Acoustic Analyses Using
MATLAB®
and ANSYS®

Acoustic Analyses Using
MATLAB®
and ANSYS®

Carl Q. Howard and Benjamin S. Cazzolato

CRC Press
Taylor & Francis Group
Boca Raton London New York

CRC Press is an imprint of the
Taylor & Francis Group, an **informa** business

A SPON PRESS BOOK

CRC Press
Taylor & Francis Group
6000 Broken Sound Parkway NW, Suite 300
Boca Raton, FL 33487-2742

First issued in paperback 2017

© 2015 by Taylor & Francis Group, LLC
CRC Press is an imprint of Taylor & Francis Group, an Informa business

No claim to original U.S. Government works

ISBN-13: 978-1-4822-2325-5 (hbk)
ISBN-13: 978-1-138-74748-7 (pbk)

Library of Congress Cataloging-in-Publication Data

Howard, Carl Q., 1970- author.
    Acoustic analyses using Matlab and Ansys / Carl Q. Howard and Benjamin S. Cazzolato.
      pages cm
    Summary: "This book describes the use of Ansys finite element analysis software and MATLAB
    Includes bibliographical references and index.
    ISBN 978-1-4822-2325-5 (hardback)
    1. Acoustical engineering--Mathematics. 2. Finite element method--Data processing. 3. MATLAB. 4. ANSYS (Computer system) I. Cazzolato, Benjamin S., 1969- author. II. Title.

TA365.H69 2015
620.20285'53--dc23   2014015094

Visit the Taylor & Francis Web site at
http://www.taylorandfrancis.com

and the CRC Press Web site at
http://www.crcpress.com

Contents

List of Figures

List of Tables

Foreword

Although ANSYS Mechanical has supported acoustic and vibroacoustic analyses for more than two decades, it was not until relatively recently that the demand for performing these simulations has grown appreciably. Over the past several years, noise considerations for rotating machinery, passenger comfort in vehicles, and acoustic performance of tablets and other ubiquitous electronic devices have taken a more dominant role in their respective design processes. These, of course, are but a few examples out of many applications where the engineer must now consider the acoustic response along with traditional structural and thermal simulations in evaluating product design.

Development of the acoustic capabilities in ANSYS Mechanical has been driven by customer feedback and requirements, and the fruit of these efforts have enabled analysts to solve challenging problems in ANSYS Workbench in a fraction of the time historically needed. That being said, however, for many engineers, acoustics may be a new field to them; conversely, acousticians may not be familiar with how certain concepts are implemented in a finite element software program.

This book provides an in-depth and practical guide on performing acoustic and vibroacoustic simulations using ANSYS Mechanical. For the engineer with limited background in acoustics, this text serves as an excellent companion to other books that cover acoustic fundamentals; the numerous examples provide many opportunities for the reader to relate finite element results with theory and learn best practices along the way. For the acoustician, the explanations focused on numerical methods as well as the comparisons with MATLAB results are illuminating, and the step-by-step instructions are invaluable for readers new to ANSYS Workbench.

It is a pleasure for me to write the introduction to this book, and I am certain it will have a wide appeal in both academic and industrial circles. Carl Howard and Ben Cazzolato have produced a well-written and practical reference that will help the reader enter the exciting world of acoustical numerical simulation.

Sheldon Imaoka
Principal Engineer
ANSYS, Inc.

Preface

The use of finite element analysis (FEA) to solve acoustic problems has enabled investigation of complex situations that would otherwise be too cumbersome or time consuming to solve using analytical methods. Many analytical methods are only suitable for solving regular-shaped objects such as ducts, hard-walled rectangular cavities, and so on.

Although finite element analysis can be used to solve complex problems, there is a steep learning curve for practitioners. One must have a good grasp of the science of acoustics. They must understand many concepts and limitations of finite element analysis. Even if they have all this knowledge, they must also know the nuances of a particular finite element analysis software package and its particular quirks. Lastly, if a practitioner has been able to calculate answers using finite element analysis, they need to have a sense of whether the answers predicted by the software are reasonable. The Garbage-In-Garbage-Out (GIGO) principle applies as it is easy to generate misleading results.

The contents of this book attempt to address only the last few of these hurdles: "how to drive" the ANSYS® finite element analysis software to solve a variety of acoustic problems. The fundamentals and applications of acoustics are covered in many other textbooks and are not the focus of this book. As for training in finite element analysis, many books are written by mathematicians or academics and their target audience seems to be for other mathematicians or academics. These books contain derivations of shape functions for various finite elements and the nuts-and-bolts of various matrix inversion algorithms so that someone with a lot of time on their hands can create their own finite element software. Whilst these are important topics, a practicing engineer has no spare time and is not going to create his or her own finite element software. Instead, an engineer wants a short sharp bullet list of instructions to get the job done.

We are both mechanical engineers and our backgrounds are in consulting engineering, academic research, and university training of students to become professional engineers. This book contains examples with flavors from these arenas. The goal was to provide instruction in solving acoustic problems starting with simple systems such as a duct, and then progressively more involved problems such as acoustic absorption and fluid–structure interaction. The theory of the acoustic problem is presented and then implemented in MATLAB® code, which is included with this book. An ANSYS finite element model of the problem is described and the completed models are included with this book. The combination of these three aspects provides the practitioner with benchmark cases that can be used as starting points for the analysis of their own acoustic problems.

At times a reader might find the instructions in this book to be verbose or repetitive. Although this might irritate some readers, it is better to provide detailed instructions rather than frustrate a new analyst, or someone that has started halfway through the book. We have tried to avoid using expressions such as "clearly," or "it is obvious," as what might be obvious for some readers is baffling to another and instead provide detailed instructions and comprehensive explanations.

The release of the ACT Acoustics extension for ANSYS Workbench has made the use of the software significantly easier for a new analyst to solve acoustic problems. The extension is essentially a toolbar that enables the user to select the relevant acoustic feature that he or she wants to include in the analysis, such as an acoustic mass source, an absorbing boundary, and so on.

We trust that you will find this book a useful resource for learning how to conduct acoustic analyses using ANSYS and MATLAB® and will enable you to solve your own acoustic problems.

Carl Howard and Ben Cazzolato
Adelaide

MATLAB® and Simulink® are registered trademarks of The MathWorks, Inc. For product information, please contact:

The MathWorks, Inc.
3 Apple Hill Drive
Natick, MA 01760-2098 USA
Tel: 508 647 7000
Fax: 508-647-7001
E-mail: info@mathworks.com
Web: www.mathworks.com

Acknowledgments

The authors wish to acknowledge the support of the following people that have enabled the completion of this book: David Roche, Sheldon Imaoka, Grama Bhashyam at ANSYS, Inc., for the numerous technical discussions; Sam Nardella; Srini Bandla, Jindong Yang, and Greg Horner at LEAP Australia, for technical assistance; Boyin Ding for proofreading the book; Colin Hansen, our mentor, for encouraging us to write this book; Chenxi Li, for assistance with the micro-perforated panel material; and John Pearse, for hosting the first author at the University of Canterbury while writing this book.

We also wish to acknowledge and thank our families for the unwavering support and encouragement: Nicole, Natalie, and Sarah; Thu-Lan, Toby, and Marcus.

1

Introduction

1.1 About This Book

The aim of this book is to provide examples of how to solve acoustic problems using MATLAB, ANSYS Workbench, and ANSYS Mechanical APDL. This book contains mathematical theory which is referenced to published works such as textbooks, journals, and conference papers. The theory is then demonstrated by example that is implemented using MATLAB and the source code is included with this book. The example is also solved using ANSYS Workbench, where an archive file .wbpz, which contains the project file .wbpj, is available with the book, and also with ANSYS Mechanical APDL, and the source code is available with this book. The software that is included with this book is listed in Appendix A.

The subjects of vibrations and acoustics encompass an enormous range of topics. The theory and examples in this book were selected in order to demonstrate how to solve problems using the ANSYS software. ANSYS has good capabilities for solving acoustics problems, and it has been the intention of the authors to cover most of these capabilities.

It is assumed that the reader has some familiarity with theory relating to vibrations and acoustics. This book is not intended to be a vibration and acoustic textbook—there are numerous excellent textbooks available that cover these topics.

In addition, it is assumed that the reader has some familiarity with Finite Element Analysis (FEA). This book only covers the very basics of finite element analysis.

1.1.1 MATLAB Code

This book includes a number of MATLAB code listings to model acoustic theoretical systems. Many of the MATLAB scripts have been written using the *publishing* feature that generates HTML output and includes mathematical equations rather than ASCII text and the graphs that have been generated by the script. To view the published output of these scripts, start the MATLAB

1

software, open the particular `filename.m` script file of interest in the editor, and click on `File | Publish filename.m`.

1.1.2   ANSYS

There are two "front-ends" or Graphical User Interfaces (GUIs) for AN-SYS, namely ANSYS Workbench and ANSYS Mechanical APDL, that enable the construction, analysis, and visualization of models and results. The ANSYS Workbench interface is intended to be used mainly by using a mouse to select menu items and the keyboard to enter values of parameters and naming objects. The ANSYS Mechanical APDL GUI can be used either by using a mouse to select menu items or by typing APDL commands into an input box.

The ANSYS GUIs use color extensively in menus, displays of models, and results. However, the printed version of this book is in grayscale and ANSYS color contour results are plotted using reverse-grayscale, where the maximum value has a white color, and the minimum value has a black color.

Both GUIs generate `ASCII` text APDL commands, which are usually hidden from the user, and are sent to the numerical solving "engine." The underlying computational engine has a user interface as shown in Figure 1.1, where it can be used interactively, or run in batch mode where it is supplied with a text file containing APDL commands to execute.

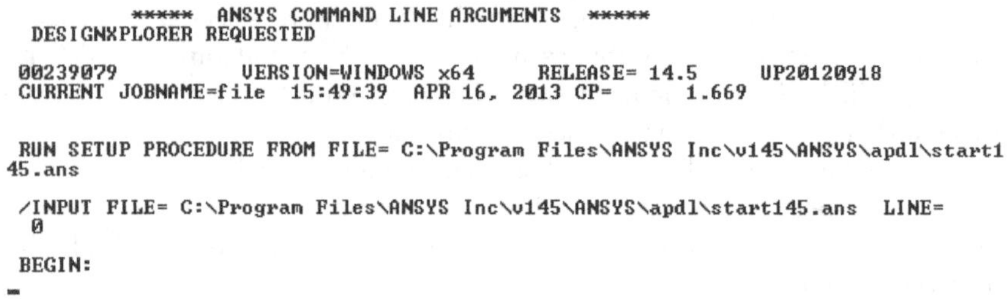

```
        ***** ANSYS COMMAND LINE ARGUMENTS *****
   DESIGNXPLORER REQUESTED

00239079              VERSION=WINDOWS x64       RELEASE= 14.5       UP20120918
   CURRENT JOBNAME=file  15:49:39  APR 16, 2013 CP=       1.669

 RUN SETUP PROCEDURE FROM FILE= C:\Program Files\ANSYS Inc\v145\ANSYS\apdl\start1
45.ans

 /INPUT FILE= C:\Program Files\ANSYS Inc\v145\ANSYS\apdl\start145.ans   LINE=
   0

 BEGIN:
 -
```

FIGURE 1.1
Text input interface for the ANSYS finite element analysis software.

1.1.3   ANSYS Workbench Models

A number of ANSYS Workbench models are included in this book which can be used to calculate the acoustic theoretical models. These models were created using ANSYS Release 14.5 and are stored as archive files with extension `.wbpz`, which removes any file path dependencies and enables the project and associated files to be restored to a computer to any directory. To restore the archive, copy the archive of interest to a local disk drive, start ANSYS Workbench, and click on File | Restore Archive....

Instructions are provided to create the ANSYS Workbench models described in this book. Some general advice about configuring and using ANSYS is in Appendix B.

Warning

When using ANSYS Workbench, if you intend to undertake detailed post-processing it is possible that you will inadvertently trigger ANSYS to request the project to be re-analyzed. If you intend to experiment with post-processing, particularly using Command (APDL) objects, it is advisable to either start with a small model so that if a re-analysis is required then it will not take long to solve, or alternatively consider using the ANSYS Mechanical APDL interface rather than ANSYS Workbench.

1.1.4 ANSYS Mechanical APDL Code

In addition to the ANSYS Workbench models, a number of ANSYS Mechanical ANSYS PARAMETRIC DESIGN LANGUAGE (APDL) scripts are included that can be used to analyze the vibration and acoustic models. These models were created using ANSYS Release 14.5 and are text files with file extensions *.inp , *.txt, *.mac. These scripts can be viewed with any text editor. The typical method for using these scripts in the ANSYS Mechanical APDL software is to either use the mouse to select File | Read Input From, or, alternatively from the command line, to type /input,myscript,inp where myscript is replaced with the filename of interest, and the inp is replaced with the appropriate extension such as inp, txt, or mac.

Although it is possible to use ANSYS Mechanical APDL by clicking menu items, it is not covered in this book.

1.2 A Philosophy for Finite Element Modeling

This section contains a qualitative discussion about the concept of networks in relation to finite element modeling. This concept is applicable to modeling many engineering systems that have interconnected parts or systems. It applies to structural and acoustic finite element models, and also thermal, fluid finite element models, electrical networks, statistical energy analysis, water piping networks, and many others.

Consider the truss shown in Figure 1.2 where all the joints are welded (not pinned) and a load is applied at the apex of the truss. One can imagine that the applied load will result in the beam members placed under compression, tension, and bending stresses. Now consider the situation where the center upright truss member has been removed, as shown in the right sketch in Figure 1.2. One would expect that there would be a redistribution of the stresses within the beams to accommodate the applied load, compared to the truss on the left in Figure 1.2. This simple example is used to highlight the concept of a network and the redistribution of stress or energy that occurs throughout the network if parts of the network are altered.

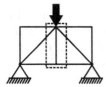 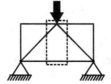

FIGURE 1.2
A truss with welded joints (left sketch), and a truss with an upright removed (right sketch).

Finite element models consist of a network of interconnected elements and nodes. An analogy that is used to describe this network is a "'sea of springs," as shown in Figure 1.3, comprising a network of interconnected springs and masses. The springs can be considered as the finite elements, and the masses can be considered as the nodes within the finite element model. Loads are applied at the nodes in the model that can be considered as forces, acoustic volume velocity, and so on. The restraints that are applied to the sea of springs, are the boundary conditions in the finite element model. One can imagine that if a load were applied to the model on the left in Figure 1.3, most of the springs would deflect. If some of the springs were removed from the model, as shown in the right-hand model in Figure 1.3, the springs would deflect differently compared to the model on left in Figure 1.3.

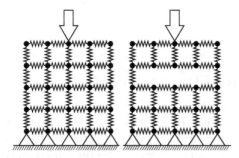

FIGURE 1.3
Sea of springs analogy.

Although these two previous examples are simple, the purpose of describing them is to change the reader's perception from considering a physical system to an analogous mathematical model applicable to finite element modeling. Analysts should consider these two *mental models* when conducting finite element analyses and how they are relevant when selecting loads, boundary conditions (constraints), material properties, speed of sound of materials, temperature gradients, impedances, the number of degrees of freedom (i.e., whether the node is capable of translational and rotational motion) at each node, and mesh density.

Here are some examples and traps for new players:

- Consider the example of an attempt to model the deflection of a cantilever beam. If you were to apply a harmonically varying displacement at the end of the beam, the resulting tip displacement will only be the value that was applied as the "load." Perhaps one intended to apply a harmonically varying force to the end of the cantilever beam, and measure the resulting tip displacement. The equivalent trap for an acoustic finite element analysis is the application of an acoustic pressure at a node causes the acoustic pressure to always be the value that was specified. This is obvious in hindsight, but consider the implication if you were modeling a reactive acoustic muffler excited with a harmonic acoustic pressure source. This will constrain the acoustic pressure at the source location to be the value that was specified. If downstream of the acoustic pressure source there are impedance changes that cause acoustic reflections, so that an incident acoustic pressure wave is reflected upstream, the acoustic pressure source provides an artificial constraint to your model. Perhaps you might have expected to see a pressure doubling near the source due to the superposition of the incident and reflected waves. However, this will not occur as the amplitude of the acoustic pressure was specified at the source location. In mathematical parlance, the application of an acoustic pressure to a node is applying a *boundary condition* to the model. An alternative is to apply an acoustic volume velocity source, which is effectively applying a load to the model.

- Following from the discussion about the redistribution of load in the modified truss in Figure 1.2, acoustic or vibrational energy can be inadvertently redistributed in a system by poor modeling practices, such as the following:

  · If the boundary conditions or constraints of a system are modeled incorrectly, this will have the effect of changing the response of the system. Consider a shaft with the ends supported by rolling element bearings. If the bearings are modeled as simple-support boundary conditions, it is likely that the shaft will deflect greater than expected. If the bearings are modeled as clamped boundary conditions, it is likely that the shaft will be over-constrained. A more accurate representation is likely to be between a simple-support and a clamped boundary condition.

  · If the mesh density is too low, meaning that the size of the elements is too large, in one region, the model might be artificially "stiff" and cause vibro-acoustic energy to divert to a different region. This effect is highlighted in Section 3.5, where a duct that has large dimensions has a poor mesh in one region and causes the generation of non-plane waves. This effect can occur for irregular or asymmetric meshes. When the duct is re-meshed with an adequate mesh density, only plane waves are observed.

- The analyst has to take care to consider whether it is worth modeling a system in detail, or whether the system can be simplified. For example, when modeling some acoustic systems, it is not necessary to model intricate details such as protrusions or small cavities, if the acoustic wavelength is so long that the acoustic wave will not be altered by the presence of the feature.

- When building a structural model for a vibro-acoustic analysis, the analyst should be aware that some 3D brick elements do not have rotational degrees of freedom at their nodes. This can cause problems if one is applying a moment loading to a structure. Another situation to be aware of is if the structure to be modeled is constrained with a fully clamped condition, which means that all translational and all rotational displacements at the node are set to zero, that many solid elements do not have rotational degrees of freedom and so the rotational constraint on the node will be ineffective and hence the model may rotate more than expected.

1.3   Analysis Types

There are several analysis types that are available to conduct acoustic and vibration investigations using the ANSYS software. These include:

Modal used to calculate the natural frequencies and mode shapes of systems.

Harmonic used to calculate the acoustic or vibration response of a system due to excitation by a sinusoidally varying driving force, displacement, acoustic pressure, and others, where the excitation is continuous at constant frequency. A number of harmonic analyses can be conducted over a frequency range.

Transient used to calculate the time-history response of a system due the application of a time-varying excitation.

Random used to calculate the response of a system due to the application of a prescribed frequency and amplitude spectrum of excitation.

Each of these analysis types are described further in the following sections. The focus of this book is covering modal and harmonic analyses for acoustic systems.

1.3.1 Modal

A modal analysis can be conducted to calculate the natural frequencies and mode shapes of an acoustic or structural system, or a combined structural-acoustic system. The results from a modal analysis conducted in ANSYS can be used to calculate a harmonic response, transient, or response spectrum analysis.

The equations of motion for an acoustic or structural system can be written as

$$\left(-\omega^2 \left[\mathbf{M}\right] + j\omega \left[\mathbf{C}\right] + \left[\mathbf{K}\right]\right) \{\mathbf{p}\} = \{\mathbf{f}\} , \tag{1.1}$$

where $[\mathbf{M}]$ is the mass matrix, $[\mathbf{C}]$ is the damping matrix, $[\mathbf{K}]$ is the stiffness matrix, $\{\mathbf{p}\}$ is the vector of nodal pressures for an acoustic system or displacements for a structural system, and $\{\mathbf{f}\}$ is the acoustic or structural load applied to the system. For a basic modal analysis, it is assumed that there is no damping and no applied loads, so the damping matrix $[\mathbf{C}]$ and the load vector $\{\mathbf{f}\}$ are removed from Equation (1.1), leaving [4, Eq. (17.46)]

$$\left(-\omega^2[\mathbf{M}] + [\mathbf{K}]\right) \{\mathbf{p}\} = \{\mathbf{0}\} . \tag{1.2}$$

For an (undamped) system, the free pressure oscillations are assumed to be harmonic of the form

$$\{\mathbf{p}\} = \{\phi\}_n \cos \omega_n t , \tag{1.3}$$

where $\{\phi\}_n$ is the eigenvector of pressures of the nth natural frequency, ω_n is the natural circular frequency (radians/s), t is time. Substitution of Equation (1.3) into (1.2) gives

$$\left(-\omega_n^2 \left[\mathbf{M}\right] + [\mathbf{K}]\right) \{\phi\}_n = \{\mathbf{0}\} . \tag{1.4}$$

The trivial solution is $\{\phi\}_n = 0$. The next series of solutions is where the determinant equates to zero and is written as [4, Eq. (17-49)]

$$\left|[\mathbf{K}] - \omega_n^2 \left[\mathbf{M}\right]\right| = 0 , \tag{1.5}$$

which is a standard eigenvalue problem and is solved to find the natural frequencies (eigenvalues) ω_n and mode shapes (eigenvectors) $\{\phi\}_n$. ANSYS will list results of the natural frequencies f_n in Hertz, rather than circular frequency in radians/s, where

$$f_n = \frac{\omega_n}{2\pi} . \tag{1.6}$$

For many finite element models, the mass and stiffness matrices are symmetric, and ANSYS has several numerical solvers that can be used to calculate the natural frequencies and mode shapes, which include the supernode, block Lanczos, and Preconditioned Conjugate Gradient PCG Lanczos methods [5]. When the finite element model has unsymmetric matrices, which can occur when the model contains fluid–structure interaction, then an unsymmetric solver must be used. When the system includes damping, it is necessary to use the damped or QR damped solver. The ANSYS online help manual [5, Table 15.1] lists which modal analysis solver is appropriate for the conditions present in the finite element model and is summarized here in Table 1.1.

TABLE 1.1
Modal Analysis Solver Types Available in ANSYS for Determining Natural Frequencies and Mode Shapes

Undamped / Damped	Symmetric / Unsymmetric	Solver Name	APDL Command
Undamped	Symmetric	Supernode	MODOPT,SNODE
Undamped	Symmetric	Block Lanczos	MODOPT,LANB
Undamped	Symmetric	PCG Lanczos	MODOPT,LANPCG
Undamped	Unsymmetric	Unsymmetric	MODOPT,UNSYM
Damped	Symmetric or Unsymmetric	Damped	MODOPT,DAMP
Damped	Symmetric or Unsymmetric	QR Damped	MODOPT,QRDAMP

Sections 3.3.1, 3.3.3, 4.4.2, and 4.4.3 describe examples of undamped modal analyses conducted using ANSYS. Section 7.4.2 describes an example of a damped modal analysis of a room that has acoustic absorptive material on the floor.

1.3.2   Harmonic

The harmonic response of a system can be calculated using two methods: full and modal summation (or superposition).

The full method involves forming the mass $[\mathbf{M}]$, damping $[\mathbf{C}]$, and stiffness $[\mathbf{K}]$ matrices and the loading vector $\{\mathbf{f}\}$ of the dynamic equations of motion, combining the matrices, then inverting the combined matrix and multiplying it with the load vector to calculate the nodal displacements $\{\mathbf{u}\}$, as follows [6]:

$$[\mathbf{M}]\{\ddot{\mathbf{u}}\} + [\mathbf{C}]\{\dot{\mathbf{u}}\} + [\mathbf{K}]\{\mathbf{u}\} = \{\mathbf{f}\}$$
$$-\omega^2[\mathbf{M}]\{\mathbf{u}\} + j\omega[\mathbf{C}]\{\mathbf{u}\} + [\mathbf{K}]\{\mathbf{u}\} = \{\mathbf{f}\}$$
$$\left(-\omega^2[\mathbf{M}] + j\omega[\mathbf{C}] + [\mathbf{K}]\right)\{\mathbf{u}\} = \{\mathbf{f}\}$$
$$\{\mathbf{u}\} = \left(-\omega^2[\mathbf{M}] + j\omega[\mathbf{C}] + [\mathbf{K}]\right)^{-1}\{\mathbf{f}\}. \quad (1.7)$$

The modal summation method involves the calculation of the mode shapes of a structural or acoustic system, and determining what portion of each mode, called the modal participation factors P_n, contributes to the overall response. The mathematical derivation of the mode superposition method is presented in the ANSYS theory manual [7] [8].

To illustrate the concept of the modal summation method, consider a simply supported beam that has vibration mode shapes ψ_n that resemble half sine waves as shown in Figure 1.4. Each mode can be multiplied by a modal participation factor P_n, then summed to calculate the total response of the beam $\sum P_n \psi_n$. Any complicated displacement shape can be represented by a weighted sum of a sufficiently large number of modes. A similar acoustic analogy exists where any complicated acoustic response of a system can be represented by a weighted sum of a sufficiently large number of mode shapes of the acoustic system.

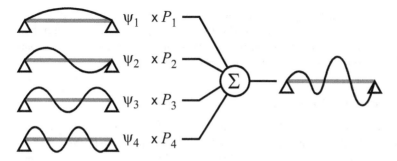

FIGURE 1.4
Schematic of the concept of modal summation, where fractions (modal participation factors) P_n of each mode shape ψ_n contribute to the total response of the system.

Harmonic analysis of acoustic systems using the modal summation method is not available using ANSYS Workbench at Release 14.5 and ACT Acoustics extension version 8, but is expected to be available at Release 15.0. This technique can be employed using ANSYS Mechanical APDL and an example is shown in Section 4.4.3.

1.3.3   Transient Dynamic Analysis

Transient dynamic analysis (sometimes called time-history analysis) is an analysis technique used to determine the response of a system to any time-dependent load. It is used when inputs into the system cannot be considered stationary (unlike a harmonic analysis). The basic equation of motion which is solved by a transient dynamic analysis is [9]

$$[\mathbf{M}]\{\ddot{\mathbf{u}}\} + [\mathbf{C}]\{\dot{\mathbf{u}}\} + [\mathbf{K}]\{\mathbf{u}\} = \{\mathbf{f(t)}\} , \tag{1.8}$$

where the terms are the same as defined for the harmonic analysis with the exception of the load vector, $\{\mathbf{f}(\mathbf{t})\}$, which represents the time-dependent loads applied to the system. As was the case for the harmonic analysis, transient dynamic analysis can be calculated using either a *full* or *mode-superposition* method. The advantages and disadvantages of these two methods when undertaking transient dynamic analysis are the same as was discussed for the harmonic analysis. However, there are other issues that only affect transient analysis. The mode-superposition method is restricted to fixed time steps throughout the analysis, so automatic time stepping is not allowed. Such a feature is often desirable when there are events of differing time scales. Without automatic time stepping, the shortest time scale needs to be used over the entire simulation, which can increase solution times. The full method accepts non-zero displacements as a form of load, whereas the mode-superposition method does not.

In Chapter 7 a full transient dynamic analysis is conducted on a model of a damped reverberation room, where the time-varying pressure at a number of locations is predicted in response to a sharp acoustic impulse. The various solver options are discussed in detail in this chapter.

1.3.4   Spectrum Analysis

The spectrum analysis in ANSYS is one in which the results from a modal analysis are used, along with a known spectrum, to calculate the response of a system. It is mainly used in place of transient dynamic analyses to determine the response of a system to either time-dependent or random loading conditions which may be characterized by an input spectrum. This method is linear, so a transient dynamic analysis must be used if the system behavior is non-linear.

There are two broad categories of spectrum analysis: deterministic and probabilistic. The deterministic methods in ANSYS (Response Spectrum and Dynamic Design Analysis Method) use an assumed phase relationship between the various modes, whereas in the probabilistic method (called Power Spectral Density method in ANSYS and also known as random vibration analysis), the way in which the response of the modes are summed is probabilistic.

Although spectrum analyses are commonly used in predicting the response of a system to shock and vibration, these methods are not suitable for acoustic models as of ANSYS Release 14.5, and therefore there are no examples of these methods in this book.

2

Background

2.1 Learning Outcomes

The learning outcomes of this chapter are to:

- understand the general concepts of conducting a fluid–structure interaction acoustic analysis,

- understand the types of pressure-formulated acoustic elements available in ANSYS,

- understand the capabilities of the ACT Acoustics extension add-on and the available acoustic boundary conditions and load types,

- understand the acoustic energy metrics that can be calculated from the results of an acoustic analysis, and

- understand the scaling of results required when only using a portion of an acoustic model to model a full system, with assumed symmetry.

2.2 Introduction

Finite Element Analysis (FEA) is a numerical method that can be used to calculate the response of a complicated structure due to the application of forcing functions, which could be an acoustic source or a distribution of mechanical forces. FEA can also be used to estimate the sound power radiated by a structure or the distribution of the sound field in an enclosed space. Estimating the sound power radiated by a structure into an acoustic region generally requires a large numerical model and the associated computer memory requirements are large. An alternative is to use FEA to calculate the vibration response of the noise-radiating structure only and then use a numerical evaluation of the Rayleigh Integral to calculate the radiated sound power. Alternatively, if the structure is excited by an external sound field, then FEA can be used to determine separately the in-vacuo (i.e., without the acoustic fluid) structural response as well as the resonance frequencies and mode shapes of the rigid-walled enclosed sound field. Then the actual sound pressure distribution in the enclosed space can be calculated using modal coupling analysis implemented with a programming tool such as MATLAB. Software for conducting this type of analysis is described in Appendix C. The underlying theory for FEA is covered in many textbooks and will not be repeated here. However, its practical implementation using a commercially available FEA package will be discussed in an attempt to help potential users apply the technique to acoustic analysis. Finite element analysis of acoustic systems involves the discretization of the acoustic volume into elements and nodes. An enclosed acoustic volume might be surrounded by rigid-walls, a flexible structure, or walls that provide acoustic damping. Alternatively, the acoustic radiation of a structure into an anechoic or free-field can also be examined.

Analytical methods can be used to calculate the acoustic fields and structural vibration for only the most simple systems, with geometries typically limited to rectangles and circles and their 3D equivalents. Any geometry more complex than these is onerous to analyze and vibro-acoustic practitioners opt for a numerical method such as finite element or boundary element analysis to solve their particular problem.

Finite element analysis of acoustic systems has numerous applications including the acoustic analysis of interior sound fields, sound radiation from structures, the transmission loss of panels, the design of resonator-type silencers and diffraction around objects. The finite element method takes account of the bi-directional coupling between a structure and a fluid such as air or water. In acoustic fluid–structure interaction problems, the equations related to the structural dynamics need to be considered along with the mathematical description of the acoustics of the system, given by the Navier–Stokes equations of fluid momentum and the flow continuity equation. The discretized structural dynamics equation can be formulated using structural finite elements. The fluid momentum (Navier–Stokes) and continuity equations are simplified to form the acoustic wave equation using the following assumptions:

- The acoustic pressure in the fluid medium is determined by the wave equation.

- The fluid is compressible where density changes are due to pressure variations.

- There is no mean flow of the fluid.

- The density and pressure of the fluid can vary along the elements and the acoustic pressure is defined as the pressure in excess of the mean pressure.

- Finite element analyses are limited to relatively small acoustic pressures so that the changes in density are small compared with the mean density.

The acoustic wave equation is used to describe the acoustic response of the fluid. When the viscous dissipation of the fluid is neglected, the equation is referred to as the lossless wave equation. Suitable acoustic finite elements can be derived by discretizing the lossless wave equation using the Galerkin method. For a derivation of the acoustic finite element, the reader is referred to Ref. [58]. There are two formulations of finite elements that are used to analyze acoustic problems: pressure and displacement. The most commonly used finite element to analyze acoustic problems is the pressure-formulated element, which is discussed in the following section.

2.3 Pressure-Formulated Acoustic Elements

The acoustic pressure p within a finite element can be written as

$$p = \sum_{i=1}^{m} N_i p_i \,, \tag{2.1}$$

where N_i is a set of linear shape functions, p_i are acoustic nodal pressures at node i, and m is the number of nodes forming the element. For pressure-formulated acoustic elements, the lossless finite element equation for the fluid in matrix form is

$$[\mathbf{M_f}] \{\ddot{\mathbf{p}}\} + [\mathbf{K_f}] \{\mathbf{p}\} = \{\mathbf{F_f}\} \,, \tag{2.2}$$

where $[\mathbf{K_f}]$ is the equivalent fluid stiffness matrix, $[\mathbf{M_f}]$ is the equivalent fluid mass matrix, $\{\mathbf{F_f}\}$ is a vector of applied fluid loads, $\{\mathbf{p}\}$ is a vector of unknown nodal acoustic pressures, and $\{\ddot{\mathbf{p}}\}$ is a vector of the second derivative of acoustic pressure with respect to time.

There are four acoustic element types available in ANSYS that are based on pressure formulation: types FLUID29, FLUID30, FLUID220, and FLUID221. Section 2.7 describes the capabilities of these elements.

2.4 Fluid–Structure Interaction

The previous section described a standard pressure-formulated acoustic element. These elements can be connected to structural elements so that the two become coupled—the acoustic pressure acts on a structure which causes it to vibrate, and so is the converse where a vibrating structure causes sound to be generated in an acoustic fluid.

This section describes the matrix equations for the coupled fluid–structure interaction problem. The purpose of describing these equations is to highlight:

- how the responses of the acoustic fluid and structure are connected,

- the unsymmetric matrices that result from fluid–structure interaction problems, and

- how the matrices can be reformulated from unsymmetric to symmetric matrices, leading to a reduction in computation time.

The equations of motion for the structure are (for more details see the ANSYS online help manual [10]):

$$[M_s]\{\ddot{U}\} + [K_s]\{U\} = \{F_s\} \,, \tag{2.3}$$

where $[K_s]$ is the structural stiffness matrix, $[M_s]$ is the structural mass matrix, $\{F_s\}$ is a vector of applied structural loads, $\{U\}$ is a vector of unknown nodal displacements and hence $\{\ddot{U}\}$ is a vector of the second derivative of displacements with respect to time, equivalent to the acceleration of the nodes. The interaction of the fluid and structure occurs at the interface between the structure and the acoustic elements, where the acoustic pressure exerts a force on the structure and the motion of the structure produces a pressure. To account for the coupling between the structure and the acoustic fluid, additional terms are added to the equations of motion for the structure and fluid (of density, ρ_0), respectively, as

$$[M_s]\{\ddot{U}\} + [K_s]\{U\} = \{F_s\} + [R]\{p\} \,, \tag{2.4}$$

$$[M_f]\{\ddot{p}\} + [K_f]\{p\} = \{F_f\} - \rho_0[R]^T\{\ddot{U}\} \,, \tag{2.5}$$

where $[R]$ is the coupling matrix that accounts for the effective surface area associated with each node on the fluid-structure interface. Equations (2.4) and (2.5) can be formed into a matrix equation including the effects of damping as

$$\begin{bmatrix} M_s & 0 \\ \rho_0 R^T & M_f \end{bmatrix} \left\{ \begin{array}{c} \ddot{U} \\ \ddot{p} \end{array} \right\} + \begin{bmatrix} C_s & 0 \\ 0 & C_f \end{bmatrix} \left\{ \begin{array}{c} \dot{U} \\ \dot{p} \end{array} \right\} + \\ \begin{bmatrix} K_s & -R \\ 0 & K_f \end{bmatrix} \left\{ \begin{array}{c} U \\ p \end{array} \right\} = \left\{ \begin{array}{c} F_s \\ F_f \end{array} \right\} \,, \tag{2.6}$$

where $[C_s]$ and $[C_f]$ are the structural and acoustic damping matrices, respectively. For harmonic analyses, this equation can be reduced to an expression without differentials as

$$\begin{bmatrix} -\omega^2 M_s + j\omega C_s + K_s & -R \\ -\omega^2 \rho_0 R^T & -\omega^2 M_f + j\omega C_f + K_f \end{bmatrix} \left\{ \begin{array}{c} U \\ p \end{array} \right\} = \left\{ \begin{array}{c} F_s \\ F_f \end{array} \right\} . \tag{2.7}$$

The important feature to notice about Equation (2.7) is that the matrix on the left-hand side is unsymmetric and solving for the nodal pressures and displacements requires the inversion of this unsymmetric matrix, which requires a significant amount of computer resources. The fluid–structure interaction method described above accounts for two-way coupling between structures and fluids. This mechanism is significant if a structure is radiating into a heavier-than-air medium such as water, or if the structure is very lightweight, such as a car cabin. In some vibro-acoustic systems, an acoustic field will be dissipated by the induced vibration of a structure, which has the effect of damping the acoustic response of the system.

When using this coupling in FEA, it is necessary to carefully construct the model to accurately represent the interface between the fluid and the structure. Figure 2.1 illustrates a finite element model of an acoustic duct with a structural partition. The left and right sides of the duct have acoustic elements with only pressure DOFs. The elements for the structural partition contain displacement DOFs. At the interface between the acoustic fluid and the structure is a single layer of acoustic elements that have pressure and displacement DOFs. It is this thin layer of elements that enables the bi-directional coupling between the vibration of the structure and the pressure response in the fluid. Although it is possible to use acoustic elements with both pressure and displacement DOFs for the entire acoustic field, this is unnecessary and would result in long solution times compared to only using this type of element at the fluid-structure interfaces and using acoustic elements with only a pressure DOF for the remainder of the acoustic field.

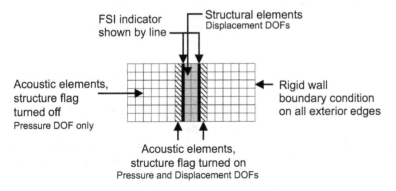

FIGURE 2.1
Schematic of a finite element model with fluid–structure interaction.

When using the ANSYS software it is necessary to explicitly define which surface of the structure and the fluid are in contact by using the Fluid-Structure-Interface (FSI) flag (meaning a switch to indicate the presence of FSI). Release 14.5 of the ANSYS software will try to identify and create the FSI flags at the interface of the structure (only solid elements, not beam or shell elements) and fluid if none are defined. However, it is good practice to manually define the interfaces rather than relying on the automated identification. The ANSYS Mechanical APDL command that is used to define the FSI flag is SF, Nlist, FSI, 1 where Nlist is either ALL to select all the nodes currently selected or P to select the nodes individually. This FSI flag is only relevant for FLUID29, FLUID30, FLUID220, and FLUID221 acoustic elements.

Modeling an acoustically rigid-wall is achieved by not defining acoustic elements on an edge, shown in Figure 2.1 on all the exterior sides of the model. Modeling a free surface can be achieved by setting the pressure to be zero on the nodes of pure acoustic elements (i.e., only pressure DOF). Alternatively, if using acoustic elements with both pressure and displacement DOFs, a free surface is modeled by not defining any loads, displacement constraints, or structure. The motion of the fluid boundary can then be obtained by examining the response of the nodes on the surface.

The matrix in Equation (2.7) is unsymmetric, which means that the off-diagonal entries are not transposes of each other. The inversion of an unsymmetric matrix takes longer to compute than a symmetric matrix. There is an option within ANSYS to use a symmetric formulation for the fluid–structure interaction [122, 11]. This can be accomplished by defining a transformation variable for the nodal pressures as

$$\dot{\mathbf{q}} = j\omega\mathbf{q} = \mathbf{p}\,, \tag{2.8}$$

and substituting this into Equation (2.7) so that the system of equations becomes

$$\begin{bmatrix} -\omega^2\mathbf{M_s} + j\omega\mathbf{C_s} + \mathbf{K_s} & -j\omega\mathbf{R} \\ -j\omega\mathbf{R}^{\mathrm{T}} & \dfrac{\omega^2\mathbf{M_f}}{\rho_0} - \dfrac{j\omega\mathbf{C_f}}{\rho_0} - \dfrac{\mathbf{K_f}}{\rho_0} \end{bmatrix} \left\{ \begin{array}{c} \mathbf{U} \\ \mathbf{q} \end{array} \right\}$$
$$= \left\{ \begin{array}{c} \mathbf{F_s} \\ \dfrac{j}{\omega\rho_0}\mathbf{F_f} \end{array} \right\}\,. \tag{2.9}$$

Equation (2.9) has a symmetric matrix which can be inverted and solved for the vectors of the structural nodal displacements $\mathbf{U}$ and the transformation variable for nodal pressures $\mathbf{q}$, faster than the unsymmetric formulation in Equation (2.7). The nodal pressures $\mathbf{p}$ can then be calculated using Equation (2.8).

An example of the use of unsymmetric and symmetric formulations is described in Section 9.4.3.

The previous discussion described how one can conduct a fluid–structure interaction analysis with bi-directional coupling. This type of analysis can require large computational resources. In some situations it may be reasonable to conduct a one-way analysis where a vibrating structure induces a pressure response in an acoustic medium, or vice versa. For this type of analysis, one must remember that some acoustic mechanisms are being neglected such as radiation damping, mass, stiffness, and damping loading of the structure. Hence one should be cautious if considering to conduct this type of analysis. A procedure that can be used to conduct this type of analysis is:

- Construct the acoustic and structural models where the nodes of the structure are coincident with the nodes on the exterior boundary of the acoustic domain.

- Suppress the nodes and elements associated with the acoustic domain.

- Conduct a vibration analysis of the structure and determine the displacements of the nodes that are in contact with the acoustic domain (i.e., write the displacement results to a file).

- Unsuppress the acoustic model and suppress the structural model.

- Use the displacement results from the structural vibration analysis as boundary conditions for the displacements of the acoustic nodes.

- Calculate the acoustic response of the acoustic model.

2.5 Displacement-Formulated Acoustic Elements

Another formulation of acoustic elements is based on nodal displacements and these elements are based on standard structural elements. A typical structural acoustic finite element model based on displacement-formulated elements is shown in Figure 2.2.

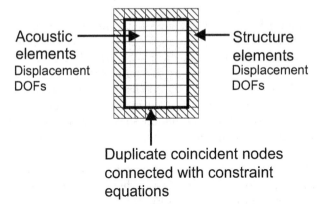

FIGURE 2.2
Finite element model with displacement-formulated acoustic elements connected to structural elements.

The difference between the solid structural elements and fluid elements is that the underlying material behavior is altered to reflect the behavior of a fluid, so that the stiffness terms associated with shear stresses are set to near zero and the Young's modulus is set equal to the bulk modulus of the fluid. What this means is that the element has no ability to resist shear stress and can lead to peculiar results. For example, a modal analysis of an acoustic space using the displacement-formulated acoustic elements will produce a large number of zero energy modes that are associated with shearing mechanisms in the fluid, and these results have no relevance.

A 3-dimensional (spatial) displacement-formulated acoustic element has 3 displacement DOFs. In comparison, a 3-dimensional (spatial) pressure-formulated acoustic element has 3 displacement DOFs and 1 pressure DOF, a total of 4 DOFs.

It is conceptually easier to understand the fluid–structure interaction mechanism by considering displacement-formulated acoustic elements and structural elements. For the pressure-formulated acoustic elements, the equations for the fluid–structure interaction in Equation (2.7) included a coupling matrix $[\mathbf{R}]$ to relate the acoustic pressure at the surface of a structure and the resulting normal displacement of the structure, which is not an obvious mathematical equation. For fluid–structure interaction systems using displacement-formulated acoustic elements, the displacements at the nodes of the structural elements can be directly coupled to displacement-formulated acoustic elements. Note that only the nodal displacements of the acoustic elements that are normal to the structure should be coupled, as it is only the out-of-plane motion of the structure that generates acoustic pressure in the fluid, as the in-plane motion of the structure does not generate any acoustic pressure in the fluid. Hence, the fluid and the structure must retain independent displacement DOFs for motions that are tangential to the fluid–structure interaction surface, which can be difficult to model.

These systems are modeled by defining separate but coincident nodes for the fluid and the structure at the fluid–structure interaction interface, and then coupling the appropriate nodal displacement DOFs, or defining mathematical relationships for the nodal displacement motion of the structure and fluid, to define the compatibility of displacements that are normal to the fluid–structure interaction surface. In many cases it is advantageous to rotate the nodal coordinate systems of the structure and fluid meshes along the interface so that there is one axis that is normal, and two axes that are tangential to the interface surface.

One of the main advantages of using displacement-formulated acoustic elements is that the matrix equation is symmetric and thus is quicker to solve than the unsymmetric matrix equation shown in Equation (2.7) for the same number of degrees of freedom in a model.

Tables 2.1 and 2.2 list the advantages and disadvantages of both the pressure and displacement formulated acoustic elements.

There are two element types available in ANSYS that are based on displacement formulation: types FLUID79 and FLUID80. However, it is not recommended to use these elements, now termed *legacy elements*. The documentation for these elements has been moved into the ANSYS Mechanical APDL Feature Archive section of the ANSYS help manual.

TABLE 2.1

Advantages and Disadvantages of Pressure-Formulated Elements

Advantages	Disadvantages
A minimum of a single pressure DOF per node.	The set of equations to be solved in a general fluid–structure inter-action analysis are unsymmetric, requiring more computational re-sources.
No zero energy fluid modes are obtained in a modal analysis.	
The pressure DOF can be associated with either the total acoustic pressure (incident plus scattered) or only the scattered component of the acoustic pressure.	
Both displacement and pressure DOFs are available at a fluid–structure interface. Hence, defining fluid–structure coupling is relatively easy and does not require the use of duplicate nodes at the interface.	
Relatively easy to define a radiation boundary condition.	
Nodal acoustic pressures are output as solution quantities for direct use in post processing.	

2.6 Practical Aspects of Modeling Acoustic Systems with FEA

The following paragraphs describe some practical considerations when modeling acoustic systems with Finite Element Analysis (FEA).

Acoustic wavelength. The acoustic wavelength in a media is related to the speed of sound and the excitation frequency by the following equation

$$\lambda = \frac{c_0}{f},\tag{2.10}$$

where λ is the acoustic wavelength, c_0 is the speed of sound in the media, and f is the excitation frequency. It is vitally important to consider the acoustic wavelength when meshing the acoustic and structural models, as this will affect the accuracy of the results.

Mesh density. The finite element method can be useful for low-frequency problems. However, as the excitation frequency increases, the number of nodes

TABLE 2.2

Advantages and Disadvantages of Displacement-Formulated Elements

Advantages	Disadvantages
The set of equations to be solved in a general fluid–structure interaction analysis are symmetric.	There are 3 displacement DOFs per node, which can result in a model with a large number of DOFs.
Displacement boundary conditions and applied loads have the same physical meaning as those used for standard structural elements.	The definition of the fluid–structure interface is complex requiring the use of duplicate nodes and expressions to couple the relevant DOFs.
Energy losses can be included in the displacement element via a fluid viscosity parameter as well as the standard techniques with solid elements.	Modal analyses can result in a large number of (near) zero frequencies associated with shearing of the fluid elements.
	The acoustic pressure at a point in the fluid cannot be expressed in terms of a known incident pressure and an unknown scattered pressure.
	The shape of the elements should be nearly square for good results.

and elements required in a model increases exponentially, requiring greater computational resources and taking longer to solve. A general rule-of-thumb is that acoustic models should contain at least 6 elements per wavelength as a starting point [150, p. 5-1]. For better accuracy, it is recommended to use 12 elements per wavelength for linear elements (i.e., FLUID29 and FLUID30) and 6 elements per wavelength for quadratic elements (i.e., FLUID220 and FLUID221). Accurate models can still be obtained for lower mesh densities; however, caution should be exercised. At regions in a model where there is a change in the acoustic impedance, for example where the diameter of a duct changes, at a junction of two or more ducts, or at the opening of the throat of a resonator into a duct, a complex acoustic field can exist with steep pressure gradients. It is important to ensure that there is sufficient mesh density in these regions to accurately model a complicated acoustic field.

Mean flow. Many finite element software packages with acoustic finite elements require that there is no mean flow of the fluid, which is a significant limitation. When there is mean flow of fluid, a different formulation of the wave equation is required, which modifies the propagation of the acoustic disturbance (due to "convection"), depending on whether the flow is rotational or irrotational. However, it is still possible to conduct finite element modeling for low-speed fluid flow, where the compressibility effects of the fluid are negligible, using "no flow" FEA software packages, but some assumptions that underpin the analysis will be violated. When there is mean flow in a duct, aero-acoustic phenomena might be important. For example, consider the situation of mean flow in a duct where the throat of a Helmholtz resonator attaches to the main duct, or over a sharp edge. It is possible that as air flows over the edge of the throat, noise will be generated, similar to blowing air over the top of a glass soda bottle. In some situations the flow over the structure might cause vortex shedding. Standard finite element models, such as those in ANSYS finite element packages, are not able to model these effects.

If the flow speed is significant or it is expected that there will be aero-acoustic phenomena, consider the use of Computational Fluid Dynamic (CFD) software to analyze the problem. However this software also has limitations for the analysis of acoustic problems. Alternatively, some Boundary Element Analysis software packages are able to model acoustic systems with mean flow, but are not able to model noise generation from shedding type phenomena.

Rigid or Flexible Boundaries. Acoustic finite element models have rigid-wall conditions at boundaries where no elements are defined. This assumption is valid in situations where it is not expected that the motion of the boundary is likely to have any significant effect on the acoustics of the system. However, consider an automobile cabin comprising flexible sheet metal panels. Depending on the stiffness of these panels, acoustic excitation within the enclosure can cause the panels to vibrate, which in turn will affect the acoustic mode shapes and resonance frequencies of the enclosure. As highlighted above, modeling fluid–structure interaction can be computationally complex and can require substantial computer resources to solve. Hence careful consideration is required to decide whether the fluid–structure interaction should or must be modeled. A second subtle point is the consideration of re-radiation of structures in a different part of the acoustic model. Consider a duct with two Helmholtz resonators attached to a duct to reduce sound radiated from its exit as shown in Figure 2.3. A simple acoustic model could be constructed assuming rigid-walls. However if parts of the system are in fact flexible, for example the wall dividing the two resonators, then high sound levels in the first resonator would vibrate the dividing wall and reduce its effectiveness and

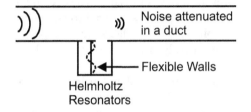

FIGURE 2.3
Duct with two Helmholtz resonators with a flexible dividing wall.

would re-radiate sound into the second Helmholtz resonator. For further discussion see Refs [55, 76]. Alternatively, if the entire system were made from lightweight sheet metal, then vibrations could be transmitted along the duct work and result in the re-radiation of sound into the main duct.

Results and Frequency Range. The results from acoustic analyses are usually the acoustic pressure at discrete locations. Sometimes this level of detail is required but often it is not; instead an indicative global sound pressure level or total sound power level may be required for assessment, which will require post-processing of the results from the analysis. This can sometimes be performed within ANSYS or may require exporting data and post-processing in another software package such as MATLAB. For higher-frequency problems, statistical energy analysis methods may be more appropriate and significantly faster in obtaining a solution.

2.7   Element Types in ANSYS for Acoustic Analyses

There are a number of pressure-formulated acoustic element types available in ANSYS for conducting acoustic analyses, which are summarized in Table 2.3.

TABLE 2.3

Summary of Element Types Available for Acoustic Analyses in ANSYS

Name	2D/3D	Nodes	Description
FLUID29	2D	4	Planar element
FLUID129	2D	2	Line element for simulating an infinite boundary
FLUID30	3D	8	Brick element
FLUID130	3D	4,8	Planar element for simulating an infinite boundary
FLUID220	3D	20	Brick element
FLUID221	3D	10	Tetrahedral element

The following sections provide a brief overview of each of these element types. For further details on these elements the reader is referred to the online ANSYS help manual included with the software; in the help index, type the name of the element type of interest. Alternatively, use the index of this book to locate uses of these elements.

All the element types listed in Table 2.3 are formulated using pressure. There are two element types available in ANSYS based on displacement formulation (FLUID79 and FLUID80) that can be used for conducting acoustic analyses. As noted in Section 2.5, it is not recommended that they are used.

2.7.1 FLUID29 2D Acoustic Fluid Element

The FLUID29 element is a 2D planar acoustic element defined by 4 nodes. The potential geometry configurations, as shown in Figure 2.4, include a planar quadrilateral or a triangle, where the last two nodes are coincident. Each node has 1 pressure degree of freedom, and two optional translational DOFs along the x and y axes. This 2D planar element can only be defined along the global x-y plane. The element is capable of modeling fluid–structure interaction and damping at the boundary interface. By changing some of the options for the element KEYOPT(3), the FLUID29 element can be configured as a 2D planar element, axi-symmetric, or axi-harmonic. Table 2.4 lists a summary of the features of the FLUID29 element.

FLUID29

Quadrilateral Triangle

FIGURE 2.4
Schematic of a FLUID29 2D, 4-node, linear, acoustic element.

TABLE 2.4
FLUID29 Element Summary

Feature	Comment
Number of nodes	4
DOFs at each node	1 pressure + optional 2 displacement
Shapes	Quadrilateral, triangle
PML capable	No
Unsymmetric FSI option	Yes
Symmetric FSI option	No
Non-uniform acoustic media	No

When the element is used for fluid–structure interaction, it is recommended that the displacement degrees of freedom that are not involved with the fluid–structure interaction interface should be set to zero to avoid warning messages about "zero pivots." This can be done using APDL commands by selecting the nodes associated with these elements, but not on the FSI interface, and using the D command to set translation in the UX and UY degrees of freedom to zero.

Table 2.5 lists the analysis types that can be conducted using the FLUID29 element. When conducting a modal analysis that involves FSI, unsymmetric matrices are generated and hence only an unsymmetric eigen-solver can be used. When there is no FSI, symmetric matrices of the equations of motion are formed and a symmetric eigen-solver can be used.

This element is used in the following examples in this book:

- to model a circular piston in an infinite plane baffle in Section 8.4 and

- to model a rectangular 2D duct in Chapter 5.

TABLE 2.5
Analysis Types That Can Be Conducted with the FLUID29 Element

Analysis Type	Method	no FSI	with FSI
Modal	Undamped or symmetric	Yes	No
Modal	Damped or unsymmetric	Yes	Yes
Harmonic	Full	Yes	Yes
Harmonic	Modal summation	Yes	No
Transient	Full	Yes	Yes
Transient	Modal summation	No	No

2.7.2 FLUID30 3D Acoustic Fluid Element

The FLUID30 element is a linear 3D brick acoustic element defined by 8 nodes. This element does not have mid-side nodes. The potential geometry configurations, as shown in Figure 2.5, include a brick (hexahedral), wedge, pyramid, and tetrahedral. Each node has one pressure degree of freedom, and three optional translational DOFs along the x, y, and z axes. The element is capable of modeling fluid–structure interaction and the three translational DOFs are only applicable on the nodes on the FSI interface.

The element is capable of modeling sound-absorbing material on the boundary and an impedance sheet inside a fluid can also be defined. The material properties associated with the element can be specified with fluid viscosity to model acoustic dissipation.

This element type has existed in the portfolio of elements in ANSYS for several decades and has undergone some evolutionary changes with regard to the method that loads are applied to the element, the results that are available, the use of Perfectly Matched Layers (PML) , capability for non-uniform material properties, and a symmetric element configuration for fluid–structure

FLUID30

Hexahedral   Wedge   Pyramid   Tetrahedral

FIGURE 2.5
Schematic of a FLUID30 3D, 8-node, acoustic element.

interaction. Before Release 14.0 of ANSYS, after conducting a harmonic analysis, the sound pressure level (in decibels) within an element could be obtained by using the APDL command PLESOL,NMISC,4, as is still the case for the 2D FLUID29 elements. The sound pressure level is now obtained by requesting that complex valued results are displayed as amplitude using the APDL command SET,,,,AMPL, and then plotting the sound pressure level with the command PLNSOL,SPL, or listing the results with the command PRNSOL,SPL.

Table 2.6 lists a summary of the features of the FLUID30 element.

TABLE 2.6
FLUID30 Element Summary

Feature	Comment
Number of nodes	8
DOFs at each node	1 pressure + optional 3 displacement
Shapes	Brick, wedge, pyramid, tetrahedral
PML capable	Yes
Unsymmetric FSI option	Yes
Symmetric FSI option	Yes
Non-uniform acoustic media	Yes

The alternatives to using this linear 3D element are to use the higher-order FLUID220 element for hexahedral shapes, or the higher-order FLUID221 for tetrahedral shaped elements.

Table 2.7 lists the analysis types that can be conducted using the FLUID30 element. The table indicates that all modal analysis methods are available, however certain conditions must be met. Modal analyses are conducted using either symmetric or unsymmetric eigen-solvers. The method that can be used depends on whether the matrices of the equations of motion of the system are symmetric or unsymmetric. The FLUID30 elements can be specified to use a symmetric formulation for fluid–structure interaction coupling, in which case a symmetric eigen-solver can be used in the modal analysis. If the unsymmetric form of the fluid–structure interaction is used, then only an unsymmetric eigen-solver can be used to conduct a modal analysis. A coupled modal analysis cannot be conducted for a symmetric matrix formulation with viscous materials.

TABLE 2.7
Analysis Types That Can Be Conducted with the FLUID30 Element

Analysis Type	Method	no FSI	with FSI
Modal	Undamped or symmetric	Yes	Yes
Modal	Damped or unsymmetric	Yes	Yes
Harmonic	Full	Yes	Yes
Harmonic	Modal summation	Yes	No
Transient	Full	Yes	Yes
Transient	Modal summation	No	No

ANSYS recommends that at least 12 elements per wavelength should be used for this linear element.

The material properties for this element can be described in terms of non-uniform temperature and pressure, and hence it is possible to accommodate temperature and pressure gradients in the system. For example, a temperature gradient along a duct will cause a change in the speed of sound of the fluid, and therefore change the resonance frequencies, impedances, and acoustic response of the system compared to a system with a uniform temperature.

2.7.3   FLUID129 2D Infinite Acoustic Element

The FLUID129 element is a 2D element intended to provide an infinite acoustic boundary around a mesh of FLUID29 2D acoustic elements, so that outgoing acoustic pressure waves are absorbed with little reflection into the fluid. Figure 2.6 shows the geometry of the FLUID129 element as a line element defined by 2 nodes, with only a single pressure degree of freedom at each node. As with the FLUID29 element, it can only be defined in the global x-y plane.

FLUID129

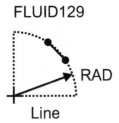

Line

FIGURE 2.6
Schematic of a FLUID129 2D, 2-node, acoustic element for modeling an infinite boundary.

Figure 2.7 shows an example use of the FLUID129 elements that are on the exterior boundary of a mesh of FLUID29 elements. The nodes of each FLUID129 element must be defined on a circle of radius RAD using a REAL constant set (as described below). Errors can be generated if the nodes of this element are not *precisely* on the surface of a circular arc. Errors can occur even when using the standard meshing operations both in Mechanical APDL and Workbench. Section D.2.4 provides more details on this error and how to fix it. ANSYS recommends that the enclosing arc containing the FLUID129 element is at least 0.2 of a wavelength from any structure in the fluid. The coordinates of the center of the arc (X0,Y0) must also be defined using a REAL constant set. If the coordinates of the center of the arc not defined, it is assumed that the center is at the global origin. The radius RAD and center of the arc (X0,Y0) are defined using a set of REAL constants using the APDL command R,NSET,RAD,X0,Y0.

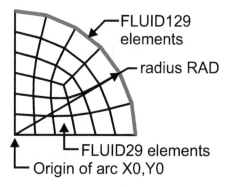

FIGURE 2.7

Example of a finite element model where FLUID129 2D elements are on a circular arc on the exterior boundary of FLUID29 elements.

Multiple Meanings of *Real*

The term *real* is used in several contexts in this book, which is likely to lead to confusion, especially for people that use English as an additional language.

real in the context of the real part of a complex number. For example, the real part of the complex number $\mathrm{Re}(2 + 3j) = 2$.

real in the context of being genuine.

REAL the APDL command REAL, NSET is used to select the set numbered NSET that contains several real-valued (i.e., not a complex number) constants just before meshing an object. The set of constants are defined using the APDL command R, NSET, R1, R2, R3, R4, R5, R6.

Table 2.8 lists a summary of the features of the FLUID129 element.

TABLE 2.8
FLUID129 Element Summary

Feature	Comment
Number of nodes	2
DOFs at each node	1 pressure
Shapes	Line
PML capable	No
Unsymmetric FSI option	Yes
Symmetric FSI option	No
Non-uniform acoustic media	No

Table 2.9 lists the analysis types that can be conducted using the FLUID129 element. Note that the only modal analysis method that is available is the damped method.

TABLE 2.9
Analysis Types That Can Be Conducted with the FLUID129 Element

Analysis Type	Method	no FSI	with FSI
Modal	Undamped or symmetric	No	No
Modal	Damped or unsymmetric	Yes	Yes
Harmonic	Full	Yes	Yes
Harmonic	Modal summation	Yes	No
Transient	Full	Yes	Yes
Transient	Modal summation	No	No

2.7.4 FLUID130 3D Infinite Acoustic Element

The FLUID130 element is a 3D element intended to provide an infinite acoustic boundary around a mesh of 3D acoustic elements, which include FLUID30, FLUID220, and FLUID221, so that outgoing acoustic pressure waves are absorbed with little reflection into the fluid. The FLUID130 element is the 3D equivalent of the 2D FLUID129 element. There is some consistency in the element type numbering scheme: the FLUID129 elements can be used with FLUID29 elements, and the FLUID130 elements can be used with FLUID30 elements. Figure 2.8 shows the geometry of the FLUID130 element as either a quadrilateral or as a degenerate triangular shape. The element can be defined with 4 nodes and will connect to the FLUID30 elements that do not have mid-side nodes, or with 8 nodes and will connect to the higher-order FLUID220 and FLUID221 elements that have mid-side nodes. Degenerate shapes of the elements are defined by assigning duplicate nodes at one of the vertices, such that the degenerate version of the 4-node quadrilateral element has a triangular shape, and the degenerate version of the 8-node octagonal element is a triangular-shaped element. Each node only has a single pressure degree of freedom.

FLUID130

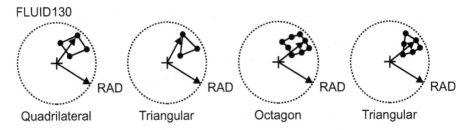

| Quadrilateral | Triangular | Octagon | Triangular |

FIGURE 2.8
Schematic of a FLUID130 3D acoustic element for modeling an infinite boundary with 4 nodes as a quadrilateral or degenerate triangular shape, or 8 nodes as an octagonal or degenerate triangular shape.

Similar to the FLUID129 element, the nodes of the FLUID130 element must be defined on a spherical surface of radius RAD using a REAL constant set (as described below). Figure 2.9 shows an example use of the FLUID130 elements

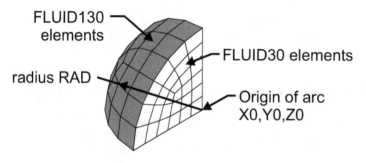

FIGURE 2.9
Example finite element model where FLUID130 3D elements are on the exterior
boundary of a spherical volume of FLUID30 elements.

where 1/8 th of a sphere has been meshed with FLUID30 acoustic elements, and
the exterior of the spherical surface at radius RAD from center point (X0,Y0,Z0)
has been meshed with FLUID130 elements. Errors can be generated if the nodes
of this element are not *precisely* on the surface of a sphere. Errors can occur
even when using the standard meshing operations both in Mechanical APDL
and Workbench. Section D.2.4 provides more details on this error and how
to fix it. ANSYS recommends that the enclosing spherical surface containing
the FLUID130 element is at least 0.2 of a wavelength from any structure in the
fluid. The coordinates of the center of the arc (X0,Y0,Z0) must also be defined
using a REAL constant set. If the coordinates of the center of the arc are not
defined, it is assumed that the center is at the global origin. The radius RAD
and center of the arc (X0,Y0,Z0) are defined using a set of REAL constants
using the APDPL command R,NSET,RAD,X0,Y0,Z0.

Table 2.10 lists a summary of the features of the FLUID130 element.

TABLE 2.10
FLUID130 Element Summary

Feature	Comment
Number of nodes	4 or 8
DOFs at each node	1 pressure
Shapes	Quadrilateral, triangle
PML capable	No
Unsymmetric FSI option	Yes
Symmetric FSI option	No
Non-uniform acoustic media	No

The formulation for the FLUID130 element assumes that the fluid has a constant density and constant speed of sound with no damping loss mechanisms
in the infinite domain. Therefore the acoustic elements in contact with the
FLUID130 elements must not be defined with viscosity, or non-uniform acoustic properties, or defined with the Johnson–Champoux–Allard model option.

The acoustic pressure wave-absorbing property of the FLUID130 element works best when the outgoing acoustic waves are normally incident to the boundary.

2.7.5   FLUID220 3D Acoustic Fluid 20-Node Solid Element

The FLUID220 element is a higher-order version of the FLUID30 element type, and has 20 nodes instead of 8 nodes. Figure 2.10 shows the geometry configurations that are possible, and that the element has mid-side nodes. The element is defined by all 20 nodes, and the degenerate shapes of the wedge, pyramid, and tetrahedral are defined by duplicate nodes at the vertices.

FLUID220

Hexahedral   Wedge   Pyramid   Tetrahedral

FIGURE 2.10
Schematic of a FLUID220 3D, 20-node, acoustic element and the degenerate shapes of a wedge, pyramid, and tetrahedral.

The characteristics and capabilities of this element are very similar to the FLUID30 element. However, the FLUID30 element has a linear shape function, which means it is assumed that pressure varies linearly through the element, whereas the FLUID220 element has a quadratic shape function, which means that pressure through the element can be described by a quadratic formulation. The following paragraphs describe the differences, advantages, and disadvantages of a quadratic shape function.

Figure 2.11 shows a pressure distribution $p(x)$ along a one-dimensional coordinate x as the curve with the solid line. The finite element method attempts to model the pressure distribution with an acoustic element. One can choose an acoustic element that has a linear shape function, such as the FLUID30 element, which is represented by the curve with the long dashes in Figure 2.11. It can be seen that this linear fit does not represent the complicated pressure distribution very well. If one wanted to use acoustic elements with linear shape functions, then it would be advisable to use at least 4 elements to model this pressure distribution. See Section 2.11 for further discussion about mesh density. If an acoustic element with a quadratic shape function, such as the FLUID220 element, were used to model this pressure distribution, it can be seen in Figure 2.11 that there is much closer approximation to the exact pressure distribution than the element with a linear shape function. The advantage of the increased accuracy comes with the penalty that more nodes (and hence

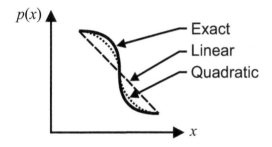

FIGURE 2.11
Comparison of fitting linear or a quadratic shape function to a pressure distribution.

degrees of freedom) are used in the model, which may take longer to solve than using linear elements.

ANSYS recommends that at least 6 elements per wavelength should be used for this quadratic element.

Examples of the use of the FLUID220 element type are described in:

- Section 3.3.2 where the resonance frequencies of a duct with rigid ends is calculated using FLUID30 and FLUID220 elements.

- Section 3.5 where it is shown that the effects of an irregular mesh of FLUID30 elements in a duct can cause the generation of non-plane waves, and the effects can be reduced by using FLUID220 elements.

- Section 3.6.3 where the sound field in a duct with a linear temperature gradient is calculated.

- Sections 6.5 and 6.5.2 where the sound field is calculated in a duct that has walls with locally reacting liners that absorb sound, using ANSYS Mechanical APDL and Workbench, respectively.

- Section 8.6.4 where the scattered acoustic pressure is calculated of an incident plane wave striking a rigid-walled cylinder.

- Section 9.4.3 where a fluid–structure interaction (FSI) analysis is conducted of a rectangular box with a flexible panel on one side.

2.7.6 FLUID221 3D Acoustic Fluid 10-Node Solid Element

The FLUID221 is similar to the FLUID220 element, however it only has one geometry configuration available as a 10-node tetrahedral with mid-side nodes, as shown in Figure 2.12.

The characteristics and capabilities of this element are identical to the FLUID30 element. This tetrahedral element is suitable for use when there is a complicated geometry and it is difficult to mesh with hexahedral (brick) elements.

FLUID221

Tetrahedral

FIGURE 2.12
Schematic of a FLUID221 3D, 10-node, acoustic element.

2.8 ACT Acoustics Extension

The ACT Acoustics extension add-on is intended to make it easier to conduct acoustic analyses using the ANSYS Workbench Graphical User Interface. The package installs a new menu bar in Mechanical, as shown in Figure 2.13, that has options specifically for conducting acoustic analyses. When the items in the menu are selected, additional objects are inserted into an analysis that contains APDL commands that are executed when the model is solved. The menu options are described in the following sections.

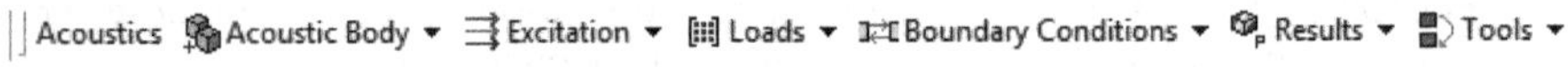

|| Acoustics 🐡 Acoustic Body ▾   ≣ Excitation ▾   🏢 Loads ▾   ≖ Boundary Conditions ▾   🐡ₚ Results ▾   📖 Tools ▾

FIGURE 2.13
ACT Acoustics extension menu bar in Mechanical.

2.8.1 Acoustic Body

The ACT Acoustics extension in ANSYS Workbench has one type of acoustic body that can be implemented as listed in Table 2.11.

TABLE 2.11
Acoustic Bodies Available in the ACT
Acoustics Extension

Method	APDL command
🐡 Acoustic Body ▾	
🐡 Acoustic Body	EMODIF

When the acoustic body branch is inserted into an analysis, bodies can be designated as acoustic domains to be meshed with acoustic elements. ANSYS Workbench will mesh bodies with structural solid elements by default and the Acoustic Body option will transform the body into acoustic fluid elements as listed in Table 2.12 using the APDL command EMODIF.

TABLE 2.12

Transforms of Solid Structural Elements to Acoustic Fluid Elements Using the Acoustic Body Option in the ACT Acoustics Extension

Original Element	Acoustic Element
SOLID185	FLUID30
SOLID186	FLUID220
SOLID187	FLUID221

An example of the Details of "Acoustic Body" window is shown below.

Details of "Acoustic Body - Acoustic Interior"	
⊟ **Scope**	
Scoping Method	Named Selection
Named Selection	acoustic_interior
⊟ **Definition**	
Temperature Dependency	No
Mass Density	1.2041 [kg m^-1 m^-1 m^-1]
Sound Speed	343.24 [m sec^-1]
Dynamic Viscosity	0 [Pa sec]
Thermal Conductivity	0 [W m^-1 C^-1]
Specific Heat Cv	0 [J kg^-1 C^-1]
Specific Heat Cp	0 [J kg^-1 C^-1]
Equivalent Fluid of Perforated Material	No
Reference Pressure	2E-05 [Pa]
Reference Static Pressure	101325 [Pa]
Acoustic-Structural Coupled Body Options	Uncoupled
Perfectly Matched Layers (PML)	Off

It is possible to define the acoustic body to have properties of the fluid that vary with temperature by changing the row Temperature Dependency to Yes. This will enable entering values in a table for density, speed of sound, dynamic viscosity, thermal conductivity, coefficient of specific heat at constant volume, and coefficient of specific heat at constant pressure that vary with temperature. The reference temperature can be defined in ANSYS Workbench by clicking in the Outline window on the analysis object such as Modal (A5) or Harmonic Response (A5), in the window Details of ... , enter the desired temperature in the row Options | Environment Temperature. The equivalent command using ANSYS Mechanical APDL is TREF.

The default fluid properties are for air at standard pressure and temperature, where the density is $\rho_0 = 1.2041\text{kg/m}^3$, speed of sound $c_0 = 343.24\text{m/s}$, and the dynamic viscosity, thermal conductivity, coefficients of specific heat at constant volume and pressure are 0.

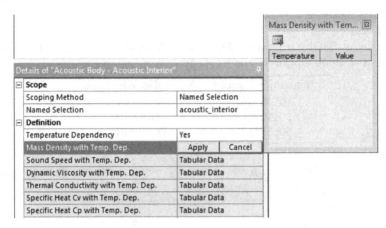

It is also possible to simulate the acoustic behavior of a porous material by changing the row `Equivalent Fluid of Perforated Material` to Yes. The equivalent fluid properties follow the Johnson–Champoux–Allard model [2]. Note that although ANSYS uses the term "perforated material," this does not mean a piece of perforated sheet metal; it is used in the context of "porous media" that absorbs incident acoustic waves. When this feature is activated it will enable entering material properties for the fluid resistivity, porosity, tortuosity, characteristic viscous length, and characteristic thermal length.

Equivalent Fluid of Perforated Material	Yes ▼
Fluid Resistivity	0 [N sec m^-1 m^-1 m^-1 m^-1]
Porosity	0
Tortuosity	0
Viscous Characteristic Length	0 [m]
Thermal Characteristic Length	0 [m]

The Johnson–Champoux–Allard model is further described in Section 6.4.4.3. Examples of the use of this porous material are described in Sections 6.5.2 and 6.5.3, and the ANSYS Verification Manual test case VM242 [12, VM242].

The row `Reference Pressure` has a default value of 2E-05 Pa, which is used for the calculation of sound pressure levels in decibels, as described in Section 2.8.5.2.

The row `Reference Static Pressure` has a default value of 1 atmosphere = 101325 Pa. This is used to model non-uniform acoustic media where the fluid properties vary with temperature and static pressure. See Section 2.8.3.1 for further details.

These two pressures are defined using the APDL command R, ,PREF, PSREF where PREF is the reference pressure, which has a default value of 2E-05 Pa, and PSREF is the reference static pressure, which has a default value of 101325 Pa.

The row `Acoustic-Structural Coupled Body Options` has three options for specifying whether the acoustic elements should include displacement and pressure degrees of freedom at the nodes, and whether the fluid–structure interaction equations should be formulated using unsymmetric or symmetric matrices.

Acoustic-Structural Coupled Body Options	Uncoupled
Perfectly Matched Layers (PML)	Uncoupled
	Coupled With Unsymmetric Algorithm
	Coupled With Symmetric Algorithm

Uncoupled the acoustic elements only have pressure degrees of freedom at the nodes and there is no coupling to structural elements.

Coupled With Unsymmetric Algorithm the acoustic elements have both pressure and displacement degrees of freedom at the nodes, and an unsymmetric formulation for the FSI matrices are used.

Coupled With Symmetric Algorithm the acoustic elements have both pressure and displacement degrees of freedom at the nodes, and a symmetric formulation for the FSI matrices are used.

Section 2.4 describes the theoretical aspects of fluid–structure interaction, and Chapter 9 contains examples of the use of this feature.

The row `Perfectly Matched Layers (PML)` can be set to On, where a row for `PML Options` is available. This row can be set to On - 3D PML or On - 1D PML, then additional options are available for defining the PML reflection coefficients in the x, y, z axes.

PML Options	On - 3D PML
Element coordinate system number	0
Reflection Coefficient in Negative X Direction	0.001
Reflection Coefficient in Positive X Direction	0.001
Reflection Coefficient in Negative Y Direction	0.001
Reflection Coefficient in Positive Y Direction	0.001
Reflection Coefficient in Negative Z Direction	0.001
Reflection Coefficient in Positive Z Direction	0.001

Examples of using PML acoustic bodies are in Sections 8.2.1 and 8.3.

2.8.2   Excitation

When applying a sound source to an acoustic finite element model, prior to ANSYS Release 14.0 there were two main types: acoustic mass acceleration sources F,NODE,FLOW and pressure sources D,NODE,PRES (which can be considered as a boundary condition). If the displacement degrees of freedom were activated on the acoustic elements, then it was possible to apply a displacement boundary condition with the APDL command D, NODE, Lab, VALUE, where NODE is the node number, Lab is the axes in which the boundary condition would be applied and is one of UX, UY, or UZ, and VALUE is the value of the displacement. The acoustic excitation types that are available with the ACT Acoustics extension are listed in Table 2.13.

TABLE 2.13
Excitation Types Available in the ACT Acoustics Extension

Method	APDL command
⊟ Excitation ▼	
⊟ Wave Sources (Harmonic)	AWAVE
⫯ Normal Surface Velocity (Harmonic)	SF,,SHLD or SFA,,SHLD
▦ Mass Source (Harmonic)	BF,,JS,value,,,phase
→_S Surface Velocity (Harmonic)	
	BF,,EF, X comp, Y comp, Z comp, phase
⫯ Normal Surface Acceleration (Transient)	SF,,SHLD or SFA,,SHLD
▦ Mass Source Rate (Transient)	BF,,JS,value
→_A Surface Acceleration (Transient)	
	BF,,EF, X comp, Y comp, Z comp, phase

Tables 2.14 and 2.15 provide a summary of the configuration options available for applying acoustic surface loads and body loads, respectively, using APDL commands. These load types will be explained in this chapter. Note that the APDL command SF applies only to area and volume elements.

2.8.2.1   Wave Sources

An incident acoustic wave can be created by selecting

⊟ Wave Sources (Harmonic)

from the Excitation menu in the ACT Acoustics menu bar. The wave originates from outside the model and a reference body, face, or vertex is not required to define the source of the wave.

The wave types that are available are incident plane wave, monopole or pulsating spherical incident, wave, incident dipole wave, back-enclosed loudspeaker, bare loudspeaker, as shown in Table 2.16. The theories for these acoustic wave sources are described in ANSYS [13] and Patronis [128].

TABLE 2.14

Options Available for Applying Acoustic Surface Loads Using APDL Commands

	Body	APDL Command			
	load	Node	Element	Line	Area
Description	label	SF	SFE	SFL	SFA
Fluid–structure interaction flag	FSI	Y	Y	N	Y
Impedance or admittance coefficient	IMPD	Y	Y	N	Y
Surface normal velocity or acceleration	SHLD	Y	Y	N	Y
Maxwell surface flag or equivalent source surface	MXWF	Y	Y	N	Y
Free surface flag	FREE	Y	Y	N	Y
Exterior Robin radiation boundary flag	INF	Y	Y	N	Y
Attenuation coefficient	CONV	Y	Y	N	Y

TABLE 2.15

Options Available for Applying Acoustic Body Force Loads Using APDL Commands

	Body	APDL Command				
	load	Node	Element	Line	Area	Volume
Description	label	BF	BFE	BFL	BFA	BFV
Mass source or mass source rate	JS	Y	N	Y	Y	Y
Impedance sheet	IMPD	Y	N	N	Y	N
Static pressure	CHRGD	Y	N	N	N	N
Temperature	TEMP	Y	N	Y	Y	Y
Velocity or acceleration	EF	Y	N	N	Y	N

TABLE 2.16

Formulation of Analytical Wave Sources Used with the APDL Command
`AWAVE`

Wave Type	Schematic
Incident plane wave `Wavetype=PLAN`	
Monopole wave `Wavetype=MONO`	
Dipole wave `Wavetype=DIPO`	
Bare loudspeaker `Wavetype=BARE`	
Back-enclosed loudspeaker `Wavetype=BACK`	

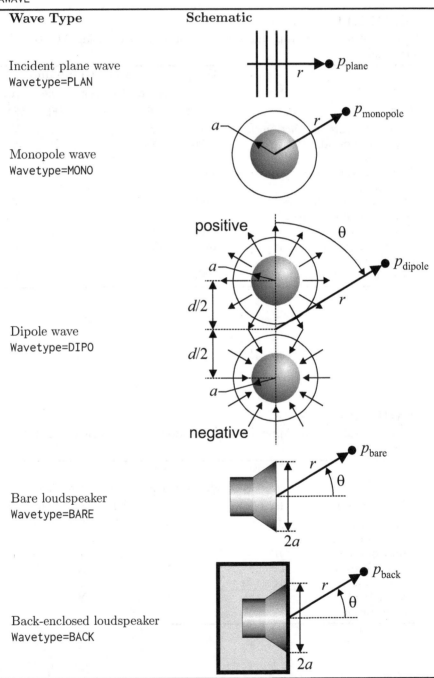

The wave is created using the APDL command

AWAVE, Wavenum, Wavetype, Opt1, Opt2, VAL1, VAL2, VAL3, VAL4, VAL5, VAL6, VAL7, VAL8, VAL9, VAL10, VAL11, VAL12

where

Wavenum is the number that the user assigns for each wave (and has no relationship to the acoustic wavenumber $k = \omega/c$)

Wavetype is one of

PLAN incident plane wave

MONO monopole incident wave, similar to a pulsating sphere

DIPO incident dipole wave

BACK back-enclosed loudspeaker

BARE bare loudspeaker

STATUS displays the status of the waves that have been defined with the AWAVE command, but is not used in the ACT Acoustics extension.

DELE deletes the specified Wavenum acoustic wave, but is not used in the ACT Acoustics extension.

Opt1 determines whether the amplitude of the wave is defined by pressure by selecting PRES, or by the amplitude of the normal velocity by selecting VELO.

Opt2 is either EXT for an incident acoustic wave outside the model, or INT for an incident acoustic wave that originates from inside the model. This is only available for pure scattered pressure formulation, where the APDL command HFSCAT, SCAT has been used. When the Wavetype is set to PLAN for an acoustic plane wave, the only option available is Opt2=EXT.

The input parameters VAL1 to VAL12 depend on the Wavetype that was selected and are described in Table 2.17.

Incident Plane Wave

If Wavetype=PLAN, the acoustic plane wave source is defined and has a spherical coordinate system as shown in Figure 2.14.

An incident plane wave can be created that originates from inside or outside the acoustic finite element model and propagates through it. The incident complex pressure is [47, p. 622] [13, Eq. (8-49)]

$$p_{\text{plane}} = P_0 \, e^{j\phi - j(k_x x + k_y y + k_z z)} , \qquad (2.11)$$

where P_0 is the amplitude of the plane wave, ϕ is the initial phase angle shift of the plane wave that is usually ignored, and k_x, k_y, k_z are the wavenumber

TABLE 2.17
Description of the VALn Input Parameters for the APDL Command AWAVE

Description	VALn	Wavetype				
		PLAN	MONO	DIPO	BACK	BARE
Amplitude of pressure or normal velocity to the sphere surface	VAL1	Y	Y	Y	Y	Y
Phase angle of the applied pressure or velocity in units of degrees, with a default value of 0	VAL2	Y	Y	Y	Y	Y
Global Cartesian x coordinate for the position of the acoustic source (*) If Wavetype=PLAN, then VAL3 is the incident angle ϕ from the x axis toward the y axis.	VAL3	Y*	Y	Y	Y	Y
Global Cartesian y coordinate for the position of the acoustic source (*) If Wavetype=PLAN, then VAL4 is the incident angle θ from the z axis toward the y axis.	VAL4	Y*	Y	Y	Y	Y
Global Cartesian y coordinate for the position of the acoustic source If Wavetype=PLAN, then VAL5 is not used.	VAL5	N	Y	Y	Y	Y
Density of the acoustic media, with a default value of 1.2041kg/m^3	VAL6	Y	Y	Y	Y	Y
Speed of sound of the acoustic media with a default value of 343.24m/s.	VAL7	Y	Y	Y	Y	Y
Radius of the pulsating sphere	VAL8	N	Y	Y	Y	Y
Length of the dipole	VAL9	N	N	Y	N	Y
x coordinate of a unit vector for the dipole axis from the positive to the negative	VAL10	N	N	Y	N	Y
y coordinate of a unit vector for the dipole axis from the positive to the negative	VAL11	N	N	Y	N	Y
z coordinate of a unit vector for the dipole axis from the positive to the negative	VAL12	N	N	Y	N	Y

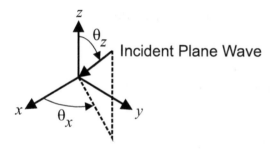

FIGURE 2.14
Spherical coordinate system for defining an incident plane wave vector propagating towards the origin of the Cartesian coordinate system.

components along the x, y, z axes, respectively, such that their magnitude is equal to the wavenumber as

$$k^2 = \left[\frac{\omega}{c_0}\right]^2 = k_x^2 + k_y^2 + k_z^2 . \qquad (2.12)$$

Using the ACT Acoustics extension, the input parameters for a `Planar Wave` source can be defined in the window `Details of "Acoustic Wave Sources"` as shown below.

Details of "Acoustic Wave Sources"	무
⊟ **Definition**	
Wave Number	1
Wave Type	Planar Wave ▼
Excitation Type	Pressure
Source Location	Outside The Model
Pressure Amplitude	1 [Pa]
Phase Angle	0 [°]
Angle Phi (From X Axis Toward Y Axis)	0 [°]
Angle Theta (From Z Axis Toward X Axis)	0 [°]
Mass Density Of Environment Media	1.2041 [kg m^-1 m^-1 m^-1]
Sound Speed In Environment Media	343.24 [m sec^-1]
Pure Scattering Options	Off

Monopole Wave

A monopole is one of the fundamental acoustic source types and is used to form more complicated acoustic wave source types. It can be modeled as a pulsating sphere where its surface expands and contracts radially. The pressure at a distance r from a monopole is [128, p. 21] [47, p. 188] [131] [102, Eq. (7.36), p. 164]

$$p_{\text{monopole}} = \frac{j\rho_0 c_0 kSu}{4\pi r} e^{j(\omega t - kr)} , \tag{2.13}$$

where S is the surface area of the sphere of radius a given by

$$S = 4\pi a^2 , \tag{2.14}$$

u is the velocity of the expanding and contracting surface of the sphere, ρ_0 is the density of the acoustic fluid, c_0 is the speed of sound in the acoustic fluid, $\omega = 2\pi f$ is the angular frequency, and $k = \omega/c_0$ is the wavenumber. In Equation (2.13) the product $Su = Q$ is the acoustic volume velocity of the source.

A monopole acoustic wave source can be defined to originate from inside or outside the acoustic finite element model. The input parameters for a Monopole source can be defined in the window Details of "Acoustic Wave Sources" as shown below.

Details of "Acoustic Wave Sources"	卫
⊟ **Definition**	
Wave Number	1
Wave Type	Monopole
Excitation Type	Pressure
Source Location	Outside The Model ▼
Pressure Amplitude	1 [Pa]
Phase Angle	0 [°]
Global X Coordinate At Source Origin	0 [m]
Global Y Coordinate At Source Origin	0 [m]
Global Z Coordinate At Source Origin	0 [m]
Mass Density Of Environment Media	1.2041 [kg m^-1 m^-1 m^-1]
Sound Speed In Environment Media	343.24 [m sec^-1]
Radius Of Pulsating Sphere	0 [m]
Pure Scattering Options	Off

The rows for Mass Density of Environment Media, and Sound Speed of Environment Media correspond to ρ_0 and c_0 in Equation (2.13), respectively. The row Radius of Pulsating Sphere corresponds to a in Equation (2.14).

Dipole Wave

A dipole comprises two monopole sources of equal strength but of opposite phase, that are separated by a distance d. Referring to Table 2.16, the far-field sound pressure at a distance r and angle θ from the axis of the dipole is [128,

p. 21] [47, p. 192]

$$p_{\text{dipole}} = -j\frac{\rho_0 c_0 k^2 S\, u\, d}{4\pi r} \cos(\theta)\, e^{j(\omega t - kr)}\,, \tag{2.15}$$

where the radiating surface area of each monopole is from the surface of a sphere of radius a and is given by

$$S = 4\pi a^2\,, \tag{2.16}$$

and the remaining terms are the same as described for a monopole source.

A dipole acoustic wave source can be defined to originate from inside or outside the acoustic finite element model. The input parameters for a Dipole source can be defined in the window Details of "Acoustic Wave Sources" shown below.

Details of "Acoustic Wave Sources"	
⊟ **Definition**	
Wave Number	1
Wave Type	Dipole
Excitation Type	Pressure
Source Location	Outside The Model
Pressure Amplitude	1 [Pa]
Phase Angle	0 [°]
Global X Coordinate At Source Origin	0 [m]
Global Y Coordinate At Source Origin	0 [m]
Global Z Coordinate At Source Origin	0 [m]
Mass Density Of Environment Media	1.2041 [kg m^-1 m^-1 m^-1]
Sound Speed In Environment Media	343.24 [m sec^-1]
Radius Of Pulsating Sphere	0 [m]
Dipole Length	0 [m]
X Component Of Unit Dipole Vector	0 [m]
Y Component Of Unit Dipole Vector	0 [m]
Z Component Of Unit Dipole Vector	0 [m]
Pure Scattering Options	Off

The row Radius of Pulsating Sphere corresponds to parameters a in Equation (2.16), and the row Dipole Length corresponds to d in Equation (2.15). The default value for the rows X Component of Unit Dipole Vector, Y Component of Unit Dipole Vector, and Z Component of Unit Dipole Vector are $(0,0,0)$ and an error will be generated unless values are defined (see page 647). Note that when defining the components of the vector, it is *not* necessary to define the components of the vector so that the magnitude $(\sqrt{x^2 + y^2 + z^2})$ equals 1—the equivalent vector of unit length will be calculated by ANSYS.

Bare Loudspeaker

A bare (unbaffled) loudspeaker has the characteristic that as its cone moves outward the front face generates a "positive" acoustic pressure, the rear face generates a "negative" acoustic pressure, and around the edges of the loudspeaker the two acoustic waves interfere, as illustrated in Figure 2.15.

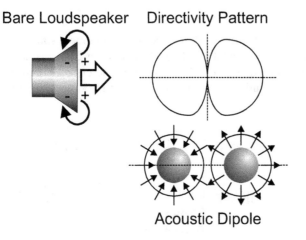

FIGURE 2.15
Schematic of a bare loudspeaker and the resulting directivity arising from the acoustic interference at the edges, and how it can be modeled as an acoustic dipole.

A bare loudspeaker can be simulated as an acoustic dipole where the radiating surface is the area of a circle of radius a is given by [128, p. 21]

$$S = \pi a^2 . \tag{2.17}$$

Note the main difference between the bare loudspeaker and the dipole sources is that for the dipole source, the radiating area is the surface of two spheres (Equation (2.14)) and for the bare loudspeaker the radiating area is the surface area of a circle (Equation (2.17)). These areas differ by a factor of 4, and hence one could expect that the sound pressure levels of a model using a bare loudspeaker would be $20 \times \log_{10}(4) = 12 \text{dB}$ less than the sound pressure levels of model using a dipole.

An example showing a comparison between a bare loudspeaker and a dipole is in Section 8.3.5.

Determining the input parameters applicable for a real loudspeaker, such as the effective radiating area, is not simple. For further information the reader is referred to Backman [43].

A bare loudspeaker acoustic wave source can be defined to originate from inside or outside the acoustic finite element model. The input parameters for a

Bare Loudspeaker source can be defined in the window Details of "Acoustic Wave Sources".

Details of "Acoustic Wave Sources"	
Definition	
Wave Number	1
Wave Type	Bare Loudspeaker
Excitation Type	Pressure
Source Location	Outside The Model
Pressure Amplitude	1 [Pa]
Phase Angle	0 [°]
Global X Coordinate At Source Origin	0 [m]
Global Y Coordinate At Source Origin	0 [m]
Global Z Coordinate At Source Origin	0 [m]
Mass Density Of Environment Media	1.2041 [kg m^-1 m^-1 m^-1]
Sound Speed In Environment Media	343.24 [m sec^-1]
Radius Of Pulsating Sphere	0 [m]
Dipole Length	0 [m]
X Component Of Unit Dipole Vector	0 [m]
Y Component Of Unit Dipole Vector	0 [m]
Z Component Of Unit Dipole Vector	0 [m]
Pure Scattering Options	Off

Back-Enclosed Loudspeaker

A loudspeaker that has the rear portion contained within a sealed enclosure, does not have the interference that occurs between the front and rear faces of the driver cone that occurs with a bare loudspeaker. This configuration of the loudspeaker can be considered to have an acoustic response similar to an acoustic monopole, where the radiating surface area is that of a circular piston of radius a given by [128, p. 21]

$$S = \pi a^2. \tag{2.18}$$

Note the main difference between the ANSYS model of a back-enclosed loudspeaker and the monopole sources is that for the monopole source the radiating area is the surface of a sphere (Equation (2.14)) and for the back-enclosed loudspeaker the radiating area is the surface area of a circle (Equation (2.18)). These areas differ by a factor of 4, and hence one could expect that sound pressure levels of a model using a back-enclosed loudspeaker would be $20 \times \log_{10}(4) = 12$dB less than the sound pressure levels of a model using a monopole for the same radius.

Note that the ANSYS model of the back-enclosed loudspeaker does not deal with:

- the coupling of the acoustic impedances between the loudspeaker enclosure and the electrical–mechanical impedance of the loudspeaker itself,

- the vibration and sound radiation from the walls of the loudspeaker, or

- the acoustic baffle effect caused by the face of the loudspeaker enclosure.

An example showing a comparison between a bare (unbaffled) loudspeaker and a monopole is in Section 8.3.3.

A back-enclosed loudspeaker acoustic wave source can be defined to originate from inside or outside the acoustic finite element model. The input parameters for a `back-enclosed loudspeaker` source can be defined in the window `Details of "Acoustic Wave Sources"`.

Details of "Acoustic Wave Sources"	
⊟ **Definition**	
Wave Number	1
Wave Type	Back Enclosed Loudspeaker ▼
Excitation Type	Pressure
Source Location	Outside The Model
Pressure Amplitude	1 [Pa]
Phase Angle	0 [°]
Global X Coordinate At Source Origin	0 [m]
Global Y Coordinate At Source Origin	0 [m]
Global Z Coordinate At Source Origin	0 [m]
Mass Density Of Environment Media	1.2041 [kg m^-1 m^-1 m^-1]
Sound Speed In Environment Media	343.24 [m sec^-1]
Radius Of Pulsating Sphere	0 [m]
Pure Scattering Options	Off

For the simulations where an acoustic source is placed within the mesh of an acoustic finite element model, it is relatively easy to understand how this would produce an acoustic pressure response in the model, as this is analogous to a structural analysis where a force is applied to a structure causing it to deflect. Acoustic wave excitation sources can also be defined to originate from outside the mesh of the finite element model. A detailed mathematical derivation of how this can be accomplished is beyond the scope of this book, and the reader is referred to Ref. [13], and will only be described qualitatively. The total pressure response comprises the incident and scattered acoustic waves. The incident incoming acoustic wave is applied on the boundary of the acoustic model, and the scattered outgoing acoustic wave is assumed to be planar and is absorbed by the PML region.

An example of the use of an acoustic plane wave source is described in Sections 8.6.4.

Care needs to be taken when using this excitation source and PML regions to investigate the transmission loss of a system. Typically one wants an incident acoustic wave field to exist *only* in a "source" room or cavity, and the vibro-acoustic energy that is transmitted into a "receiver" room or cavity is calculated. When the AWAVE command is issued, the incident acoustic wave can propagate through *all* the acoustic domains when they are bounded by PML regions, which is not the desired model. The required behavior can be simulated as two free-field regions separated by a planar baffle with a test partition, where the acoustic domain on the "source" side can be surrounded with PML elements, and the acoustic domain on the "receiver" side can be a hemispherical region surrounded by FLUID130 infinite elements. This configuration will permit an incident plane wave to exist only in the "source" room, but not in the "receiver" room.

2.8.2.2 Normal Surface Velocity

A normal surface velocity load suitable for harmonic response analyses can be simulated in ANSYS Workbench by selecting

Normal Surface Velocity (Harmonic)

from the Excitation menu in the ACT Acoustics menu bar.

An acoustic source can be simulated by applying a velocity normal to a face on the exterior of the acoustic domain, as shown in Figure 2.16. A limitation

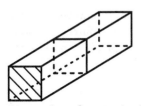

Normal surface velocity applied to an end face will work

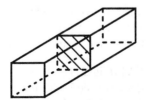

Normal surface velocity applied to an interior face will not work

FIGURE 2.16
A normal surface velocity applied to a face on the exterior of the fluid domain will work correctly.

of this excitation type in ANSYS is that it will not work if the normal surface velocity load is applied to a face on the interior of the acoustic domain. The vibrating surface causes acoustic particles adjacent to the surface to move and therefore will generate an acoustic pressure. This wave source is applicable in harmonic response analyses.

The APDL command that is issued to create the normal surface velocity on a face is

SF, Nlist, SHLD, VALUE, VALUE2

which applies a surface velocity to the selected nodes, where VALUE is the magnitude of the surface normal velocity, and VALUE2 is the phase angle of the normal surface velocity (defaults to zero).

If an analysis is conducted using Mechanical APDL, then it is also possible to apply surface loads to areas using the APDL command

SFA, AREA, SHLD, Lab, VALUE, VALUE2

Note that ANSYS Workbench does not create areas that can be translated into Mechanical APDL. The DESIGNMODELER module in ANSYS Workbench creates faces, which are only recognized within the Workbench environment. To translate regions from Workbench to Mechanical APDL, it is necessary to define Named Selections with the elements or nodes of interest.

2.8.2.3 Mass Source

A mass source suitable for harmonic response analyses can be simulated in ANSYS Workbench by selecting

Mass Source (Harmonic)

from the Excitation menu in the ACT Acoustics menu bar.

A mass source is a mass flow rate and is defined as a (normalized) value per unit volume as [14]

$$\frac{\text{mass}}{\text{length}^3 \times \text{time}}. \tag{2.19}$$

A mass source can be applied to a vertex, edge, face, or body, however the units have to be adjusted. Table 2.18 lists the units of the mass source depending on the object to which the source is attached. For example, if a mass source

TABLE 2.18
Mass Source Units for a
Vertex, Edge, Face and Body

Geometry	Units
Vertex	kg / (s)
Edge	kg / (m × s)
Face	kg / (m^2 × s)
Body	kg / (m^3 × s)

acoustic excitation was applied to a face, the value that is entered is the mass flow rate in kg/s divided by the surface area of the face in m^2 and hence the units are kg/m^2/s , even though the units in the row Amplitude of Mass Source will always be written as [kg m^-1 m^-1 m^-1 sec^-1].

The APDL command that is issued to create the mass source is

```
BF, NODE, JS, VAL1, , , VAL4
```

which defines a nodal body force load. The parameter VAL1 is the magnitude of the mass source in units of $kg/m^3/s$ for a harmonic response analysis, or mass source rate with units $kg / m^3 / s^2$ for a transient analysis (as discussed in Section 2.8.2.6), and VAL4 is the phase angle in degrees.

Table 2.19 shows examples of the application of a mass source to a vertex, edge, face, and body. A finite element model was created to simulate a "free-field" with a region of acoustic elements surrounded by a Perfectly Matched Layer of elements that act to absorb outgoing acoustic waves. The left column shows the object type and where the mass source was applied in the acoustic model. The right column shows the corresponding sound pressure level that was generated.

2.8.2.4 Surface Velocity

A surface velocity for harmonic response analyses can be simulated in ANSYS Workbench by selecting

$\overrightarrow{\rightarrow}_5$ Surface Velocity (Harmonic)

from the Excitation menu in the ACT Acoustics menu bar.

The APDL command that is issued is

```
BF, NODE, EF, VAL1, VAL2, VAL3, VAL4
```

which defines a nodal body force load on node NODE, where VAL1, VAL2, and VAL3 are the velocity components of the node in the x, y, z directions, respectively, and VAL4 is the phase angle in degrees. The equivalent command for applying a surface velocity to an area is the command BFA, AREA, EF, VAL1, VAL2, VAL3, VAL4.

A limitation of this excitation type in ANSYS is that it will not work if the surface velocity load is applied to a face on the interior of the acoustic domain, as shown in Figure 2.16.

This excitation type allows one to define the velocity components of a vibrating surface, however only vibration that is normal to the surface will generate an acoustic pressure.

2.8.2.5 Normal Surface Acceleration

A normal surface acceleration, which is applicable for acoustic transient simulations, can be simulated in ANSYS Workbench by selecting

$\overrightarrow{\Vert_{\overline{H}}}$ Normal Surface Acceleration (Transient)

from the Excitation menu in the ACT Acoustics menu bar.

The APDL command that is issued is

```
SF, Nlist, SHLD, VALUE
```

which applies a surface acceleration to the selected nodes, where VALUE is the magnitude of the surface normal acceleration. Note that VALUE2, which is used

TABLE 2.19

Contour Plots of the Sound Pressure Level Arising from an Acoustic
Mass Source Applied to a Vertex, Edge, Face, and Body

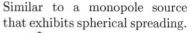

Location of Mass Source	Radiation Pattern

Vertex:

One vertex at the center of the acoustic model.

Similar to a monopole source that exhibits spherical spreading.

Edge:

Two edges near the center of the acoustic model.

Similar to a line source that exhibits cylindrical spreading.

Face:

Four faces in a vertical plane in the acoustic model.

Sound radiates out from both sides of a face with the same phase.

Body:

One body in the shape of a cube near the center of the acoustic model.

Depends on the shape of the body. In this example, sound radiates from each face of the cubic body.

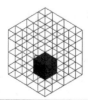

when applying a normal surface velocity for a harmonic analysis, is not used for a transient analysis.

A limitation of this excitation type in ANSYS is that it will not work if the normal surface acceleration load is applied to a face on the interior of the acoustic domain, as shown in Figure 2.16.

2.8.2.6 Mass Source Rate

A mass source rate, which is applicable for acoustic transient simulations, can be simulated in ANSYS Workbench by selecting

Mass Source Rate (Transient)

from the Excitation menu in the ACT Acoustics menu bar.

The APDL command that is issued is

BF, NODE, JS, VAL1

which defines a nodal body force load. The parameter VAL1 is the mass source in units of kg / m^3 / s^2 for a transient analysis. Note that the parameter VAL4, which is used in harmonic analyses to define the phase angle, is not used in a transient analysis.

2.8.2.7 Surface Acceleration

A surface acceleration, which is applicable for acoustic transient simulations, can be simulated in ANSYS Workbench by selecting

$\overrightarrow{\Rightarrow}_R$ Surface Acceleration (Transient)

from the Excitation menu in the ACT Acoustics menu bar.

The APDL command that is issued is

BF, NODE, EF, VAL1, VAL2, VAL3, VAL4

which defines a nodal body force load, where VAL1, VAL2, and VAL3 are the acceleration components in the x, y, z directions, respectively, and VAL4 is the phase angle in degrees.

A limitation of this excitation type in ANSYS is that it will not work if the surface acceleration load is applied to a face on the interior of the acoustic domain, as shown in Figure 2.16.

2.8.3 Body Force Loads

The ACT Acoustics extension in ANSYS Workbench has three body force load types that are listed in Table 2.20. Table 2.15 provides a summary of the configuration options available for applying body loads, respectively, using APDL commands.

TABLE 2.20

Load Types Available in the ACT Acoustics Extension

Method	APDL command
⊞ Loads ▾	
🔧 Static Pressure	BF,,CHRGD, value
⊞ Impedance Sheet	BF,,IMPD, Res, Rea or BFA,,IMPD, Res, Rea
≋ Temperature	BF,,TEMP, val,

2.8.3.1 Static Pressure

A static pressure can be applied to the acoustic fluid by selecting

🔧 Static Pressure

from the Loads menu in the ACT Acoustics menu bar.

The 3D acoustic fluid elements FLUID30, FLUID220, FLUID221 can be defined with non-uniform acoustic fluid properties, where the static pressure and the temperature can vary throughout the model. A static pressure can be applied to the fluid, which will cause an increase in density.

The speed of sound is calculated using the ideal gas law as [15, Eqs. (8-76), (8-79)]

$$c^2(x) = \gamma RT(x) , \tag{2.20}$$

$$p_{\text{state}} = \rho(x)\, RT(x) , \tag{2.21}$$

where $\gamma = C_p/C_v$ is the ratio of specific heats, C_p is the coefficient of specific heat at constant pressure per unit mass, C_v is the coefficient of specific heat at constant volume per unit mass, R is the universal gas constant, T is temperature, and p_{state} is the absolute pressure of the gas measured in atmospheres.

The acoustic fluid density ρ at a position x in the fluid is given by

$$\rho(x) = \frac{p_{\text{state}}(x)}{T(x)} \frac{\rho_0 T_0}{p_{\text{state},0}} , \tag{2.22}$$

where ρ_0 is the density at the reference temperature T_0 and reference pressure $p_{\text{state},0}$.

The speed of sound will vary with position as

$$c(x) = c_0 \sqrt{\frac{T(x)}{T_0}} . \tag{2.23}$$

The APDL command that is issued to define the static pressure is

BF, NODE, CHRGD, VAL1

which defines a nodal body force load. The parameter VAL1 is the static pressure for a non-uniform acoustic media.

2.8.3.2 Impedance Sheet

A thin impedance sheet can be applied to a face within an acoustic fluid by selecting

[⊞] Impedance Sheet

from the Loads menu in the ACT Acoustics menu bar. This boundary condition can be used when both sides of the sheet are in contact with the acoustic fluid. An example application for using this type of boundary condition is where a porous material is placed mid-length in an acoustic impedance tube measurement device. If only one side of the impedance sheet is in contact with the acoustic fluid, then use an Impedance Boundary boundary condition instead.

The mathematical description for this boundary condition is cast in terms of the acoustic admittance of the impedance sheet $Y(x) = 1/Z(x)$, the inverse of acoustic impedance $Z(x)$, as [16]

$$Y(x) = \frac{v_{n,F+} - v_{n,F-}}{p(x)}, \qquad (2.24)$$

where $v_{n,F+}$, $v_{n,F-}$ are the acoustic particle velocities normal to the impedance sheet on the $+$ top and $-$ bottom sides of the sheet, and $p(x)$ is the acoustic pressure on the impedance sheet. Although the mathematical description is cast in terms of acoustic admittance, values of acoustic impedance are input into ANSYS.

The APDL command that is issued is

BF, NODE, IMPD, VAL1, VAL2

which defines a nodal body force load. The parameters VAL1 and VAL2 are the acoustic resistance and reactance, respectively, with units of Pa s / m.

The values of acoustic impedance are sometimes available from product manufacturers, textbooks, or they can be measured using an impedance tube, or if the material properties such as porosity, tortuosity, density, etc. are known, the impedance can be calculated [47, Appendix C, p. 679].

2.8.3.3 Temperature

The temperature of the acoustic fluid can be altered from the reference value, which will cause a change in the speed of sound of the fluid, by selecting

≋ Temperature

from the Loads menu in the ACT Acoustics menu bar.

The APDL command that is issued is

BF,NODE,TEMP,VAL

which defines a nodal body force load for the temperature at the nodes. The parameter VAL is the temperature in units of Kelvin.

An example application using this load is shown in Section 3.6, where the temperature of an acoustic fluid is varied along the length of a duct.

2.8.4 Boundary Conditions

The ACT Acoustics extension in ANSYS Workbench can be used to insert acoustic boundary conditions into a model and these are listed in Table 2.21 along with the corresponding APDL command that is executed.

TABLE 2.21

Boundary Conditions Available in the ACT Acoustics Extension and the Corresponding APDL Command

Method	APDL Command
Boundary Conditions ▾	
Acoustic Pressure	D,,PRES, Real, Imag
Impedance Boundary	SF,,IMPD, Res, React or SFA,,IMPD, Res, React
Thermo-viscous BLI Boundary (beta)	SF,,BLI
Free Surface	SF,ALL,FREE
FSI Interface	SF,,SHLD or SF,,FSI,1
Radiation Boundary	SF,ALL,INF
Absorbing Elements (Exterior To Enclosure)	ESURF with FLUID130 elements
Equivalent Source Surface	SF,ALL,MXWF
Attenuation Surface	SF,,CONV, alpha

Note that the term "boundary condition" is used in the context of a *numerical* boundary condition, and does not necessarily have to be applied on an outer surface or boundary of the model. Numerical boundary conditions can be defined on interior parts of an acoustic model, such as an acoustic pressure boundary condition that is described in the following section.

2.8.4.1 Acoustic Pressure

An acoustic pressure boundary condition can be applied to an vertex, edge, face, or body by selecting

 Acoustic Pressure

from the Boundary Conditions menu in the ACT Acoustics extension toolbar.

The APDL command that is issued is

D,,PRES, Real, Imag

which defines constraints on the pressure degrees of freedom of the selected nodes.

Note that one should not confuse an acoustic pressure boundary condition as an acoustic excitation source, as noted in Section 1.2. By specifying an acoustic pressure boundary condition, one is defining the value of pressure in the model and this will influence the acoustic response throughout the entire model. This could lead to unexpected results where perhaps a forced acoustic particle (or volume) velocity excitation should have been specified, rather than using a pressure boundary condition. For duct acoustic problems, the noise source can be represented as either a pressure or velocity source [119, Fig. 2.5, p. 54], and it is important to select the correct representation in order to model the system accurately.

2.8.4.2 Impedance Boundary

An impedance boundary can be applied to an exterior face of an acoustic body, where only one side in contact with the acoustic fluid as shown in Figure 2.17, by selecting

⬛ Impedance Boundary

from the Boundary Conditions menu in the ACT Acoustics menu bar. An example application for this type of boundary condition is where a porous material is adhered to the interior surface of an acoustic cavity. If both sides of the acoustic porous material are in contact with the acoustic fluid, then consider using an Impedance Sheet from the Loads menu instead. In future releases of ANSYS there will be a Trim element that can be inserted into the acoustic domain.

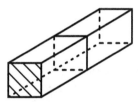

Impedance BC
applied to an end face
will work

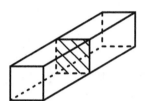

Impedance BC
applied to an interior face
will not work

FIGURE 2.17
Impedance boundary condition can only be applied to the exterior face of an acoustic domain.

The formulation for an impedance boundary condition [17] is similar to an impedance sheet, where the relationship between acoustic admittance, normal particle velocity, and acoustic pressure is given by Equation (2.24).

The APDL command that is issued is

SF, NODE, IMPD, VAL1, VAL2

which defines *surface* loads on the selected nodes. The parameters VAL1 and VAL2 can be used to define the acoustic impedance or frequency varying acoustic admittance, depending on whether the value is positive or negative, as listed in Table 2.22. Table 2.23 lists the terms that describe the real and imaginary parts of impedance and admittance (which is the inverse of impedance).

The most common use is where VAL1 and VAL2 are defined with positive values, and hence the command SF, NODE, IMPD, VAL1, VAL2 will define the impedance on the selected nodes where VAL1 is the resistance—the real part of the impedance $\mathrm{Re}(Z)$, and VAL2 is the reactance—the imaginary part of the impedance $\mathrm{Im}(Z)$. The window of Details of "Acoustic Impedance Boundary" has a row Definition | Impedance or Admittance which can be changed to define whether impedance or admittance values are entered as shown in the two images below.

Details of "Acoustic Impedance Boundary"	
☐ **Scope**	
Scoping Method	Geometry Selection
Geometry	1 Face
☐ **Definition**	
Impedance Or Admittance	Impedance
Resistance	0 [Pa m^-1 sec]
Reactance	0 [Pa m^-1 sec]

Details of "Acoustic Impedance Boundary"	
☐ **Scope**	
Scoping Method	Geometry Selection
Geometry	1 Face
☐ **Definition**	
Impedance Or Admittance	Admittance
Conductance	0 [Pa^-1 m sec^-1]
Product Of Susceptance And Angular Frequency	0 [Pa^-1 m sec^-1 rad sec^-1]

TABLE 2.22

Interpretation of the Impedance Boundary Condition That Is Defined Using the APDL Command SF,,IMPD,VAL1,VAL2 Depending on Whether VAL1 Is Positive or Negative

Sign	Interpretation	
VAL1	**VAL1=**	**VAL2=**
VAL1 $\geq$ 0	$\mathrm{Re}(Z)=R$ resistance [Pa.s/m]	$\mathrm{Im}(Z)=X$ reactance [Pa.s/m]
VAL1 $<$ 0	$-\mathrm{Re}(Y) = -G$ negative conductance [m/(Pa.s) $\equiv$ mho]	$\mathrm{Im}(Y)\,\omega$ susceptance $\times$ angular freq. [m/(Pa.s^2)]

Note: Z is the complex specific impedance and Y is the complex specific admittance. These definitions apply to harmonic analysis only. For transient analyses, VAL2 is not used.

TABLE 2.23

Definitions of the Real and Imaginary Parts of Acoustic Impedance and Acoustic Admittance

Term	Equation	Real Part	Imaginary Part
Impedance	$Z = R + jX$	R resistance	X reactance
Admittance	$Y = 1/Z = G + jB$	G conductance	B susceptance

Which version of impedance?

There are four variants of impedance that are often used in vibro-acoustic textbooks that can cause confusion. The variants are listed in Table 2.24 [47, Table 1.3, p. 53] and it is important to be consistent when using formulations of impedance in the 4-pole theory. See Fahy [65, p. 56] for an extensive discussion on the various forms of impedance.

TABLE 2.24

The Four Types of Vibro-Acoustic Impedances Described in Textbooks

Impedance Type	Equation	Units
Mechanical impedance	$Z_{\text{mechanical}} = \dfrac{F}{u} = \dfrac{pS}{u}$	$[\text{M} / \text{T}]$
Specific acoustic impedance	$Z_{\text{specific acoustic}} = \dfrac{p}{u}$	$[\text{M} / (\text{T L}^2)]$
Acoustic impedance	$Z_{\text{acoustic}} = \dfrac{p}{v} = \dfrac{p}{uS}$	$[\text{M} / (\text{T L}^4)]$
Characteristic impedance	$Z_{\text{characteristic}} = \rho_0 c_0$	$[\text{M} / (\text{T L}^2)]$

Note: F is the force, u is the acoustic particle velocity, p is the acoustic pressure, v is the acoustic volume velocity, S is the cross-sectional area of the duct. The generalized units are M=mass, T=time, L=length.

Many acoustic problems have impedances that vary with frequency. Unfortunately, with Release 14.5 of ANSYS Workbench, there is no simple way to define a "lookup" table of impedance values versus frequency for use in acoustic analyses. Instead, one has to use APDL commands either by using Mechanical APDL or inserting Command Objects containing APDL commands into an ANSYS Workbench simulation that defines the value of impedance at each analysis frequency. An example of how this can be implemented is shown in Section 3.3.7.4. Although not included in this book, a lookup table could be implemented using APDL commands with an array containing columns for frequency, real, and imaginary impedance values, that are used by the APDL command SF, NODE, IMPD, VAL1, VAL2 at each harmonic analysis frequency.

Other examples of the use of an impedance boundary in ANSYS are in Sections 6.5.2 and 6.5.3, where an acoustic absorbing material placed on the walls of a duct acts as a muffler to absorb an incident acoustic wave.

2.8.4.3   Thermo-viscous BLI Boundary

A new feature that has been implemented in ANSYS Workbench is a thermo-viscous boundary layer impedance model that can be applied to a face. This can be implemented by selecting

🔍 Thermo-viscous BLI Boundary (beta)

from the Boundary Conditions menu in the ACT Acoustics menu bar.

The APDL command that is issued is

SF,NODES,BLI

which defines a surface impedance on the selected NODES.

The theory for this model is described in Ref. [50]. An example of this "beta" feature is not covered in this book.

2.8.4.4   Free Surface

A free surface can be simulated in ANSYS Workbench by selecting

📐 Free Surface

from the Boundary Conditions menu in the ACT Acoustics menu bar.

The APDL command that is issued is

SF,ALL,FREE

which defines the free surface flag on the select nodes of the acoustic fluid. When using the command in ANSYS Mechanical APDL, the nodes belonging to the free surface should be selected, then issue the APDL command SF,ALL,FREE.

This boundary condition can be used to simulate "sloshing" problems, where the pressure at the surface of fluid is zero and the pressure at depth z is given by

$$p = \rho g z \,, \tag{2.25}$$

where ρ is the density of the fluid, g is the acceleration due to gravity, and z is the depth from the free surface.

The sloshing surface must be parallel to the coordinate plane of the global Cartesian system.

In earlier releases of ANSYS before Release 14.5, the displacement-formulated acoustic elements FLUID79 and FLUID80 were recommended for modeling 2D and 3D fluids contained in a vessel with a free surface. These elements are now termed *legacy* elements and the documentation has been moved into the Mechanical APDL Feature Archive section.

2.8.4.5 Radiation Boundary

A radiation boundary is a boundary condition that can be applied to the exterior faces of an acoustic body to absorb outgoing acoustic waves. It can be simulated in ANSYS Workbench by selecting

📦 Radiation Boundary

from the Boundary Conditions menu in the ACT Acoustics menu bar.

The APDL command that is issued is

SF,ALL,INF

which defines a "Robin" radiation boundary flag on the selected nodes of the acoustic fluid.

The relationship for impedance in Equation (2.24) can be rearranged, where the difference in the normal acoustic particle velocities is written as a pressure gradient normal to an absorbing boundary as

$$\frac{\partial p(x)}{\partial n} + j\omega\rho_0 Y_0 p(x) = 0 \,, \tag{2.26}$$

where $Y_0 = 1/Z_0 = 1/(\rho_0 c_0)$ is the characteristic admittance (the inverse of the characteristic acoustic impedance) of the fluid. This expression means that the impedance on the boundary of the acoustic fluid will cause outgoing acoustic pressure waves that are *normal* to the boundary to be absorbed, and will not be reflected back into the acoustic domain. However, for acoustic waves that do not strike the boundary at 90 degrees, there will be some reflection.

An example that demonstrates the use of this boundary condition is in Section 3.3.6.

2.8.4.6 Absorbing Elements

Another method of absorbing outgoing acoustic pressure waves is to define Absorbing Elements overlaid onto the outside of a spherical-shaped body. This can be implemented in ANSYS Workbench by selecting

▯▯▯ Absorbing Elements (Exterior To Enclosure)

from the Boundary Conditions menu in the ACT Acoustics menu bar.

The APDL command that is issued is ESURF, which meshes FLUID130 elements on the exterior faces of a spherical-shaped body, such as a sphere, hemisphere, 1/4 or 1/8 of a sphere.

The boundary condition that must be satisfied is the Sommerfeld radiation condition [133], which qualitatively means that outgoing acoustic waves that result from scattering by an object, or from acoustic sources within the acoustic domain, should continue to propagate outward not inward.

It is recommended that when developing a model that will use the FLUID130 (or FLUID129) elements, the exterior spherical surface should be more than 0.2× of the acoustic wavelength from the nearest object.

The parameters that need to be defined in the window Details of "Absorbing Elements (Exterior to Enclosure)" are the faces of the spherical surface, the speed of sound and density of the fluid, the radius of the spherical body, and the origin of the spherical body. By default it is assumed that the origin of the spherical body is at the global Cartesian origin.

Details of "Absorbing Elements"	⊓
⊟ **Scope**	
Scoping Method	Geometry Selection
Geometry	1 Face
⊟ **Definition**	
Sound Velocity	343 [m sec^-1]
Mass Density	1.21 [kg m^-1 m^-1 m^-1]
Radius Of Enclosure	0.7 [m]
X Coordinate Of Enclosure Origin	0 [m]
Y Coordinate Of Enclosure Origin	0 [m]
Z Coordinate Of Enclosure Origin	0 [m]

An example that demonstrates the use of this boundary condition in 3D is in Section 3.3.7.2. Another example in Sections 8.4.4 and 8.4.5 show how this boundary condition can be implemented for a 2D analysis using FLUID129 elements.

An error message sometimes occurs in ANSYS (and the analysis will halt) when using the FLUID129 and FLUID130 elements, indicating that the nodes of the element are not precisely on the curved surface of the defined radius. This error is discussed further in Appendix D.2.4 as well as how to fix the issue.

2.8.4.7 Attenuation Surface

An attenuation surface can be applied to a face that absorbs incident acoustic waves in ANSYS Workbench by selecting

      🖝 Attenuation Surface

from the Boundary Conditions menu in the ACT Acoustics menu bar.

The APDL command that is issued is

SF, Nlist, CONV, VALUE

which defines an attenuation surface on the selected nodes, where VALUE is the attenuation coefficient of the surface for harmonic and transient analyses.

The material properties of acoustic absorbing materials are often defined in terms of the attenuation coefficient α, which is the ratio of the sound power that is absorbed by the material I_{absorbed} to the sound power incident on the material I_{incident},

$$\alpha = \frac{I_{\text{absorbed}}}{I_{\text{incident}}}, \tag{2.27}$$

with a range $0 < \alpha < 1$. The absorption coefficient can be written in terms of a reflection coefficient [47, Eq. (C.46)]

$$|R| = \sqrt{1 - \alpha}\,, \tag{2.28}$$

and the acoustic impedance of the material is given by

$$Z = \rho_0 c_0 \, \frac{1 + \sqrt{1 - \alpha}}{1 - \sqrt{1 - \alpha}}, \tag{2.29}$$

which will result in a real value for impedance and the complex part will be zero $(\mathrm{Im}(Z) = 0)$.

The normal impedance of a material is often measured in an impedance tube test device, where acoustic plane waves exist in the tube and strike the test sample at 90 degrees angle of incidence, and hence the impedance normal to the surface of the material is measured.

Often when the material is installed in an application, sound will strike the material at random angles of incidence and it is necessary to determine the statistical absorption coefficient which can be estimated from the normal impedance as described in Ref. [47, Appendix C]. Alternatively, the acoustic absorption coefficient can be determined by installing a large area of material in an acoustic reverberation chamber and measuring the time it takes for the sound level in the room to reduce by 60 dB after a noise source is switched off [47, p. 315]. A simulation of this test procedure using ANSYS is described in Chapter 7.

Furthermore, the absorption and impedance of porous materials vary with frequency, often having poor absorption at low frequencies. As described in Section 2.8.4.2, ANSYS Release 14.5 does not have a simple means of defining a lookup table of attenuation coefficient versus frequency, although it could be implemented using APDL commands. Refer to Section 2.8.4.2 for further details.

2.8.4.8 Equivalent Source Surface

It is mathematically possible to calculate the acoustic pressure from an acoustic source at a position beyond the extent of an acoustic finite element mesh by using the surface equivalent principle, which is used in electro-magnetic simulations [18, Section 4.3.2.6.].

An equivalent source surface is used in combination with Perfectly Matched Layer acoustic bodies to enable the calculation of various acoustic results at locations beyond the finite element model, using the Acoustic Far Field and Acoustic Far Field Microphone objects from the Results menu of the ACT Acoustics extension menu bar, or within the finite element model using the Acoustic Near Field object from the Results menu.

The equivalent source surface should be defined between the acoustic source (or an object that causes acoustic scattering) and the perfectly matched layer acoustic body, as shown in Figure 2.18. By defining the equivalent source surface close to the acoustic source, accurate near-field and far-field results can be calculated. It is recommended that there should be more than half a wavelength separation between the acoustic source (or scattering object) and the equivalent source surface. There should also be another acoustic body defined between the perfectly matched layer acoustic body and the equivalent source surface that acts as a buffer or spacer.

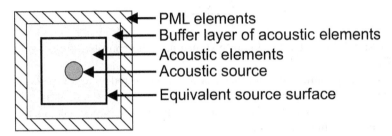

FIGURE 2.18
Schematic showing an equivalent source surface defined between an acoustic source and Perfectly Matched Layer acoustic body.

This can be implemented in ANSYS Workbench by selecting

         Equivalent Source Surface

from the Boundary Conditions menu in the ACT Acoustics menu bar.
The APDL command that is used to define the equivalent source surface is SF,ALL,MXWF.

See Section 8.2.1 for further details about Perfectly Matched Layers and Equivalent Source Surfaces. Section 8.3 has an example that demonstrates the use of an Equivalent Source Surface to calculate the sound pressure level versus angle for several acoustic sources.

2.8.5   Results

The ACT Acoustics extension in ANSYS Workbench can be used to request acoustic results and these are listed in Table 2.25. These results are discussed more in detail in the following sections.

2.8.5.1   Acoustic Pressure

The calculated pressure response can be displayed as a contour plot by selecting

         Acoustic Pressure

from the Results menu in the ACT Acoustics menu bar.
The APDL command that is issued is

             PLNSOL,PRES

which will plot the results of the pressure degree of freedom of the selected geometries.

TABLE 2.25

Results Available in the ACT Acoustics Extension and the Corresponding
APDL Command

Method	APDL Command
Results ▼	
Acoustic Pressure	PLNSOL,PRES
Acoustic SPL	PLNSOL,SPL
Acoustic Velocity X	PLNSOL,PG,X
Acoustic Velocity Y	PLNSOL,PG,Y
Acoustic Velocity Z	PLNSOL,PG,Z
Acoustic Velocity SUM	PLNSOL,PG,SUM
Acoustic Velocity Vectors	PLNSOL,PG,VECTORS
Acoustic Pressure Gradient X	PLNSOL,PG,X
Acoustic Pressure Gradient Y	PLNSOL,PG,Y
Acoustic Pressure Gradient Z	PLNSOL,PG,Z
Acoustic Pressure Gradient SUM	PLNSOL,PG,SUM
Acoustic Pressure Gradient Vectors	PLNSOL,PG,VECTORS
Acoustic Far Field	PLFAR
Acoustic Far Field Microphone	PLFAR and extract results at a node
Acoustic Near Field	PLNEAR
Acoustic Time_Frequency Plot	PLVAR
Muffler Transmission Loss	

In the window Details of "Acoustic Pressure", one can select the
Geometry of interest, the Expression is usually left as =PRES to request the
pressure DOF results, and the Output Unit should be left as Pressure. The
row By has a drop-down menu when the small triangle at the end of the row
is clicked.

By	Frequency ▼
Frequency	Frequency
Phase Angle	Set
	Maximum Over Frequency
Coordinate System	Frequency Of Maximum
Identifier	Maximum Over Phase
Suppressed	Phase Of Maximum

The options for the row By and the corresponding descriptions are listed
in Table 2.26.

TABLE 2.26

List of Options for the Row By in the `Acoustic Pressure` Result

By Option	Description
`Frequency`	A value can be entered into the following row to request the results at a desired frequency. Note that it is possible to request results at a frequency that was not analyzed. Enter a value of frequency that is listed in the `Tabular Data` window.
`Set`	Enter a value for the `Set` that is listed in the `Tabular Data` window.
`Maximum Over Frequency`	Plots the maximum pressure over the analysis frequency range. The phase angle is held constant and each node / element / sample point is swept through the analysis frequency range to find its maximum result. This result is applicable in harmonic analyses only.
`Frequency Of Maximum`	The contour plot is in units of frequency Hz, where the frequency at which the maximum pressure occurs. The phase angle is held constant and each node / element / sample point is swept through the analysis frequency range to find its maximum result. This is only available in harmonic analyses.
`Maximum Over Phase`	The frequency is selected and each node / element / sample point is swept through a phase angle of 0 to 360 degrees at specified increments to find its maximum result. A row containing `Phase Increment` will appear where a value can be entered. This is only applicable in harmonic analyses.
`Phase of Maximum`	Plots the phase angle at which the maximum pressure occurs. The frequency is selected and each node / element / sample point is swept through a phase angle of 0 to 360 degrees at specified increments to find its maximum result. A row containing `Phase Increment` will appear where a value can be entered. This is only applicable in harmonic analyses.

Note that a small error can occur in the legend for the `Acoustic Pressure` results where `Frequency: 0. Hz` might be displayed, regardless of what frequency was selected to be displayed in the `Details of "Acoustic Pressure"` window. The images below show that the `Last` frequency has been selected, yet the legend shows `Frequency: 0. Hz`.

Details of "Acoustic Pressure"	
⊟ **Scope**	
Scoping Method	Geometry Selection
Geometry	All Bodies
⊟ **Definition**	
Type	User Defined Result
Expression	= PRES
Input Unit System	Metric (m, kg, N, s, V, A)
Output Unit	Pressure
By	Frequency
Frequency	Last
Phase Angle	0. °
Coordinate System	Global Coordinate System
Identifier	
Suppressed	No
⊟ **Results**	
☐ Minimum	-403.82 Pa
☐ Maximum	820.68 Pa
Minimum Occurs On	Solid
Maximum Occurs On	Solid

A: Harmonic Response
Acoustic Pressure
Expression: PRES
Frequency: 0. Hz
Phase Angle: 0. °
Unit: Pa

An alternative way to determine the analysis frequency of the pressure contour plot is to inspect the `Tabular Data` window to look up the result set number and the corresponding analysis frequency. Either select an analysis frequency listed in the table in the `Tabular Data` window, or select the `Set` number.

Tabular Data		
	Set	☑ Frequency [Hz]
1	1.	500.
2	2.	600.

2.8.5.2 Acoustic Sound Pressure Level

The sound pressure level in decibels (dB) can be displayed as a contour plot by selecting

🔲 Acoustic SPL

from the `Results` menu in the ACT Acoustics menu bar.

The APDL command that is issued is

PLNSOL,SPL

which will plot the sound pressure level in decibels of the selected geometries.

Sound pressure level is calculated as

$$L_p = 20 \log_{10} \left[\frac{p_{\text{RMS}}}{p_{\text{ref}}} \right], \tag{2.30}$$

where p_{RMS} is the root-mean-square (RMS) pressure, and p_{ref} is the reference pressure level, which is defined in the `Details of "Acoustic Body"` window with a default value of 20 μPa. A common mistake is forgetting to recognize that sound pressure level always uses RMS of the complex acoustic pressure [47, p. 38] and one will find that the results calculated by ANSYS are $20 \log_{10}(\sqrt{2}) = 3$ dB lower than expected. The likely explanation is that the results calculated using ANSYS are correct and the analyst did not use RMS pressure in the calculation of the sound pressure level.

Note that similar to the `Acoustic Pressure` result, a minor error can occur in the legend for the `Acoustic SPL` result, where `Frequency: 0. Hz` might be written, regardless of what frequency was selected to be displayed in the `Details of "Acoustic SPL"` window. The images below show that the results at the `Last` frequency was selected, although the legend shows `Frequency: 0. Hz`.

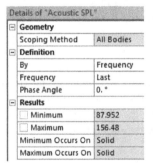

2.8.5.3 Acoustic Velocity

The acoustic particle velocity along the x, y, z axes, or magnitude can be displayed as a contour plot by selecting one of

> Acoustic Velocity X
>
> Acoustic Velocity Y
>
> Acoustic Velocity Z
>
> Acoustic Velocity SUM

from the `Results` menu in the ACT Acoustics menu bar.

The APDL command that is issued is

```
PLNSOL,PG,Comp
```

which will plot the continuous contours of the acoustic particle velocities of the nodes, where `Comp` is either `X`, `Y`, `Z`, or `SUM`. This result is only available for modal and harmonic analyses.

It is also possible to display the results as a vector plot, by selecting

🔍 Acoustic Velocity Vectors

from the Results menu in the ACT Acoustics menu bar.

The APDL command that is issued is

PLNSOL,PG,VECTORS

which will plot the vectors of the acoustic particle velocity calculated at the nodes. This result is only available for modal and harmonic analyses.

2.8.5.4 Acoustic Pressure Gradient

Similar to the results for Acoustic Velocity, which were only available for modal and harmonic analyses, for acoustic transient analyses, contour plots of the acoustic pressure gradient at the nodes can be displayed by selecting one of

🔲PG Acoustic Pressure Gradient X

🔲PG Acoustic Pressure Gradient Y

🔲PG Acoustic Pressure Gradient Z

🔲PG Acoustic Pressure Gradient SUM

from the Results menu in the ACT Acoustics menu bar.

The APDL command that is issued is

PLNSOL,PG,Comp

which will plot the continuous contours of the acoustic pressure gradient at the nodes, where Comp is either X, Y, Z, or SUM. This result is only available for acoustic transient analyses.

It is also possible to display the results as a vector plot, by selecting

🔲PG Acoustic Pressure Gradient Vectors

from the Results menu in the ACT Acoustics menu bar.

The APDL command that is issued is

PLNSOL,PG,VECTORS

which will plot the vectors of the acoustic pressure gradient calculated at the nodes. This result is only available for acoustic transient analyses.

2.8.5.5 Acoustic Far Field

If an acoustic analysis is conducted that uses acoustic bodies with Perfectly Matched Layers (PMLs) or a Radiation Boundary around the acoustic domain, then it is possible to calculate acoustic far-field results, beyond the acoustic domain, using the equivalent source principle. This feature requires that an equivalent source surface is defined (see Section 2.8.4.8), which can be done manually or automatically created by ANSYS. According to the ANSYS manual [19, FLUID30], the equivalent source surface "may be automatically applied to a PML-acoustic medium interface or exterior surface with the label INF (i.e., an acoustic radiation boundary), if MXWF surfaces (i.e., equivalent source surfaces) have not been flagged manually." The acoustic far-field results can be obtained by selecting

⌖ **Acoustic Far Field**

from the `Acoustic Results` menu in the ACT Acoustics menu bar.

The APDL command that is issued is

`PRFAR, PRES, Option, PHI1, PHI2, NPH1, THETA1, THETA2, NTHETA, VAL1, VAL2, VAL3`

which calculates and prints the far-field pressure and parameters. For directivity plots, where the sound pressure level is plotted versus angle, the APDL command that is issued is `PLFAR`, which has similar input parameters to the `PRFAR` command. The `Option` field is one of the items listed in Table 2.27. The remaining input parameters and descriptions are listed in Table 2.28.

TABLE 2.27

Options Available for the `Acoustic Far Field` Result in the ACT Acoustics Extension

Option	Description
SUMC	Maximum pressure (default)
PHSC	Pressure phase angle
SPLC	Sound pressure level
DGCT	Acoustic directivity
PSCT	Maximum scattered pressure
TSCT	Target strength
PWL	Sound power level

TABLE 2.28

Input Parameters for the APDL Command `PRFAR,PRES` to Calculate the Acoustic Far-Field Results

Parameter	Description
PHI1, PHI2	Starting and ending ϕ angles in degrees in the spherical coordinate system.
NPHI	Number of divisions between the starting and ending ϕ angles where the results will be calculated.
THETA1, THETA2	Starting and ending θ angles in degrees in the spherical coordinate system. Defaults to 0 in 3-D and 90 in 2-D.
NTHETA	Number of divisions between the starting and ending θ angles where the results will be calculated.
VAL1	Radius of an imaginary spherical surface over which the results are to be calculated for the options SUMC, PHSC, SPLC, PSCT, or TSCT.
VAL2	A reference value that defaults to the reference RMS sound pressure level 20×10^{-6} Pa when Lab = PRES and Option = SPLC, and the reference sound power level 1×10^{-12} W when Lab = PRES and Option = PWL.

A spherical coordinate system is used to define the angles as shown in Figure 2.14. Note that to calculate the results around a full circle, define the Starting Angle Phi (From X Axis Toward Y Axis) as 0, and the Ending Angle Phi as 360—defining the start and end angles as 0 degrees will only calculate the result at 0 degrees.

The input parameters for the APDL command are entered into the Details of "Acoustic Far Field" window. Most of the options require entry of the angles for the spherical coordinate system.

Details of "Acoustic Far Field"	
⊟ **Properties**	
Result Set	1
Boundary Condition On Model Symmetric Plane	No
Result	SPL In Cartesian Plot ▼
Starting Angle Phi (From X Axis Toward Y Axis)	0 [°]
Ending Angle Phi	360 [°]
Number Of Divisions Phi	36
Starting Angle Theta (From Z Axis Toward X Axis)	0 [°]
Ending Angle Theta	360 [°]
Number Of Divisions Theta	36
Sphere Radius	4 [m]
Reference RMS Sound Pressure	2E-05 [Pa]
Model Thickness in Z Direction (2D extension)	0 [m]
Spatial Radiation Angle	Full Space

The row Result Set is used to select the analysis frequency (or time) in which the particular Result type (e.g., SPL In Cartesian Plot, Sound Power Level, etc) should be displayed.

The row Boundary Condition on Model Symmetric Plane is used to indicate if it is assumed that the model is a symmetric representation of the actual system under investigation, and the calculated results will be adjusted accordingly. If no assumption about symmetry has been made about the model, then keep this option set to the default value of No. If it is assumed that symmetry exists, then change this row to Yes, which will reveal further options where the planes of symmetry can be defined as shown in the following image.

Details of "Acoustic Far Field"	
⊟ **Properties**	
Result Set	1
Boundary Condition On Model Symmetric Plane	Yes
Coordinate System Number For Symmetric Plane	0
YZ Plane	None ▼
ZX Plane	None
XY Plane	Sound Soft Boundary
	Sound Hard Boundary
Result	SPL In Cartesian Plot

The row `Coordinate System Number For Symmetric Plane` has a default value of `0`, which corresponds to the global Cartesian coordinate system. If a local coordinate system has been defined for the planes of symmetry, change this row to the appropriate coordinate system number. The next 3 rows have options to indicate whether the `YZ`, `ZX`, and `XY` planes have no symmetry (`None`), a `Sound Soft Boundary`, or a `Sound Hard Boundary`. A `Sound Soft Boundary` condition in ANSYS corresponds to a Dirichlet boundary condition [17], and should be selected when the pressure on the plane has been defined using the ACT Acoustics extension option `Boundary Conditions | Acoustic Pressure` or the APDL command `D,node,PRES,value,value2`. The option `Sound Hard Boundary` in ANSYS corresponds to a Neumann boundary condition, which should be selected when it is assumed that the pressure is symmetric about the plane and that the acoustic particle velocity (i.e., gradient of acoustic pressure) normal to the plane is zero. If you try to select a value for the radius such that it is inside the equivalent source surface, then unexpected results will be generated.

If the row for `Result` is set to `Sound Power Level`, the value of the `Reference Sound Power` level can be entered and has a default value of 10^{-12} W.

Details of "Acoustic Far Field"	中
⊟ **Properties**	
Result Set	1
Boundary Condition On Model Symmetric Plane	No
Result	Sound Power Level ▼
Reference Sound Power	1E-12 [W]
Model Thickness in Z Direction (2D extension)	0 [m]
Spatial Radiation Angle	Full Space

The sound power level is calculated as

$$L_w = 10 \log_{10} \left[\frac{W}{W_{\text{ref}}} \right], \tag{2.31}$$

where W is the sound power (Watts), and W_{ref} is the reference power, which is usually 10^{-12} W. The sound power level is usually expressed in units of [dB re 10^{-12} W].

2.8.5.6 Acoustic Near Field

Similar to the `Acoustic Far Field`, the results in the region within the acoustic domain in the acoustic near field can be obtained by selecting

△ Acoustic Near Field

from the `Acoustic Results` menu in the ACT Acoustics menu bar.

The APDL command that is issued is

PLNEAR, Lab, Opt, KCN, VAL1, VAL2, VAL3, VAL4, VAL5, VAL6, VAL7, VAL8, VAL9

which plots the acoustic pressure near to the *exterior* of the equivalent source surface. There is a similar APDL command, PRNEAR, that can be used to print the value of near field acoustic results. The input parameters are:

Lab where the keyword SPHERE can be entered to plot the pressure response on an imaginary spherical surface, or PATH to calculate the pressure response along a path.

Opt is one of three keywords:

  PSUM Maximum complex pressure.

  PHAS Phase angle of complex pressure.

  SPL Sound pressure level.

KCN is the number of a coordinate system which is used to define the spherical surface or path where the results will be calculated.

The parameters for VAL1 to VAL9 are only applicable for defining the angles of an imaginary spherical surface (LAB=SPHERE) where results should be calculated and are:

VAL1 the radius of the imaginary spherical surface, with the origin defined by the coordinate system in KCN.

VAL2 the starting angle ϕ from the x axis toward the y axis, in the spherical coordinate system.

VAL3 the ending angle ϕ in the spherical coordinate system.

VAL4 the number of divisions between the start and end angle ϕ.

VAL5 the starting angle θ from the z axis toward the y axis, in the spherical coordinate system

VAL6 the ending angle θ in the spherical coordinate system.

VAL7 the number of divisions between the start and end angle θ.

VAL8 the RMS reference sound pressure, which has a default value of 20 μPa.

VAL9 the thickness of a 2-D model extension in the z direction.

A spherical coordinate system is used to define the angles as shown in Figure 2.14.

All the input parameters can be entered into the window Details of "Acoustic Near Field".

Details of "Acoustic Near Field"	
⊟ **Properties**	
Result Set	1
Boundary Condition On Model Symmetric Plane	No
Near Field Position	On Sphere
Result	Sound Pressure Level ▼
Coordinate System Number For Near Field Position	0
Starting Angle Phi (From X Axis Toward Y Axis)	0 [°]
Ending Angle Phi	0 [°]
Number Of Divisions Phi	0
Starting Angle Theta (From Z Axis Toward X Axis)	0 [°]
Ending Angle Theta	0 [°]
Number Of Divisions Theta	0
Sphere Radius	0 [m]
Reference RMS Sound Pressure	2E-05 [Pa]
Model Thickness in Z Direction (2D extension)	0 [m]

If the results are requested over a path LAB=PATH, it is defined using the APDL command PATH. The window Details of "Acoustic Near Field" has entries to define the x, y, z coordinates for the start and end points of the straight line path.

Details of "Acoustic Near Field"	
⊟ **Properties**	
Result Set	1
Boundary Condition On Model Symmetric Plane	No
Near Field Position	Along Path ▼
Result	Sound Pressure Level
Position 1: X Coordinate	0 [m]
Position 1: Y Coordinate	0 [m]
Position 1: Z Coordinate	0 [m]
Position 2: X Coordinate	0 [m]
Position 2: Y Coordinate	0 [m]
Position 2: Z Coordinate	0 [m]
Model Thickness in Z Direction (2D extension)	0 [m]

If you try to select a value for the radius such that it is inside the equivalent source surface, then unexpected results will be generated.

2.8.5.7 Acoustic Time Frequency Plot

The acoustic results can be plotted over the analysis frequency range for harmonic response analyses or time for transients analyses by selecting

> 🗋 Acoustic Time_Frequency Plot

from the `Acoustic Results` menu in the ACT Acoustics menu bar.

There are two APDL commands that are issued. The first is

> `NSOL, NVAR, NODE, Item`

which is used to store the results in a table (confusingly called a variable). The input parameter `NVAR` is an integer, in the range from 2 to 10 (but can be increased by using the `NUMVAR` command) to define the variable containing the pressure results. The input parameter `Item` is either `PRES` to store the complex pressure results, or `SPL` to store the sound pressure level results.

Once the results have been stored into a table (`NVAR` variable), a second APDL command is issued, `PLVAR,NVAR`, that will plot the table of results over the analysis frequency range or time.

The window `Details of "Acoustic_Time Frequency"` has options to select the geometry, whether the complex pressure or sound pressure level is plotted, and the `Spatial Resolution`, which is either the average, minimum, or maximum.

Details of "Acoustic Time_Frequency Plot"	
⊟ **Scope**	
Scoping Method	Geometry Selection
Geometry	1 Face
⊟ **Definition**	
Result	Pressure
Spatial Resolution	Use Average ▼
Display	Bode

If the acoustic pressure results are plotted `Item=PRES` for the `NSOL` command, or `Result = Pressure` in the ACT Acoustics extension, they can be displayed as the real part, imaginary part, real and imaginary parts, the amplitude of the complex pressure, the phase angle, or as a Bode diagram which will plot the amplitude and phase.

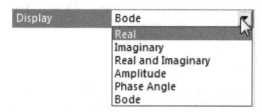

Display	Bode ▶
	Real
	Imaginary
	Real and Imaginary
	Amplitude
	Phase Angle
	Bode

If the sound pressure level results are plotted, there are no display options.

2.8.5.8   Muffler Transmission Loss

The transmission loss of a silencer that has plane wave conditions can be calculated by conducting a harmonic analysis and selecting

♧ Muffler Transmission Loss

from the Acoustic Results menu in the ACT Acoustics extension menu bar.

In order to use this feature, the model must be set up such that there are anechoic terminations on the inlet and outlet of the duct. This can be achieved using Radiation Boundary to the end faces, or defining acoustic bodies as Perfectly Matched Layer (PML) on the ends of the duct.

The window Details of "Muffler Transmission Loss" requires several features to be defined.

Details of "Muffler Transmission Loss"		
⊟ **Outlet**		
	Scoping Method	Named Selection
	Named Selection	NS_OUTLET
⊟ **Inlet**		
	Scoping Method	Named Selection
	Named Selection	NS_INLET
⊟ **Definition**		
	Inlet Source	Acoustic Normal Surface Velocity
	Inlet Pressure	0.206647642 [Pa]
	Mass Density Of Environment Media	1.2041 [kg m^-1 m^-1 m^-1]
	Sound Speed In Environment Media	343.24 [m sec^-1]

The Outlet and Inlet faces can be defined using either Geometry Selection, where the selection filter should be changed to Faces, or alternatively using Named Selection.

The acoustic excitation also needs to be defined in the row Definition | Inlet Source. It is assumed that under the branch Harmonic Response (A5), either an excitation source of a Normal Surface Velocity or a Mass Source (Harmonic) has been defined. By clicking on the row Definition | Inlet Source you can select the appropriate excitation source. The following row Definition | Inlet Pressure will automatically calculate the equivalent incident acoustic pressure excitation at the inlet of the duct. For a Mass Source (Harmonic) excitation, the equivalent incident pressure is

$$p_{\text{incident, mass source}} = \frac{Qc_0}{2}, \qquad (2.32)$$

where Q is the volume velocity, and c_0 is the speed of sound of the fluid. The term $1/2$ in Equation (2.32) comes from the fact that half the volume velocity source propagates downstream and half propagates upstream. It is only the downstream propagating volume velocity that is used to determine the transmission loss. For a `Normal Surface Velocity` excitation, the equivalent incident pressure is

$$p_{\text{incident, normal velocity}} = \frac{v_n \rho_0 c_0}{2},$$ (2.33)

where v_n is the normal velocity of the inlet, and ρ_0 is the density of the fluid.

The reason that the equivalent incident acoustic pressure is calculated rather than measuring the average sound pressure at the inlet face is because reactive silencers cause an impedance change and acoustic energy is reflected back upstream toward the inlet and can cause the sound pressure level to increase compared to a straight duct without the reactive silencer. Hence if the average sound pressure at the inlet was used in the calculations for transmission loss, the results would be artificially higher than if the incident acoustic sound pressure was used.

The transmission loss is calculated as

$$\text{TL} = 20 \log_{10} \left[\frac{p_{\text{incident}}}{p_{\text{transmitted}}} \right] + 20 \log_{10} \left[\frac{S_{\text{inlet}}}{S_{\text{outlet}}} \right],$$ (2.34)

where p_{incident} is the equivalent incident acoustic pressure calculated using either Equation (2.32) or (2.33), $p_{\text{transmitted}}$ is the average pressure at the outlet face, and S_{inlet} and S_{outlet} are the areas of the inlet and outlet faces, respectively.

The limitation of this feature is that it is assumed that plane wave conditions exist at the inlet and outlet faces.

Where non-plane wave conditions exist at these faces, the calculation of transmission loss is more complicated. It is necessary to export the real and imaginary parts of the pressure and acoustic particle velocity and the area at each node and at each frequency on the inlet and outlet faces. These results are then post-processed to calculate the acoustic intensity at each node, then multiplied by the effective nodal area and summed for all the nodes on each face to determine the sound power. The sound power at the inlet and outlet faces can be used to calculate the transmission loss.

Examples of the use of the `Muffler Transmission Loss` object are described in Sections 3.4.2.3 and 3.4.3.

2.8.5.9 Tools

The ACT Acoustics extension in ANSYS Workbench has tools that aid in the creation of acoustic-related loads, boundary conditions, results, and others. Figure 2.19 shows the menu options that are available. These tools are discussed further in the following sections.

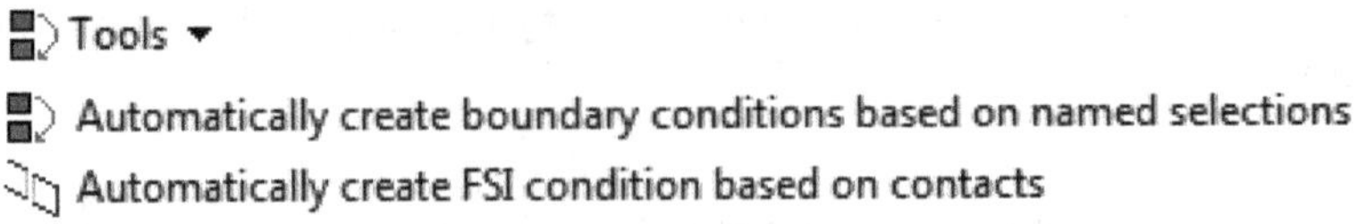

⬛⟩ Tools ▾

⬛⟩ Automatically create boundary conditions based on named selections

⬛⟩ Automatically create FSI condition based on contacts

FIGURE 2.19
Tools to automatically insert boundary conditions and FSI interfaces in the
ACT Acoustics extension.

2.8.5.10   Insertion of Boundary Conditions Based on Named Selections

The tool Automatically create boundary conditions based on named
selections can be used to create boundary conditions and loads by using
Named Selections that correspond to the keywords listed in Table 2.29. If
the Named Selection corresponds to a keyword, the boundary condition or
load is inserted into the analysis, and the scope of application are the group
of bodies or faces that correspond to the Named Selection.

TABLE 2.29
Named Selection Keywords Used to Automatically Insert
Objects into an Analysis

Keyword	Object Created
acousticbody	Acoustic Body
normalvelocity	Normal Surface Velocity
normalacceleration	Normal Surface Acceleration
masssource	Mass Source
massrate	Mass Source Rate
surfacevelocity	Surface Velocity
surfaceacceleration	Surface Acceleration
staticpressure	Static Pressure
impsheet	Impedance Sheet
temperature	Temperature
pressure	Acoustic Pressure
impedance	Impedance Boundary
thermovisc	Thermo-viscous BLI Boundary
free	Free Surface
fsi	FSI Interface
radiation	Radiation Boundary
absorbingelem	Absorbing Elements
attenuation	Attenuation Surface
plot	Acoustic Time Frequency Plot

For example, a model can be created with a face involved in Fluid-
Structure-Interaction (FSI) between air and a vibrating plate. A Named

Selection can be created for this face and called `plate_fsi` as shown in the following image.

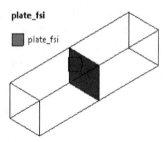

It can be seen that `fsi` is a recognized keyword that is listed in Table 2.29. When the tool `Automatically Create Boundary Conditions based on Named Selection` is selected, a new `Acoustic FSI Interface` object is inserted into the analysis tree. In the window `Details of "Acoustic FSI Interface"`, the `Selection Method` is changed to `Named Selection` and the `Named Selection` row is changed to `plate_fsi`, as shown in the following image.

Details of "Acoustic FSI Interface"	무
⊟ **Scope**	
Scoping Method	Named Selection
Named Selection	plate_fsi

Other names such as `plateFSI`, `platefsiair` would also be recognized and converted.

2.8.5.11 Insertion of FSI Interfaces Based on Contacts

The tool `Automatically create FSI conditions based on contacts` can be used to insert Fluid-Structure Interaction interface objects into an analysis by using the `Contact` regions that have been defined either automatically during the transfer of the solid geometry from DESIGNMODELER to Mechanical, or manually created in Mechanical. An example of the use of this tool is shown on page 573.

2.9 Other Acoustic Loads

The previous sections covered the acoustic excitation, loads, boundary conditions, and results that can be obtained by using the ACT Acoustics extension toolbar. Other excitation types that can also be defined are listed in the following sections.

2.9.1   Displacement

It is possible to define an applied displacement to the displacement DOFs of a structure, or the displacement DOFs of acoustic elements that have the FSI flag switched on. In the ANSYS Workbench interface, a displacement excitation can be defined by selecting Supports | Displacement from the Environment toolbar, and specifying non-zero values for the x, y, z components in the window Details of "Displacement".

This can be implemented using the APDL command

D, NODE, Lab, VALUE, VALUE2, NEND, NINC, Lab2, Lab3, Lab4, Lab5, Lab6

which defines constraints on the selected nodes where Lab is a translational degree of freedom UX, UY, or UZ, or a rotational degree of freedom ROTX, ROTY, or ROTZ. Note that the rotational degrees of freedom are not applicable to acoustic elements, only some structural elements.

Remember that only motion normal to the surface causes an acoustic pressure response in the acoustic fluid.

The displacement constraint can be applied to the nodes of acoustic elements that have the displacement degrees of freedom activated, which can be achieved by changing the option in the Acoustic Body in the row Acoustic - Structural Coupled Body Options from Uncoupled, where the displacement DOFs at the nodes are not included, to Coupled With Unsymmetric Algorithm or Coupled With Symmetric Algorithm so that the displacement DOFs at the nodes are included.

A displacement constraint can also be applied to the nodes of a structure that is in contact with an acoustic fluid, which will cause the structure to vibrate and generate an acoustic pressure in the fluid.

2.9.2   Flow

Another type of acoustic excitation that is available in ANSYS is a FLOW load. To a newcomer to ANSYS, the term FLOW would seem to imply a mean motion of the fluid, however this is not the case. The FLOW load is used to apply a volume acceleration source (i.e., the rate of change of acoustic volume velocity), using the APDL command F,node,FLOW. Although this type of acoustic load source can still be used via APDL commands, the documentation about this feature has been removed from the ANSYS manuals. An old ANSYS acoustic tutorial from 1992 [150, p. 3-4] states that

> A FLOW fluid load is equal to the negative of the fluid particle acceleration normal to the mesh boundary (+ outward), times an effective surface area associated with the node, times the mean fluid density.

Consequently, the FLOW source on a boundary creates an inward acceleration (normal to the surface). The tutorial also mentions that

> A FLOW fluid load defined at a fluid mesh interior node is represen-

tative of a point sound source. For 3-D analyses the sound source is spherical and for 2-D analyses the source is cylindrical. The effective surface area associated with the sound source is dependent on the local fluid element size.

The FLOW load can be written mathematically for a harmonic source as

$$
\text{FLOW} = \overbrace{[-\omega^2 x]}^{\text{acceleration}} A\rho_0
$$
$$
= [j\omega\, j\omega\, x]\, A\rho_0
$$
$$
= [j\omega\rho_0]\,[j\omega x A]\,, \tag{2.35}
$$

where ω is the circular frequency in radians / s, x is the particle displacement, A is the effective surface area associated with the node, and ρ_0 is the density of the acoustic fluid. The volume velocity of a source is equal to the particle velocity times the effective nodal area $Q = (j\omega x)A$. (Note that the ANSYS theory manual [20, Eq. (8.1)] unfortunately uses Q as a mass source, whereas many acoustic textbooks define Q as a volume velocity source.) Hence the ANSYS FLOW load can be written in terms of an equivalent acoustic volume velocity as

$$
\text{FLOW} = j\omega\rho_0 Q\,. \tag{2.36}
$$

A mass source excitation applied to a vertex has units of kg/s (see Table 2.18) and is defined as

$$
\text{Mass Source} = \rho_0 A v = \rho_0 Q\,, \tag{2.37}
$$

where v is the particle velocity of the node. Hence the relation between a flow load and a mass source is

$$
\text{FLOW} = j\omega \times [\text{Mass Source}]\,. \tag{2.38}
$$

2.10 Other Measures of Acoustic Energy

The main acoustic results from an ANSYS analysis is acoustic pressure, acoustic pressure gradient, or acoustic particle velocity. By further post-processing, these results can be transformed into other measures of acoustic energy such as

- sound intensity
- sound power
- acoustic potential energy
- acoustic energy density

and these are further discussed in the following sub-sections.

2.10.1   Sound Intensity

The *sound intensity* of a wave is the average rate of flow of energy per unit area that is perpendicular to the direction of the propagation of the wave, as shown in Figure 2.20.

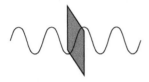

FIGURE 2.20
Sketch showing the area that is normal to the direction of wave propagation.

The *instantaneous sound intensity* I_i at time t describes the sound power per unit area at a given location and is calculated as the product of the pressure $p(r,t)$ and acoustic particle velocity vector $\vec{v}(r,t)$ at a point r as [47, Eq. (1.64)]

$$I_i(r,t) = p(r,t)\vec{v}(r,t) \, . \tag{2.39}$$

As sound pressure and velocity vary with time and location, it is more useful to describe the sound energy at a point by the time-averaged sound intensity as [47, Eq. (1.65)] [102, p. 125, Eq. (5.9.1)]

$$I(r) = \langle I_i(r,t)\rangle_T = \langle p\vec{v}\rangle_T = \lim_{T\to\infty} \frac{1}{T}\int_0^T p(r,t)\vec{v}(r,t)dt \, . \tag{2.40}$$

For a monofrequency wave, T is the period. It can be shown [64, p. 48] that the time-averaged *active* sound intensity is given by [91, p. 53, Eq. (6.13)]

$$I = \frac{1}{2}\mathrm{Re}\{pv^*\} \, , \tag{2.41}$$

where the superscript $*$ indicates the complex conjugate. The active intensity corresponds with the local net transport of sound energy. The time-averaged *reactive* intensity is calculated as

$$I_{\text{reactive}} = \frac{1}{2}\mathrm{Im}\{pv^*\} \, , \tag{2.42}$$

The reactive intensity is a measure of the energy stored in the sound field during each cycle but is not transmitted.

A harmonic sound wave with pressure p and acoustic particle velocity v can be defined as

$$p = P_{\max}\cos(\omega t + \theta_p) \tag{2.43}$$

$$v = V_{\max}\cos(\omega t + \theta_v) \, , \tag{2.44}$$

where P_{max} and V_{max} are the peak amplitude of the sound pressure and acoustic particle velocity, ω is the circular frequency, and θ_p and θ_v are the phase angles of the pressure and particle velocity, respectively. The corresponding active and reactive intensities are given by [47, Eqs. (1.72) and (1.73)]

$$I = \frac{1}{2}\text{Re}\{pv^*\} = \frac{P_{max}V_{max}}{2}\cos(\theta_p - \theta_v) \tag{2.45}$$

$$I_{reactive} = \frac{1}{2}\text{Im}\{pv^*\} = \frac{P_{max}V_{max}}{2}\sin(\theta_p - \theta_v). \tag{2.46}$$

These equations indicate that the difference in phase angles between the pressure and acoustic particle velocity $(\theta_p - \theta_v)$ are crucial in determining the sound intensity. There are two cases of interest that will be discussed further: (1) a progressive traveling wave, where it is assumed that the difference in phase angles is zero, and (2) for a standing wave configuration, where the difference in phase angles is 90°.

For a traveling progressive harmonic sound wave, such as from a plane, cylindrical, or spherical spreading wave, the pressure is defined as

$$p(x, t) = P_{max}\cos(k(x - ct)), \tag{2.47}$$

and the corresponding particle velocity in the far field is assumed to be in phase with the pressure and defined as

$$v(x, t) = \frac{p(x, t)}{\rho_0 c_0}. \tag{2.48}$$

The maximum intensity is given by [46, p. 33, Eq. (2.24)] [91, p. 53, Eq. (6.15)]

$$I_{max} = \frac{P_{max}^2}{2\rho_0 c_0} = \frac{p_{RMS}^2}{\rho_0 c_0}, \tag{2.49}$$

where p_{RMS} is the square root of the mean (time) square value of $p(x, t)$. However for the general case, where the sound intensity is not related to only the sound pressure, both sound pressure and particle velocity must be evaluated at the same instant of time and location.

The second case of interest is a standing wave configuration, such as an undamped duct with rigid ends, which is examined in more detail in Section 3.3. For this case, the pressure and acoustic particle velocity are in quadrature, which means $|\theta_p - \theta_v| = 90°$, and therefore when this is substituted into Equation (2.45), the time-averaged *active* sound intensity is zero [65, p. 80]. The sound intensity field is characterized by the *reactive* sound intensity from Equation (2.46), where sound energy oscillates locally during each cycle but is not transmitted along the duct.

When a duct has some acoustic damping installed, the difference in the phase angle between pressure and velocity is not 90°, which results in a nonzero value of active sound intensity indicating the net transport of sound energy along the duct. This situation is further described in Section 5.5.

The sound intensity level is calculated as [46, Eq. (1.17)]

$$L_I = 10 \log_{10} \left[\frac{I}{I_{\text{ref.}}} \right] \quad [\text{dB re } 10^{-12} \text{W/m}^2] \,, \qquad (2.50)$$

where I is the sound intensity in units of W/m^2, and $I_{\text{ref.}}$ is the reference sound intensity that has a value of 10^{-12} W/m^2.

For further information about sound intensity, see [47, p. 33] [64] [65, p. 76] [91, p. 51].

It has been shown that sound intensity is a function of the pressure and acoustic particle velocity, and these results are available from ANSYS simulations. For pressure-formulated acoustic elements (FLUID29, FLUID30, FLUID220, FLUID221) the pressure at each node is one of the degrees of freedom of the element, and this result is always available. The estimate of the acoustic particle velocity can be obtained from the pressure gradient results, or if the displacement degrees of freedom of the acoustic elements are activated, by multiplying the displacement at the nodes of the acoustic elements by $j\omega$ (for harmonic waves).

2.10.2   Sound Power

The *sound power* W radiated by a source can be evaluated by integrating the sound intensity over a surface that encloses the sound source. Figure 2.21 illustrates the concept where an oscillating piston installed in an infinite baffle generates sound that radiates outward. A hypothetical hemispherical surface is shown that encloses the sound source. The sound power can be calculated

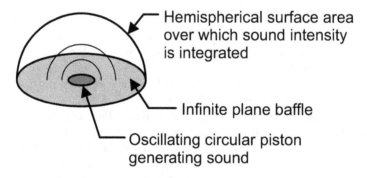

FIGURE 2.21
Example of the hypothetical surface that encloses a sound source for evaluation of sound power. An oscillating piston in an infinite plane baffle radiates sound, and the sound power is evaluated by integrating the sound intensity over a hemispherical surface.

as [46, p. 81, Eq. (4.5)]

$$W = \int_S I\,dS , \qquad (2.51)$$

where I is the sound intensity that is *normal* to an element of surface area dS, and S is the surface area that encloses the sound source. This can be converted into an expression suitable for finite element analysis as

$$W = \sum_n I_n S_n , \qquad (2.52)$$

where I_n is the sound intensity evaluated at node n in the direction that is normal to S_n, the effective surface area of a node. Methods of determining the effective area associated with a node in ANSYS are described in Appendix E.1.

The sound power level in decibels is then calculated using Equation (2.31).

There is a subtle but important point that the intensity is calculated in a direction *normal* to a surface, which requires further discussion. When determining the acoustic particle velocity for estimating the sound intensity and sound power, the nodal coordinate system is aligned with the global Cartesian system by default, and results for particle velocity are reported along each axis of the global Cartesian system. This would present problems for the example shown in Figure 2.21, where the nodes on the exterior of the hemisphere are aligned with the global Cartesian system, whereas what is required is the sound intensity in the direction that is perpendicular to the surface of the hemisphere. There are two suggested ways to resolve this difficulty:

1. Rotate the nodal coordinate system so that one of the axes is perpendicular to the face of the surface. In the example of the hemisphere, it is possible to rotate the nodal coordinate system for the nodes of the exterior surface of the hemisphere so that they are aligned with a global spherical coordinate system, and then the acoustic particle velocity in the radial direction can be determined.

2. Define a *local* coordinate system where one of the axes is aligned with the normal of the hypothetical surface, and determine the acoustic particle velocity along this local axis.

In order to make it easier to determine the sound intensity and sound power using ANSYS, it is recommended that the surface that is used for the integration of sound intensity should be a face of a solid body that is aligned with the global Cartesian system. For example, when analysing the acoustic power flow in a duct, the face that should be used for the integration of the sound intensity should be a plane that is perpendicular to the axis of the duct which is also aligned with the global Cartesian coordinate system, as shown in Figure 2.22(a). If the surface is not aligned with the global Cartesian coordinate system, as shown in Figure 2.22(b), it is necessary to define a *local* coordinate system (x', y') where one axis is normal to the face of the "sliced" surface, and determine the acoustic particle velocity in the x' direction.

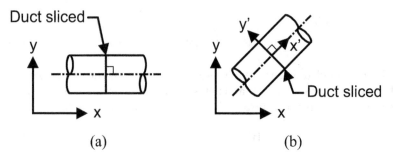

(a)   (b)

FIGURE 2.22
(a) Surface for estimating sound intensity in a duct is recommended to be
"sliced" normal to the axis of the duct and aligned with global Cartesian
coordinate system. (b) Local coordinate system (x', y') defined on axis of duct
and normal to the "sliced" face.

2.10.3   Acoustic Potential Energy

The acoustic potential energy $E_p(\omega)$ is a useful measure of the acoustic energy contained within a cavity at frequency, ω. This measure can be used to
evaluate the effectiveness of noise control in an enclosure and is given by [52]

$$E_p(\omega) = \frac{1}{4\rho_0 c_0^2} \int_V |p(\mathbf{r}, \omega)|^2 \, dV \, , \tag{2.53}$$

where $p(\mathbf{r}, \omega)$ is the pressure at a location in the cavity, and V is the volume of
the cavity. This expression can be implemented in a finite element formulation
as

$$E_p(\omega) = \frac{1}{4\rho_0 c_0^2} \sum_{n=1}^{N_a} p_n^2(\omega) V_n \, , \tag{2.54}$$

where p_n is the acoustic pressure at the nth node and V_n is the volume associated with the nth node. This equation can be rearranged so that the acoustic potential energy is calculated in terms of the modal pressure amplitudes
as:

$$E_p(\omega) = \sum_{n=1}^{N_a} \Lambda_n |p_n(\omega)|^2 = \mathbf{p_n}^H \boldsymbol{\Lambda_n} \mathbf{p_n} \, , \tag{2.55}$$

where $\boldsymbol{\Lambda_n}$ is a $(N_a \times N_a)$ diagonal matrix for which the diagonal terms are

$$\boldsymbol{\Lambda_n}(n, n) = \frac{\Lambda_n}{4\rho_0 c_0^2} \tag{2.56}$$

where Λ_n is the modal volume of the nth cavity mode. Determining the acoustic modal volume from ANSYS is an advanced topic and is explained in Section 9.3.2.

This method of determining the modal volume is used in ANSYS Mechanical APDL and MATLAB scripts that are described in Section 9.4.5 and Appendix C.

To enable comparison of acoustic potential energy from different analyses with varying frequency spacing, the acoustic potential energy can be "normalized" using the following expression

$$\overline{\text{APE}} = 10 \log_{10} \left(\frac{\Delta f_1 \times \sum\limits_{i=1}^{i_{\text{max}}} \text{APE}(f_i)}{\Delta f_2} \right) , \qquad (2.57)$$

where $\text{APE}(f)$ is the acoustic potential energy evaluated at frequency f, Δf_1 is the frequency spacing using for the analysis, and Δf_2 is the frequency range over which the analysis was conducted.

2.10.4 Acoustic Energy Density

The *acoustic energy density* at a point is equal to the sum of the acoustic potential energy density and the kinetic energy density, and quantifies the total acoustic energy at a point. The acoustic potential energy density is related to the acoustic pressure at a point. The acoustic kinetic energy density is a function of the acoustic particle velocity. In terms of characterizing the acoustic field, acoustic energy density exhibits lower spatial variance in reactive sound fields compared to the acoustic potential energy estimate offered by microphones, making it a more robust measure of the acoustic energy within an enclosure [67] [51]. Acoustic energy density also has application in active noise control in enclosures [54], the free-field [98] and random sound fields [113].

The instantaneous acoustic energy density, $E_D(t, \vec{x})$, at some point $\vec{x}$ is given by [63, Eq. (4.7)][52, Eq. (3.1)]

$$E_D(t, \vec{x}) = \frac{p(t, \vec{x})^2}{2\rho_0 c_0^2} + \frac{\rho_0 v(t, \vec{x})^2}{2} , \qquad (2.58)$$

where $p(t, \vec{x})$ and $v(t, \vec{x})$ are the instantaneous pressure and particle velocity at $\vec{x}$ respectively. This can also be written in terms of the time-averaged acoustic energy density

$$\bar{E}_D(\vec{x}) = \frac{\bar{p}(\vec{x})^2}{2\rho_0 c_0^2} + \frac{\rho_0 \bar{v}(\vec{x})^2}{2} , \qquad (2.59)$$

where $\bar{p}$ and $\bar{v}$ are the time-averaged acoustic pressure and acoustic particle velocity at $\vec{x}$, respectively.

In practice, the acoustic pressure is measured by a pressure microphone. However, the acoustic particle velocity may be measured by a number of means including directly using particle velocity sensors or pressure gradient microphones, or indirectly by estimating the acoustic pressure gradient from a number of pressure microphones [53]. Cazzolato [52] and Cazzolato and Benjamin [54] demonstrate how acoustic energy density may be calculated in ANSYS using pressure gradients obtained from microphones in common 3D configurations, as well as how to obtain the energy density estimate from four closely spaced arbitrary nodes in the acoustic field.

See Cazzolato et al. [56] for further discussion on the use of acoustic energy density and the potential errors that can occur when attempting to conduct a harmonic analysis using modal summation due to using an insufficient number of modes in the calculations.

2.10.5　Structural Kinetic Energy

Structural kinetic energy is the energy an object has due to its motion. For a rigid (lumped) body the kinetic energy is calculated as

$$E_k = \frac{1}{2}mv^2 \,, \tag{2.60}$$

where m is the mass of the rigid body, and v is the velocity of the rigid body. This simple equation can be re-written using integrals over the surface or volume of an object, where the lumped body is discretized into small areas, and the mass is described as the density ρ times the elemental volume. As an example, consider a harmonically vibrating plate of dimensions $L_x \times L_y$, and thickness h; the time-average kinetic energy is given by [66, p. 174, Eq. (3.73)]

$$E_k = \frac{1}{2} \int_S \frac{1}{T} \int_0^T \rho h v^2(x,y,t) dt dS \tag{2.61}$$

$$= \frac{\rho_{\text{plate}} h}{4} \int_0^{L_y} \int_0^{L_x} |v|^2 dx dy \,, \tag{2.62}$$

where T is a suitable period of time, S is the surface of the structure, and $v(x, y, z)$ is the velocity of the element at location (x, y) at time t. This can be written in an equivalent form for finite element models as

$$E_k = \frac{1}{4} [\dot{q}]^T [\mathbf{M}][\dot{q}] \,, \tag{2.63}$$

where $[\dot{q}]$ is the velocity of an element, and $[\mathbf{M}]$ is the mass of an element. This can be written in terms of the structural modal participation factors $\mathbf{w_m}$ as

$$E_k = \frac{\omega^2}{4} \mathbf{w_m}^H \mathbf{\Lambda_m} \mathbf{w_m} \,, \tag{2.64}$$

where $\mathbf{\Lambda_m}$ is a diagonal matrix of the modal mass of the structure.

Examples of the use of structural kinetic energy of a plate are described in Sections 9.4.5 and 9.5.4.

2.11   Mesh Density

An important decision that must be made when modeling a structure or acoustic domain using finite element analysis is to consider the mesh density. The selection of an insufficient mesh density can lead to inaccurate or misleading results and the selection of an excessive mesh density can cause long computation times. Hence, the analyst must make a balanced and informed decision.

For stress analysis simulations, the required mesh density is based on the stress gradients and deformation of the elements. In regions of high-stress gradients, it is prudent to increase the mesh density so that there are sufficient nodes and elements to depict the stress field.

Note that regions of stress concentrations can result in stress *singularities*, where increasing the mesh density (i.e., reducing the area of the element) will result in ever-increasing values of the calculated stress. In simple terms, stress (σ) is calculated as the ratio of force (F) divided by the area (A) over which the force is spread, as

$$\sigma = F/A . \tag{2.65}$$

As the mesh density is increased, so that the area of an element decreases, the calculated stress will increase. This effect can occur at stress concentrations such as re-entrant corners, point loads, point boundary conditions, and others.

In regions where there is large distortion of the structural elements and nodes due to an applied load, it is recommended to increase the mesh density.

Consider the "sea of springs" analogy described in Section 1.2. If there is an insufficient mesh density, it is possible that the model of the structure is artificially "stiff," or that the load path through the structure is re-directed to another region, and the results that are generated are inaccurate.

For dynamic structural and acoustic analyses, the recommended mesh density is based on the considerations for static analyses and also the number of *elements per wavelength*. Consider the simply supported beam shown in Figure 2.23 that is vibrating at its second mode (the first mode shape would resemble a half-sine wave). The *minimum* number of finite element nodes required to identify that the beam is vibrating at the second vibrational mode is 5, as shown by the circular dots in the figure. This model of the beam has 4 elements (count the number of links between the nodes). Hence this model has (4 elements / 2nd mode=) 2 elements per wavelength (EPW). This is analogous to the Nyquist–Shannon Sampling Theory in signal processing methods [120, 121, 136, 137].

Figure 2.24 shows the same beam as in Figure 2.23, which is modeled using 2, 4, and 6 beam elements where the nodes are aligned with the deflected mode shape in the left column and where they have been offset by half and element spacing in the right column. The worst case is shown in the top left diagram where there are only 2 beam elements and it can be seen that the results from a finite element analysis would be unable to resolve the deflected mode shape of

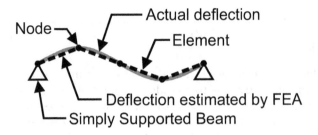

FIGURE 2.23
Second mode shape of a simply supported beam modeled with 4 elements and 5 nodes.

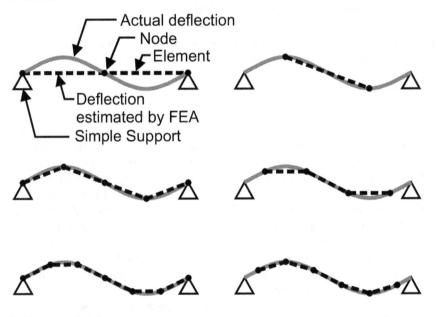

FIGURE 2.24
Second mode shape of a simply supported beam modeled with 2, 4, and 6 elements with the elements aligned with the mode shape and offset by half an element spacing.

the beam. Referring to the top right diagram, if the same element spacing were used, but the location of the elements were offset by half an element spacing, then the results from a finite element analysis would be able to resolve the deflected mode shape of the beam. As one is not always able to predict the response in advance (which is why the analysis is being conducted!), it would be a gamble as to whether the nodes and elements were in suitable locations to resolve the response of the system. It can be seen that as the number of elements is increased there is closer agreement between the results from a finite element analysis and the actual deflected shape, and that the FEA results are insensitive to location of the nodes and elements.

To obtain accurate results it is recommended that between 6 and 12 elements per wavelength are used [42, 111]. ANSYS recommends using 12 elements per wavelength when using FLUID30 elements and 6 elements per wavelength when using the quadratic FLUID220 elements that have mid-side nodes. Note that the guideline of using 6 to 12 elements per wavelength should be re-assessed when there are discontinuities in the model. For the same reason that the mesh density should be increased for structural discontinuities, the mesh density around features that cause acoustic pressure discontinuities should also be increased so that the pressure gradients can be simulated with a sufficient number of elements.

Figure 2.25 shows a schematic of the cross section of a vibrating plate that radiates acoustic pressure. Adjacent regions of the plate will generate "positive" and "negative" acoustic pressure that will combine and interfere with each other. Consider the left-hand side of the unbaffled plate in Figure 2.25. The upward movement of the plate will generate a positive acoustic pressure, but as the plate is unbaffled the pressure can circulate and equalize with the negative acoustic pressure on the underside side of the plate, with the result of acoustic cancellation.

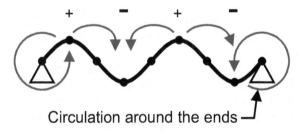

Circulation around the ends

FIGURE 2.25
Cross-sectional view of a vibrating plate showing the interaction of radiated acoustic pressure.

A similar effect occurs with an unbaffled loudspeaker as shown in Figure 2.15. In summary, there is a complicated acoustic radiation from a plate. The consequence is that when conducting a coupled fluid–structure interaction finite element analysis, the mesh density must be adequate in the region near this complicated acoustic radiation pattern to correctly simulate the pressure interactions. Consider the case if the mesh density of the acoustic domain were insufficient, then the cancellation effect due to the interaction of the positive and negative acoustic pressures would not be modeled, and the GIGO principle would prevail.

The influence of variation of mesh density on the results from simulations is shown by example in

- Section 3.3.2, where the resonance frequencies and sound pressure distribution in a duct with rigid end terminations are calculated;

- Section 3.4.3, where the mesh density is varied in an expansion chamber resonator silencer, and shows that 6 elements per wavelength of FLUID30 elements is sufficient to calculate the transmission loss;

- Section 3.5, where 6 elements per wavelength of FLUID30 elements are used in a duct where non-plane waves can exist and causes variations in sound pressure levels and re-directs acoustic energy;

- Section 4.4.4, where a 3D rigid-walled enclosure is meshed with 6 elements per wavelength with FLUID30 elements and the order of the mode shapes is incorrect, and is corrected by increasing the mesh density to 12 elements per wavelength; and

- Section 8.4.4, where the radiation from a baffled piston is simulated using 12 and 20 elements per wavelength, and it is shown that the near-field results approach theoretical estimates as the mesh density is increased.

2.12 Use of Symmetry

The use of symmetry in finite element models is expedient for reducing the number of nodes and elements in a model, and hence will reduce the time taken to solve a model. However, it has to be used with caution in acoustic and vibration finite element models.

The learning outcome of this section is to understand that if symmetry is assumed to exist in a finite element model of an acoustic system, then depending on the excitation source that is applied, the results may need to be scaled.

Before further discussion on this topic, it is worthwhile illustrating what is meant by symmetry and asymmetry for a vibrating structural system where the elements have nodes with displacement degrees of freedom. Consider a vibrating beam that can exhibit asymmetric and symmetric mode shapes as shown, respectively, in the upper and lower graphs in Figure 2.26. The asymmetric mode shapes in the upper graph resemble sine waves, and symmetric mode shapes in the lower graph resemble cosine waves. A plane of symmetry (or asymmetry) can be defined using ANSYS Workbench or ANSYS Mechanical APDL which imposes boundary conditions on the nodes (that have displacement degrees of freedom) that lie on the plane of symmetry (or asymmetry). The ANSYS Mechanical APDL command DSYM, Lab, Normal can be used to specify a symmetric or asymmetric boundary condition.

However for models comprising acoustic elements (that have pressure degrees of freedom), ANSYS does not have any facility to define planes of symmetry or asymmetry—an analyst creates a model where symmetry is implied and it is up to the analyst to interpret the results correctly, or alternatively a

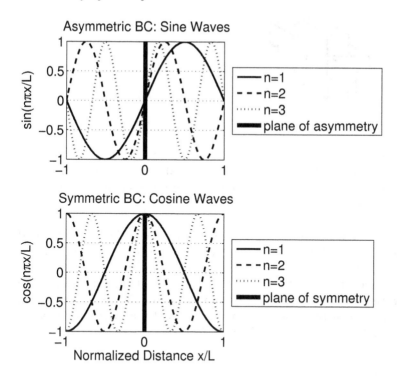

FIGURE 2.26

Plane of asymmetry at $x/L = 0$ with sine wave mode shapes, and plane of symmetry at $x/L = 0$ with cosine wave mode shapes.

zero pressure boundary condition can be applied to a face to simulate an asymmetric boundary condition for a modal analysis. Examples of these symmetry conditions are described below.

Modal Analysis Symmetric Model

The ANSYS manual has written [21],

> The naturally occurring boundary condition in acoustics is a symmetry boundary condition, so no specification on the acoustic elements is necessary to designate a symmetry plane.

The literal interpretation of this statement is that every rigid-wall is effectively a plane of symmetry. Figure 2.27(a) shows a sketch of a rigid-walled rectangular cavity of dimensions 1 m × 1 m × 2 m that has acoustic elements with only pressure degrees of freedom. A symmetric model of this rectangular cavity is shown in Figure 2.27(b) where the single cavity has been mirrored about the six exterior faces. An acoustic modal analysis was conducted using ANSYS Workbench and the resonance frequency of the $(0, 0, 1)$ mode is

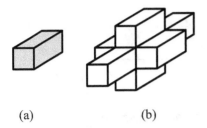

(a)   (b)

FIGURE 2.27
(a) Model of a rigid-walled rectangular duct. (b) Model of system if the rigid-walls of the model in (a) are assumed to have symmetric boundary conditions for the pressure degree of freedom.

85.8 Hz, as shown in Figure 2.28. This is the second mode that was calculated in ANSYS, where the first mode is the "bulk compression" mode at 0 Hz.

A modal analysis of the full model can be conducted where the single rectangular cavity has been mirrored about each face. Figure 2.29 shows the pressure in the cavity for mode 7 that corresponds to 85.8 Hz, where it can be seen that the acoustic pressure is symmetric about the exterior faces of the single rectangular cavity shown in Figure 2.28.

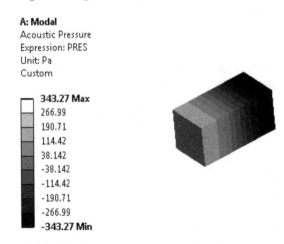

A: Modal
Acoustic Pressure
Expression: PRES
Unit: Pa
Custom

343.27 Max
266.99
190.71
114.42
38.142
-38.142
-114.42
-190.71
-266.99
-343.27 Min

FIGURE 2.28
Contour plot of the acoustic pressure for mode 2 corresponding to 85.5 Hz for the base model of the single block.

This example has shown that

- it is possible to assume a symmetric response of an acoustic system, where planes of symmetry exist about planar rigid-walls;

- the analyst has to be cautious about the interpretation of the mode shapes—the first axial mode of the single rectangular cavity shown in Figure 2.28

which corresponded to mode 2, has equivalent symmetric mode shape shown in Figure 2.29 which corresponds to mode 7; and

- there is no special definition or facility in ANSYS to define planes of symmetry for acoustic systems–it is up to the analyst to assume symmetry and interpret the results correctly.

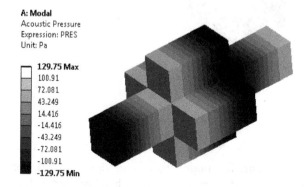

FIGURE 2.29

Contour plot of the acoustic pressure for mode 7 corresponding to 85.5 Hz for the full model where the base model has been mirrored about the faces.

Modal Analysis Asymmetric Models

An asymmetric boundary condition can be simulated for undamped modal analyses by defining a zero pressure boundary condition on a face. An asymmetric model of the rigid-walled rectangular cavity, with dimensions 1 m × 1 m × 2 m examined in the previous section, can be simulated by applying a zero pressure boundary condition on the $z = 0$ face, as shown in Figure 2.30.

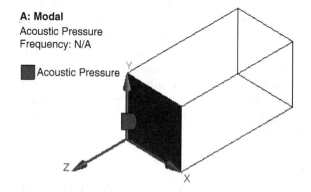

FIGURE 2.30

Half model of the cavity where face at $z = 0$ m has a zero pressure boundary condition applied.

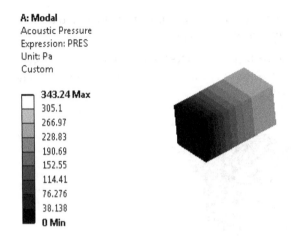

A: Modal
Acoustic Pressure
Expression: PRES
Unit: Pa
Custom

343.24 Max
305.1
266.97
228.83
190.69
152.55
114.41
76.276
38.138
0 Min

FIGURE 2.31
Contour plot of the acoustic pressure for mode 1 corresponding to 42.9 Hz for the base model of the single block and a pressure of 0 Pa has been defined on the face at $z = 0$ m.

The first acoustic mode of the system calculated in ANSYS is at 42.9 Hz as shown in Figure 2.31, where the pressure is zero at $z = 0$ m and is a maximum at $z = -2$ m. To prove that this can be considered as an asymmetric model, the full system was simulated with a rigid-walled rectangular cavity with dimensions 1 m × 1 m × 4 m and a modal analysis was conducted. Figure 2.32 shows the second acoustic mode calculated in ANSYS at 42.9 Hz (the first mode is at 0 Hz and is the bulk compression mode), where the maximum pressure is at $z = 2$ m, the pressure is zero at $z = 0$ m, and the minimum pressure is at $z = -2$ m.

Figure 2.33 shows the normalized acoustic modal pressure versus the z axis coordinate calculated using ANSYS Workbench for the full model, from the results shown in Figure 2.32, and the results from the half model shown in Figure 2.31, where a zero pressure boundary condition was applied at $z = 0$ m. The normalized modal pressure is calculated by dividing the modal pressure results by the maximum value in the cavity. It can be seen that the two sets of results overlay each other. Hence, the use of a zero pressure boundary condition applied to a face can be used to simulate an asymmetric modal response for an undamped system.

Harmonic Response Analyses: Symmetric Models

Two finite element models were created using ANSYS Workbench of a rigid-walled circular duct that was 3 m in length, 0.1 m in diameter, and filled with air. One end of the duct had a simulated piston compressing the air in the duct, and the far end of the duct had a rigid end cap. Figure 2.34(a) shows a

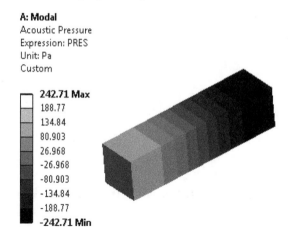

A: Modal
Acoustic Pressure
Expression: PRES
Unit: Pa
Custom

242.71 Max
188.77
134.84
80.903
26.968
-26.968
-80.903
-134.84
-188.77
-242.71 Min

FIGURE 2.32
Contour plot of the acoustic pressure for mode 2 corresponding to 42.9 Hz for the full model where the base model has been mirrored about the faces.

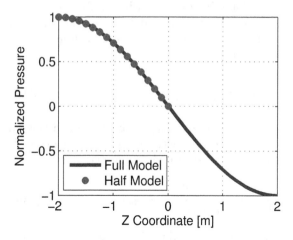

FIGURE 2.33
Normalized modal pressures within the full and half models of the cavity, where the half model had a zero pressure boundary condition applied to the face at $z = 0$.

full model of the duct, and Figure 2.34(b) shows where only a quarter section of the duct was modeled.

It was shown previously that for a modal analysis of a rigid-walled cavity, the rigid walls effectively create a symmetric boundary condition. When a harmonic response analysis is conducted, where there is a forced excitation of the acoustic system, some caution needs to be exercised as at Release 14.5 of

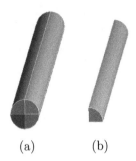

(a)   (b)

FIGURE 2.34
Full and quarter models of an air-filled duct, 3 m long, 0.1 m in diameter,
created using ANSYS Workbench.

ANSYS there is no specific mechanism for defining symmetric or asymmetric
boundary conditions for acoustic systems.

The acoustic response of the duct will depend on the excitation applied to
the piston. It will be shown in the following example where a force is applied
to the piston that it is necessary to scale the applied force by the reduction
in area, or scale the results. If the pressure generated by the piston due the
application of a force in the full model is

$$p_{\text{full}} = \frac{F}{A_{\text{full}}}, \tag{2.66}$$

where F is the force applied to the piston, and A_{full} is the area of the piston
in the full model. In the 1/4 model of the duct, the area of the piston is
$A_{\text{quarter}} = A_{\text{full}}/4$ and hence the pressure generated by the piston is

$$p_{\text{quarter}} = \frac{F}{A_{\text{quarter}}} = \frac{4 \times F}{A_{\text{full}}} = 4 \times p_{\text{full}}, \tag{2.67}$$

which is 4 times the pressure compared to the full model. This will be explored
further in the following example. An alternative to using a force excitation
could be to use a pressure boundary condition on the piston, but this also
has to be used with caution as described in Section 1.2, or specify an oscillating displacement or velocity of the piston face, in which case the pressure
developed by the piston will be identical for the full and quarter models.

The piston end of the duct was modeled as having fluid–structure interaction, and was driven with a force of $F_z = 1 \times 10^{-3}$ N at the vertex on the axis
of the cylinder, as shown in Figure 2.35. The same location, magnitude, and
direction of the force was used for the full model. All the nodes on the piston
end of the cylinder had their displacement degrees of freedom coupled in the
axis of the cylinder (z axis), so that all the nodes moved together, essentially
creating a rigid-faced piston.

FIGURE 2.35
Force applied to the end of the duct along the axis of the cylinder.

Figure 2.36 shows the pressure distribution along the duct at 70 Hz for the full model and the quarter model. It can be seen that the pressure for the quarter model is ×4 the value of the full model. This intuitively makes sense as the volume of the quarter-duct model is 1/4 of the volume of the full duct model, and the piston is driven with the same amplitude for both models. Hence, to get the correct results, the pressure for the quarter model needs to be *divided* by 4.

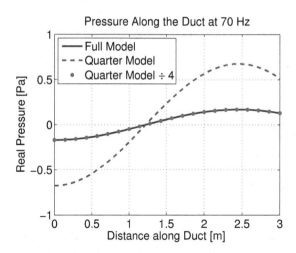

FIGURE 2.36
Real part of acoustic pressure of a piston attached to the end of a duct 3 m in length, 0.1 m in diameter, with a rigid end cap, calculated using ANSYS Workbench with a full model and a quarter model.

We now consider the mechanical impedance of the piston, which is the applied force to the piston divided by its velocity, and is defined as

$$Z_{\mathrm{m}} = \frac{\text{Force}}{\text{Velocity}} = \frac{F}{v}. \tag{2.68}$$

The theoretical expression for the mechanical impedance of the piston attached to the end of a duct at $x = 0$ is calculated as [102, Eq. (10.2.7), p. 273]

$$Z_{m0} = (\rho_0 c S) \times [-j \cot(kL)] \,, \qquad (2.69)$$

where $\rho_0 = 1.21$ kg/m^3 is the density, $c_0 = 343$ m/s is the speed of sound, $k = \omega/c_0$ is the wavenumber, $L = 3.0$ m length of the duct, $S = \pi a^2$ is the cross-sectional area of the circular duct, and $a = 0.05$ m is the radius of the circular duct. In a lossless duct, the mechanical impedance in Equation (2.69) is entirely imaginary, and the real part is zero. Figure 2.37 shows the imaginary part of the mechanical impedance for the theoretical, full, and quarter finite element model results. This figure of the impedance was generated using the MATLAB script `impedance_driven_closed_pipe.m` that is included with this book. The mechanical impedance for the quarter model needs to be *multiplied* by 4 to obtain the correct theoretical mechanical impedance.

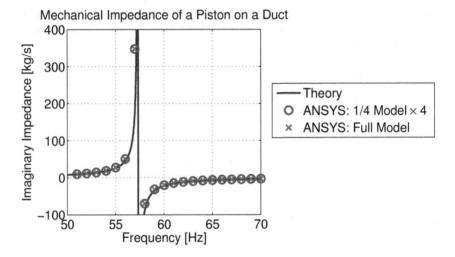

FIGURE 2.37
Imaginary part of the mechanical impedance of a piston attached to the end of a duct 3 m in length, 0.1 m in diameter, with a rigid end-cap, calculated theoretically, and using ANSYS Workbench with full and quarter models.

If mechanical power is to be calculated using 1/4 models, then the results have be scaled accordingly. Mechanical power is proportional to the inverse of the mechanical impedance [66, Eq. (2.4), p. 77]

$$\text{Power} \propto \frac{1}{Z_{\text{mechanical}}} \,. \qquad (2.70)$$

The mechanical power calculated for a 1/4 model needs to be *divided* by 4 to obtain the results for the full model. This result is further illustrated in Section 3.3.7.2, where the results of mechanical power of a 1/4 model is compared with the results from a full model.

These simple examples show how the results from finite element models of acoustic systems that exploit symmetry can be scaled to obtain the correct values for full models.

However, this can only be used where there is no structure involved. When a vibro-acoustic analysis is to be conducted where there is a vibrating structure, the use of symmetry can lead to additional complications due to symmetric and asymmetric vibration about the plane(s) of symmetry. It is recommended that the full system be modeled.

3

Ducts

3.1 Learning Outcomes

The learning outcomes of this chapter are to:

- learn how to calculate the natural frequencies of a 3D duct with rigid-walls with various end conditions;

- examine the influence of mesh density and the accuracy of natural frequencies and pressure distribution;

- understand how to model acoustic systems quadratic acoustic finite elements (FLUID220) and compare the results to those obtained using linear acoustic finite elements (FLUID30);

- understand how to model a semi-infinite acoustic domain using FLUID130 infinite acoustic elements;

- understand how the results from an analysis of a 1/4 model needs to be scaled to be applicable to a full model for certain loading conditions;

- understand how to apply a frequency varying impedance to a face in a model;

- understand four-pole or transmission line analysis method for analyzing ducts and reactive silencer elements; and

- understand the influence of a temperature variation in a duct and how to model it in ANSYS.

3.2 Theory

The following sections describe how to calculate the natural frequencies of ducts with various end conditions, the four-pole transmission line method for predicting the acoustic response of ducts that have plane-wave conditions, and a discussion of the various acoustic metrics that can be used to assess the performance of silencers or mufflers.

3.2.1 Natural Frequencies

The natural frequencies and mode shapes of undamped tubes (pipes or ducts) are listed in Table 3.1 [46, Table 6.2].

TABLE 3.1

Natural Frequencies and Axial Mode Shapes of Pipes with Various End Conditions

Configuration	Schematic	Mode Index $n =$	Natural Frequencies f_n [Hz]	Mode Shape ψ_n [no units]
rigid–rigid		$0, 1, 2 \cdots$	$\dfrac{nc_0}{2L}$	$\cos\left[\dfrac{n\pi x}{L}\right]$
open–rigid		$1, 3, 5 \cdots$	$\dfrac{nc_0}{4L}$	$\cos\left[\dfrac{n\pi x}{2L}\right]$
open–open		$1, 2, 3 \cdots$	$\dfrac{nc_0}{2L}$	$\sin\left[\dfrac{n\pi x}{L}\right]$

For a duct with rigid–rigid end conditions, the first mode index is $n = 0$, which is called the *bulk compression* mode and occurs at 0 Hz.

For the open–rigid end condition, the natural frequencies occur at odd-numbered harmonics $n = 1, 3, 5 \cdots$. This open–rigid configuration, sometimes called a quarter-wavelength tube, is often used in reactive silencers when it is attached to a main duct and the length of the quarter-wavelength tube is adjusted so that its natural frequency coincides with the frequency of an unwanted harmonic noise. This topic is further discussed in Section 3.4.

For the open–open end condition, the natural frequencies are the same as a rigid–rigid end condition, and only the mode shapes are different. In reality, when there is an open-ended pipe as in the cases of the open–rigid and open–open end conditions, the effective length of the pipe is slightly longer than the physical length of the pipe, as a small amount of gas external to the pipe is entrained in the motion of the gas. The slightly longer pipe causes the natural frequencies of the pipe to be reduced slightly. This can be modeled by modifying the termination impedance at the outlet of the duct and is described in Section 3.3.7.

The natural frequencies of ducts with various end conditions will be calculated using the expressions in Table 3.1 in MATLAB, and will be compared with the results from finite element analysis conducted using ANSYS in Section 3.3.

3.2.2 Four-Pole Method

The four-pole, or transmission line method, is a useful theoretical tool for estimating the acoustic performance of resonator silencers. See Munjal [119, Section 2.18], and Beranek and Vér [46, Chapter 10] for further details on this topic.

Figure 3.1 shows a typical configuration for a silencer. An acoustic source, such as a reciprocating engine, is attached to an upstream duct that is connected to the inlet of a silencer. The geometry of the silencer could consist of expansion chambers, transverse tube resonators, perforated tubular elements, and so on. The outlet of the silencer is connected to a tail-pipe section that radiates sound into a free-field. Each of these components has an acoustic impedance and can be represented by 4-pole transmission line matrices. The acoustic source has an impedance Z_s. The end of the acoustic duct has a termination impedance Z_T, which in the example shown in Figure 3.1, is the radiation impedance of an unflanged duct radiating into a free-field.

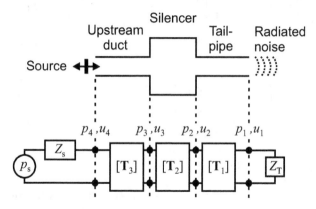

FIGURE 3.1
Schematic of a typical silencer configuration and the equivalent 4-pole transmission matrix representation.

The pressure and mass velocity upstream and downstream of an element are related by a 4-pole transmission matrix as [46, p. 377, Eq. (10.14)]

$$\begin{bmatrix} p_2 \\ \rho_0 S_2 u_2 \end{bmatrix} = \begin{bmatrix} T_{11} & T_{12} \\ T_{21} & T_{22} \end{bmatrix} \begin{bmatrix} p_1 \\ \rho_0 S_1 u_1 \end{bmatrix} \tag{3.1}$$

$$\begin{bmatrix} p_2 \\ V_2 \end{bmatrix} = \mathbf{T} \begin{bmatrix} p_1 \\ V_1 \end{bmatrix}, \tag{3.2}$$

where p_i is the acoustic pressure at point i along the system. The mass flow velocity V_i is the density of the gas times cross-sectional area of the duct times the acoustic particle velocity and is calculated as

$$V_i = \rho_0 S_i u_i, \tag{3.3}$$

where ρ_0 is the density of the gas, S_i is the cross-sectional area of the duct at point i, and u_i is the acoustic particle velocity (not the mean flow velocity) at point i.

The 4-pole transmission matrix for a straight segment of duct of length L is given by [46, p. 377, Eq. (10.15)]

$$\mathbf{T} = \begin{bmatrix} \cos(kL) & j\dfrac{c_0}{S}\sin(kL) \\ j\dfrac{S}{c_0}\sin(kL) & \cos(kL) \end{bmatrix}, \tag{3.4}$$

where $k = \omega/c_0$ is the wavenumber, $\omega = 2\pi f$ is the circular frequency, f is the frequency of excitation, and c_0 is the speed of sound.

The equations describing the response of the system shown in Figure 3.1 can be written as

$$\begin{bmatrix} p_s \\ \rho_0 S_s u_s \end{bmatrix} = \begin{bmatrix} 1 & Z_s \\ 0 & 1 \end{bmatrix} \begin{bmatrix} p_4 \\ \rho_0 S_4 u_4 \end{bmatrix} \tag{3.5}$$

$$\begin{bmatrix} p_4 \\ \rho_0 S_4 u_4 \end{bmatrix} = [\mathbf{T}_3] \begin{bmatrix} p_3 \\ \rho_0 S_3 u_3 \end{bmatrix} \tag{3.6}$$

$$\begin{bmatrix} p_3 \\ \rho_0 S_3 u_3 \end{bmatrix} = [\mathbf{T}_2] \begin{bmatrix} p_2 \\ \rho_0 S_2 u_2 \end{bmatrix} \tag{3.7}$$

$$\begin{bmatrix} p_2 \\ \rho_0 S u_2 \end{bmatrix} = [\mathbf{T}_1] \begin{bmatrix} p_1 \\ \rho_0 S_1 u_1 \end{bmatrix} \tag{3.8}$$

$$\begin{bmatrix} p_1 \\ \rho_0 S u_1 \end{bmatrix} = \begin{bmatrix} 1 & Z_T \\ 0 & 1 \end{bmatrix} \begin{bmatrix} 0 \\ \rho_0 S_1 u_1 \end{bmatrix}, \tag{3.9}$$

where the 4-pole transmission matrices $[\mathbf{T}_i]$ depend on the configuration of each duct segment. These equations can be written in matrix form as

$$\begin{bmatrix} p_s \\ \rho_0 S u_s \end{bmatrix} = \begin{bmatrix} 1 & Z_s \\ 0 & 1 \end{bmatrix} \mathbf{T}_3 \mathbf{T}_2 \mathbf{T}_1 \begin{bmatrix} 1 & Z_T \\ 0 & 1 \end{bmatrix} \begin{bmatrix} 0 \\ \rho_0 S u_1 \end{bmatrix} \tag{3.10}$$

$$= \begin{bmatrix} T_{11} & T_{12} \\ T_{21} & T_{22} \end{bmatrix} \begin{bmatrix} 0 \\ \rho_0 S u_1 \end{bmatrix}. \tag{3.11}$$

The impedance of an *unflanged* pipe radiating into a free-field is given by [102, Eq. (10.2.14), p. 274]

$$Z_T = R_0 + jX_0 \tag{3.12}$$

$$= \left[(\rho_0 c_0 S)\frac{(ka)^2}{4} \right] + j\left[(\rho_0 c_0 S)0.61ka \right], \tag{3.13}$$

where a is the radius of the duct at the exit, R_0 is the real part of the impedance called the *resistance*, and X_0 is the imaginary part of the impedance called the *reactance*. The impedance of a *flanged* pipe radiating into a free-field is given by [102, Eq. (10.2.10), p. 274]

$$Z_T = R_0 + jX_0 \tag{3.14}$$

$$= \left[(\rho_0 c_0 S) \frac{(ka)^2}{2} \right] + j \left[(\rho_0 c_0 S) \frac{8}{3\pi} ka \right] . \tag{3.15}$$

The 4-pole transmission matrix method relies on the principle of plane-wave propagation inside the duct network. To ensure that the transmission matrix method can be used with validity, it is important to estimate the cut-on frequency, which is defined as the frequency below which only plane waves propagate inside the duct. The cut-on frequency for circular ducts is given by [47, Eq. (9.137), p. 490]

$$f_{\text{cut-on: circular}} = \frac{1.8412 \times c_0}{2\pi a} = \frac{0.293 \times c_0}{a}, \tag{3.16}$$

and for ducts with a rectangular cross-section [47, Eq. (9.136), p. 490]

$$f_{\text{cut-on: rectangular}} = \frac{c_0}{2H}, \tag{3.17}$$

where H is the largest cross-sectional dimension.

3.2.3   Acoustic Performance Metrics

There are a number of metrics that can be used to evaluate the acoustic performance of a silencer:

Insertion Loss is defined as the reduction in radiated sound power level due to the replacement (insertion) of a section of duct with the proposed silencer and is calculated as

$$\text{IL} = L_{w:\text{ before}} - L_{w:\text{ after}}, \tag{3.18}$$

where $L_{w:\text{ before}}$ is the sound power level of the system without the silencer installed, and $L_{w:\text{ after}}$ is the sound power level after the silencer has been installed by replacing a section of the duct.

Transmission Loss is the difference between the sound power *incident* on the silencer ($L_{w:\text{incident}}$) and the sound power that continues to be transmitted after the silencer ($L_{w:\text{transmitted}}$), when the system has an infinite (anechoic) end condition [119, p. 58]. When a reactive silencer is installed in a duct, it tends to reflect sound power back upstream and the sound pressure level can increase upstream of the silencer compared with a system without the silencer installed. The calculation of transmission loss uses the sound power *incident* on the silencer, and not the total sound power that exists upstream of the silencer, to quantify the baseline acoustic performance.

Similarly, the sound power that is *transmitted* into an anechoic termination after the silencer, is used to evaluate transmission loss. Transmission loss is expressed mathematically as

$$\text{TL} = L_{w:\text{incident}} - L_{w:\text{transmitted}} \, . \tag{3.19}$$

Transmission loss is independent of the source impedance, and can be calculated using the four-pole transmission matrix formulation as [46, Eq. (10.10), p. 374]

$$\text{TL} = 20 \log_{10} \left| \frac{T_{11} + \dfrac{S}{c} T_{12} + \dfrac{c}{S} T_{21} + T_{22}}{2} \right| , \tag{3.20}$$

where $T_{11}, T_{12}, \dots$ are the elements of a 4-pole transmission matrix for an individual duct segment such as in Equation (3.4), or for a combined network of duct segments such as Equation (3.11).

Noise Reduction (also known as Level Difference) is the difference between the sound pressure level measured upstream and downstream of the silencer [119, p. 59] and is calculated as

$$\text{NR} = \text{LD} = L_{\text{upstream}} - L_{\text{downstream}} \, . \tag{3.21}$$

The measurement of noise reduction does not require anechoic duct terminations.

3.3 Example of a Circular Duct

We will now consider a specific example of circular duct shown in Figure 3.2.

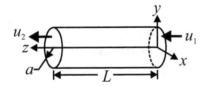

FIGURE 3.2
Schematic of a circular duct of radius a, length L, u_1 and u_2 velocities of the face at each end of the duct.

The analyses that will be conducted are the calculation of the:

- natural frequencies,

- sound pressure distribution along the duct for a harmonic volume velocity excitation at one end of a duct with a finite length,

- pressure distribution along an infinitely long duct, and

- pressure distribution along a duct with a finite length that has a frequency-varying impedance at one end to simulate the end of the duct radiating into free space.

The parameters used in the example are listed in Table 3.2.

TABLE 3.2

Parameters Used in the Analysis of a Circular Duct

Description	Parameter	Value	Units
Diameter	$2a$	0.1	m
Length	L	3	m
Speed of sound	c_0	343	m/s
Density	ρ_0	1.21	kg/m^3
Velocity at piston	u_2	0.0	m/s
Velocity at rigid end	u_1	0.0	m/s

3.3.1 ANSYS Workbench

This section describes instructions for modeling a circular rigid duct, with a diameter of 0.1 m and a length of 3 m. A modal analysis will be conducted using ANSYS Workbench.

The completed ANSYS Workbench archive file called res_freqs_duct. wbpz , which contains the .wbpj project file, is included with this book.

Instructions

- Start ANSYS Workbench.

- In the Toolbox window, under the Analysis Systems, left-click and hold the mouse button down on the Modal icon and then drag it into the Project Schematic window.

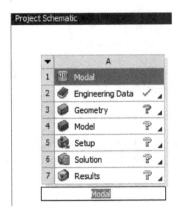

- Double-click on row 3 Geometry to start DESIGNMODELER.

- Select Meter as the desired length unit, and click the OK button.

- Click on the XYPlane icon in the Tree Outline window.

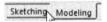

- Click on the New Sketch icon.

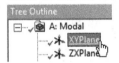

- Click on the Sketch1 icon.

- Click on the Sketching tab.

- Before we start to create a circular area for the duct, the Auto Constraint Cursor must be turned on to ensure that the cursor will "snap" to points and edges. In the Sketching Toolbox window, click on the Constraints tab. To scroll through the Constraint menu options, click on the downward-pointing triangle next to the Setting tab until the Auto Constraints option is visible, then left-click on it. Click in the box next to Cursor: to activate the generation of automatic constraints.

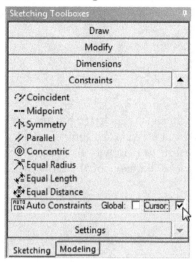

- The next step is to create a circular area to represent the duct. Click on the Draw tab in the Sketching Toolbox window and select the Circle tool.

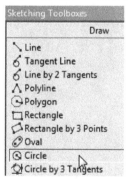

- Move the cursor so that it is over the origin of the axes, and make sure that the origin point changes to red and that the letter P is shown at the cursor (to indicate that the cursor will snap to a coincident point at the origin), then left-click the mouse button to start drawing the circle. It is very important that the center of the circle is selected to be at the origin. In later steps, the circular area will be divided into 4 sectors about the XZ and YZ planes, and if the center of the circle is not coincident with the origin, then the areas for each of the quadrants of the circle will differ, and there will also be issues when trying to use symmetry to model a quarter section of the duct.

- Move the mouse cursor away from the origin so that a circle appears. Notice that in the status bar at the bottom of the screen, the dimension of the radius is shown. Left-click the mouse button to complete drawing a circle at any radius. The correct dimension for the diameter of the circle will be assigned in a later step.

- Click on the Dimension tab in the Sketching Toolboxes window.

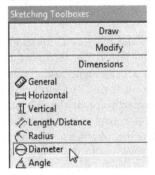

- Click on the Diameter tool.

- In the Graphics window, left-click once on the perimeter of the circle to start drawing the dimension. Move the cursor away from the perimeter and left-click again to indicate the location for the diameter dimension. A diameter D1 will be shown.

- In the window Details View, click the mouse in the square box next to the label D1.

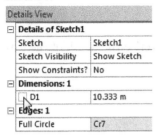

- The letter D will appear in the square box, and a dialog window will open. In the text area for Parameter Name: type duct_diam, then click the OK button.

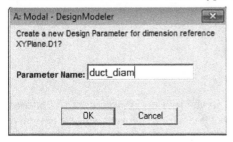

- The next step is to extrude the circle along the Z axis to create a volume for the circular duct. Click the Modeling tab.

- Click on the Sketch1 icon.

- Click on the Extrude icon.

- The geometry that is to be extruded was already selected as Sketch1, so click the Apply button next to the row Geometry. After pressing the Apply button, Sketch1 should appear in the Geometry row.

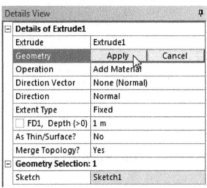

- Click in the square box next to FD1, to parameterize the depth of the extrusion.

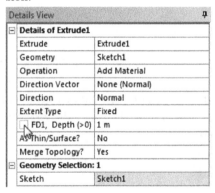

- In the dialog window that appears, in the text entry area next to Parameter Name: type duct_length. The correct length of the duct will be assigned in a later step.

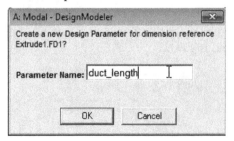

- Notice that there is a lightning bolt next to the `Extrude1` icon in the `Tree Outline` window. Click on the `Generate` icon to create the extrusion.

- An extruded shape will be created in the `Graphics` window.

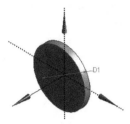

- You might be feeling uneasy that the extruded shape does not resemble a duct. Although it is not necessary to fix this issue at the moment, we will update the parameters so that the model resembles a duct. First, in the menu bar, click on `File | Save Project`, and give your project a filename such as `res_freqs_duct.wbpj`.

- Return to the `Project Schematic` window and double-click on the `Parameter Set` box. Several tables will appear with the parameter names that were defined earlier. We will return to the `Parameter Set` window several times to insert new variables and calculated expressions.

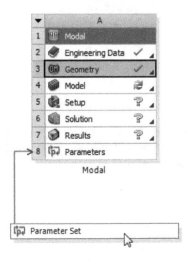

- Click in cell `C4` for the value of the `duct_diam` and type in the value `0.1`, then press the `<Enter>` key on the keyboard.

Outline: No data				
	A	B	C	D
1	ID	Parameter Name	Value	Unit
2	⊟ Input Parameters			
3	⊟ 🔲 Modal (A1)			
4	⏚ P9	duct_diam	0.1	
5	⏚ P2	duct_length	1	
*	⏚ New input parameter	New name	New expression	
7	⊟ Output Parameters			
*	⏚ New output parameter		New expression	
9	Charts			

- Repeat this step to assign the `Parameter Name duct_length` a value of 3.

Outline: No data				
	A	B	C	D
1	ID	Parameter Name	Value	Unit
2	⊟ Input Parameters			
3	⊟ 🔲 Modal (A1)			
4	⏚ P1	duct_diam	0.1	
5	⏚ P2	duct_length	3	
*	⏚ New input parameter	New name	New expression	
7	⊟ Output Parameters			
*	⏚ New output parameter		New expression	
9	Charts			

- Click on the `Refresh Project` icon.

- The solid model in DESIGNMODELER will have been updated with the new values for the parameters. Return to the DESIGNMODELER window. The model might not be visible, so click on the `Zoom to Fit` icon.

- The solid model should now resemble the shape of a duct. It is likely that the placement of the dimension text will be awkward. Although it not necessary to fix this, if you wish to move the placement of the dimension text, click on the `Sketching` tab, click on the `Dimensions` tab, and click on the `Move` icon. If you cannot see the `Move` icon, you might need to click on the triangles to the right of `Constraints` or `Dimensions` to scroll the window to reveal the `Move` icon. Click on the diameter dimension to select it, move the mouse cursor to where you would like to place the dimension text, and then click the mouse cursor again.

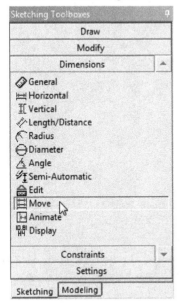

- The full model of the duct will be dissected into 4 slices to create a quarter model. Select `Create | Slice` from the menu bar.

- You need to select the plane that will be used to slice the model. Click on the icon for the YZPlane and then click the Apply button in the row for Base Plane.

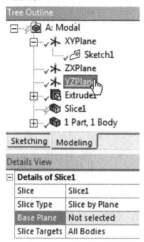

- Make sure that the row for Slice Targets says All Bodies, then click the Generate icon. The cylinder will be split into two halves.

- Repeat these steps to slice all the bodies along the ZXPlane, so that there are 4 bodies.

- Check that each of the 4 bodies have the same Volume and Surface Area by clicking on the first Solid body in the tree beneath 4 Parts, 4 Bodies.

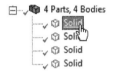

In the window `Details View`, note the values for the `Volume` and `Surface Area` and then click on the other 3 `Solid` bodies and ensure that the values are identical. If the values are not the same, then you should consider fixing this issue as the results from the simulation will be different from those shown in this book.

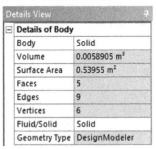

- The next step involves merging the 4 parts and 4 bodies into a single part. The reason for doing this is to ensure that the nodes on the faces between parts are shared and hence there is continuity of pressure between two connected volumes. In the `Tree Outline` window, click on the plus sign next to the `4 Parts, 4 Bodies` branch to expand the list. Click on the first `Solid` body.

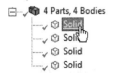

- Hold down the `Shift` key on the keyboard and click on the `Solid` body icon at the bottom of the list, so that all 4 `Solid` bodies are highlighted. Right-click the mouse button to open a context menu and select `Form New Part`.

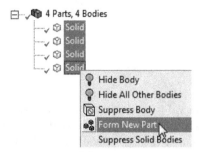

- The 4 separate parts will have been merged into a single part, which will be indicated by the tree branch showing 1 Part, 4 Bodies.

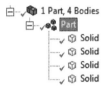

- Save the model again by selecting File | Save Project.

- We will create 3 Named Selections of features of this solid model for the faces on the inlet and outlet of the duct, and the edge along the center axis of the duct. This will make it easier in the later steps when defining the loads and boundary conditions on the model, where the Named Selections can be used. Click on the Faces selection filter icon or press <Ctrl> f on the keyboard so that the Faces icon appears depressed.

- Select the 4 faces on the end of the cylinder by holding down the <Ctrl> key and left-clicking on each face.

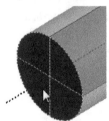

- In the menu bar, click on Tools | Named Selection.

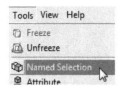

- Make sure the 4 faces are still selected, which will be highlighted in green and the status bar at the bottom of the screen will say 4 Faces. Click the Apply button in the Geometry row in the Details View window.

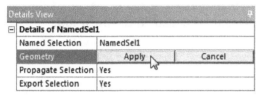

- In the cell next to Named Selection, type NS_outlet. Click the Generate icon.

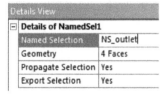

- Repeat these steps to define a named selection for the 4 faces on the opposite end of the cylinder at the XY Plane, and call it NS_inlet.

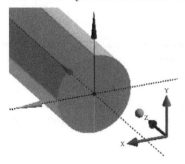

- Click on the Edges selection filter icon. Select the edge along the axis of the cylinder, which is also along the Z axis.

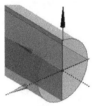

- Create a Named Selection for this edge and call it NS_duct_axis. Click on the Generate icon.

- The completed Tree Outline should be similar to the list below.

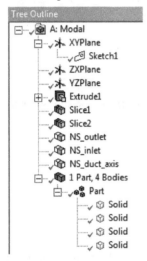

- Click on File | Save Project.

That completes the creation of the solid model. The next stage is to develop the finite element model using ANSYS Mechanical.

- This example (along with most examples in this book using ANSYS Workbench) will make use of the ACT ACOUSTIC extension. Make sure that it is loaded by clicking on Extensions | Manage Extensions from the Project Schematic window.

- Make sure there is a tick in the column Load for the row ExtAcoustics. Click on the Close button when completed.

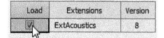

- In the Project Schematic window, double-click on row 4 Model to start Mechanical. You may be presented with a dialog box to "Read Upstream Data"; you should click on the Yes button.

- The solid model from DESIGNMODELER should be transferred to Mechanical, and under the Geometry branch there should be 1 Part, and 4 Solid bodies listed under Model (A4) in the Outline window.

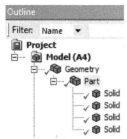

- Make sure that the tab Geometry is selected so that the model is displayed in the window. The other two tabs for Print Preview and Report Preview are not discussed in this book.

- The next steps will involve selecting the method that will be used to mesh the bodies. The Sweep method will generate a mesh pattern that is repeated along the axis of the duct. Right-click on the Mesh branch and select Insert | Method.

- For the geometry selection, right-click in the Geometry window and from the context menu, left-click on Select All, which will highlight all the solid bodies in green.

Click on the Apply button in the Geometry row and the cell should have written that 4 Bodies have been selected.

- In the window Details of "Automatic Method" - Method, change the Method from Automatic to Sweep.

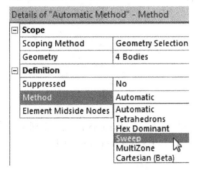

- Some further options will be presented for the Sweep method. Change the row labeled Type from Number of Divisions to Element Size.

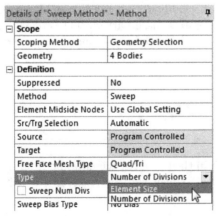

- Click on the square cell next to Sweep Element Size and the letter P will appear in the cell to indicate that this value will be defined in the Parameter Set.

Type	Element Size
P Sweep Element Size	Please Define
Sweep Bias Type	No Bias

- Right-click on the Mesh branch and select Insert | Sizing.

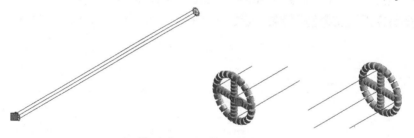

- Change the filter selection to Edges. Change the selection method from Single Select to Box Select. Select all 16 edges on the faces on both ends of the cylinder, and then click the Apply button in the Geometry row.

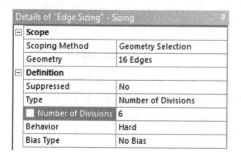

- Change the Type from Element Size to Number of Divisions. Change the value in the row for Number of Divisions to 6. Change the Behavior from Soft to Hard. When trying to mesh bodies, there can be multiple constraints that have been applied and the meshing algorithm may adjust some of the constraints. By selecting Hard means that this meshing constraint will not be over-ridden.

Details of "Edge Sizing" - Sizing	
Scope	
Scoping Method	Geometry Selection
Geometry	16 Edges
Definition	
Suppressed	No
Type	Number of Divisions
Number of Divisions	6
Behavior	Hard
Bias Type	No Bias

- For this model the FLUID30 acoustic elements will be used that do not have mid-side nodes. Left-click on the Mesh branch, click in the Details of "Mesh" window, and click on the plus sign next to Advanced. In the row for Element Midside Nodes, change the value from Program Controlled to Dropped.

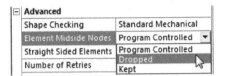

⊟ Advanced	
Shape Checking	Standard Mechanical
Element Midside Nodes	Program Controlled ▼
Straight Sided Elements	Program Controlled
	Dropped
Number of Retries	Kept

- Before the model is meshed, it is necessary to define the size of the elements. At a later time we will explore the effect of changing the size of the elements on the results. To make it easier to change the size of elements, we will define size as a parameter in the Parameter Set window. Return to the Workbench Project Schematic window and double-click on the box for Parameter Set.

- Click on the cell that is labeled New name, type c_speed_sound, and then press the <Tab> key to move to the cell to the right.

	A	B	C	D
	ID	Parameter Name	Value	Unit
1	ID	Parameter Name	Value	Unit
2	⊟ Input Parameters			
3	⊟ 🔳 Modal (A1)			
4	ℂₚ P1	duct_diam	0.1	
5	ℂₚ P2	duct_length	3	
6	ℂₚ P3	Sweep Method Sweep Element Size	0	m ▼
*	ℂₚ New input parameter	New name	New expression	
8	⊟ Output Parameters			
*	🄿ᴶ New output parameter		New expression	
10	Charts			

Outline: No data

- The cursor should be in the cell for the Value of c_speed_sound, so type 343 and press the <Enter> key on the keyboard, which assigns the Input Parameter ID P4.

7	ℂₚ P4	c_speed_sound	343

- Repeat these steps to define new Parameter Name entries for n_index = 1 for the mode index and epw = 12 for the number of elements per wavelength, which will be assigned Input Parameter IDs P5 and P6, respectively.

	A	B	C	D
1	ID	Parameter Name	Value	Unit
2	☐ Input Parameters			
3	☐ ⊞ Modal (A1)			
4	⟨p P1	duct_diam	0.1	
5	⟨p P2	duct_length	3	
6	⟨p P3	Sweep Method Sweep Element Size	0	m ▾
7	⟨p P4	c_speed_sound	343	
8	⟨p P5	n_index	1	
9	⟨p P6	epw	12	
*	⟨p New input parameter	New name	New expression	

- Create a new `Parameter Name` called `max_freq`. In the cell, to define its value, we will enter a mathematical expression for the natural frequency of a rigid–rigid duct, which is $f_n = nc_0/(2L)$. Type in the cell to define its `Value` as `P5*P4/2/P2`. After you press the `<Enter>` key, it will evaluate to `57.167`.

10	⟨p P7	max_freq	P5*P4/2/P2

- Repeat this process to define a new `Parameter Name` called `esize`, which will represent the element size for the acoustic elements. The element size will be defined as

$$\text{esize} = \frac{\lambda}{\text{epw}} = \frac{c_0}{f} \times \frac{1}{\text{epw}} = \frac{\text{P4}}{\text{P7}} \times \frac{1}{\text{P6}}, \tag{3.22}$$

where λ is the acoustic wavelength, c_0 is the speed of sound, and f is the frequency of interest. Hence, in the `Value` cell, enter the expression `P4/P7/P6` and press the `<Enter>` key, which will evaluate to `0.5`.

- Click in the `Value` cell for `P3 Sweep Method Sweep Element Size`. In the lower window labeled `Properties of Outline C6: P3`, click in the cell next to `Expression` and type `P8*1[m]`, which will set the element size to the value that was calculated previously. Note that it is necessary to define the parameter with units by typing `*1[m]` at the end of the expression to assign the units of meters. Press the `<Enter>` key on the keyboard and the `Value` should evaluate to `0.5`, the same as `esize`.

	A	B
	Properties of Outline C6: P3	
1	Property	Value
2	☐ General	
3	Expression	P8*1[m]
4	Description	

- Click on the `Refresh Project` icon, which will transfer the calculated values from the `Parameter Set` into Mechanical.

- Click on the `Save` icon to save the project.

- Click on the `Workbench Mechanical` window.

- Click on the `Sweep Method` in the `Mesh` branch and you should notice that the row labeled `Sweep Element Size` has been updated from `Please Define` to `0.5m`.

P Sweep Element Size	0.5 m

- Now that the element sizes have been defined, the solid bodies can be meshed. However, before doing so, always save your model by clicking on `File | Save Project` from the menu bar. Right-click on the `Mesh` branch and select `Generate Mesh`.

- Once the meshing has been completed, you should notice that there are only 6 elements along the axis of the cylinder. If you zoom to show the mesh at the end of the cylinder, you will see that there are 6 elements on each of the edges, as was defined. The peculiar feature of this mesh is that the patterns in the quadrants are not the same. In ANSYS Release 14.5, it is not easy to create a repeated mesh pattern in ANSYS Workbench compared to using Mechanical APDL. The mesh has resulted in very elongated brick elements, which is usually not advisable. However for this analysis, where acoustic plane wave conditions will exist, the elements with poor aspect ratio will still provide accurate results.

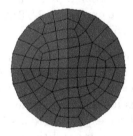

- Up to this point, ANSYS Workbench has meshed the solid model with the default structural `SOLID186` elements. However, we want to conduct an acoustic analysis using acoustic elements `FLUID30`. By using the ACT Acoustics extensions toolbox, it is easy to make the conversion from the structural to acoustic element types. In the ACT Acoustics extension toolbar, select `Acoustic Body | Acoustic Body`.

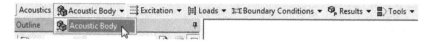

- In the Graphics window, right-click with the mouse and click on Select All, so that all 4 bodies are highlighted.

- In the window for Details of "Acoustic Body", click in the cell next to Geometry and click on the Apply button.

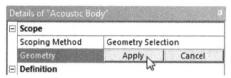

- In the window for Details of "Acoustic Body", change the Mass Density to 1.21, change the Sound Speed to 343, and leave the rest of the entries unaltered. Note this step requires that the units are set to MKS in ANSYS Mechanical.

Details of "Acoustic Body"	
□ **Scope**	
Scoping Method	Geometry Selection
Geometry	4 Bodies
□ **Definition**	
Temperature Dependency	No
Mass Density	1.21 [kg m^-1 m^-1 m^-1]
Sound Speed	343 [m sec^-1]
Dynamic Viscosity	0 [Pa sec]
Thermal Conductivity	0 [W m^-1 C^-1]
Specific Heat Cv	0 [J kg^-1 C^-1]
Specific Heat Cp	0 [J kg^-1 C^-1]
Reference Pressure	2E-05 [Pa]
Reference Static Pressure	101325 [Pa]
Acoustic-Structural Coupled Bo...	Uncoupled

- Click on the Analysis Settings branch. In the window Details of "Analysis Settings", click on the plus sign next to Output Controls. Change the rows for Nodal Forces, Calculate Reactions, General Miscellaneous to Yes.

- Click on the plus sign next to the row for Analysis Data Management, and change Save MAPDL db to Yes. This step is important because the finite element model database (.db) is used by the ACT Acoustics extension when post-processing to calculate results.

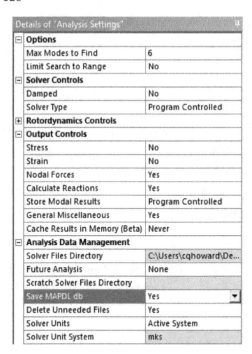

- In the Acoustics toolbar, click on Results | Acoustic Pressure. Leave the Geometry row as All Bodies.

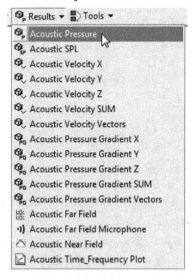

- Repeat these steps to request results for another Acoustic Pressure, but this time change the Scoping Method to Named Selection, and the Named

Selection to NS_duct_axis, which will calculate the acoustic pressure along the axis of the duct.

- Click on the icon for User Defined Result.

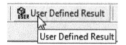

- Change the Scoping Method to Named Selection and the Named Selection to NS_duct_axis. In the cell next to Expression type LOCZ. This will determine the Z axis coordinates of the nodes on the axis of the cylinder.

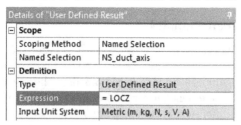

Note that it is also possible to define a Path where the results will be calculated. However, the locations of coordinates will be relative to the starting point of the path instead of displaying the location in the global Cartesian coordinate system. The advantage of using a Path is that a graph of the result versus the distance along the path can be displayed in ANSYS Mechanical.

- Right-click on the branch for User Defined Result and select Rename Based on Definition.

- From the menu bar select File | Save Project.

That completes the setup of the analysis. Click on the Solve icon and wait for the computations to finish.

- Once the results have been calculated, click on the Acoustic Pressure result to show the response in the duct at mode 1. You should notice that there is a constant pressure for mode 1, and in the window labeled Tabular Data, the natural frequencies will be listed.

- Right-click in the window labeled Tabular Data and select Export. Type an appropriate filename, such as res_freqs_duct.txt and press the <Enter> key on the keyboard. These results can be analyzed using a spreadsheet or MATLAB.

Tabular Data		
	Mode	✔ Frequency [Hz]
1	1.	4.3752e-005
2	2.	57.822
3	3.	119.6
4	4.	189.11
5	5.	267.44
6	6.	343.07

Export
Select All

The following section shows the comparison of these ANSYS results with theoretical predictions.

3.3.2   Results: Effect of Mesh Density

The finite element model was created with element sizes of esize=0.5 m, which correspond to 12 elements per wavelength (epw) at the first natural frequency of the rigid–rigid duct which is 57 Hz. Table 3.3 lists the natural frequencies calculated from the finite element analysis compared with the theoretical values, and the percentage difference between the two. The last column in the table lists the effective elements per wavelength, which is calculated as

$$\text{epw}_{\text{effective}} = \frac{\lambda_n}{\text{esize}} = \frac{c}{f_n} \times \frac{1}{\text{esize}}. \tag{3.23}$$

For the mode $n = 1$, where the mesh density was epw=12, there is only a 1% difference between the theoretical and finite element results. For $n = 2$, when there is only epw=6, the results are still reasonable with only a 5% difference in the predicted natural frequencies. However, for epw < 6, the percentage difference increases rapidly, which is to be expected as there is an insufficient number of elements and nodes to cover an acoustic wavelength.

Figure 3.3 shows the mode shapes of the normalized pressure (the modal pressure calculated in ANSYS divided by the maximum pressure at each mode) of the rigid–rigid duct for modes $n = 3, 4, 5$ calculated theoretically and using ANSYS Workbench. The mode shapes predicted using ANSYS correlate well with the theoretical mode shapes, despite the natural frequencies having poor correlation with the theoretical natural frequencies, as indicated in Table 3.3.

In the previous section, the ANSYS Workbench model of the duct used linear FLUID30 acoustic elements that have 8 nodes per element. In this section, the FLUID30 elements will be replaced with the quadratic FLUID220 acoustic elements that have 20 nodes per element. The analyses will be repeated and the results using the two element types will be compared. The completed ANSYS Workbench project is contained in the archive file called res_freqs_duct_FLUID220.wbpz and is available with this book.

TABLE 3.3
Results of Modal Analyses of a Rigid–Rigid Duct

Mode n	MATLAB Nat. Freq. [Hz]	ANSYS Nat. Freq. [Hz]	Diff. [%]	Eff. epw
0	0	0.0	nil	n/a
1	57.2	57.8	1.1	12.0
2	114.3	119.6	4.6	6.0
3	171.5	189.1	10.3	4.0
4	228.6	267.4	17.0	3.0
5	285.8	343.1	20.0	2.4

Note: Results were calculated theoretically using MATLAB and
ANSYS Workbench where FLUID30 elements were used. The
columns show the mode number n, the natural frequencies calculated using MATLAB and ANSYS, the percentage difference
between the results, and the effective elements per wavelength.

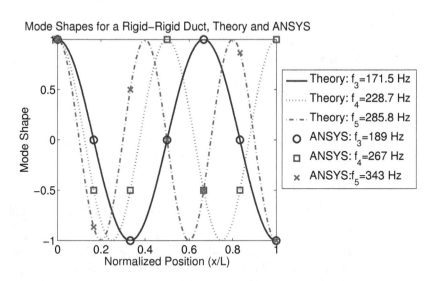

FIGURE 3.3
Mode shapes of the normalized pressure of a rigid–rigid duct for modes $n =$
3, 4, 5 calculated theoretically and using ANSYS Workbench at epw=4.0, 3.0,
2.4, respectively.

Instructions

- Open the Workbench project file res_freqs_duct.wbpj.

- Click File | Save As and type a new filename such as res_freqs_duct_FLUID220.wbpj.

- Double-click on the Parameter Set box in the Project Schematic window.

- Change the value in row P6 epw to 6, to set 6 elements per wavelength.

- Click on the Refresh Project icon.

- Click on the Save icon.

- Click on the Return to Project icon.

- In the Project Schematic window, double-click on row 4 Model to start Mechanical.

- Click on the Mesh branch, click on the plus + sign next to Advanced, and in the row Element Midside Nodes change it to Kept, which will ensure that the quadratic FLUID220 acoustic elements are used that have 20 nodes per element. As described previously, if this option is selected as Dropped, then the linear FLUID30 acoustic elements will be used that have 8 nodes per element.

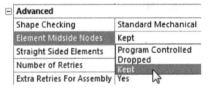

- Click on the Sweep Method under the Mesh branch and check that the row Element Midside Nodes is set to Use Global Setting.

- Select File | Save Project.

- Right-click on the Mesh branch and select Clear Generated Data, which will clear the existing mesh.

- Right-click on the Mesh branch and select Generate Mesh which will generate the new finite element mesh. There will only be 3 elements along the length of the duct, so it will be interesting to see the accuracy that can be achieved with such few elements. Click on the plus sign next to the Statistics and there will be 1459 nodes and 264 elements. Note that it does not matter if the number of nodes and elements created in your model is slightly different.

• Click File | Save Project.

That completes the setup of the finite element model. Click the Solve icon and wait for the computations to complete.

Click on the Solution Information branch under the Solution (A6) tree. Scroll down until you find the entries that are similar to the following:

```
*** MASS SUMMARY BY ELEMENT TYPE ***

TYPE       MASS
   5   0.285097E-01

Range of element maximum matrix coefficients in global
  coordinates
Maximum = 1.06840588 at element 75.
Minimum = 0.775643317 at element 199.

  *** ELEMENT MATRIX FORMULATION TIMES
 TYPE    NUMBER    ENAME      TOTAL CP  AVE CP

   5       264    FLUID220     0.062   0.000236
```

A simple "sanity" check of the model can be done by comparing the estimated mass of the fluid with what is reported in the Solution Information. The estimated mass of air in the duct is

$$\text{Mass} = \rho_0(\pi a^2 L) \tag{3.24}$$
$$= 1.21 \times (\pi \times 0.05^2 \times 3) \tag{3.25}$$
$$= 0.0285 \text{ kg}, \tag{3.26}$$

which is almost the same as reported in the Solution Information, and therefore the volume and mass of the model are correct.

Also notice in the Solution Information that element type 5 corresponds to FLUID220 elements, and hence the model was meshed with the desired element type.

Table 3.4 lists the comparison of natural frequencies calculated theoretically and the ANSYS modal analysis results where the model was meshed using FLUID30 elements at epw=12 and FLUID220 elements at epw=6. The table shows that the natural frequencies are nearly identical, despite having half the number of elements per wavelength along the axial direction.

The interesting statistic from the analysis of these two models is that the model using the FLUID220 elements had fewer elements and more nodes than the model with the FLUID30 elements, however the maximum wavefront when using the FLUID220 elements was larger. The wavefront statistics are listed in the Solution Information and is the number of degrees of freedom retained by the solver during the formation of the matrices. The maximum wavefront of

TABLE 3.4

Results of Modal Analyses of a Rigid–Rigid Duct

	Theory	ANSYS	ANSYS
Element Type	n.a.	FLUID30	FLUID220
epw	n.a.	12	6
Nodes	n.a.	707	1459
Elements	n.a.	528	264
Max. Wavefront	n.a.	112	320
Total CPU Time [s]	n.a.	2.4	2.6
Elapsed Time [s]	n.a.	4.0	4.0
Mode n			
0	0.0	0.0	0.0
1	57.2	57.8	57.2
2	114.3	119.6	115.6
3	171.5	189.1	189.1
4	228.7	267.4	261.5
5	285.8	343.1	360.9

Note: Results were calculated theoretically using MAT-LAB and using ANSYS Workbench with FLUID30 elements (no mid-side nodes) and FLUID220 (with mid-side nodes).

a model directly affects the memory required to solve a model. The root-mean-square (RMS) wavefront, is indicative of how long it will take to solve a model. The lower the value of the RMS wavefront, the less time it will take to solve the model. For this simplistic model of a one-dimensional duct, based on the values of the maximum wavefront, the computations would be completed faster and use less memory than using the model comprising FLUID30 elements, which had more elements and fewer nodes than the model comprising FLUID220 elements. This is not generally the case, and these wavefront statistics will vary depending on the geometry of the model.

3.3.3 Natural Frequencies of Open–Rigid and Open–Open Ducts

The previous finite element model will be further developed to show how to calculate the natural frequencies of a duct with idealized open end conditions.

The learning outcome is to show how to simulate an open end condition by applying a boundary condition of zero acoustic pressure on the face of an acoustic body. This is not an accurate way to model an open-ended duct and a more accurate method is described in Section 3.3.7.

- Return to the Mechanical window, and restore the model where the FLUID30 elements were used to mesh the model (i.e., the res_freqs_duct.wbpj model).

- In the ACT Acoustic Extensions toolbar, click on Boundary Conditions | Acoustic Pressure.

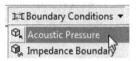

- In the window Details of "Acoustic Pressure", change the Scoping Method to Named Selection, and the Named Selection to NS_outlet. Leave the Pressure (Real) entry as 0 [Pa].

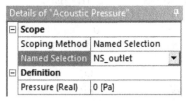

- Right-click on the Acoustic Pressure icon and select Rename. Change the name to Acoustic Pressure_open_rigid.

- Insert another Boundary Conditions | Acoustic Pressure, and this time for the Geometry, change the selection filter to Faces, and select the 4 faces on the end of the cylinder for NS_outlet, and the 4 faces on the end of the cylinder for NS_inlet, then click the Apply button. The cell next to Geometry should indicate that 8 Faces were selected.

- Right-click on this Acoustic Pressure, select Rename, and change the name to Acoustic Pressure_open_open.

- The first analysis that will be conducted is to determine the natural frequencies for an open–rigid duct, so the boundary conditions for the open–open case will be suppressed. Right-click on the icon for
Acoustic Pressure_open_open and select Suppress. Note the X next to label of the object to indicate it has been suppressed.

- Under the Solution (A6) branch, click on each of the entries for Acoustic Pressure and make sure that the row for Mode is 1.

- From the menu bar select File | Save Project.

  That completes the setup of the analysis. Click on the Solve icon and wait for the computations to finish.

The results from the analysis can be exported as described in the previous section. Table 3.5 lists the comparison between the natural frequencies predicted using ANSYS and theoretical predictions using MATLAB. The finite element model had element sizes of 0.5 m along the axis of the cylinder. The results show that the natural frequencies predicted using ANSYS are reasonably accurate up to $n = 3$ where there are 8 elements per wavelength (epw=8), and for higher mode indices the accuracy is poor.

TABLE 3.5

Results of Modal Analyses of an Open–Rigid Duct

Mode n	MATLAB Nat. Freq. [Hz]	ANSYS Nat. Freq. [Hz]	Diff. [%]	Eff. epw
1	28.6	28.7	0.3	24
3	85.8	88.0	2.6	8
5	142.9	153.2	7.2	4.8
7	200.1	227.4	13.6	3.4
9	257.3	307.3	19.5	2.7
11	314.4	368.8	17.3	2.2

Note: Results were calculated theoretically using MATLAB and using ANSYS Workbench where FLUID30 elements were used. The columns show the mode number n, the natural frequencies calculated using MATLAB and ANSYS, the percentage difference between the results, and the effective elements per wavelength.

Figure 3.4 shows the mode shapes of the open–rigid duct calculated theoretically using MATLAB and numerically using ANSYS Workbench. The FLUID30 elements had a length of 0.5 m, the same as the previous analysis for the rigid–rigid end conditions. The mode shapes predicted using ANSYS Workbench correlate well with the theoretical mode shapes, even for the mode $n = 7$ where there was a 14% difference in the predicted natural frequency.

Similar analyses can be conducted for the open–open end conditions by changing the boundary conditions under the Modal (A5) branch so that Acoustic Pressure_open_rigid is Suppressed (which is indicated an X next to the object), and the Acoustic Pressure_open_open boundary condition is Unsuppressed (which is indicated by a green tick next to the object). Table 3.6 shows the comparison of the natural frequencies predicted theoretically using MATLAB and numerically using ANSYS Workbench for the open–open duct.

In summary, it is recommended by ANSYS that when using the linear FLUID30 elements that the mesh density should have at least 12 elements per wavelength. The modal analysis results presented in this section indicate that this can be stretched to about epw=6, but only with caution.

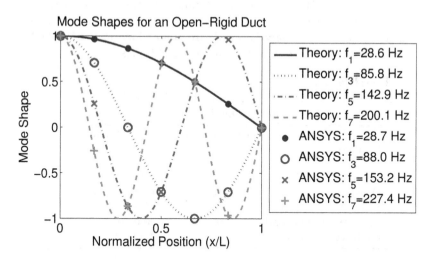

FIGURE 3.4

Mode shapes of an open–rigid duct for modes $n = 1, 3, 5, 7$ calculated theoretically and using ANSYS Workbench with element size of 0.5 m.

TABLE 3.6

Results of a Modal Analysis of an Open–Open Duct

Mode	MATLAB	ANSYS	Diff.	Eff.
n	Nat. Freq. [Hz]	Nat. Freq. [Hz]	[%]	epw
1	57.2	57.8	1.1	12.0
2	114.3	119.6	4.6	6.0
3	171.5	189.1	10.3	4.0
4	228.7	267.4	17.0	3.0
5	285.8	343.1	20.0	2.4
6	343.0	2029.6	491.7	2.0

Note: Results were calculated theoretically using MATLAB and
ANSYS Workbench.

3.3.4 Pressure and Velocity Distribution along the Duct

The model that was created in Section 3.3.1 will be further developed to
enable the calculation of the pressure distribution along the length of the duct,
due to the forced velocity excitation at one end of the duct. Table 3.7 lists
the parameters used in the analysis, which are almost identical to Table 3.2,
except that the particle velocity at the inlet is $u_2 = 1.0$ m/s.

TABLE 3.7
Parameters Used in the Analysis of a Circular Duct
with Forced Excitation at One End

Description	Parameter	Value	Units
Diameter	$2a$	0.1	m
Length	L	3	m
Speed of sound	c_0	343	m/s
Density	ρ_0	1.21	kg/m^3
Velocity at piston	u_2	1.0	m/s
Velocity at rigid end	u_1	0.0	m/s
Excitation frequency	f	200	Hz

Instructions

The following instructions describe how to modify the previous model res_
freqs_duct.wbpj to model a duct with a rigid end that is driven by a piston
with a velocity excitation at the other end. The completed ANSYS Workbench
archive file called driven_duct_pres_dist.wbpz , which contains the .wbpj
project file, is included with this book.

- Start ANSYS Workbench.

- Open the model created in the previous section called res_freqs_duct.wbpj.

- In the menu bar click on File | Save As, and give the project a new
  name such as driven_duct_pres_dist. Note that it is also possible to link
  a Harmonic Response from the Analysis Systems window to the Modal
  analysis, so that the project contained both modal and harmonic analyses.
  However, they will be kept as separate projects for this example.

- In the Project Schematic window left-click on the small down pointing triangle next in the top corner of the Modal cell and select Replace With | Harmonic Response.

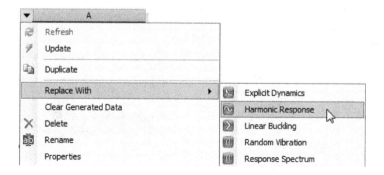

- In the Project Schematic window, the Modal Analysis System will be replaced with Harmonic Response.

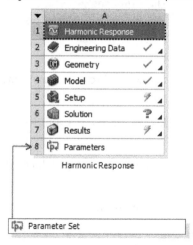

- Double-click on the box for Parameter Set.

- You will see that two entries in cells C 10 and C 11, corresponding to Input Parameters P7 and P8, respectively, are highlighted in red, as some of the Input Parameters have been reassigned new identity numbers.

	A	B	C
	Outline: No data		▾ ⏗ ✕
	ID	Parameter Name	Value
1			
2	⊟ Input Parameters		
3	⊟ 🖼 Harmonic Response (A1)		
4	📭 P11	Sweep Method Sweep Element Size	0.5
5	📭 P9	duct_diam	0.1
6	📭 P10	duct_length	3
7	📭 P4	c_speed_sound	343
8	📭 P5	n_index	1
9	📭 P6	epw	12
10	📭 P7	max_freq	name 'P2' is not defined
11	📭 P8	esize	referenced 'P7' could not be successful
*	📭 New input parameter	New name	New expression
13	⊟ Output Parameters		
*	📭 New output parameter		New expression
15	Charts		

- Update the expressions for P7 max_freq with the value 200.

10	📭 P7	max_freq	200

- Update the Expression for P11 Sweep Method Sweep Element Size with the expression P8*1[m].

- Click on the Refresh Project icon to transfer the updated value to Mechanical.

- Click on the Return to Project icon, then click on row 5 Setup, which will start Mechanical.

- Right-click on the Mesh branch and select Generate Mesh.

- Left-click on Analysis Settings under the branch Harmonic Response (A5). In the window Details of "Analysis Settings", change the Range Minimum to 199, Range Maximum to 200, and Solution Intervals to 1. This will calculate a single harmonic response at 200 Hz when the model is solved. Change Solution Method to Full, click on the plus sign next to Output Controls and change all the options to Yes, and click on the plus sign next to Analysis Data Management and change Save MAPDL db to Yes.

Details of "Analysis Settings"	
Options	
Range Minimum	199. Hz
Range Maximum	200. Hz
Solution Intervals	1
Solution Method	Full
Variational Technology	Program Controlled
Output Controls	
Stress	Yes
Strain	Yes
Nodal Forces	Yes
Calculate Reactions	Yes
General Miscellaneous	Yes
Damping Controls	
Analysis Data Management	
Solver Files Directory	C:\Users\cqhoward\Desktop\D...
Future Analysis	None
Scratch Solver Files Di...	
Save MAPDL db	Yes
Delete Unneeded Files	Yes
Solver Units	Active System
Solver Unit System	mks

- The next step is to define a velocity excitation on the end of the duct. In the ACT Acoustic Extensions toolbar, click on Excitation | Normal Surface Velocity (Harmonic).

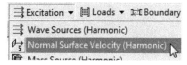

- In the window Details of "Acoustic Normal Surface Velocity", change the Scoping Method to Named Selection, Named Selection to NS_inlet, and Amplitude Of Normal Velocity to -1.0.

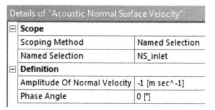

- Click on the Acoustic Pressure entries under the Solution (A6) branch and change the Set Number to 1, as the harmonic analysis is only being conducted at a single frequency of 200 Hz and we want the first and only solution.

- In the ACT Acoustics extension toolbar, click on `Results | Acoustic SPL`. Leave the `Scoping Method` as `All Bodies`, so that the sound pressure level through the duct will be shown.

- Click on `Results | Acoustic Velocity Z`, to request that the acoustic particle velocity along the Z axis, which corresponds to the axis of the duct, is calculated. Change the `Scoping Method` to `Named Selection`, and `Named Selection` to `NS_duct_axis`. Rename this result to `Acoustic Velocity Z_real`, for the real part of the complex valued particle velocity.

- Insert another `Results | Acoustic Velocity Z` entry to calculate the acoustic velocity along `NS_duct_axis`, and change the `Phase Angle` to `-90°`. Rename this result to `Acoustic Velocity Z_imag`, for the imaginary part of the complex valued particle velocity. Note that this is an error with the ACT Acoustics extension that the imaginary part of a result is retrieved by specifying a phase angle of `-90°`, whereas one would typically specify a phase angle of `+90°`. See also Ref. [22].

Details of "Acoustic Velocity Z_imag"	ᴨ
Scope	
Scoping Method	Named Selection
Named Selection	NS_duct_axis
Definition	
Type	User Defined Result
Expression	= PGZ
Input Unit System	Metric (m, kg, N, s, V, A)
Output Unit	Pressure
By	Frequency
Frequency	Last
Phase Angle	-90. °
Coordinate System	Global Coordinate System
Identifier	
Suppressed	No

- Insert another `Results | Acoustic Pressure`, and change the `Scoping Method` to `Named Selection`, `Named Selection` to `NS_duct_axis`, `Phase Angle` to `-90°`, and then right-click on this `Acoustic Pressure` branch and select Rename, and name it `Acoustic Pressure_imag`, for the imaginary part of the complex valued acoustic pressure.

Details of "Acoustic Pressure_imag"	⏚
⊟ **Scope**	
Scoping Method	Named Selection
Named Selection	NS_duct_axis
⊟ **Definition**	
Type	User Defined Result
Expression	= PRES
Input Unit System	Metric (m, kg, N, s, V, A)
Output Unit	Pressure
By	Frequency
Frequency	Last
Phase Angle	-90. °
Coordinate System	Global Coordinate System
Identifier	
Suppressed	No

- Rename the other entry for Acoustic Pressure that has the Named Selection as NS_duct_axis, to Acoustic Pressure_real. Leave the Phase Angle as 0°, for the real part of the complex valued particle velocity.

- In the menu bar click on File | Save Project.

That completes the setup of the model for analysis. Click on the Solve icon and wait for the computations to complete, which should not take long as the model is small and only a single analysis frequency has been requested.

If attempting to solve the ANSYS Workbench model and an error message is generated such as one or more features used are beta, this can be addressed by turning on the Beta Options. In the Project Schematic window, click on Tools | Options. Click on the Appearance branch, and use the scroll bar to move the window to the bottom to reveal the Beta Options box. Left-click in the box so that a tick appears and then left-click on the OK button.

3.3.5   Results: Pressure and Velocity along the Duct

The MATLAB script spl_along_duct_4pole.m included with this book can be used to calculate the sound pressure and acoustic particle velocity along a circular duct.

Figure 3.5 shows the sound pressure level along the length of the 3 m circular duct, calculated theoretically using the 4-pole method, and using ANSYS Workbench. The results show good correlation.

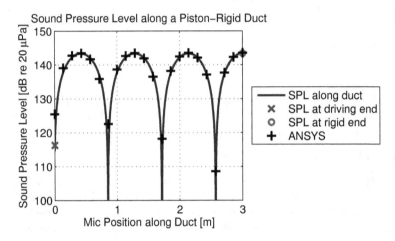

FIGURE 3.5
Sound pressure level along a 3 m circular duct driven by a piston at one end and a rigid-wall boundary condition at the other, calculated theoretically using the 4-pole method and using ANSYS Workbench.

The real pressure along the duct is zero, both for the theoretical predictions and the ANSYS results and therefore are not plotted. Figure 3.6 shows the imaginary component of the sound pressure calculated theoretically and using ANSYS Workbench, and again there is good correlation between the two results.

Figure 3.7 shows the real part of the complex particle velocity calculated using the 4-pole method and ANSYS Workbench. The imaginary part of the complex particle velocity is zero and is not shown. The results within the main body of the duct show good correlation, however the points at either end of the duct slightly differ from the expected results. The piston was driven with a velocity of 1 m/s, however the velocity calculated using ANSYS was 0.92 m/s. Similarly, the particle velocity at the rigid end should be 0 m/s, however ANSYS calculated the velocity as -0.25 m/s. The particle velocity is calculated in ANSYS for modal and full harmonic analyses using the pressure gradient and is evaluated at nodes and the centers of each acoustic element [23]. At the end of the duct, the estimates of pressure gradient (and particle velocity) are discontinuous, as there are no results beyond the extent of the finite element model to enable the correct calculation at the end of the duct. By increasing the elements per wavelength from 12 to 36 will provide a higher mesh density in the model, or alternatively using quadratic FLUID220 acoustic elements, and the results predicted using ANSYS will approach the theoretical values.

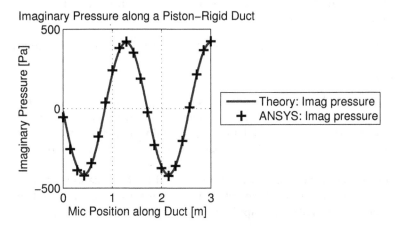

FIGURE 3.6
Imaginary pressure along a 3 m circular duct driven by a piston at one end and a rigid-wall boundary condition at the other, calculated theoretically using the 4-pole method and using ANSYS Workbench.

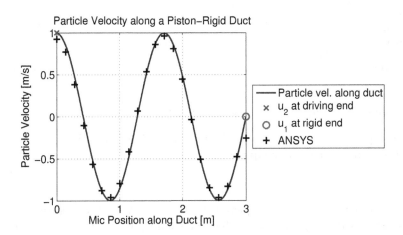

FIGURE 3.7
Real part of the particle velocity along a circular duct 3 m in length calculated theoretically using the 4-pole method, and using ANSYS Workbench.

3.3.6   Infinite and Semi-Infinite Loss-Less Ducts

The learning outcomes from this section are:

- how to simulate an acoustic infinite end condition in a duct,

- sound that propagates along a duct does not attenuate with distance as there is no spreading, and

- sound pressure level can be constant but the acoustic pressure will oscillate.

Instructions

In this section we will apply an absorbing boundary to the outlet of the duct to simulate a semi-infinite duct. The upstream inlet end provides an acoustic excitation as a surface velocity of 1 m/s, and the downstream outlet end has a radiation boundary applied, which is one method of specifying an absorbing boundary.

The model used in Section 3.3.4 will be re-used for this example. There is only one step that needs to be done, which is to change the rigid end condition into an anechoic termination by applying a Radiation Boundary as follows:

- In the ACT Acoustic Extensions toolbar, click on Boundary Conditions | Radiation Boundary. Change the Scoping Method to Named Selection and Named Selection to NS_outlet.

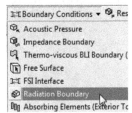

- In the menu bar, click on File | Save Project.

That completes the steps to set up the model. Click on the Solve icon and wait for the computations to complete.

Results

When the computations have completed, click on the Acoustic SPL branch under the Solution (A6) tree to see that the sound pressure level along the duct is (nearly) constant at 143.5 dB re 20 μPa, as shown in Figure 3.8.

The theoretical impedance of an infinite duct with a uniform cross-section is [47, Eq. (9.102), p. 467]

$$Z_{\text{semi-inf duct}} = \frac{p}{Q} == \frac{p}{Su} = \frac{\rho_0 c_0}{S}. \tag{3.27}$$

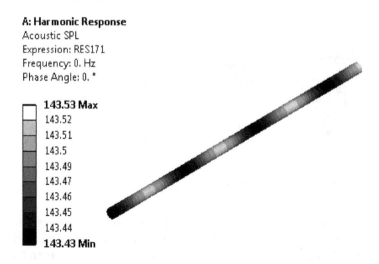

A: Harmonic Response
Acoustic SPL
Expression: RES171
Frequency: 0. Hz
Phase Angle: 0. °

143.53 Max
143.52
143.51
143.5
143.49
143.47
143.46
143.45
143.44
143.43 Min

FIGURE 3.8
Sound pressure level in a duct with a piston at the right end and an anechoic termination at the left end.

Note that the volume velocity $Q = Su$ is the product of S the cross-sectional area of the duct, since the piston fills the entire cross-section, and u the velocity of the piston. If an infinite duct were modeled, then only half the volume velocity would propagate downstream, and half would propagate upstream. By rearranging this equation, the pressure in the duct is given by

$$p_{\text{semi-inf duct}} = \rho_0 c_0 u , \tag{3.28}$$

so therefore for a piston velocity of $u = 1$ m/s the sound pressure level in decibels is

$$L_{p\text{- semi-inf duct}} = 20\log_{10}\frac{p_{\text{semi-inf duct}}}{\sqrt{2} \times 20 \times 10^{-6}} \tag{3.29}$$

$$= 20\log_{10}\frac{1.21 \times 343 \times 1.0}{\sqrt{2} \times 20 \times 10^{-6}}$$

$$= 143.3 \text{ dB re } 20 \ \mu\text{Pa} ,$$

where the $\sqrt{2}$ comes from the conversion of peak sound pressure to RMS. Hence the prediction using ANSYS of 143.5 dB re 20 μPa compares well with the theoretical value of 143.3 dB re 20 μPa.

3.3.7 Radiation from an Open-Ended Duct

The aim of this section is to determine the mechanical impedance of a piston on the end of a duct that radiates into a baffled free space, as shown in

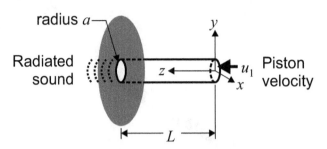

FIGURE 3.9
Schematic of a duct radiating into a plane baffle.

Figure 3.9. The mathematical theory is presented, followed by instructions on how to simulate the situation using ANSYS Workbench where a hemispherical infinite acoustic domain is used. Another implementation is demonstrated using ANSYS Workbench where the infinite hemispherical domain is replaced with an impedance at the end of the duct that varies with frequency.

3.3.7.1   Theory

The theory for this example comes from Kinsler et al. [102, p. 273–276]. The radiation impedance from an open end of a pipe radiating into a plane baffle is given by [102, Eq. (10.2.10), p. 274]

$$Z_{mL} = (\rho_0 c_0 S) \left[\frac{1}{2}(ka)^2 + j \frac{8}{3\pi}(ka) \right] , \tag{3.30}$$

where ρ_0 is the density of the acoustic medium, c_0 is the speed of sound of the acoustic medium, S is the cross-sectional area of the duct and piston, $k = \omega/c_0$ is the wavenumber, $\omega = 2\pi f$ is the circular frequency, f is the frequency of excitation, and a is the radius of the duct and piston.

The input mechanical impedance for a piston driving at the end of a duct of length L is [102, Eq. (10.2.4), p. 273]

$$Z_{m0} = (\rho_0 c_0 S) \frac{\dfrac{Z_{mL}}{(\rho_0 c_0 S)} + j \tan(kL)}{1 + j \dfrac{Z_{mL}}{(\rho_0 c_0 S)} \tan(kL)}. \tag{3.31}$$

Substitution of Equation (3.30) into Equation (3.31) gives

$$Z_{m0} = (\rho_0 c_0 S) \frac{\left[\frac{1}{2}(ka)^2 + j \frac{8}{3\pi}(ka) \right] + j \tan(kL)}{1 + j \left[\frac{1}{2}(ka)^2 + j \frac{8}{3\pi}(ka) \right] \tan(kL)}. \tag{3.32}$$

This expression for the mechanical impedance can be used to derive an expression for the time-averaged power that is delivered by the piston in terms of the applied force as [66, Eq. (2.4), p. 77]

$$\text{Power} \quad = \quad \frac{1}{2}|\tilde{F}|^2 \operatorname{Re}\left[\frac{1}{\tilde{Z}}\right] \tag{3.33}$$

$$= \quad \frac{|\tilde{F}|^2 R}{2(R^2 + X^2)}, \tag{3.34}$$

where the mechanical impedance is defined as [66, Eq. (2.2), p. 77]

$$\tilde{Z} = \frac{\tilde{F}}{\tilde{u}} = R + jX, \tag{3.35}$$

and $\tilde{F}$ and $\tilde{u}$ are the amplitude of the complex force and velocity, respectively.

Note there is a small error in the equation for power in Kinsler et al. [102, p. 276]. See Section D.1.1 for more details.

3.3.7.2 ANSYS Workbench

A model of the previously described system will be created in ANSYS Workbench of a duct with a hemispherical free-field that simulates a plane baffle, as shown in Figure 3.10. Note the change in the location of the origin of the coordinate system between the theoretical model shown in Figure 3.9 and Figure 3.10. It is convenient in the theoretical model to define the piston at $z = 0$, whereas when creating a geometric model of a sphere in ANSYS, the default location is at the origin and hence the location of the piston is placed at $z = -L$.

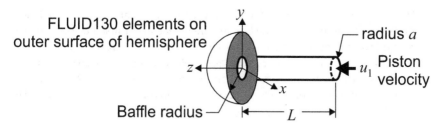

FIGURE 3.10

Schematic of the finite element model that will be created in ANSYS Workbench of a circular duct radiating into a plane baffle, which is modeled with FLUID130 infinite acoustic elements on the surface of a hemispherical acoustic volume.

Instructions

The ANSYS Workbench archive model radiation_open_duct.wbpz , which contains the .wbpj project file, is included with this book. The following instructions provide an overview of the steps required to modify the previous model.

- Start ANSYS Workbench and load the project driven_duct_pres_dist. wbpj.

- To ensure that the original model is not corrupted, save the project by selecting File | Save As and type a filename such as radiation_open_duct. wbpj.

- Start DESIGNMODELER.

- For this model the starting and finishing locations of the duct will be reversed compared to the previous model, so that the duct finishes at the origin, and a hemisphere that represents the baffled acoustic region is centered at the origin and extends into the +Z axis, as shown in Figure 3.10. In the Tree Outline window, click on the branch for Extrude1. In the row Direction, change it to Reversed and click the Generate icon. In the Tree Outline window click on the icon for XYPlane to show the XY axes and notice that the duct now extends in the −Z direction.

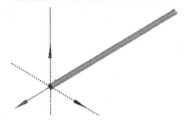

- As the orientation of the duct has been altered, the definitions for the Named Selections of the inlet and outlets are incorrect and must be fixed. In the Tree Outline window, right-click on the entry for NS_outlet and left-click on Edit Selections. Make sure the Faces selection filter is active, then hold down the <Ctrl> key and left-click on the four faces on the XY plane at the exit of the duct. Left-click on the Apply button and then left-click the Generate icon.

- Repeat these steps to redefine the Named Selection for NS_inlet as the 4 faces at the far end of the duct at $z = -3$m.

- The next step will be to create a hemisphere to represent the free-field acoustic region which will have its origin at the exit of the duct. In the menu bar, left-click on Create | Primitives | Sphere. Change the Operation to Add Frozen. Keep the coordinates of the origin as (0,0,0). Make the radius a parameter by clicking on the box next to FD6 and name it sphere_r. Press the Generate icon. The radius will be set as a function of the wavelength in the Parameter Set.

- A second smaller sphere will be created at the exit of the duct that will be used as a transition region for the acoustic finite elements. The finite element mesh in the duct has a swept or "mapped mesh," and the large spherical region for the free-field will have an inflation mesh. This smaller spherical region at the exit of the duct enables one to have a transition zone between the two mesh regions. Create another sphere at the same location. In the row FD6, Radius (>0), enter the value 0.05 to make it easier to see the two spheres. Define the radius as a parameter called sphere_duct_r, which will be set as the radius of the main duct in the Parameter Set.

- Click on Create | Boolean, and change the Operation to Intersect. For the Tool Bodies select the two spheres that were just created. Change Preserve Tool Bodies? to Yes, Sliced. Change the Intersect Result to Union of All Intersections.

- Insert Create | Slice three times to slice all bodies along the XYPlane, ZXPlane, YZPlane.

- The next step is to delete the 8 unwanted bodies of the sphere that are in the −Z axis. Insert a `Create | Body Operation`, and change the row `Type` to `Delete`. Change the `Select Mode` to `Box Select` and select the 8 bodies that are the 4 large and 4 small one-eighth spherical bodies comprising the hemisphere that overlaps the duct. Click the `Generate` icon.

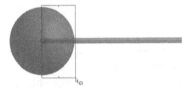

- In the `Outline` window, in the `Parts` branch, select all the solid bodies, and right-click and select `Form New Part`. There should be `1 Part, 12 Bodies` listed in the `Tree Outline`.

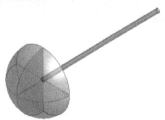

- Click on the `Save Project` icon.

    That completes the creation of the solid model. There are two new parameters that were created and need to be assigned values.

- In the `Workbench Project` window, double-click on the `Parameter Set` box.

- The first parameter that will be defined is the radius of the outer hemisphere. In the row for `P12 sphere_r`, in the `Value` cell, enter `0.7`.

- The next parameter that will be defined is the radius of the duct. Click on the row `P13 sphere_duct_r`. In the window `Properties of Outline B9:P14`, click in the box for `Expression` and type `P9/2`, which will define it as the duct diameter divided by 2.

- Click on the `Refresh Project` icon, which will update the model with the dimensions that have just been defined.

  The next step is to set up the harmonic analysis.

- Start Workbench Mechanical.

- Under the `Mesh` branch, check that the `Sweep Method` is applied to the 4 bodies for the duct.

- Check that the `Edge Sizing` is applied to the `Geometry` for the 16 edges on the ends of the cylinder. The `Number of Divisions` should be 6 and the `Behavior` should be `Hard`.

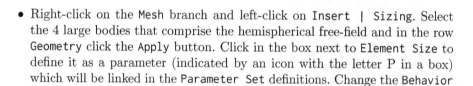

- Right-click on the `Mesh` branch and left-click on `Insert | Sizing`. Select the 4 large bodies that comprise the hemispherical free-field and in the row `Geometry` click the `Apply` button. Click in the box next to `Element Size` to define it as a parameter (indicated by an icon with the letter P in a box) which will be linked in the `Parameter Set` definitions. Change the `Behavior` to `Hard`.

- In the `Parameter Set`, alter the expression for the `P14 Body Sizing Element Size` to `P8*1[m]`, and click on `Refresh Project`. This will define the element size as a variable that can be altered depending on the desired number of elements per wavelength, excitation frequency, and speed of sound. Initially this will equate to an element size of 0.142 m.

- Under the `Mesh` branch, right-click `Insert | Sizing`. Select the 5 edges that comprise the hemisphere on the end of the duct. The edges on the end of the cylinder already have edge divisions defined. Change the Type to Number of Divisions, Number of Divisions to 6, and Behavior to Hard.

- The next step is to create an inflation mesh around the region of the exit of the duct. Make sure the Body selection filter is active and right-click on the Mesh branch and select Insert | Inflation. Select the 4 large bodies that comprise the hemisphere for the free-field, and right-click and select Hide All Other Bodies. In the Details of "Inflation" - Inflation window, in the Geometry row select the 4 large bodies for the hemispherical free-field. In the row for Boundary, select the 4 faces for the outer surface of the small hemisphere. The Inflation Option should be Smooth Transition, the Transition Ratio as 0.8, Maximum Layers as 5, Growth Rate as 1.2, and Inflation Algorithm as Pre.

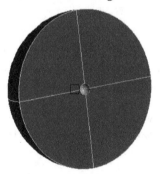

- Right-click on the Mesh branch and select Insert | Method. For the Geometry select the 4 small bodies on the end of the duct that comprise a hemisphere. Change the Method to Automatic.

- Right-click on the Mesh branch and select Insert | Sizing. Select the same 4 bodies as above. Change the Type to Element Size, and in the row for Element Size enter 8e-3.

- The next step is to apply a force to the piston face. There are several ways this can be achieved. One way is to apply a force to the face NS_inlet and couple all the nodal displacements in the Z axis by inserting an object Conditions | Coupling, which is a beta feature in ANSYS Release 14.5. Instead of using the beta feature, the way it will be done for this example is to apply a force to the vertex on the piston face and use a Commands (APDL) object to couple the displacement of nodes. This method is also instructive to see how components can be selected in ANSYS Workbench using APDL code. Right-click on the Harmonic Response (A5) branch and select Insert | Force. Select the vertex on the axis of the duct on the inlet face. Set the Magnitude to 1.e-003. Click in the cell next to Direction and then click on an edge that is along the axis of the duct, so that the red arrow that indicates the direction of the force is pointing into the duct.

- Check that the Acoustic Body is defined for the 4 bodies comprising the duct. The Acoustic-Structural Coupled Body Options should be Coupled With Unsymmetric Algorithm. The Mass Density should be 1.21, and the Sound Speed should be 343. Note that it is also possible to use the option Coupled With Symmetric Algorithm, provided that *all* Acoustic Body objects in the model are set to the option Coupled With Symmetric Algorithm.

- Insert another Acoustic Body, and select the 8 bodies on the end of the duct that model the hemispherical free-field region using the Box selection filter. Change the Mass Density to 1.21, and the Sound Speed to 343. Leave the Acoustic-Structural Coupled Body Options as Uncoupled, and Perfectly Matched Layers (PML) should be Off.

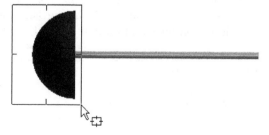

- The next step is to create the absorbing conditions on the exterior of the hemisphere to simulate the free-field. In the ACT Acoustic Extensions toolbar, select Boundary Conditions | Absorbing Elements (Exterior to Enclosure). In the Geometry row, select the 4 faces on the exterior of the hemisphere and then click the Apply button. In the row Radius of Enclosure type the value 0.7, which is the radius of the hemisphere and the same value of the parameter sphere_r. This step will lay FLUID130 elements on the surface of the hemisphere that provide the acoustic absorption of the outgoing waves.

- The next step is to enable the fluid–structure interaction at the inlet to the duct so that the displacement degrees of freedom become active and hence the acoustic particle velocity can be determined. In the ACT Acoustics toolbar click on Boundary Conditions | FSI Interface. In the window Details of "Acoustic FSI Interface" change the row Scoping Method to Named Selection and Named Selection to NS_inlet.

- As mentioned earlier, a force is applied to a vertex belonging to the piston, and all the nodes belonging to the piston will have their displacement degree of freedom along the Z axis coupled. The coupling of the nodal displacements can be achieved using an APDL code snippet. Right-click on Harmonic Response (A5) and select Insert | Commands.

- Click on the branch for Commands (APDL) and enter the following commands which will couple all the nodes associated with NS_INLET in the UZ axis, which will essentially create a rigid piston face, and motion in the UX and UY axes will be disabled.

```
1  ! Select all the nodes associated with NS_INLET
2  CMSEL,S,NS_INLET,NODE
3  ! Couple all the UZ DOFs to create a rigid piston face.
4  ! The following command will use the NEXT available set number
5  ! for the coupling equations.
6  CP,NEXT,UZ,ALL
7  D,ALL,UX,0
8  D,ALL,UY,0
9  ALLS
```

- One of the results that we want to obtain is the displacement versus frequency of the piston due to the applied force. Once the displacement result is calculated, it is possible to determine the mechanical impedance, which will be compared with the theoretical value. Right-click on Solution (A6) and select Insert | Frequency Response | Deformation. Click in the row for Geometry and select the vertex on the axis of the cylinder on the inlet face for the piston. Change the row Definition | Orientation to Z Axis.

- Although we have modeled the entire system, we will only analyze 1/4 of the model so that we can reduce the number of nodes and elements in the model. On the triad in the lower right of the screen, click on the -Z axis to change the view of the model. Change the selection filter to Bodies, and Select Mode to Box Select. Select all the bodies in the +X and +Y region, so that only 1/4 of the model is selected. Right-click and select Suppress All Other Bodies.

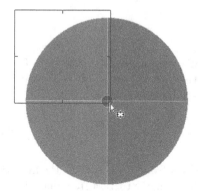

At a later stage, if you wish to confirm that the results obtained using the 1/4 model are the same as the full model, you can select Unsuppress All Bodies and re-run the analyses.

- The model should comprise 3 bodies.

- Right-click on the Mesh branch and select Generate Mesh. Once the meshing has completed, in the Statistic branch in the Details of "Mesh", there will be about 1555 nodes, and 3092 elements (do not be concerned if the statistics of your mesh are not exactly these values). The mesh around the outlet of the duct should have a fine mesh and there should a transition layer of fine to coarse elements.

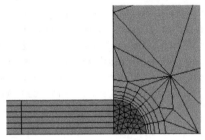

- Click on Analysis Settings and change the Range Minimum to 0, Range Maximum to 200, and the Solution Intervals to 100. This will result in harmonic analyses conducted at frequency increments Δf of

$$\Delta f = \frac{(\text{Range Maximum}) - (\text{Range Minimum})}{(\text{Solution Intervals})}. \qquad (3.36)$$

Hence, these settings will provide solutions from 2 Hz to 200 Hz in 2 Hz increments. Note that an analysis at the Range Minimum frequency is not conducted. Under the Output Controls, make sure everything is set to Yes. In the Analysis Data Management, make sure that Save MAPDL db is set to Yes.

- The tree for Harmonic Response (A5) should look like the following figure, and the other entries from the previous analyses can be deleted.

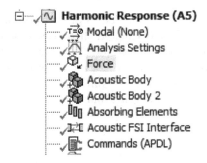

- Click on File | Save Project.

That completes the setup of the model. Click the Solve icon and wait for the computations to complete.

3.3.7.3 Results

Once the analysis has completed, click on the Frequency Response branch under Solution (A6). In the Tabular Data window, click on the cell in the top left corner of the table so that all the entries in the table are highlighted. Right-click and select Copy Cell. These results for the displacement of the piston can be pasted into MATLAB or a spreadsheet for processing. The results that will be calculated are the mechanical impedance and the mechanical power.

	Frequency [Hz]	✔ Real [m]	✔ Imaginary [m]
1	2.	-8.842e-004	-1.3421e-008
2	4.	-2.1829e-004	-6.7943e-009

If the displacement results are pasted into MATLAB as a variable ansys_uz, then the velocity ansys_vel is calculated as

$$\mathsf{ansys\_vel} = j(2\pi f) \times \mathsf{ansys\_uz} \,. \tag{3.37}$$

The mechanical impedance ansys_Z_m0 can be calculated as

$$\mathsf{ansys\_Z\_m0} = \frac{F}{\mathsf{ansys\_vel}} = \frac{1 \times 10^{-3} \text{ N}}{\mathsf{ansys\_vel}} \,. \tag{3.38}$$

The mechanical power can be calculated using Equation (3.33).
The MATLAB code to calculate these parameters is

```
1  ansys_vel=1i*2*pi*ansys_uz(:,1).*(ansys_uz(:,2)+1i*ansys_uz(:,3));
2  ansys_Z_m0=1e-3*ones(size(ansys_vel))./ansys_vel;
3  ansys_power=0.5*1e-3^2*real(ones(size(ansys_Z_m0))./ansys_Z_m0);
```

Figures 3.11 and 3.12 show the real and imaginary parts of the mechanical impedance of a piston attached to the duct calculated theoretically and using ANSYS Workbench with a quarter model, respectively. The results calculated using ANSYS were *multiplied* by 4 (because a 1/4 model was used in the simulation), to calculate the correct values for the full model. For further discussion on the use of symmetry when conducting analyses of acoustic systems, see Section 2.12.

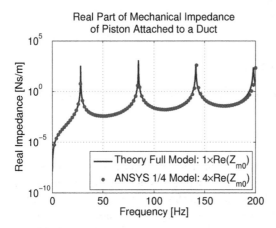

FIGURE 3.11
Real part of mechanical impedance of a piston attached to a 3 m long circular duct of radius 0.05 m that radiates into a baffled plane, calculated theoretically and using ANSYS Workbench with a 1/4 model.

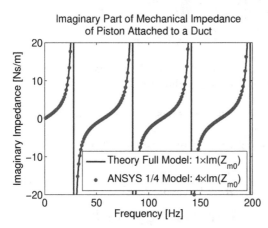

FIGURE 3.12
Imaginary part of mechanical impedance of a piston attached to a 3 m long circular duct of radius 0.05 m that radiates into a baffled plane, calculated theoretically and using ANSYS Workbench with a 1/4 model.

Figure 3.13 shows the mechanical power that was delivered to the piston calculated theoretically and using ANSYS Workbench with a quarter model. The results of power calculated using ANSYS were *divided* by 4, to calculate the correct values for the full model.

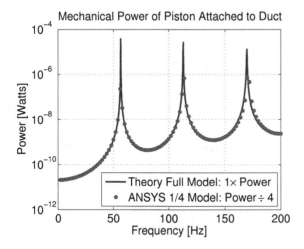

FIGURE 3.13
Mechanical power delivered to a piston attached to a 3 m long circular duct of radius 0.05 m that radiates into a baffled plane, calculated theoretically and using ANSYS Workbench with a 1/4 model.

The theoretical values were calculated using the MATLAB script `radiation_end_of_pipe.m` that is included with this text book.

3.3.7.4 Impedance Varying with Frequency

The next development of the ANSYS Workbench model is to replace the 1/4 hemisphere (1/8th sphere) that is used to simulate a baffled free-field with an equivalent impedance on the exit face of the duct.

The mechanical impedance caused by the radiation at the outlet of the duct into the hemispherical baffle is given by Equation (3.30). Referring to Table 2.24, mechanical impedance is force divided by velocity and specific acoustic impedance is pressure divided by the acoustic particle velocity. To convert the mechanical impedance $Z_m = pS/u$ for the radiation of the end of the duct into the baffled plane into an equivalent specific acoustic impedance $Z_s = p/u$, it is necessary to divide by the cross-sectional area of the duct. Hence the specific acoustic impedance is

$$Z_{sL} = \frac{Z_{mL}}{S} = (\rho_0 c_0)\left[\frac{1}{2}(ka)^2 + j\frac{8}{3\pi}(ka)\right].\qquad(3.39)$$

This equation needs to be rearranged so that it is suitable for insertion as a formula into ANSYS Workbench. The complex number needs to be split into real and imaginary parts, and the constants need to be separated from the frequency varying components. The real part of the specific acoustic impedance can be written as

$$\text{Re}(Z_{sL}) = (\rho_0 c_0) \left[\frac{1}{2}(ka)^2 \right] \tag{3.40}$$

$$= \frac{\rho_0 c_0}{2} \left[\frac{2\pi f a}{c_0} \right]^2 \tag{3.41}$$

$$= \left[\frac{2\rho_0 \pi^2 a^2}{c_0} \right] f^2. \tag{3.42}$$

The imaginary part of the specific acoustic impedance can be written as

$$\text{Im}(Z_{sL}) = (\rho_0 c_0) \left[\frac{8}{3\pi}(ka) \right] \tag{3.43}$$

$$= \rho_0 c_0 \frac{8}{3\pi} \left(\frac{2\pi f}{c_0} \right) a \tag{3.44}$$

$$= \left[\frac{16\rho_0 a}{3} \right] f. \tag{3.45}$$

Equations (3.42) and (3.45) will be used in a command object that is inserted into the Workbench model.

The completed ANSYS Workbench archive file `freq_depend_impedance.wbpz`, which contains the `.wbpj` project file, is included with this book.

Figure 3.14 shows a schematic of the finite element model that replaces the hemispherical infinite acoustic domain in Figure 3.10 with an acoustic impedance on the face of the outlet of the duct that varies with frequency.

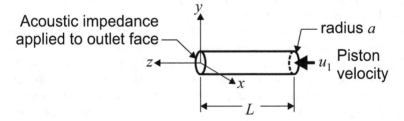

FIGURE 3.14
Schematic of the finite element model that will be created in ANSYS Workbench of a circular duct radiating into a plane baffle modeled with an acoustic impedance, which varies with frequency, on the face of the outlet of the duct.

The following instructions describe how the previously generated model radiation_open_duct.wbpj can be modified to create a new model.

- Open the previously generated ANSYS Workbench project radiation_open_duct.wbpj, click on File | Save As, and choose an appropriate filename such as freq_depend_impedance.wbpj.

- Double-click on row 4 Model to start Mechanical. There is no need to alter any of the solid geometry in DESIGNMODELER.

- For this model the volumes that were used to simulate the plane baffle (hemispherical free-field) are not required. Change the filter selection to Body, select the two 1/8 spheres on the end of the duct, right-click, and then left-click on Suppress. There should only be 1 body remaining that is active, which is the 1/4 section of the duct.

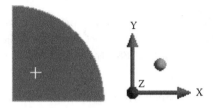

- The next step is to insert a command object of APDL code under the Harmonic Response (A5) branch. If you started creating this model using the radiation_open_duct.wbpj project, it already has a branch labeled Commands (APDL), which can either be used and overwritten with the following commands. Alternatively, the existing branch Commands (APDL) can be suppressed and a new command object created by right-clicking on Harmonic Response (A5) and select Insert | Commands. Click on the Command (APDL) branch under Harmonic Response (A5) tree. Copy and paste the APDL code from the file freq_depend_imp_commands.txt that is included with this book, into the Commands window. The APDL code couples all the UZ displacement degrees of freedom of the nodes on the named selection NS_INLET, essentially creating a rigid piston face. The harmonic response analysis is set up, the constants for the real and imaginary parts of the specific acoustic impedance in Equations (3.42) and (3.45) are calculated, an impedance surface is applied to the named selection NS_OUTLET using the APDL command SF,,IMPD, and the harmonic response analysis is conducted over the frequency range defined in the Input Parameters.

- The next step is to define values of parameters for the APDL code. Before entering the values, make sure that the units are set to MKS. In the Outline window, left-click on the branch Analysis Settings under Harmonic Response (A5). Expand the Analysis Data Management tree and make sure that the row for Solver Unit System is mks. If it is not, then change the row `Solver Units` to `Manual`, and then change the row `Solver Unit System` to `mks`. Once the use of the MKS system of units is confirmed, the values of the parameters can be entered. In the `Outline` window, left-click on the `Commands (APDL)` branch under `Harmonic Response (A5)`. In the window `Details of "Commands (APDL)"`, enter the following values in the rows beneath the branch for `Input Arguments` as follows

ARG1	1.21	the density of air in kg/m^3
ARG2	343	the speed of sound of air in m/s
ARG3	0.05	the radius of the duct in m
ARG4	2	the start frequency for the harmonic analysis in Hz
ARG5	200	the end frequency for the harmonic analysis in Hz
ARG6	2	the frequency increment for the harmonic analysis in Hz

Note that an error will occur if the starting analysis frequency in `ARG4` is set to `0`. See Section D.2.1.3 for more details.

- Click on the `Analysis Setting` branch. These values will *not* be used to conduct the analysis, since it is handled by the `APDL` code in the `Commands (APDL)` branch. Change the `Range Minimum` to `0`, `Range Maximum` to `1`, `Solution Intervals` to `1`.

- The completed `Harmonic Response (A5)` branch will look like the following figure.

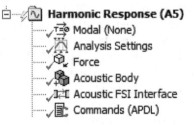

- Click on `File | Save Project`.

That completes the set up of the analysis. Click the `Solve` icon to calculate the results.

3.3.7.5 Results

Figures 3.15 and 3.16 show the real and imaginary parts of the mechanical impedance of the piston attached to the duct, respectively. The results calculated using ANSYS overlay the theoretical predictions, and are essentially the same as Figures 3.11 and 3.12, which is the desired outcome. As the mechanical impedance values are calculated correctly, it follows that the mechanical power will match theory and this result is not presented.

What has been shown is that it is possible to simulate a duct radiating into a plane baffle by using a frequency varying impedance applied to the outlet of a duct, instead of needing to model a semi-infinite acoustic space for the plane baffle condition. This simplification reduces the size and complexity of the finite element model, and accurate results are still obtained.

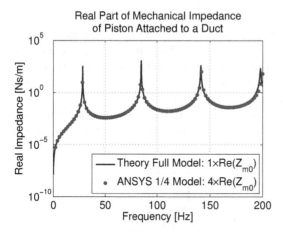

FIGURE 3.15
Real part of mechanical impedance of a piston attached to a 3 m long circular duct of radius 0.05 m that radiates into a baffled plane, calculated theoretically and using ANSYS Workbench with a 1/4 model and an impedance for the radiation.

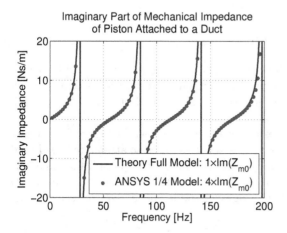

FIGURE 3.16
Imaginary part of mechanical impedance of a piston attached to a 3 m circular duct that radiates into a baffled plane, calculated theoretically and using ANSYS Workbench with a 1/4 model and an impedance for the radiation.

3.4   Resonator Silencers

3.4.1   Geometries

Resonator silencers function by providing a high reactive impedance causing an incident acoustic wave to reflect upstream. They usually have no or little acoustic absorption material within the device. This is in contrast to absorptive silencers that rely on the use of acoustic absorbing material to attenuate incident acoustic waves and is further discussed in Chapter 5.

Common geometries of resonator-type silencers are shown in Figure 3.17 and include (a) quarter-wavelength tube, (b) Helmholtz resonator, (c) expansion chamber, (d) contraction.

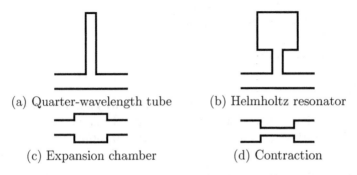

(a) Quarter-wavelength tube   (b) Helmholtz resonator

(c) Expansion chamber   (d) Contraction

FIGURE 3.17
Geometries of resonator-type silencers.

The acoustic response of these silencers can be modeled using the 4-pole (or transmission line) method described in Section 3.2.2. Examples of a quarter-wavelength tube and expansion chamber are shown in Sections 3.4.2 and 3.4.3, respectively.

Although simple reactive resonator types are shown in Figure 3.17, many of these silencers can be formed into an array and used to attenuate broadband noise. Figure 3.18 shows a sketch of a complicated silencer "splitter" that was designed using ANSYS Mechanical APDL where the model was built using 2D FLUID29 acoustic elements. The design was built and underwent scale-model

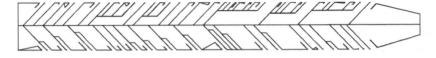

FIGURE 3.18
Design of a large exhaust silencer for a power station [55, 76].

testing, then built at full size, where the size of each silencer splitter was 45 cm
wide by 7.1 m tall, and installed in a 980 MW coal-fired power station [55, 76].

3.4.2   Example: Quarter-Wavelength Tube Silencer

Figure 3.19 shows a schematic of a quarter-wavelength tube (QWT) resonator
silencer attached to a circular main exhaust duct and Table 3.8 lists the relevant parameters.

The acoustic response of the quarter-wavelength tube system can be modeled using the 4-pole method described in the following section. A similar
analysis can be conducted to model a Helmholtz resonator using the 4-pole
method as described by Singh [138].

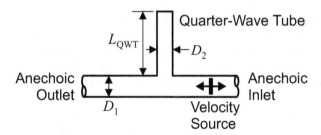

FIGURE 3.19
Schematic of a quarter-wavelength tube attached to a circular main exhaust
duct.

TABLE 3.8
Parameters Used in the Analysis of a Circular Duct
with a Quarter-Wavelength Tube

Description	Parameter	Value	Units
Diameter main duct	D_1	0.1	m
Diameter QWT	D_2	0.05	m
Length QWT	L_{QWT}	1.5	m
Speed of sound	c_0	343.24	m/s
Density	ρ_0	1.2041	kg/m^3
Velocity at inlet	u_1	0.001	m/s

3.4.2.1   Theory

The four-pole method described in Section 3.2.2 can be used to calculate
the acoustic pressure, velocity, and transmission loss of quarter-wavelength
tube resonator silencer attached to a duct. The expansion silencer component
shown in Figure 3.1 is replaced with the quarter-wavelength tube shown in
Figure 3.19, and the corresponding four-pole matrix $[\mathbf{T}_2]$ for a side-branch

resonator is given by [46, Eq. (10.20), p. 379]

$$[\mathbf{T}_2] = \begin{bmatrix} 1 & 0 \\ \dfrac{1}{Z_r} & 1 \end{bmatrix} \tag{3.46}$$

$$Z_r = Z_t + Z_c, \tag{3.47}$$

where Z_r is the impedance of the resonator that is the sum of the impedance of the throat Z_t, and the cavity of the resonator Z_c. For the quarter-wavelength tube resonator, the impedance of the cavity is given by [46, Eq. (10.21), p. 380]

$$Z_c = -j \frac{c_0}{S_{\text{QWT}}} \cot(kL_{\text{QWT}}), \tag{3.48}$$

where S_{QWT} is the cross-sectional area of the quarter-wavelength tube, c_0 is the speed of sound of the acoustic medium, and $k = \omega/c_0$ is the wavenumber, and L_{QWT} is the length of the quarter-wavelength tube which is the physical length of the tube plus a small end correction δ. The end correction length that should be added is the subject of ongoing research and references present different expressions as described in the following paragraphs.

Ji [95] shows that the additional length depends on the excitation frequency and the ratio of the diameter of the quarter-wavelength tube and the main duct. Figures are shown in Ji [95] for end correction lengths for a cylindrical quarter-wavelength tube and cylindrical main duct, where the ratio of the radii of the quarter-wavelength tube and main duct is $a_{\text{QWT}}/a_{\text{duct}} = 1$ and 0.5, as shown in Figure 3.20.

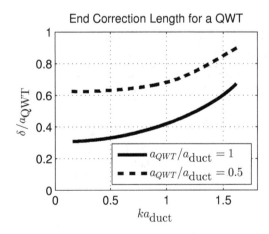

FIGURE 3.20
End correction δ for a cylindrical quarter-wavelength tube where the ratio of the radii of the quarter-wavelength tube and the cylindrical main duct is $a_{\text{QWT}}/a_{\text{duct}} = 1$ and 0.5 [95, Figs. 2–3].

Kurze and Riedel [104, p. 287] suggest an expression for the end correction length based only on the ratio of the cross-sectional areas of the duct, and not related to the excitation frequency as

$$\delta = \frac{\pi}{2}d\left(1 - 1.47\epsilon^{0.5} + 0.47\epsilon^{1.5}\right), \tag{3.49}$$

where $\epsilon = (d/b)^2$, d is the diameter of the narrower pipe, and b is the diameter of the wider pipe.

Some acoustic textbooks suggest end correction lengths that are independent of frequency, and area ratios that are between $0.6\times$ the radius of the quarter-wavelength tube a_{QWT} [47, p. 447] and $0.85\times$ [83, p. 379] where

$$\pi a_{\text{QWT}}^2 = S_{\text{QWT}}, \tag{3.50}$$

and S_{QWT} is the cross-sectional area of the quarter-wavelength tube.

The impedance of the throat Z_t is a function of the grazing flow speed and has been the subject of considerable research. However, an equation for the impedance of a circular orifice with grazing flow is problematic. There is considerable research reported for small-diameter holes, such as in perforated plates, however, there is no expression available for large-diameter holes. It has been found experimentally that the acoustic response varies considerably for variations in hole diameter, side-branch geometry, and ratio of cross-sectional areas between the main duct and side-branch resonator. For the simplest case, where there is no flow, it is assumed that there is no additional impedance at the throat ($Z_t = 0$). This enables comparison of the theoretical predictions with the finite element analysis results. For further discussion the reader is referred to Howard and Craig [77].

In summary, it is the opinion of the authors that further research is required in this area to address the effective end correction length for a variety of geometric configurations and operating conditions. As an "engineering" approximation (reminiscent of the jokes comparing engineers, physicists, and mathematicians) the reader should choose one reference with the expectation that their predictions may not match exactly with real-world measurements.

The 4-pole transmission matrices for the upstream $\mathbf{T}_3$ and downstream $\mathbf{T}_1$ duct segments are given by Equation (3.4). The total transmission matrix is given by

$$\begin{bmatrix} T_{11} & T_{12} \\ T_{21} & T_{22} \end{bmatrix} = \mathbf{T}_3\mathbf{T}_2\mathbf{T}_1, \tag{3.51}$$

The transmission loss is calculated as [46, Eq. (10.10), p. 374]

$$\text{TL} = 20\log_{10}\left|\frac{T_{11} + \dfrac{S_{\text{duct}}}{c_0}T_{12} + \dfrac{c_0}{S_{\text{duct}}}T_{21} + T_{22}}{2}\right|, \tag{3.52}$$

where S_{duct} is the cross-sectional area of the inlet duct. In this example of the silencer and duct system, there are upstream and downstream straight lossless ducts that have anechoic terminations either side of the quarter-wavelength tube. These straight ducts provide no acoustic attenuation and hence the 4-pole transmission matrices $\mathbf{T}_3$ and $\mathbf{T}_1$ can be ignored in the calculations, in which case it is only necessary to determine the elements of the four-pole matrix in Equation (3.46). If the upstream and downstream ducts provide attenuation, or if the complex values of the pressure and particle velocities at the terminations are to be calculated, then it is necessary to include the four-pole transmission matrices for these ducts, as was shown in Section 3.2.2.

3.4.2.2   MATLAB

The theoretical model was implemented using MATLAB and the script beranek_ver_fig10_11_quarter_wave_tube_duct_4_pole.m is included with this book. The aim was to reproduce the results shown in Beranek and Vér [46, Fig 10.11, p. 384].

Figure 3.21 shows the predicted transmission loss of a quarter-wavelength tube silencer, sometimes called a transverse tube silencer, for a range of area ratios of the quarter-wavelength tube to main duct $N = S_{\text{QWT}}/S_{\text{duct}}$.

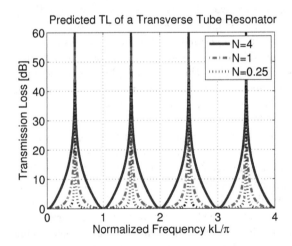

FIGURE 3.21
Transmission loss versus normalized frequency of a transverse tube resonator of length L for a range of area ratios of the transverse tube to main duct $N = S_{\text{QWT}}/S_{\text{duct}}$.

3.4.2.3   ANSYS Workbench

An ANSYS Workbench model was created of a quarter-wavelength tube attached to a main duct and was used to conduct a harmonic analysis and calculate the transmission loss of the reactive silencer. The ANSYS Workbench archive file called `quarter_wave_tube.wbpz` , which contains the `.wbpj` project file, is available with this book.

The model was created with some of the dimensions parameterized so that ANSYS Workbench could be used to calculate the results shown in Figure 3.21.

The learning outcomes from this section are:

- to obtain experience with the use of the `Muffler Transmission Loss` feature in the ACT Acoustics extension,

- to inspect the acoustic particle velocity results to ensure that there are no errors in the model and the results,

- to use the mesh `Refinement` feature when meshing the solid model around the throat region of the quarter-wavelength tube, and

- to verify that removing the silencer element from the model results in a transmission loss of the straight duct of 0 dB.

Creation of the Model

By now the reader should have experience using ANSYS Workbench to create solid models using DESIGNMODELER and conducting a `Harmonic Analysis` from the `Analysis Systems` toolbox. Hence instructions for the creation of the solid model will not be described in this section and the reader is referred to the completed model in the file `quarter_wave_tube.wbpj` that is provided with this book.

The completed solid model is shown in Figure 3.22 and looks like a T-branch. Several of the dimensions in the model have been parameterized and can be inspected by double-clicking on the `Parameter Set` cell in the `Project Schematic` window. The diameters of the quarter-wavelength tube and the

FIGURE 3.22
Solid model of a circular duct with a circular quarter-wavelength tube attached.

main exhaust duct can be altered by changing the values for QWT_Diam and Duct_Diam, respectively. For this model the initial diameters have been set as QWT_Diam=0.05 m and Duct_Diam=0.1 m, so that the ratio of areas is $N = (0.05/0.1)^2 = 0.25$. These dimensions can be easily changed to calculate the results for the ratio $N = 1$. The figure in Beranek and Vér [46, Fig. 10.11, p. 384] also includes a curve for $N = 4$ where the area the quarter-wavelength tube is 4× the area of the main exhaust duct. The parameterized solid model created in ANSYS DESIGNMODELER is not set up to handle geometries where the diameter of the quarter-wavelength tube is larger than the diameter of the main exhaust duct, and minor adjustments would be required. The length of the quarter-wavelength tube can be altered by changing the parameter QWT_Length and has an initial value of 1.5 m.

Two Named Selections were created for this solid model: NS_INLET was defined for the two faces on the right end of the main circular duct shown in Figure 3.22, and NS_OUTLET was defined on the left end.

The following section describes the setup of the harmonic analysis.

Analysis Setup

- The solid model of the quarter-wavelength tube should have been created in DESIGNMODELER. In the Project Schematic window, there should be a green tick in row 3 Geometry.

- In the Project Schematic window, double-click on row 4 Model, which will start ANSYS Mechanical.

- In the Outline window, check that the object Model (A4) | Connections *does not* have a plus sign next to it, which would indicate that there are entries for contact conditions and that the model has multiple parts, instead of a single part comprising multiple bodies.

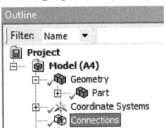

- The next few steps will involve defining the size of the mesh of the bodies, and the meshing method that will be used. In the Outline window, right-click on Model (A4) | Mesh and left-click on Insert | Sizing.

- In the window Details of "Body Sizing", left-click in the row Scope | Geometry. Change the filter selection to Body and select the 6 bodies shown in the following figure, and *do not* select the 4 bodies for the throat of the quarter-wavelength tube that connects to the main exhaust duct. Click the

Apply button. In the row Definition | Element Size, type in the value
3e-2. Change the row Definition | Behavior to Hard.

- Repeat these steps to define the size of the elements in the 4 bodies for
  the throat of the quarter-wavelength tube that connects to the main exhaust duct. In the window Details of "Body Sizing 2", click in the row
  Definition | Element Size, and type in the value 2e-2. Change the row
  Definition | Behavior to Hard. Note that we will define a finer mesh
  around the throat. This is necessary to visualize the complicated pressure
  distribution in this region.

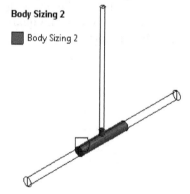

- The next steps are to define the methods used to perform the meshing. The
  mesh will be created using a tetrahedral mesh. The reason for selecting this
  type is that ANSYS Mechanical has difficulty in accommodating a mixture
  of mesh methods such as sweep, mesh refinement, and free-meshing. In the
  Outline window, right-click on Model (A4) | Mesh and left-click on Insert
  | Method. In the window Details of "Automatic Method" - Method, click
  in the row Scope | Geometry. Make sure that the Body selection filter is still
  active. In the Geometry window, right-click and then left-click on Select

All, then click the Apply button. The row should show that 10 Bodies were
selected. In the row Definition | Method, left-click in the cell for Automatic,
which will reveal a triangle icon on the right-hand side, and then left-click on
this icon to open the drop-down menu. Left-click on the entry Tetrahedrons.
This will change the name of the object in the Outline window listed under
Model (A4) | Mesh to Patch Conforming Method.

Details of "Automatic Method" - Method	무
⊟ **Scope**	
Scoping Method	Geometry Selection
Geometry	10 Bodies
⊟ **Definition**	
Suppressed	No
Method	Automatic ▼
Element Midside Nodes	Automatic
	Tetrahedrons
	Hex Dominant
	Sweep
	MultiZone
	Cartesian (Beta)

- The next step will involve increasing the mesh density around the throat
  region where the quarter-wavelength tube connects to the main exhaust
  duct. In this case, it is not necessary to increase the mesh density in this
  region as the results will still be accurate. The reason for including this step
  is to highlight the complicated acoustic particle velocity and sound pressure
  level in this region. To make it easier to select the faces of interest, first
  change the selection filter to Body, and select the 2 bodies on the quarter-
  wavelength tube that connect to the main exhaust duct. Right-click and
  left-click on Hide Body.

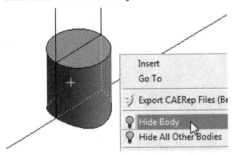

- In the Outline window, right-click on Model (A4) | Mesh and left-click on
  Insert | Refinement. Change the filter selection to Face. In the window
  Details of "Refinement" - Refinement, click in the row Scope | Geometry.
  Select the two faces at the junction between the quarter-wavelength tube
  and the main exhaust duct, and then click the Apply button. The Geometry
  row should indicate that 2 Faces were selected. Change the row Definition

| Refinement to 3, which determines the level of mesh refinement that will be attempted—the higher the number the finer the mesh.

Refinement

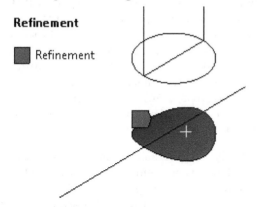

- In the menu bar, click on File | Save Project. The reason for saving the model before meshing is that a separate computer process for the meshing module is started and sometimes this crashes.

- In the Outline window, right-click on Model (A4) | Mesh, and left-click on Generate Mesh. The following figure shows the mesh around the throat of the quarter-wavelength tube.

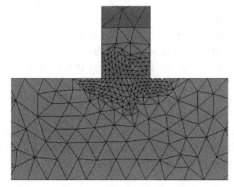

- Next, check that there are two entries listed under Model (A4) | Named Selection for NS_INLET and NS_OUTLET. Each named selection should have in the row Scope | Geometry, that 2 Faces are selected.

That completes the definition of the steps listed under Model (A4). The next set of steps involves setting up objects listed under Harmonic Response (A5) to define the analysis settings and acoustic parameters.

Analysis Options

- The harmonic response analysis will be conducted from 2 Hz to 300 Hz in increments of 2 Hz, so that there are 150 steps. In the `Outline` window, left-click on `Harmonic Response (A5) | Analysis Settings`. In the window `Details of "Analysis Settings"`, change the row `Options | Range Minimum` to `0`, `Options | Range Maximum` to `300`, and `Options | Solution Intervals` to `150`. Note that if you are experimenting with various setup options, a good practice is to set `Options | Solution Intervals` to 1, and use a low mesh density so that it does not take a long time to solve models. Change the row `Options | Solution Method` to `Full`.

- Click on the plus sign next to `Analysis Data Management` to expand the options. Change the row `Analysis Data Management | Save MAPDL db` to `Yes`. It is necessary to select this option for the post-processing of the results to calculate sound pressure levels, transmission loss, and other acoustic results.

- In the ACT Acoustics extension menu bar, click on `Acoustic Body | Acoustic Body`. Make sure that the Body selection filter is active. In the Geometry window, right-click to open the context menu and left-click on `Select All`. In the window `Details of "Acoustic Body"`, click in the row `Scope | Geometry`, and then click the `Apply` button. The row should indicate that `10 Bodies` were selected. Leave all the other parameters as the default values.

- In order to calculate the `Muffler Transmission Loss` using the ACT Acoustics extension, it is necessary to ensure that the inlet and outlet of the duct have anechoic terminations. There are several ways to achieve an anechoic termination. One of the simplest methods is to insert a `Radiation Boundary` object and select the appropriate faces that act to absorb outgoing waves. In the ACT Acoustics extension menu bar, click on `Boundary Conditions | Radiation Boundary`. In the window `Details of "Acoustic Radiation Boundary"`, change the row `Scope | Scoping Method` to `Named Selection`. Change the row `Scope | Named Selection` to `NS_INLET`. This will apply an anechoic termination to the inlet to the main exhaust duct.

Details of "Acoustic Radiation Boundary"	
⊟ **Scope**	
Scoping Method	Named Selection
Named Selection	NS_INLET

- Repeat this process to define an `Acoustic Radiation Boundary` for the `NS_OUTLET` named selection.

- The next step is to apply an acoustic excitation at the inlet of the duct to effectively create a piston with a volume velocity source. This can be achieved in a number of ways. One could use a `Mass Source`, a `FLOW` source, apply a force or displacement to the nodes on the inlet face, and couple the displacement degrees of freedom, as was shown in a previous example. For this example, the piston source will be created using an `Acoustic Normal Surface Velocity` from the `Excitation` menu in the ACT Acoustics extension menu, where the face of the inlet will be defined to have a harmonic normal velocity. Click on `Excitation | Normal Surface Velocity (Harmonic)`. In the window `Details of "Acoustic Normal Surface Velocity"`, change the row `Scope | Scoping Method` to `Named Selection`, and change the row `Named Selection` to `NS_INLET`. In the row `Definition | Amplitude of Normal Velocity` type the value `0.001`, which will define that the face has a harmonic velocity of 1 mm/s.

Details of "Acoustic Normal Surface Velocity"	
Scope	
Scoping Method	Named Selection
Named Selection	NS_INLET
Definition	
Amplitude Of Normal Velocity	0.001 [m sec^-1]
Phase Angle	0 [°]

- That completes the definitions of the steps under `Harmonic Response (A5)`. The next steps are to define how the results will be requested and displayed.

- Click on the branch `Solution (A6)`. In the ACT Acoustics extension menu, select `Results | Acoustic Pressure`, and also select `Results | Acoustic SPL`. This will request that the acoustic pressure and acoustic sound pressure level will be displayed, initially at the last analysis frequency at 300 Hz. Once we have determined the frequency at which the maximum transmission loss occurs, we will plot the sound pressure level at this frequency.

- The next result that will be requested is the (absolute) acoustic particle velocity along the Y-axis, which is aligned with the quarter-wavelength tube. The reason the absolute acoustic particle velocity will be calculated is because it is easier to see on a contour plot where the particle velocity is zero, compared with calculating the signed value of acoustic particle velocity. It is also instructive to show how to calculate a user-defined result that involves a mathematical operation. Click on the icon `User Defined Result`. In the window `Details of "User Defined Result"`, make sure that the row `Scope | Geometry` has `All Bodies` selected. In the row `Definition | Expression` type `abs(PGY)`. Note that `abs` must be in lower case. For the moment we will keep the frequency of the result that is displayed as `Last`.

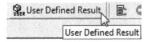

- So that it is easier to recognize the results, we will change the name of this object. Right-click on Solution (A6) | User Defined Result and then left-click on Rename Based on Definition. The object name will be changed from User Defined Result to abs(PGY).

- Repeat these steps to request that the absolute acoustic particle velocity in the Z-axis is displayed, which corresponds to the axis of the main duct. In the row Definition | Expression type in abs(PGZ). Rename this object using the Rename Based on Definition feature.

- The next result that will be calculated is the transmission loss. From the ACT Acoustics menu, select Results | Muffler Transmission loss. In the window Details of "Muffler Transmission Loss" it is necessary to define the faces for the inlet and outlet, which are used to calculate the average sound pressure levels, and the acoustic excitation that was defined in Harmonic Response (A5). Change the row Outlet | Scoping Method to Named Selection, Outlet | Named Selection to NS_OUTLET. Change Inlet | Scoping Method to Named Selection, and Inlet | Named Selection to NS_INLET. Change the row Definition | Inlet Source to Acoustic Normal Surface Velocity. The equivalent acoustic pressure for this surface velocity will be displayed in the following row in Definition | Inlet Pressure, as 0.2 Pa. It is assumed that half of the volume velocity will propagate upstream away from the muffler and half will propagate downstream toward the muffler. The equivalent acoustic pressure is calculated using Equation (2.33) as

$$
\begin{aligned}
p_{\text{incident, normal velocity}} &= \frac{u_n \rho_0 c_0}{2} \\
&= \frac{0.001 \times 1.2041 \times 343.24}{2} \\
&= 0.2066 \text{ Pa}.
\end{aligned}
$$

Details of "Muffler Transmission Loss"	
Outlet	
Scoping Method	Named Selection
Named Selection	NS_OUTLET
Inlet	
Scoping Method	Named Selection
Named Selection	NS_INLET
Definition	
Inlet Source	Acoustic Normal Surface Velocity
Inlet Pressure	0.206647642 [Pa]
Mass Density Of Environment Media	1.2041 [kg m^-1 m^-1 m^-1]
Sound Speed In Environment Media	343.24 [m sec^-1]

- The last result that will be requested is the acoustic pressure at the outlet of the duct versus the frequency of the acoustic excitation. This result can be used to confirm that the sound pressure at the exit is minimized at the frequency that corresponds to the quarter-wavelength of the side-branch resonator. In the ACT Acoustics extension menu, click on `Results | Acoustic Time_Frequency Plot`. In the window `Details of "Acoustic Time_Frequency Plot"`, change the row `Scope | Scoping Method` to `Named Selection`, `Scope | Named Selection` to `NS_OUTLET`, `Definition | Display` to `Amplitude`.

Details of "Acoustic Time_Frequency Plot"	
⊟ **Scope**	
Scoping Method	Named Selection
Named Selection	NS_OUTLET
⊟ **Definition**	
Result	Pressure
Spatial Resolution	Use Average
Display	Amplitude

- That completes the setup of the analysis. Click on `File | Save Project` to save the model before solving it.

- Click the `Solve` icon and wait for the calculations to complete, which may take 10 minutes. If you receive an error message that the number of nodes and elements exceeds the amount allowed for your license (for example, ANSYS Teaching licenses have a limit of 32,000 nodes), then you can change the element size of the upstream, downstream, and quarter-wavelength tube from `3e-002` m to `5e-002` m. Leave the element size at `2e-002` m for the 4 bodies at the junction of the main duct and the attachment of the quarter-wavelength tube.

Although for this example it is known that the results are correct, it is suggested that when developing a new model that some verification steps should be conducted, which follows the "crawl, walk, run" philosophy described on page 617. It is suggested that initially the bodies for the quarter-wavelength tube, or silencer elements should be `Suppressed`, and verify that the transmission loss of a simple straight duct with anechoic terminations is 0 dB.

The next section describes the post-processing of the results.

Results

This section describes instructions for the post-processing to view and export the results.

- The result listed under `Solution (A6) | Muffler Transmission Loss` should have a green tick next to it indicating that the results were calculated successfully. If there is a red lightning bolt next to the entry, it means that an error occurred. Check the `Messages` window for the error message. There might be error messages listed such as

An error occurred when the post processor attempted to load a
specific result. Please review all messages.

Unable to create user defined result. PRES is not a
recognized result in: 1041 = PRES

There might also be an error message which indicates that you should inspect
the log file. If this occurs, open the log file and search for the cause of the
error. You might see an error message such as the following

```
*** ERROR ***                          CP =       1.576   TIME= 10:30:16
The license is currently in use by another application in this
Workbench session such as Mechanical or another Mechanical APDL
application.  You must wait until the other application has finished
its task (for example meshing) or manually PAUSE the other Mechanical
APDL application.
```

This error could be caused by a license issue where the ACT Acoustics extension is trying to request an additional license. The work-around is to request
that these results are calculated again. Right-click on the object Solution
(A6) | Muffler Transmission Loss and left-click on Clear Generated
Data, and there should be a yellow lightning bolt next to the object. Right-
click on Muffler Transmission Loss again, and left-click on Generate to
re-calculate the transmission loss. If necessary, follow the same steps to re-
calculate the results for the Acoustic Time_Frequency Plot.

- Click on the object Solution (A6) | Muffler Transmission Loss and
  the following graph will be shown. The results indicate that the maximum
  transmission loss occurs at 58 Hz and at odd-numbered harmonics. Inspect
  the table of values in the window Data View to confirm that the transmission
  loss at 58 Hz has a local maximum.

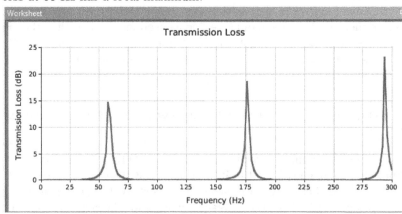

The expected frequency at which the maximum transmission loss occurs is based on the speed of sound and the effective length of the quarter-wavelength tube. The theory described in Section 3.2.2 assumes that the acoustics can be modeled as a one-dimensional system, where only acoustic plane waves propagate along the duct segments. The one-dimensional assumption also means that it is assumed that the quarter-wavelength tube is effectively attached at the center of the main duct. In reality, the quarter-wavelength tube is attached at the external diameter of the main duct, and hence an adjustment has to be made for the length of the quarter-wavelength tube. In the solid model of the quarter-wavelength tube that was created in DESIGNMODELER, a circular area was extruded perpendicularly from the central axis of the main duct for a distance of 1.5 m. Therefore, the physical length of the quarter-wavelength tube is 1.5 m minus the radius of the main duct. As described in Section 3.4.2.1, when determining the effective length of the quarter-wavelength tube it is necessary to add an additional length called the end correction. For the quarter-wavelength tube in this example, the end correction can be estimated using Figure 3.20 as $\delta = 0.61 \times a_{QWT} = 0.61 \times 0.025 = 0.01525$ m. Hence the effective length of the quarter-wavelength of the tube is

$$L_{\text{effective}} = 1.5 - a + \delta$$
$$= 1.5 - 0.05 + 0.01525$$
$$= 1.465 \text{ m}.$$

The lowest frequency at which the maximum transmission loss is expected to occur is at

$$f_{\text{QWT}} = \frac{c_0}{4 \times L_{\text{effective}}} = \frac{343.24}{4 \times 1.465} = 58.56 \text{ Hz}. \tag{3.53}$$

Hence the theoretical prediction of 58.56 Hz is close to the result from the finite element analysis that predicted 58 Hz, noting that the harmonic analysis was conducted in 2 Hz increments. If the settings for the analysis frequency range and increments were altered (found under `Harmonic Response (A5) | Analysis Settings`, in the window `Details of "Analysis Settings")` to `Options | Range Minimum` to 58, `Options | Range Maximum` to 59, `Options | Solution Intervals` to 20, so that frequency increment is $(59 - 58)/20 = 0.05$ Hz, the frequency at which the maximum transmission loss occurs is 58.55 Hz. It is left to the keen reader to attempt this analysis with the small frequency increment.

Another point to mention is that damping is not included in these analyses, and so the theoretical and finite element analysis estimates of the transmission loss at the tuned frequency of the quarter-wavelength tube will be infinite! The graph shown above of the transmission loss calculated using ANSYS Workbench might give the misleading impression that greater transmission loss is achieved at the higher harmonics of the quarter-wavelength tube tuned frequency. However, this is not the case—the closer the analysis frequency is to the theoretical resonance frequency, the higher the value of the transmission loss. Figure 3.23 shows the theoretical transmission loss of a quarter-wavelength tube with dimensions corresponding to this example, that was calculated using the MATLAB script `beranek_ver_fig10_11_quarter_wave_tube_duct_4_pole.m`. It can be seen that the transmission loss approaches infinity at 58.56 Hz. If the analysis frequency increment used in ANSYS Workbench is $\Delta f = 2$ Hz, then one might expect the peak transmission loss to be 18.4 dB at 58.0 Hz. However if the frequency increment is changed to $\Delta f = 0.1$ Hz, the peak value of the transmission loss increases to 42.1 dB at 58.6 Hz. In summary, care should be exercised when examining transmission loss (or any frequency response) results that the frequency increment used in the harmonic analysis is appropriate.

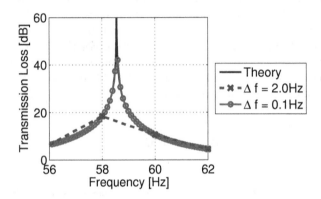

FIGURE 3.23
Influence of analysis frequency increment on perceived transmission loss results.

The following instructions continue the post-processing of the results from the ANSYS Workbench analysis to export the data and examine the sound pressure versus frequency, sound pressure level, and particle velocity.

- The transmission loss results can be exported by right-clicking on `Muffler Transmission Loss` and left-clicking on `Export`.

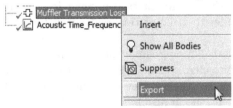

- Click on the object `Solution (A6) | Acoustic Time_Frequency Plot`, which should show that the acoustic pressure at the outlet is minimized at 58 Hz.

Also confirm that at the frequencies where the quarter-wavelength tube is not responsive, that the acoustic pressure is 0.206 Pa and corresponds to the value listed in Solution (A6) | Muffler Transmission Loss | Definition | Inlet Pressure.

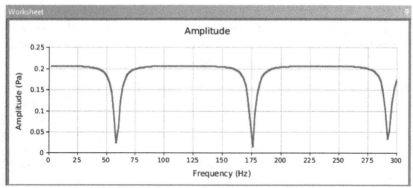

- Now that we know the frequency at which the transmission loss is maximized (58 Hz), we will change the frequency that is requested in the other results objects. Click on the object Solution (A6) | Acoustic SPL. In the window Details of "Acoustic SPL" in the row Definition | Frequency, type the number 58.

- Repeat these steps to change the frequency of the result that is displayed for the results abs(PGY), abs(PGZ), and Acoustic Pressure.

- Right-click on Solution (A6) | Acoustic SPL and left-click on Evaluate All Results. The following figure shows the sound pressure level at 58 Hz in the duct.

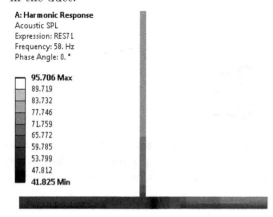

The results show that the sound pressure level:

- · in the quarter-wavelength tube is high at the closed end;
- · downstream from the quarter-wavelength tube (left side) is constant as there is no impedance change;
- · between the inlet (right side) and the quarter-wavelength tube varies, as the quarter-wavelength tube presents an impedance to the incident acoustic wave and is reflected back towards the inlet.

- • The following figure shows the absolute value of the acoustic particle velocity (pressure gradient) along the Y-axis (along the axis of the quarter-wavelength tube) at 58 Hz in the vicinity of the junction between the main duct and the quarter-wavelength tube. Note in order to see the cross-sectional view, it is necessary to select the two bodies in the side of the duct, right-click in the graphics window, and left-click on Hide Bodies. In this image the results are plotted using a logarithmic scale, which can be obtained by right-clicking on the color legend, and in the context menu that opens, left-click on Logarithmic Scale. The results show that at the interface between the quarter-wavelength tube and the main exhaust duct the fluid is moving. Hence the *effective length* is slightly longer than the length of the quarter-wavelength tube.

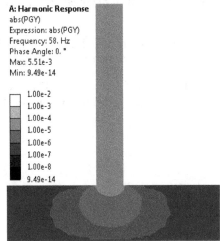

- • The following figure shows the corresponding acoustic particle velocity along the Z-axis (along the axis of the duct). The results show that downstream of the quarter-wavelength tube (left side) the acoustic particle velocity is low.

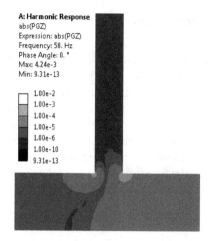

Note that acoustic particle velocity is a complex value with real and imaginary components. It is possible to inspect the imaginary part of the result by changing the row Definition | Phase Angle to -90, in the result object such as Acoustic Pressure.

Figure 3.24 shows the comparison of the transmission loss results calculated using the MATLAB code beranek_ver_fig10_11_quarter_wave_tube_duct_4_pole.m, and from the ANSYS Workbench analysis. The ANSYS analysis was repeated to calculate the transmission loss when the parameter QWT_Diam was set to 0.1 m, so that the ratio of areas of the quarter-wavelength tube to the main exhaust duct was $N = 1$. The figure shows that ANSYS results overlay the theoretical predictions.

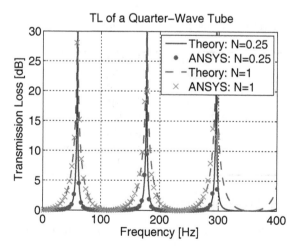

FIGURE 3.24
Transmission loss of a quarter-wavelength tube calculated theoretically and using ANSYS Workbench, for ratios of areas between the quarter-wavelength tube and the main exhaust duct of $N = 0.25$ and $N = 1.0$.

3.4.3 Example: Expansion Chamber Silencer

This example covers the use of the 4-pole transmission matrix method to calculate the transmission loss of an expansion chamber silencer, shown in Figure 3.25. An inlet duct of diameter D_3 is connected to a larger-diameter tube of diameter D_2. The interface between the inlet and the expansion chamber is a sudden expansion. The expansion chamber has length $L_{expansion}$ and is terminated with a sudden contraction to the outlet duct of diameter D_1. For this example, the inlet and outlet diameters are identical ($D_3 = D_1$). The acoustic excitation on the inlet duct is a harmonic acoustic particle velocity. The inlet and outlet ducts are assumed to have anechoic end conditions. Table 3.9 lists the parameters used in the example of the expansion chamber silencer.

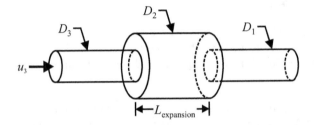

FIGURE 3.25
Schematic of an expansion chamber silencer.

TABLE 3.9
Parameters Used in the Analysis of an Expansion Chamber Silencer

Description	Parameter	Value	Units
Diameter inlet duct	D_3	0.05	m
Diameter expansion chamber	D_2	0.2	m
Diameter outlet duct	D_1	0.05	m
Length of expansion chamber	$L_{expansion}$	0.5	m
Speed of sound	c_0	343.24	m/s
Density	ρ_0	1.2041	kg/m^3
Velocity at inlet	u_1	1.0	m/s

The learning outcomes from this example are:

- exposure to transmission line (4-pole) theory for predicting the transmission loss of an expansion chamber silencer, and

- examination of the effect of mesh density on the predicted transmission loss, and the pressure field.

3.4.3.1 Theory

Figure 3.26 shows schematics of sudden expansion, straight duct, and sudden contraction duct segments that can be used to model an expansion chamber silencer. A section of duct of length L_2 extends into the expansion section.

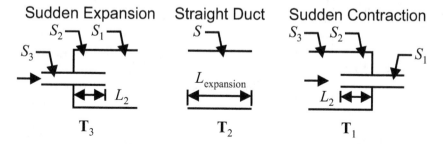

FIGURE 3.26
Sudden expansion, straight, and sudden contraction duct segments used to model an expansion chamber silencer.

The 4-pole transmission matrix for the expansion chamber silencer can be calculated as the sequence of segments (starting from the right side in Figures 3.25 and 3.26): (1) sudden contraction, (2) a straight duct, and (3) sudden expansion. The transmission matrix for these duct transitions is given by [46, Eq. (10.18), p. 379]

$$\mathbf{T} = \begin{bmatrix} 1 & KM_1Y_1 \\ \dfrac{C_2S_2}{C_1S_2Z_2 + S_2M_3Y_3} & \dfrac{C_2S_2Z_2 - M_1Y_1(C_1S_1 + S_3K)}{C_2S_2Z_2 + S_3M_3Y_3} \end{bmatrix} \qquad (3.54)$$

$$Z_2 = -j\frac{c}{S_2}\cot kL_2 \qquad (3.55)$$

$$Y_i = c/S_i , \qquad (3.56)$$

where the parameters are listed in Table 3.10, V_i is the mean flow velocity through cross-section of area S_i, $M_i = V_i/c_0$ is the Mach number through cross-section S_i.

TABLE 3.10
Parameters for the Evaluation of the Transmission Matrix for Ducts with Cross-Sectional Discontinuities [46, Table 10.1, p. 378]

Type	C_1	C_2	K
Expansion	-1	1	$[(S_1/S_3) - 1]^2$
Contraction	-1	-1	$\frac{1}{2}[1 - (S_1/S_3)]$

For this example, there are no pipe extensions within the expansion chamber ($L_2 = 0$), and there is no mean flow ($M_i = 0$). It can be shown that the transmission matrices for the sudden expansion and sudden contraction are identity matrices. The 4-pole matrices for each of these segments are

$$\text{contraction: } \mathbf{T}_1 = \begin{bmatrix} 1 & 0 \\ 0 & 1 \end{bmatrix} \tag{3.57}$$

$$\text{straight duct: } \mathbf{T}_2 = \begin{bmatrix} \cos(kL_{\text{expansion}}) & \dfrac{jc_0}{S_{\text{expansion}}}\sin(kL_{\text{expansion}}) \\ \dfrac{jS_{\text{expansion}}}{c_0}\sin(kL_{\text{expansion}}) & \cos(kL_{\text{expansion}}) \end{bmatrix} \tag{3.58}$$

$$\text{expansion: } \mathbf{T}_3 = \begin{bmatrix} 1 & 0 \\ 0 & 1 \end{bmatrix}, \tag{3.59}$$

where $S_{\text{expansion}}$ is the cross-sectional area of the expansion chamber. The 4-pole matrix for the straight duct in Equation (3.58) is the same as Equation (3.4). The combined 4-pole matrix for the expansion chamber silencer is

$$\mathbf{T}_{\text{expansion}} = \mathbf{T}_3\mathbf{T}_2\mathbf{T}_1 \tag{3.60}$$

$$= \begin{bmatrix} \cos(kL_{\text{expansion}}) & \dfrac{jc_0}{S_{\text{expansion}}}\sin(kL_{\text{expansion}}) \\ \dfrac{jS_{\text{expansion}}}{c_0}\sin(kL_{\text{expansion}}) & \cos(kL_{\text{expansion}}) \end{bmatrix}. \tag{3.61}$$

The transmission loss is calculated as [46, Eq. (10.10), p. 374]

$$\text{TL} = 20 \times \log_{10} \left| \frac{T_{11} + \dfrac{S_{\text{duct}}}{c_0}T_{12} + \dfrac{c_0}{S_{\text{duct}}}T_{21} + T_{22}}{2} \right|, \tag{3.62}$$

where S_{duct} is the cross-sectional area of the inlet duct.

The previous equations can be reduced to a single-line expression for the transmission loss as [47, Eq. (9.99), p. 464]

$$\text{TL} = 10 \times \log_{10} \left[1 + \frac{1}{4}\left(\frac{S_{\text{duct}}}{S_{\text{expansion}}} - \frac{S_{\text{expansion}}}{S_{\text{duct}}} \right)^2 \sin^2(kL_{\text{expansion}}) \right]. \tag{3.63}$$

3.4.3.2   MATLAB

The MATLAB script beranek_ver_fig10_12_single_chamber_4_pole.m included with this book can be used to calculate the transmission loss of an expansion chamber. Figure 3.27 shows the predicted transmission loss for two area ratios of $N = 64$ and $N = 16$ calculated using the 4-pole method using Equation (3.62) (from Beranek and Vér [46, p. 379]) and using Equation (3.63) (from Bies and Hansen [47, Eq. (9.99), p. 464]).

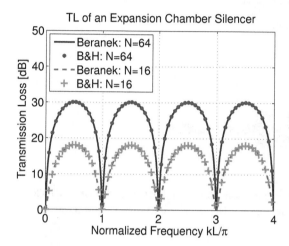

FIGURE 3.27

Transmission loss of a single expansion chamber silencer predicted theoretically using the 4-pole method using Equations (3.62) (from Beranek and Vér [46]) and (3.63) (from Bies and Hansen [47, Eq. (9.99), p. 464]) for area ratios of $N = 64$ and $N = 16$.

3.4.3.3   ANSYS Workbench

A model of the expansion chamber was created using ANSYS Workbench and is included with this book in the archive file `duct_expansion_chamber.wbpz` that contains the `.wbpj` project file.

Figure 3.28 shows the model and the associated mesh of linear `FLUID30` elements. The use of the `FLUID30` elements is selected in the window `Details of "Mesh"`, and the option `Advanced | Element Midside Nodes`, is changed to `Dropped`. The mesh density was 10 elements over an axial length of 0.1 m in the expansion chamber (=68 EPW at 500 Hz), and 10 elements through the diameter of 0.2 m.

Figure 3.29 shows the SPL at 500 Hz. Notice the complicated sound field at the entrance of the expansion chamber on the left side of the figure, which indicates that there are non-plane waves in this local region. Further along the expansion chamber the sound pressure level contours are straight lines, indicating the progression of a plane wave.

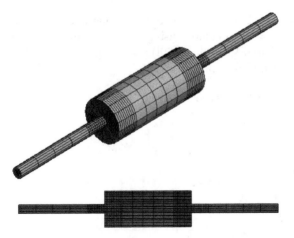

FIGURE 3.28
Finite element mesh of the expansion chamber with a mesh of `FLUID30` elements at 68 EPW at the inlet and outlet of the expansion chamber.

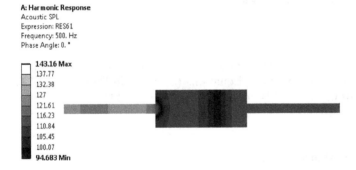

FIGURE 3.29
Sound pressure level in an expansion chamber silencer at 500 Hz predicted using ANSYS Workbench where there were 68 EPW at the inlet and outlet of the expansion chamber.

The model of the expansion chamber silencer was re-meshed with `FLUID30` elements with a density of 6 EPW as shown in Figure 3.30. Figure 3.31 shows the SPL at 500 Hz. The complicated sound field at the entrance to the expansion chamber is less obvious than in Figure 3.29, as the mesh density is coarser. This example highlights that it may be important to increase the mesh density in regions where the pressure field is expected to be complicated. However, it will be shown in the following section that for this example, there is little difference in the predicted transmission loss using either a fine or coarse mesh.

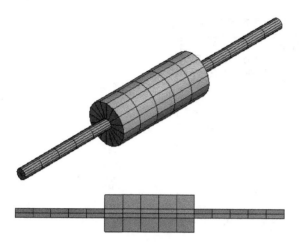

FIGURE 3.30

Finite element mesh of the expansion chamber with a mesh density of 6 EPW of `FLUID30` elements.

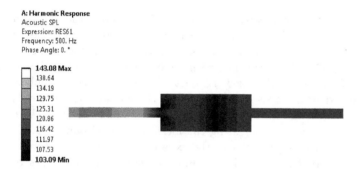

FIGURE 3.31

Sound pressure level in an expansion chamber silencer at 500 Hz predicted using ANSYS Workbench with 6 EPW of `FLUID30` elements.

3.4.3.4   Results

Figure 3.32 shows the predicted transmission loss using theory and ANSYS Workbench for the fine mesh and the coarse mesh. The results are nearly identical, indicating that there was only a marginal benefit in increasing the mesh density if calculation of the transmission loss was the purpose of the analysis. If the purpose was to investigate the local sound pressure field around the expansion segment, then a fine mesh would be required.

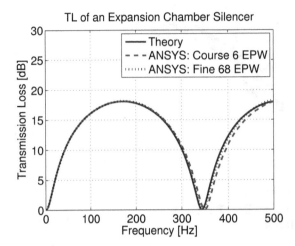

FIGURE 3.32

Transmission loss of an expansion chamber predicted using theory, and ANSYS Workbench for the fine mesh and coarse mesh of 6 EPW.

Figure 3.33 shows the sound pressure level at 174 Hz along the axis of a duct with an expansion chamber, calculated using the transmission line theory (calculated using the MATLAB script beranek_ver_fig10_12_single_chamber_4_pole.m) and using ANSYS Workbench with a mesh density of 6 EPW, and for a fine mesh of 68 EPW at the inlet and outlet of the expansion chamber. The results predicted using ANSYS indicate that the sound pressure levels at the inlet and outlet of the duct system were the same, and therefore there was no increase in accuracy by increasing the mesh density. However, it can be seen that at $z = 0.5$ m at the inlet to the expansion chamber, the results predicted using 6 EPW has insufficient number of nodes and elements to represent the local response and appears to be nearly constant 110 dB re 20 μPa at $z = 0.5 \cdots 0.6$ m.

Referring to Figure 1.3 and the "sea of springs" analogy, a finite element model of the system could be used to correctly predict the reaction force and displacement at the base of the structure and at the location of the load. However, the model could not be used to predict the local response in the region of the missing springs. Similarly for this example, the model was able to be used to predict the mean response at the inlet and outlet of the duct, and yet was unable to predict the local response in the expansion chamber as there was an insufficient number of nodes and elements.

The outcome from this example is as follows:

- The use of 6 elements per wavelength for the linear FLUID30 elements was adequate for the purpose of estimating the transmission loss, but was insufficient to accurately portray the local sound pressure field.

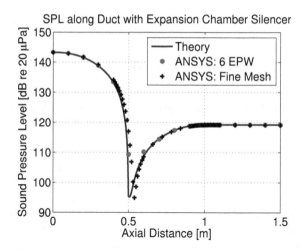

FIGURE 3.33

SPL at 174 Hz along the axis of a duct with an expansion chamber predicted using theory, and ANSYS Workbench for the coarse mesh of 6 EPW and the fine mesh with 68 EPW at the inlet and outlet of the expansion chamber.

- For 1-dimensional-type acoustic systems, where the acoustic energy can only propagate along a "waveguide," there was no increase in accuracy of the predicted sound pressure at the inlet and outlet of the duct, by increasing the mesh density in regions of impedance changes.

It will be shown in the following section where non-plane wave conditions occur in the duct, that the effect of a low mesh density can cause the re-direction of acoustic energy.

3.5 Non-Plane Waves

The previous analyses of the sound field in ducts were conducted at frequencies below cut-on, such that the acoustic field was plane-wave. When the analysis frequency is above cut-on, as calculated by Equation (3.16) or Equation (3.17), the acoustic field can have acoustic modes perpendicular to the axis of the duct.

This section contains an example of a harmonic analysis of a rectangular duct where the excitation frequency is above cut-on. The learning outcomes from this example are:

- highlight the existence of cross-modes in a duct at analysis frequencies above cut-on,

- demonstrate how poor mesh density can cause the re-direction of acoustic energy,

- determine whether the mesh density needs to be considered in regions where high pressure gradients or sudden changes in impedance are expected, and

- highlight why it is impractical to undertake theoretical analyses with complicated sound fields.

The following examples show the effect of altering the mesh density on the predicted sound field in a duct. A rectangular rigid-walled duct with anechoic terminations at each end is excited with a 1 m/s acoustic particle velocity applied across the inlet of the duct (the right side in the following figures). The dimensions of the duct are listed in Table 3.11. The cut-on frequency for the 0.4 m tall duct is calculated using Equation (3.17) as $f_{\text{cut-on}} = 343.23/(2 \times 0.4) = 429$ Hz. The following harmonic analyses are conducted at 600 Hz and hence it is possible for non-plane wave conditions to exist in the duct. The ANSYS Workbench archive file nonplane_wave_duct.wbpz, which contains the .wbpj project file and model used in the following discussion, is included with this book.

TABLE 3.11

Parameters Used in the Analysis of a Rectangular Duct

Description	Parameter	Value	Units
Duct height	L_y	0.4	m
Duct depth	L_x	0.1	m
Duct length	L_z	1.5	m
Speed of sound	c_0	343.23	m/s
Density	ρ_0	1.2041	kg/m^3
Velocity of piston	u_1	1.0	m/s

A) Regular FLUID30 Mesh 6 EPW

The rectangular duct was initially meshed with linear FLUID30 elements with approximately 6 elements per wavelength (EPW), as shown in Figure 3.34. The FLUID30 elements were selected by specifying that the mid-side nodes should be Dropped in the window Details of Mesh | Advanced. The actual mesh size that was generated was 0.087 m corresponding to 6.6 EPW.

A harmonic analysis was conducted at 600 Hz and the sound pressure level in the duct is shown in Figure 3.35. It can be seen that the contours are perpendicular to the walls of the duct and there is only a $(137.62 - 137.23 =)$ 0.4 dB variation in sound pressure level along the length of the duct. Theoretically, as there are no impedance changes along the infinite duct, the sound

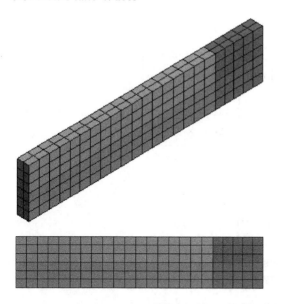

FIGURE 3.34
Case (A): isometric and side views of an ANSYS Workbench model of a rectangular duct regular swept mesh at 6.6 EPW, using linear FLUID30 elements.

A: Harmonic Response
Acoustic SPL
Expression: RES71
Frequency: 0. Hz
Phase Angle: 0. °

137.62 Max
137.58
137.53
137.49
137.45
137.4
137.36
137.31
137.27
137.23 Min

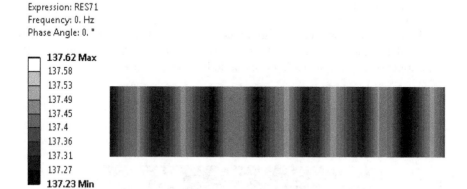

FIGURE 3.35
Case (A): Sound pressure level at 600 Hz using a FLUID30 mesh shown in
Figure 3.34.

pressure level should be constant at

$$\text{SPL} = 20 \times \log_{10} \left[\frac{\rho_0 c_0 \left(\frac{u_n}{2} \right)}{\sqrt{2} \times 20 \times 10^{-6}} \right] \tag{3.64}$$

$$= 20 \times \log_{10} \left[\frac{1.2041 \times 343.23 \times \frac{1}{2} \times 1}{\sqrt{2} \times 20 \times 10^{-6}} \right]$$

$$= 137.27 \text{ dB} .$$

B) Modified FLUID30 Mesh 6 EPW

The mesh of `FLUID30` elements was modified slightly by the introduction of a couple of tetrahedral elements in the upper portion of the duct, as shown in Figure 3.36. The mesh density is still approximately 6 EPW.

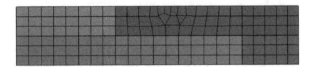

FIGURE 3.36
Case (B): ANSYS Workbench model of a rectangular duct with mostly regular swept mesh and a couple of tetrahedral elements at 6.6 EPW, using linear `FLUID30` elements.

Figure 3.37 shows the sound pressure level at 600 Hz for the mesh in Figure 3.36 is not regular and there is a variation of $(138.23 - 136.26 =) 2$ dB, and appears that an acoustic cross-mode has been excited.

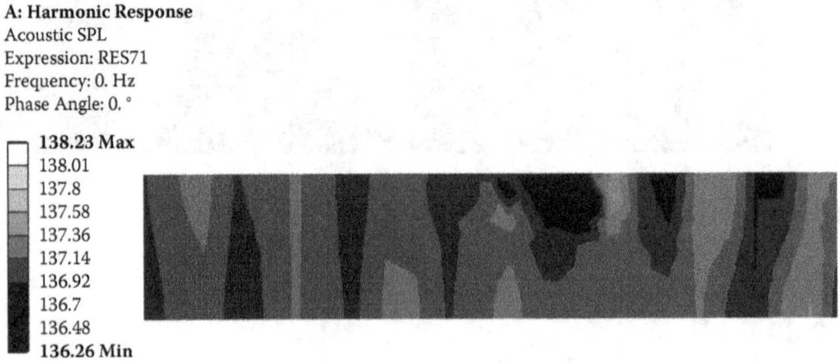

A: Harmonic Response
Acoustic SPL
Expression: RES71
Frequency: 0. Hz
Phase Angle: 0. °

138.23 Max
138.01
137.8
137.58
137.36
137.14
136.92
136.7
136.48
136.26 Min

FIGURE 3.37
Case (B): Sound pressure level at 600 Hz using a `FLUID30` mesh shown in Figure 3.36.

The effect of the non-regular mesh effectively caused a pseudo impedance change in the duct, and excited a cross-mode. Another way to consider this effect is that a region of the duct was artificially stiffer or softer than the surrounding region and caused the sound to be distorted, thereby "tripping" the incident planar sound field, causing the wavefront to bend and hit the upper and lower walls of the duct.

C) Modified FLUID220 Mesh 6 EPW

The accuracy of the results can be improved by increasing the mesh density, or by using the quadratic FLUID220 elements. The mesh was cleared and in the window Details of Mesh | Advanced, the options for the mid-side nodes was selected as Kept. The model was re-meshed and Figure 3.38 shows the new mesh, which is similar to the mesh in Figure 3.36.

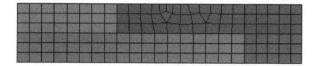

FIGURE 3.38

Case (C): ANSYS Workbench model of a rectangular duct with mostly regular swept mesh and a couple of tetrahedral elements at 6.6 EPW, using quadratic FLUID220 elements.

Figure 3.39 shows the sound pressure level at 600 Hz and there is only a 0.14 dB variation in sound pressure level. Hence, the use of the quadratic FLUID220 elements improved the accuracy of the predicted sound field.

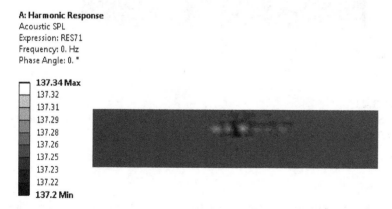

FIGURE 3.39

Case (C): Sound pressure level at 600 Hz using a mesh of FLUID220 elements shown in Figure 3.38.

D) Modified FLUID30 Mesh 12 EPW

An alternative to using the quadratic FLUID220 elements with 6 EPW is to use linear FLUID30 elements at 12 EPW. The mesh of the model was cleared, the mid-side nodes were Dropped, and the element size was reduced to 0.045 m. The resulting mesh is shown in Figure 3.38.

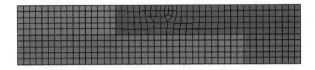

FIGURE 3.40
Case (D): ANSYS Workbench model of a rectangular duct with mostly regular swept mesh and a couple of tetrahedral elements at 12 EPW, using linear FLUID30 elements.

Figure 3.41 shows the sound pressure level at 600 Hz using the mesh shown in Figure 3.40 and the variation is $(137.68 - 136.77 =) 0.9$ dB.

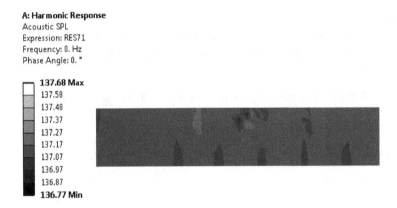

A: Harmonic Response
Acoustic SPL
Expression: RES71
Frequency: 0. Hz
Phase Angle: 0. °

137.68 Max
137.58
137.48
137.37
137.27
137.17
137.07
136.97
136.87
136.77 Min

FIGURE 3.41
Case (D): Sound pressure level at 600 Hz using a mesh of FLUID30 elements at 12 EPW shown in Figure 3.40.

Summary

Table 3.12 lists a summary of cases (A) to (D), the number of nodes, elements, and the wavefront. It can be seen that by comparing cases (B) and (C), where the element type was changed from the linear FLUID30 elements to the quadratic FLUID220 elements, that the number of nodes and wavefront increased substantially, and the accuracy improved.

TABLE 3.12

Summary of Mesh Quality for Rectangular Duct with Non-plane Wave Conditions

Case	Element	EPW	Nodes	Elements	Wavefront	Δ dB
A)	FLUID30	6	504	276	92	0.4
B)	FLUID30	6	525	317	80	2.0
C)	FLUID220	6	1947	368	380	0.1
D)	FLUID30	12	2541	1987	96	0.9

E) Rectangular Duct with Quarter-Wavelength Tube

The following example is the same rectangular duct described in the previous cases, only a quarter-wavelength tube is attached to the upper side of the duct. The rectangular quarter-wavelength tube has a length of 0.143 m, 0.1 m deep, and 0.05 m along the axis of the duct. The model was meshed with the quadratic FLUID220 elements, with an element size of 0.09 m, and 0.045 m in the region of the quarter-wavelength tube.

Figure 3.42 shows the sound pressure level in the duct at 408.5 Hz, which is the frequency where the transmission loss is high. The ACT Acoustics extension has a Muffler Transmission Loss object that is only intended to be used when plane wave conditions exist in the duct. It can be seen in Figure 3.42 that the sound pressure level varies across the height of the duct, and hence the acoustic field is not plane wave, and therefore the Muffler Transmission Loss object should not be used. In order to calculate the transmission loss of the muffler for non-plane wave conditions, it is necessary to export the real and imaginary parts of the acoustic pressure and particle velocity to calculate the acoustic intensity at each node at the outlet, then multiply the nodal intensity by the area associated with each node to calculate the sound power at each node, and then sum all the nodal power results to calculate the transmitted sound power at the outlet of the duct. The incident sound power at the inlet

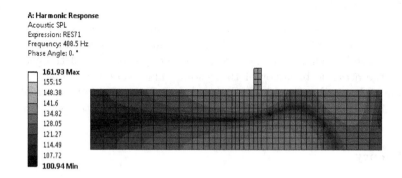

A: Harmonic Response
Acoustic SPL
Expression: RES71
Frequency: 408.5 Hz
Phase Angle: 0. °

161.93 Max
155.15
148.38
141.6
134.82
128.05
121.27
114.49
107.72
100.94 Min

FIGURE 3.42

Case (E): Sound pressure level at 408.5 Hz.

can be calculated using an Acoustic Normal Surface Velocity (Harmonic) excitation at the inlet, as described in Sections 2.8.2.2.

The sound field in Figure 3.42 is complicated, and the learning outcome from this analysis is that it would be difficult to model this relatively simple system analytically, whereas it is relatively easily accomplished using ANSYS.

Figure 3.43 shows the sound pressure level in the duct at 408.5 Hz with the same mesh, only the quarter-wavelength tube has been removed. The results show that sound pressure level is constant throughout the duct at 137.27 dB and the wave field is regular. This result highlights that the addition of the quarter-wavelength tube causes a complicated wave field to be generated.

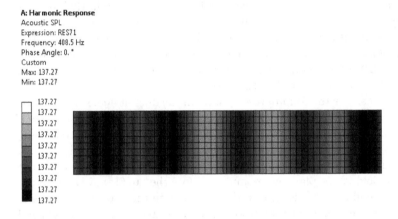

FIGURE 3.43
Case (F): Sound pressure level at 408.5 Hz.

3.6 Gas Temperature Variations

There are many practical applications where the temperature of a gas varies along the length of a duct, such as an exhaust system. As the temperature of gas changes there is a change in the speed of sound and density of the gas, that affects the acoustic behavior of the system. This section explores this phenomenon and describes the relevant 4-pole transmission matrix theory and how to use ANSYS to model a duct where the gas has a linear temperature gradient.

3.6.1 Theory

Sujith [142] derived the four-pole transmission matrix for a duct with linear and exponential temperature gradients, based the work from a previous paper [143]. The equations for the four-pole transmission matrix presented in the paper [142] are incorrect and have been corrected here. Further details can be found in Section D.1.3, and Howard [74, 75].

Figure 3.44 shows a schematic of a linear temperature distribution in a circular duct of radius a and length L. The ends of the duct have acoustic particle velocities u_1 and u_2, and gas temperatures T_1 and T_2 at axial locations $z_1 = L$ and $z_2 = 0$, respectively. The locations of the inlet and outlet have been defined in this way to be consistent with Sujith [142] and Howard [74].

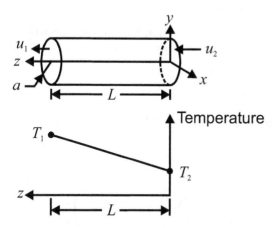

FIGURE 3.44
Schematic of a duct segment with a linear temperature gradient.

The linear temperature distribution in the duct is given by

$$T(z) = T_2 + mz,$$ (3.65)

where m is the gradient of the temperature distribution given by

$$m = \frac{T_1 - T_2}{L}.$$ (3.66)

The temperature-dependent speed of sound and density of the gas can be calculated as [47, Eq. (1.8), p. 17–18]

$$c = \sqrt{\gamma RT/M}$$ (3.67)

$$\rho = \frac{MP_{\text{static}}}{RT},$$ (3.68)

where γ is the ratio of specific heats, $R = 8.314$ J.mol^{-1}°K^{-1} universal gas constant, T temperature in Kelvin, $M = 0.029$ kg.mol^{-1} molecular weight of air, and assuming the gas in the duct is not pressurized, $P_{\text{static}} = 101325$ Pa is atmospheric pressure.

The definition for the four-pole transmission matrix with a temperature gradient differs from Equation (3.2), as the density and speed of sound of the gas changes with temperature and position along the duct. The pressure and acoustic particle velocities at the ends of the duct are related by the four-pole transmission matrix as

$$\begin{bmatrix} p_2 \\ u_2 \end{bmatrix} = \begin{bmatrix} T_{11} & T_{12} \\ T_{21} & T_{22} \end{bmatrix} \begin{bmatrix} p_1 \\ u_1 \end{bmatrix} \tag{3.69}$$

$$\begin{bmatrix} p_2 \\ u_2 \end{bmatrix} = \mathbf{T} \begin{bmatrix} p_1 \\ u_1 \end{bmatrix} , \tag{3.70}$$

where p_i is the acoustic pressure at ends of the duct, u_i is the acoustic particle velocity at ends of the duct, and the four-pole transmission matrix is

$$\mathbf{T} = \begin{bmatrix} T_{11} & T_{12} \\ T_{21} & T_{22} \end{bmatrix} , \tag{3.71}$$

where the elements of the transmission matrix are [74]

$$T_{11} = \left[\frac{\pi \omega \sqrt{T_1}}{2\nu} \right] \\ \times \left[J_1 \left(\frac{\omega \sqrt{T_1}}{\nu} \right) Y_0 \left(\frac{\omega \sqrt{T_2}}{\nu} \right) - J_0 \left(\frac{\omega \sqrt{T_2}}{\nu} \right) Y_1 \left(\frac{\omega \sqrt{T_1}}{\nu} \right) \right] , \tag{3.72}$$

$$T_{12} = 1j \times \left[\frac{\pi \omega \sqrt{T_1}}{2\nu} \right] \times \left[\frac{|m|}{m} \right] \times \left[\rho_1 \sqrt{\gamma R_s T_1} \right] \\ \times \left[J_0 \left(\frac{\omega \sqrt{T_2}}{\nu} \right) Y_0 \left(\frac{\omega \sqrt{T_1}}{\nu} \right) - J_0 \left(\frac{\omega \sqrt{T_1}}{\nu} \right) Y_0 \left(\frac{\omega \sqrt{T_2}}{\nu} \right) \right] , \tag{3.73}$$

$$T_{21} = 1j \times \left[\frac{\pi \omega \sqrt{T_1}}{2\nu} \right] \times \left[\frac{m}{|m|} \right] \times \left[\frac{1}{\rho_2 \sqrt{\gamma R_s T_2}} \right] \\ \times \left[J_1 \left(\frac{\omega \sqrt{T_2}}{\nu} \right) Y_1 \left(\frac{\omega \sqrt{T_1}}{\nu} \right) - J_1 \left(\frac{\omega \sqrt{T_1}}{\nu} \right) Y_1 \left(\frac{\omega \sqrt{T_2}}{\nu} \right) \right] , \tag{3.74}$$

$$T_{22} = \left[\frac{\pi \omega \sqrt{T_1}}{2\nu} \right] \times \left[\frac{\rho_1 \sqrt{\gamma R_s T_1}}{\rho_2 \sqrt{\gamma R_s T_2}} \right] \\ \times \left[J_1 \left(\frac{\omega \sqrt{T_2}}{\nu} \right) Y_0 \left(\frac{\omega \sqrt{T_1}}{\nu} \right) - J_0 \left(\frac{\omega \sqrt{T_1}}{\nu} \right) Y_1 \left(\frac{\omega \sqrt{T_2}}{\nu} \right) \right] . \tag{3.75}$$

The symbols used in these equations are defined in Table 3.13. The constant ν is defined as

$$\nu = \frac{|m|}{2} \sqrt{\gamma R_s} , \tag{3.76}$$

and the specific gas constant R_s is defined as

$$R_s = R/M . \tag{3.77}$$

TABLE 3.13

Symbols Used for the Four-Pole Transmission Matrix of a
Duct with a Linear Temperature Gradient

Symbol	Description
a	Radius of the duct
c_0	Speed of sound at ambient temperature
j	Unit imaginary number $= \sqrt{-1}$
J_n	Bessel function of the nth order
k	Wavenumber
L	Length of the duct
m	Linear temperature gradient
p_1, p_2	Pressure at the ends of the duct
P_{static}	Static pressure in the duct
R	Universal gas constant
R_s	Specific gas constant
S	Cross-sectional area of the duct
T	Temperature of the fluid
T_1, T_2	Temperatures of fluid at the ends of the duct
u_1, u_2	Particle velocities at the ends of the duct
Y_n	Neumann function of the nth order
z	Axial coordinate along the duct
ρ_1, ρ_2	Density of fluid at ends of duct
ω	Angular frequency
ν	Constant defined in Equation (3.76)
γ	Ratio of specific heats (C_P/C_V)

Note that if one were to define a constant temperature profile in the duct, such that $T_1 = T_2$ and $m = 0$, one would expect that this four-pole matrix would equate to the expressions in Equation (3.4). However, the terms $(|m|/m)$ and $(m/|m|)$ in Equations (3.73) and (3.74) equate to 0/0, and Equation (3.76) equates to zero, which causes numerical difficulties.

An example is used to demonstrate the use of the theory and conduct an ANSYS Workbench analysis, where a circular duct has a piston at one end, a rigid termination at the other end, and the gas has a linear temperature gradient. The parameters used in this example are listed in Table 3.14.

3.6.2 MATLAB

The MATLAB code `temp_gradient_spl_along_duct_4pole_sujith.m` included with this book can be used to calculate the sound pressure and acoustic particle velocity in a duct with a temperature gradient, using the four-pole transmission matrix method described in Section 3.6.1.

As described in Section 3.6.1, numerical difficulties occur with Equations (3.73) and (3.74) when attempting to analyze a system where there is a constant temperature profile in the duct, such that $T_1 = T_2$ and $m = 0$.

TABLE 3.14

Parameters Used in the Analysis of a Piston–Rigid Circular Duct with
a Linear Temperature Gradient

Description	Parameter	Value	Units
Radius of duct	a	0.05	m
Length of duct	L	3.0	m
Velocity at rigid end	u_1	0.0	m/s
Velocity of piston	u_2	1.0	m/s
Temperature at rigid end	T_1	673	K
Temperature at piston end	T_2	293	K
Excitation frequency	f	200	Hz
Ratio of specific heats	γ	1.4	
Universal gas constant	R	8.3144621	J.mol^{-1} K^{-1}
Molar mass of air	M	0.029	kg.mol^{-1}
Atmospheric pressure	P	101325	Pa

The work-around is to approximate a constant temperature profile by defining
one end of the duct to have a small temperature offset, say 0.1°C. It can be
shown that this will generate nearly identical results to the predictions using
the MATLAB script spl_along_duct_4pole.m described in Section 3.3.5.

Another software package called DELTAEC [109], which is intended for
the analysis of thermoacoustic systems, can be used to predict the sound level
inside ducts with temperature gradients.

3.6.3   ANSYS Workbench

Introduction

This section describes the instructions to create an ANSYS Workbench model
of a duct with a piston at one end and a rigid termination at the other. Three
analyses are conducted where the temperature of the gas inside the duct

- is at ambient temperature of 22°C,

- is at an elevated temperature of 400°C, and

- has a linear temperature distribution, where it is 400°C at the rigid end and
  20°C at the piston end.

A harmonic analysis at 200 Hz is conducted to calculate the sound pressure
levels, real and imaginary acoustic pressures, and real and imaginary acoustic
particle velocities along the duct. The analyses where the gas temperature
is constant could be done by following the instructions in Section 3.3.4 and
changing the speed of sound and density of the gas. However for this example,
the ACT Acoustics extension feature of the Temperature body force load, as
listed in Section 2.8.3.3, will be used to alter the acoustic properties of the
gas in the duct to illustrate the process.

It is more complicated to conduct an analysis where the gas has a temperature gradient along the duct. Instructions are provided to conduct this analysis in ANSYS Workbench where the following steps are performed:

- A static thermal analysis is conducted to determine the temperature profile of the gas in the duct. Although it is not necessary to conduct a thermal analysis for this problem as the temperature profile is known and could be directly defined using a command object, the instructions provided here enable the analysis of more complicated thermal problems.

- The temperatures at each node are stored in an array and then exported to disk.

- A harmonic acoustic analysis is set up. There are three analyses with three different gas temperature profiles that will be analyzed:

  1. Ambient temperature at 22°C: the default temperature for an acoustic body is 22°C, and no special conditions will be applied.
  2. Elevated temperature at 400°C: the temperature of the gas will be elevated by using the Loads | Temperature object from the ACT Acoustics extension menu bar.
  3. Linear temperature gradient across the duct from 20°C to 400°C: the nodal temperatures calculated from a static thermal analysis will be imported and applied as nodal body force loads to the nodes of the acoustic elements.

- The sound pressure levels, real and imaginary acoustic pressures, and real and imaginary acoustic particle velocities along the axis of the duct are calculated.

The results calculated using ANSYS Workbench are compared with theoretical predictions. The completed ANSYS Workbench archive file temp_grad_duct.wbpz, which contains the .wbpj project file, is available with this book.

The learning outcomes from this example are:

- demonstrate the use of the Temperature boundary condition from the ACT Acoustics extension,

- demonstrate that the wavelength of sound changes as the temperature changes, and

- demonstrate how a thermal analysis can be used to determine the temperature profile of the acoustic medium, and the temperature results can be transferred to the properties of the acoustic fluid.

Instructions

- Start ANSYS Workbench.

- In the `Toolbox` window, double-click on `Analysis Systems | Steady-State Thermal`.

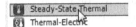

- The ANSYS Workbench model of the duct with the piston and rigid end previously generated in Section 3.3.4, called `driven_duct_pres_dist.wbpj`, will be used as a starting point for this example. In the `Project Schematic` window, right-click on the row 3 `Geometry` and select `Import Geometry | Browse`. Using the file explorer dialog box, change the path as required to select the DESIGNMODELER geometry file `xxxx\driven_duct_pres_dist_files\dp0\SYS-1\DM\SYS-1.agdb`. Click the `Open` button. If the operation was successful, there will be a green tick in the row 3 `Geometry`.

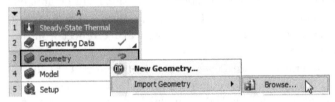

- Make sure that the ACT Acoustics extension is loaded by clicking on `Extensions | Manage Extensions` from the `Project Schematic` window.

- Make sure there is a tick in the column `Load` for the row `ExtAcoustics`. Click on the `Close` button when completed.

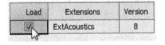

- Click on `File | Save` and type the filename `temp_grad_duct.wbpj`.

- Double-click on row 4 `Model` to start ANSYS Mechanical.

- The geometry of the duct should be shown in the `Graphics` window in ANSYS Mechanical.

- Left-click on `Model (A4) | Part`. In the window `Details of "Part"`, notice that the row `Definition | Assignment` is currently set to `Structural Steel` and needs to be changed to `Air`.

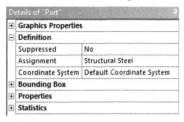

- Click in the cell for `Structural Steel` and a small triangle will appear. Click on the triangle and then click on `New Material`. A dialog box will appear with a message to remind you to click on the `Refresh Project` button after defining the new material. Press the `OK` button.

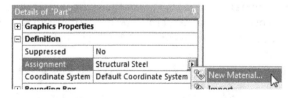

- In the window `Outline of Schematic A2: Engineering Data`, right-click in the `Click here to add a new material` row, and left-click on `Engineering Data Sources`.

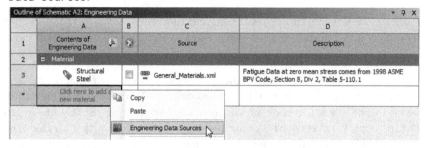

- In the window `Outline of General Materials`, scroll to the top of the table and left-click on the plus sign in column B to add the material property to

the engineering data. You should notice that an icon of a book appears in
the column next to the plus sign.

	A	B	C	D	E	
1	Contents of General Materials		Add	Source	Description	
2	⊟ Material					
3	🏷 Air			📄 General_Materials.xml	General properties for air.	
4	🏷 Aluminum Alloy			📄 General_Materials.xml	General aluminum alloy. Fatigue properties come from MIL-HDBK-5H, page 3-277.	
5	🏷 Concrete			📄 General_Materials.xml		

- In the window Toolbox, scroll down until you can see the region for Thermal
  and click on the plus sign to expand the list of material properties.

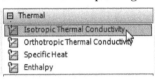

- The next step is to define the thermal conductivity of air at the initial
  temperature of the analysis, which defaults to 22°C (which can be found
  in the ANSYS Mechanical Outline window by clicking on Steady State
  Thermal (A5) | Initial Temperature, but don't do this now). The thermal
  conductivity of air at 22°C is approximately 0.0257W/m K [71, p. 643].
  Right-click on Isotropic Thermal Conductivity and left-click on Include
  Property. In the window Properties of Outline Row 3: Air in the row 2
  Isotropic Thermal Conductivity, type the value 0.0257 into column B for
  Value.

	A	B	C	D	E
1	Property	Value	Unit		
2	🖹 Isotropic Thermal Conductivity	0.0257	W m^-1 C^-1		
3	🖹 Isotropic Relative Permeability	1			

- Click on the Return to Project icon, and then click on the Refresh Project
  icon, which will update Engineering Data in the ANSYS Mechanical model.
- Click in the ANSYS Mechanical window.
- The next step is to change the material property of the parts from
  Structural Steel to Air. Click on Model (A4) | Part. In the window
  Details of "Part", click in the cell for Structural Steel in the row
  Definition | Assignment. A small triangle will appear. Click on this trian-
  gle, and then click on Air, which will change the material properties of the
  parts to air.

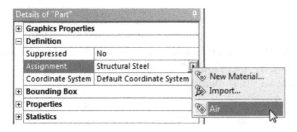

- The next step is to define the size of the mesh. Although the default settings for the mesh size are adequate for both the thermal and the acoustic analyses, we will explicitly specify them. In the Outline window, right-click on Model (A4) | Mesh and left-click on Insert | Sizing. In the toolbar, change the filter selection type to Edges. Change the selection mode to Box Select. Click on the lower right side of the cylinder and with the left mouse button still held down, move the mouse cursor so that it is over the upper left side of the cylinder, so that the 5 axial lines are selected. You should notice that a box is drawn over the cylinder where there are lines drawn through the middle of each side of the selection rectangle. This indicates that all edges within and those that cross the selection rectangle will be selected. In the status line at the bottom of the screen, it should indicate 5 Edges Selected: Length = 15. m. In the window Details of "Sizing" – Sizing, click in the cell next to Scope | Geometry and click on the Apply button.

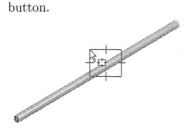

- Change the row Definition | Type to Number of Divisions. Change the row Number of Divisions to 30. Change the row Behavior to Hard.

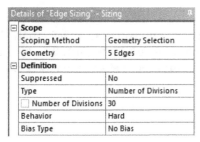

- Click on File | Save Project.

- Right-click on Model (A4) | Mesh and left-click on Generate Mesh. For this Steady-State Thermal analysis, the element type used by default in ANSYS Workbench is SOLID90, which is a 3D 20-node thermal solid element that has mid-side nodes by default. The SOLID90 element is the higher-order version of the 3-D 8-node thermal element SOLID70.

- The next step is define the temperature boundary conditions at each end of the duct. Left-click on Steady-State Thermal (A5). The Environment toolbar at the top of the screen should show options relevant for thermal analyses. Click on the icon for Temperature.

- In the window for Details of "Temperature", change the row Scope | Scoping Method to Named Selection. Change the row Named Selection to NS_inlet. In the row Definition | Magnitude, enter the value 20.

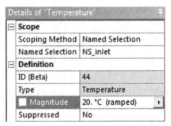

- Repeat this process to define the temperature at NS_outlet as 400.

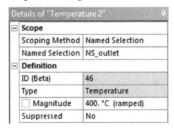

- The next step is to select the results that we want to display, which is only the temperature. In the window Outline, click on Solution (A6). The toolbar at the top of the screen will change to Solution. Left-click on Thermal | Temperature.

- The next step involves creating an array to store the nodal temperatures and then exporting the array to disk. This stored array will be read during the second phase of the analysis where the harmonic analysis of the acoustic model is conducted. Insert a command object by right-clicking on Solution (A6) and left-clicking on Insert | Commands. Type the following commands into the window Commands. These commands will save the nodal temperatures to a file called allparams.txt.

```
 1   SET,FIRST
 2
 3   /COM,--------------------------------------------
 4   /COM,Write all the nodal temperatures to an array
 5   *GET,num_nodes,NODE,0,COUNT
 6   *DIM,n_array,ARRAY,num_nodes,2
 7
 8   ALLS                        ! select all everything
 9   n_array(1,1)=NDNEXT(0)       ! first node number
10   ! Get the temperature at the first node
11   *GET, n_temp, NODE, n_array(1,1), TEMP
12   ! insert temperature at first node into first element of array
13   n_array(1,2)=n_temp
14
15   ! Put the remaining nodal temperatures into the array.
16   *DO,nn,2,num_nodes
17       n_array(nn,1)=NDNEXT(n_array(nn-1,1))
18       *GET, n_temp, NODE, n_array(nn,1), TEMP
19       n_array(nn,2)=n_temp
20   *ENDDO
21
22   ! Save all the parameters to disk
23   PARSAV,ALL,allparams,txt
```

- Click on File | Save Project.

That completes the setup of the thermal analysis. Click on the Solve icon. Once the analysis has been completed there should be a green tick next to Temperature under the Solution (A6) tree. Click on this object to show the temperature profile in the duct.

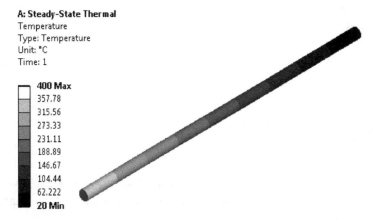

A: Steady-State Thermal
Temperature
Type: Temperature
Unit: °C
Time: 1

400 Max
357.78
315.56
273.33
231.11
188.89
146.67
104.44
62.222
20 Min

The command object that was inserted under Solution (A6) created an array containing the nodal temperatures and stored them to a file called allparams.txt. Use the windows file explorer to find the file in the directory xxxx\temp_grad_duct_files\dp0\SYS\MECH\allparams.txt. Alternatively, from the Project window, click on View | Files, and scroll down until you find the file allparams.txt. Right-click in the cell containing allparams.txt and left-click on Open Containing Folder.

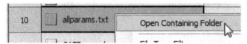

You can use a text editor to inspect the contents of the file allparams.txt. Later, this file will be copied into another directory used in the harmonic analysis.

The next stage involves setting up a harmonic analysis to calculate the acoustic results. As mentioned in the introduction, this analysis will be set up to investigate the acoustic response in the duct for three gas temperature profiles: ambient temperature, a constant (elevated) temperature profile, and a linear gradient temperature profile.

- In the Workbench window, click and hold the left mouse button on Toolbox | Harmonic Response. With the mouse button still held down, drag it into the window Project Schematic, on top of Steady-State Thermal, row 4 Model and then release the mouse button.

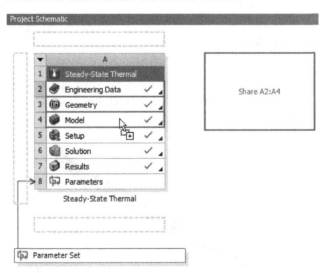

- This will create a new Harmonic Response object and will transfer the Engineering Data, Geometry, and Model data from the Steady-State Thermal analysis.

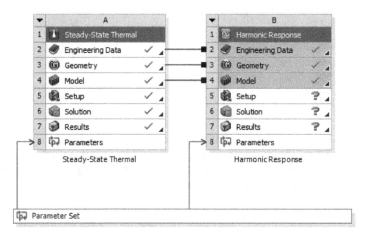

- The next step is to define the duct as `Acoustic Bodies`. Left-click on `Harmonic Response (A5)`. In the ACT Acoustics extension toolbar, select `Acoustic Body | Acoustic Body`. In the window `Details of "Acoustic Body"`, click in the cell next to `Scope | Geometry`. Right-click in the `Geometry` window and left-click on `Select All`. Click the `Apply` button.

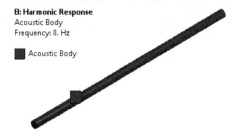

- The next step is to define the acoustic particle velocity of the nodes at the inlet to the duct. In the ACT Acoustics extension, left-click on `Excitation | Normal Surface Velocity (Harmonic)`.

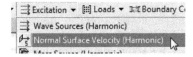

- In the window `Details of "Acoustic Normal Surface Velocity"`, change the row `Scope | Scoping Method` to `Named Selection`. Change the row `Named Selection` to `NS_inlet`. In the row `Definition | Amplitude of Normal Velocity` type the value 1.

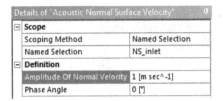

- The next step involves defining a constant temperature "load" on the nodes. The ACT Acoustics extension (at Release 8) is only able to apply a constant temperature "load" to an entire body. In the ACT Acoustics extension toolbar, left-click on Loads | Temperature.

- In the window Details of "Acoustic Temperature", left-click on the cell next to Scope | Geometry. In the Geometry window, right-click to open the context menu and then left-click on Select All. The status window at the bottom of the screen should show 4 Bodies Selected. Left-click on the Apply button. In the row Definition | Temperature, type the value 400.

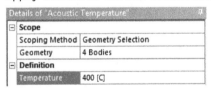

- For the moment, we will Suppress this temperature load so that the acoustic response at ambient temperature is calculated. Right-click the object Harmonic Response (B5) | Acoustic Temperature and then left-click on Suppress.

- The next step involves reading the array of the stored nodal temperatures and setting the temperatures on each node of each FLUID220 element as nodal body force loads. Insert a command object by right-clicking on Harmonic Response (B5) and left-clicking on Insert | Commands. Type the following commands into the window Commands.

```
1  /COM,Read in the parameters and array of nodal temps from disk
2  PARRES,CHANGE,allparams,txt
3
4  ! Apply the thermal gradient to the nodes
5  *DO,nn,1,num_nodes
6      BF,n_array(nn,1),TEMP,n_array(nn,2)
7  *ENDDO
```

- Right-click on the object Harmonic Response (B5) | Commands (APDL) and left-click on Rename. Change the name to Commands (APDL) - temp grad.

- For the moment, we will Suppress this temperature load so that the acoustic response at ambient temperature is calculated. Right-click the object Harmonic Response (B5) | Commands (APDL) - temp grad and then left-click on Suppress.

- The settings for the harmonic analysis will be defined. Click on the branch Harmonic Response (B5) | Analysis Settings. In the window Details of "Analysis Settings", change the rows Range Minimum to 199, Range Maximum to 200, Solution Intervals to 1, and Solution Method to Full. In the branch Output Controls, change the row General Miscellaneous to Yes. In the tree Analysis Data Management, change the row Save MAPDL db to Yes, as the database file is needed for post-processing of the acoustic results.

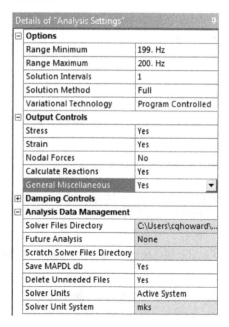

Details of "Analysis Settings"	⋥
⊟ **Options**	
Range Minimum	199. Hz
Range Maximum	200. Hz
Solution Intervals	1
Solution Method	Full
Variational Technology	Program Controlled
⊟ **Output Controls**	
Stress	Yes
Strain	Yes
Nodal Forces	No
Calculate Reactions	Yes
General Miscellaneous	Yes ▼
⊞ **Damping Controls**	
⊟ **Analysis Data Management**	
Solver Files Directory	C:\Users\cqhoward\...
Future Analysis	None
Scratch Solver Files Directory	
Save MAPDL db	Yes
Delete Unneeded Files	Yes
Solver Units	Active System
Solver Unit System	mks

- That completes the set up of the objects under Harmonic Response (B5). The next steps are to set up the acoustic results that will be displayed. Left-click on Solution (B6).

- In the ACT Acoustics extension toolbar, click on Results | Acoustic SPL, which will show the sound pressure level in the duct.

- The next steps involve inserting results objects for the sound pressure level, acoustic pressure, and acoustic particle velocity along the axis of the duct, and these results will be exported. Insert another object Results | Acoustic

SPL, and in the window Details of "Acoustic SPL", change the row
Geometry | Scoping Method to Named Selection. Change the row Named
Selection to NS_duct_axis. Right-click on this object and select Rename.
Type the name Acoustic SPL - axis, so that it is differentiated from the
Acoustic SPL result for all the bodies.

- Repeat this process to add an object for Results | Acoustic Pressure and
  change the scoping method to Named Selection and select NS_duct_axis.
  This will calculate the real part of the acoustic pressure. Right-click on the
  object Solution (B6) | Acoustic Pressure and left-click on Rename. Type
  the name Acoustic Pressure - real.

- It is also necessary to calculate the imaginary part of the acoustic pressure.
  Right-click on the object Solution (B6) | Acoustic Pressure - real and
  then left-click on Duplicate Without Results.

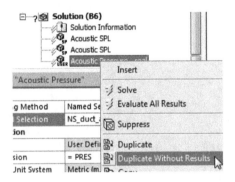

- Rename the object to Acoustic Pressure - imag. In the window Details
  of "Acoustic Pressure - imag", change the row Definition | Phase
  Angle to -90.

- Repeat this process to insert objects for the real and imaginary parts of the
  Results | Acoustic Velocity Z.

- The completed Outline window should look like the following figure.

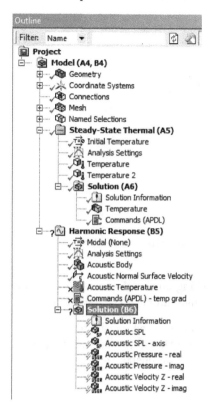

- Click on File | Save Project.

At this point one should be able to solve the model. However, an error would be generated warning that Air contains invalid property data. The work-around for this is to change the material property for the geometry back to Structural Steel.

- In the Outline window, click on Model (A4,B4) | Geometry | Part. In the window Details of "Part", change the row Definition | Assignment from Air to Structural Steel.

- Click on File | Save Project.

- Right-click on Solution (B6) and left-click on Solve.

- Once the computations have completed, click on the object Solution (B6) | Acoustic SPL to show the sound pressure level in the duct at ambient temperature.

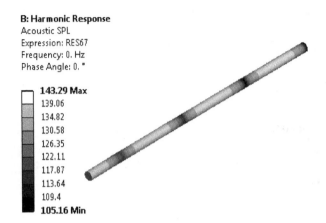

B: Harmonic Response
Acoustic SPL
Expression: RES67
Frequency: 0. Hz
Phase Angle: 0. °

143.29 Max
139.06
134.82
130.58
126.35
122.11
117.87
113.64
109.4
105.16 Min

- Click on the other acoustic results under the `Solution (B6)` branch to confirm that the results were calculated.

- The results can be exported by right-clicking on the object and left-clicking on `Export`. Type in an appropriate filename and click the `Save` button.

- The next step is to conduct the harmonic analysis where the temperature of the gas in the duct is elevated. Right-click on the object `Harmonic Response (B5) | Acoustic Temperature` and left-click on `Unsuppress`.

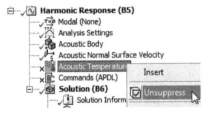

- Click on `File | Save Project`.

- Right-click on `Solution (B6)` and left-click on `Solve`.

- The results can be exported by right-clicking on the object and left-clicking on `Export`. Type in an appropriate filename and click the `Save` button.

- The next step is to calculate the acoustic response in the duct where there is a linear temperature gradient of the gas. Right-click on the object `Harmonic Response (B5) | Acoustic Temperature` and left-click on `Suppress`.

- Right-click on the object `Harmonic Response (B5) | Commands (APDL) - temp grad` and left-click on `Unsuppress`.

- Before solving this model, it is necessary to copy the file `allparams.txt`, which contains the nodal temperature data, into the directory where the files for the harmonic response analysis are stored. Use the windows file explorer (or equivalent) to copy the file .\temp_grad_duct_files\dp0\SYS\MECH\allparams.txt to the directory: .\temp_grad_duct_files\dp0\SYS-1\MECH.

- In the ANSYS Mechanical window, click on File | Save Project.

- Right-click on Solution (B6), and left-click on Solve.

- The results can be exported by right-clicking on the object and left-clicking on Export. Type in an appropriate filename and click the Save button.

Comparison of Results

Figure 3.45 shows the comparison of the sound pressure level in the duct calculated using theory, which was implemented in the MATLAB model described in Section 3.6.2, and ANSYS Workbench for the cases where the temperature of the gas in the duct was:

- at an ambient temperature of 22°C,

- at an elevated temperature of 400°C,

- a linear temperature gradient of 400°C at the rigid end and 20°C at the piston end.

The ANSYS Workbench results agree with the theoretical predictions.

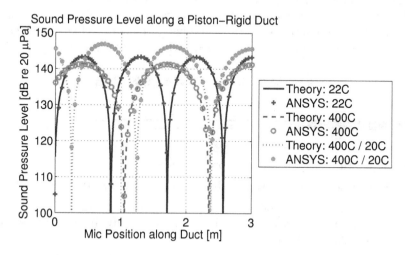

FIGURE 3.45
Sound pressure level versus axial location in a piston–rigid duct at 22°C, at 22°C, and with a linear temperature gradient from 400°C to 20°C.

3.6.4   ANSYS Mechanical APDL

The system shown in Figure 3.44 was modeled using ANSYS Mechanical APDL and the complete script duct_temp_grad.inp is included with this book. The script performs the following steps:

- Creates a solid cylinder and initially meshes the solid body with FLUID30 3D acoustic elements, where the displacement DOFs have been turned off. The following figure shows the finite element mesh of the FLUID30 elements.

- The 8-node FLUID30 elements are swapped for 8-node SOLID70 thermal solid elements.
- The temperatures at each end of the duct are defined as boundary conditions.
- A static thermal analysis is conducted to calculate the temperature at each node. The result will be a temperature profile with a linear distribution. The following figure shows the temperature profile (in units of Kelvin) of the gas in the duct where the model was meshed with SOLID70 elements.

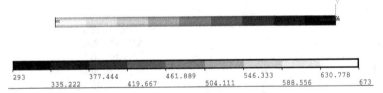

- The temperature at each node is stored in an array.
- The SOLID70 elements are swapped back to the original FLUID30 elements.
- The temperature at each node calculated during the static thermal analysis is retrieved from the array and applied as a nodal body force load to each node of the FLUID30 elements.
- The nodes at the piston end of the duct are defined to have a velocity.
- A harmonic analysis is conducted at a single frequency.
- A path is defined on the axis of the cylinder, starting at the piston and finishing at the rigid end. The acoustic pressure, acoustic particle velocity,

and sound pressure level are calculated along the path and the results are stored in a binary file.

- The temperature profile that was applied to the nodes is deleted.

- The harmonic analysis is repeated.

- The results from the analysis without the temperature gradient are retrieved.

- The acoustic pressure, acoustic particle velocity, and sound pressure level are calculated along the path again.

- The sound pressure level results are plotted for the analyses with and without the temperature gradient.

Figure 3.46 shows the graph that is generated using ANSYS Mechanical APDL of the sound pressure levels in the duct with the temperature gradient (thick line) and with a constant temperature of 22°C (thin line).

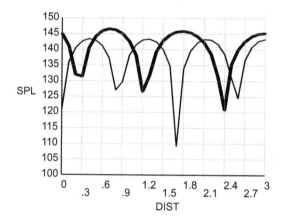

FIGURE 3.46
Graph of sound pressure level at 200 Hz generated by ANSYS Mechanical APDL of a piston–rigid duct with a linear temperature gradient (thick line) and with a constant temperature of 22°C (thin line).

In order to calculate the temperature distribution in the duct by conducting a static thermal analysis, it is necessary to define the thermal conductivity of the gas. The APDL code that is used to define the thermal conductivity is MP,KXX,matid,value. Note that an issue occurs when using ANSYS Release 14.5, that if the thermal conductivity of the gas is defined when conducting a harmonic analysis using acoustic elements, then the acoustic particle velocity is not calculated. This issue has been fixed in subsequent releases. See Section D.2.1.2 for more details about this issue.

The ANSYS Mechanical APDL script was used to calculate the sound pressure level, real and imaginary components of the acoustic pressure, and the real and imaginary components of the acoustic particle velocity. These results were compared with the theoretical predictions calculated using the MATLAB code described in Section 3.6.2.

Figure 3.47 shows the sound pressure level along the axis of the duct $(x = 0, y = 0, z = 0 \cdots L)$ with a linear temperature gradient calculated theoretically and using ANSYS Mechanical APDL. The results show that there is good agreement between the two models. At the rigid end of the duct ($z_1 = 3$ m) where the temperature is $T_1 = 400°C = 673$ K, the corresponding speed of sound at this elevated temperature is higher than at the piston end $z_2 = 0$ where $T_2 = 20°C = 293$ K and hence the wavelengths are longer. It can be seen that the spacing between two acoustic nodes (where the pressure is close to zero) is large at the hot end of the duct ($z_2 = 3$ m), compared to the colder end of the duct ($z_1 = 0$ m).

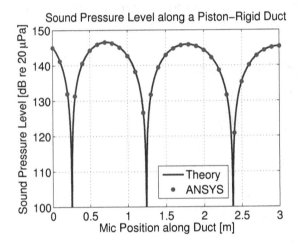

FIGURE 3.47
Sound pressure level inside a piston-rigid duct with a linear temperature gradient calculated theoretically and using ANSYS Mechanical APDL.

Figure 3.48 shows the real and imaginary parts of the sound pressure calculated theoretically and using ANSYS Mechanical APDL. The imaginary part of the acoustic pressure varies along the length of the duct, while the real part of the pressure is zero since the system contains no damping. The results calculated using ANSYS agree with theoretical predictions.

Figure 3.49 shows the real and imaginary parts of the acoustic particle velocity calculated theoretically and using ANSYS Mechanical APDL. The real part of the particle velocity varies along the length of the duct, while the imaginary part is zero. The results calculated using ANSYS agree with theoretical predictions.

Another example of using ANSYS to evaluate the transmission loss of a duct with a quarter-wavelength tube reactive silencer that has linear temperature gradients in each duct segment is shown in Howard [75].

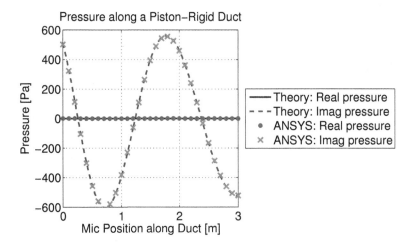

FIGURE 3.48
Real and imaginary parts of the sound pressure inside a piston-rigid duct with a linear temperature gradient calculated theoretically and using ANSYS Mechanical APDL.

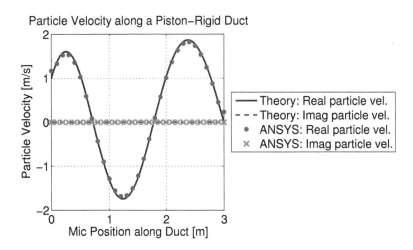

FIGURE 3.49
Real and imaginary parts of the acoustic particle velocity inside a piston-rigid duct with a linear temperature gradient calculated theoretically and using ANSYS Mechanical APDL.

4

Sound Inside a Rigid-Walled Cavity

4.1 Learning Outcomes

The learning outcomes from this chapter are:

- ability to calculate the undamped natural frequencies of a rigid-walled cavity using MATLAB and ANSYS,

- ability to use the FLUID30 fluid element in ANSYS to model a rigid-walled cavity,

- recognize that a rigid-wall in ANSYS is obtained at the outer boundary of the fluid elements with only pressure degrees of freedom active,

- application of a volume velocity acoustic source in ANSYS,

- ability to conduct a harmonic response analysis in ANSYS, and

- ability to calculate the modal forcing vector for an acoustic source.

4.2 Description of the System

The system under investigation is a rectangular acoustic cavity that is bounded by rigid-walls, as shown in Figure 4.1. A sound source is placed within the cavity and the sound pressure is measured at a receiver within the cavity.

4.3 Theory

The theory that is described in the following sections includes:

- natural frequencies of a rigid-wall rectangular cavity,

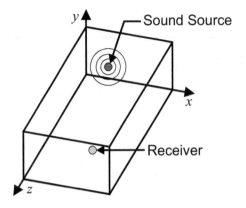

FIGURE 4.1
Rigid-walled rectangular cavity with an acoustic volume velocity source and
a receiver to measure the resulting sound pressure.

- mode shapes of a rigid-wall rectangular cavity, and

- sound pressure level at a receiver inside a rigid-wall rectangular cavity due
  to an acoustic volume velocity source, using the modal summation method.

4.3.1   Natural Frequencies and Mode Shapes

The natural frequencies of a rigid-walled rectangular cavity can be expressed
in terms of the cavity dimensions L_x, L_y, L_z as [46, Eq. 6.13]

$$f_n = \frac{c_0}{2} \sqrt{\left[\frac{n_x}{L_x}\right]^2 + \left[\frac{n_y}{L_y}\right]^2 + \left[\frac{n_z}{L_z}\right]^2} \qquad \text{(Hz)}, \qquad (4.1)$$

where c_0 is the sound velocity and n_x, n_y, and n_z denote the modal indices
that have a range from 0 to infinity (i.e. $n_x, n_y, n_z = 0, 1, 2, 3, \cdots, \infty$). An
interesting feature of the natural frequencies of a cavity is that they do not
depend on the density of the fluid, only the speed of sound and the dimensions
of the cavity.

The mode shapes of a rigid-walled rectangular cavity ψ_n are given by [46,
Eq. 6.13]

$$\psi_n (x, y, z) = \cos\left(\frac{n_x \pi x}{L_x}\right) \cos\left(\frac{n_y \pi y}{L_y}\right) \cos\left(\frac{n_z \pi z}{L_z}\right). \qquad (4.2)$$

When conducting a harmonic response analysis using modal superposition,
only a finite number of modes are used in the summation, which will be further
described in the following section. The number of modes should be sufficient
to accurately cover the frequency range of interest. Once the mode shapes are
calculated, they should be sorted into order of increasing frequency and only

the first N_a (number of acoustic) modes are retained for conducting further analyses using modal superposition or structural-acoustic modal-coupling.

4.3.2 Harmonic Response

The sound pressure inside a cavity due to a sound source can be calculated using a modal summation method. The general process is as follows:

- calculate the natural frequencies,

- sort the natural frequencies (and the modal indices) into increasing values,

- retain only the first selected modes up to a frequency that is at least double the frequency range of interest,

- calculate the value of the mode shape at the sound source location(s) and the receiver location(s), for each mode of interest,

- form the matrix equation for the response of the system,

- invert the matrix and calculate the modal participation factors at each frequency to be analyzed, and

- calculate the pressure response at the receiver location(s).

This process is formalized mathematically below.

The sound pressure at any point in the cavity is [46, Eq. 6.14]

$$p(x, y, z) = \sum_n^\infty P_n \psi_n(x, y, z). \tag{4.3}$$

In layperson terms, each acoustic mode ($\psi_n(x, y, z)$) in the cavity contributes a fraction (P_n) toward the total sound pressure at a point in the cavity. See page 8 for a further discussion.

Consider a monopole point sound source located inside the cavity at (x_s, y_s, z_s) with a volume velocity amplitude of Q_s. The sound pressure at location (x, y, z) inside the cavity can be calculated as [46, Eq. 6.28]:

$$p(x, y, z) = \rho_0 c_0^2 Q_s \sum_{n=0}^{N_a} \frac{\omega \, \psi_n(x, y, z) \psi_n(x_s, y_s, z_s)}{V_n(\omega^2 - \omega_n^2)}, \tag{4.4}$$

where V_n is the modal volume and is calculated as

$$V_n = V \epsilon_{nx} \epsilon_{ny} \epsilon_{nz} \quad \text{where } \epsilon_i = \begin{cases} 1 & \text{for } i = 0 \\ \frac{1}{2} & \text{for } i \geq 1 \end{cases}, \tag{4.5}$$

$V = L_x L_y L_z$ is the volume of the rectangular cavity, and $\omega_n = 2\pi f_n$ is the nth natural frequency in radians/second.

TABLE 4.1

Parameters of a Rigid-Walled Cavity

Description	Parameter	Value	Units
Cavity:			
Length X	L_x	0.5	m
Length Y	L_y	0.3	m
Length Z	L_z	1.1	m
Acoustic Source:			
Location X	x_s	0.15	m
Location Y	y_s	0.12	m
Location Z	z_s	0.0	m
Acoustic Receiver:			
Location X	x_b	0.3	m
Location Y	y_b	0.105	m
Location Z	z_b	0.715	m
Fluid:			
Speed of sound	c_0	343	m/s
Density	ρ_0	1.21	kg/m^3
Number of modes	N_a	500	no units

4.4   Example

Consider the rigid-walled rectangular box shown in Figure 4.1 that has parameters as listed in Table 4.1. This example is similar to the example in [47, Chapter 12, p. 646].

The locations of the source and receiver were selected because it is known that the mesh in the finite element model will create nodes at those locations. It is also possible to use *hard keypoint points* in the mesh, which will create a keypoint at a desired location, and then when the solid model is meshed, a node will exist at the location of the keypoint. However, mapped meshing is not supported when hard keypoints are used, so it is not possible to create a regular mesh with brick elements.

4.4.1   MATLAB

The MATLAB script rigid_wall_cavity.m included with this book can be used to calculate the natural frequencies, mode shapes, and the acoustic pressure at a point within the rigid-walled rectangular enclosure.

4.4.2 ANSYS Workbench

This section describes the instructions to create an ANSYS Workbench model of a rigid-walled enclosure and conduct:

- a modal analysis to calculate the natural frequencies and mode shapes, and

- a harmonic analysis to calculate the acoustic pressure response within the cavity using the full method and an acoustic mass source. ANSYS Release 14.5 does not support modal superposition for harmonic response analyses using acoustic elements, but has been implemented in Release 15.0.

The completed ANSYS Workbench archive file `rigid_cav.wbpz`, which contains the `.wbpj` project file, is available with this book.

It is assumed that you have the ACT Acoustics extensions installed and operating correctly. This can be checked in the Workbench project view by selecting the `Extensions | Manage Extensions` menu. You should see the extension `ExtAcoustics` listed in the table and a tick in the `Load` column.

Instructions

- In the `Project` window, from the `Toolbox | Analysis Systems` window on the left-hand side, select a `Modal Analysis` and drag it into the `Project Schematic` window.

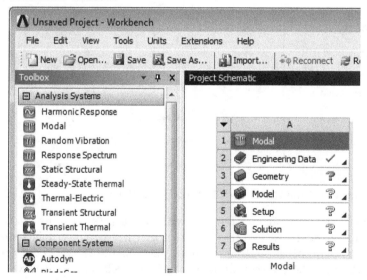

- In the `Project Schematic` window, double-click on the icon 3 `Geometry` to start the `Design Modeler`. In the dialog box that asks `Select desired length unit:` it is recommended that you select `Meter` and keep everything in SI units. Click on the `OK` button.

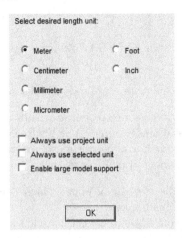

- In the `Tree Outline` window, click on the `XYPlane` icon, and axes will appear in the `Graphics` window.

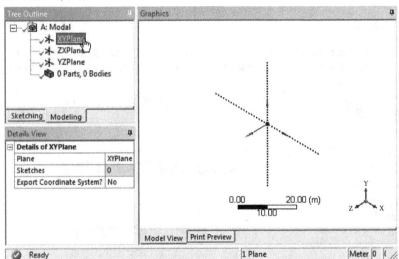

- Click on the `Sketching` tab in the `Sketching Toolboxes`.

- Before we start to create an area for the cavity, the `Auto Constraint Cursor` must be turned on to ensure that the cursor will "snap" to points and edges. In the `Sketching Toolbox` window, click on the `Constraints` tab. To scroll through the `Constraint` menu options, click on the downward-pointing triangle next to the `Setting` tab until the `Auto Constraints` option is visible then left-click on it. Click in the box next to `Cursor:` to activate the generation of automatic constraints.

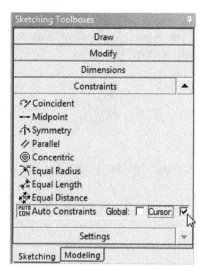

- Click on the Draw tab and then select the Rectangle tool.

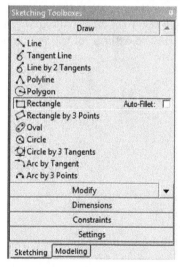

- The cursor will change into a pencil shape. Move the cursor to the center (origin) of the axes and the small red cube at the origin should change in color to red, and the letter P will be shown at the origin. Click the left mouse button on this point to start the creation of the rectangle.

- Move the mouse cursor upward and to the right and click the left mouse button to create a rectangle. The size of the rectangle does not matter at this stage as you will define the dimensions shortly.

- Click on the Dimensions tab in the Sketching Toolboxes window. Select the General dimension option.

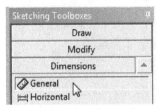

- Hover the mouse cursor over the vertical line of the rectangle that is on the Y-axis until the line color changes to red. Click on the line. An expanding dimension line will appear. Click the mouse on the exterior of the rectangle to place the dimension line.

- Define the horizontal dimension by following the same process.

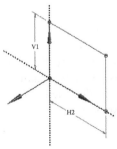

- In the Details View window, which is beneath the Sketching Toolboxes window, you can see the dimensions labeled H2 and V1 with arbitrary lengths.

Details View	
Details of Sketch1	
Sketch	Sketch1
Sketch Visibility	Show Sketch
Show Constraints?	No
Dimensions: 2	
☐ H2	13.618 m
☐ V1	11.336 m
Edges: 4	
Line	Ln7
Line	Ln8
Line	Ln9
Line	Ln10

- Click on the numbers and change the dimensions so that H2 = 0.5 m and V1 = 0.3 m. After changing the value of the dimensions, it is likely that the model will not be clear.

- Resize the model by clicking on the Zoom to Fit icon, which looks like a magnifying glass over a cube. The placement of the dimensions could be improved, although this does not affect the results.

- In the Dimensions tab, click on the Move icon, left-click on the dimension to select it, move the cursor to a new location, and left-click to place the dimension at the new location. Repeat this process as necessary.

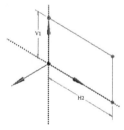

- Click on the Modeling tab in the Tree Outline window. Click on the + sign next to the XYPlane icon so that the Sketch1 icon appears and select it so that the lines of the rectangle change to yellow.

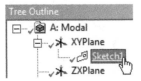

- Click on the Extrude icon. In the window Details of View, click in the number next to FD1, Depth (>0) and enter a dimension of 1.1. Click the Generate icon to create the box.

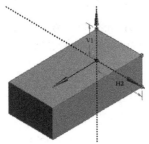

That completes the creation of the solid model. Save the project by clicking on File | Save Project, type an appropriate filename, and click the Save button.

The next step involves the creation of the mesh of the finite element model.

Meshing

- In the Project Schematic window, double-click on row 4 Model to start ANSYS Mechanical.

- The next step is to define the properties of the finite element mesh. This involves defining the element sizes, the method used to perform the meshing, and the element type that will be used. In this step the number of divisions along each edge will be explicitly defined. In the Selection Filter menu bar, change the selection type to Edge by clicking on the cube with the one green edge. In the Outline window, select Mesh and then right-click with the mouse and select Insert | Sizing.

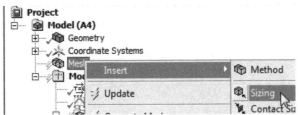

- In the window labeled `Details of "Sizing" - Sizing`, the row labeled `Geometry` is highlighted in yellow. Click the mouse in this row. Move the mouse cursor into the window with the rectangular acoustic body and right-click with the mouse and in the menu that appears click on `Select All`. All the edges of the model will be highlighted in green. Click on the `Apply` button in the `Geometry` row and there should be `12` `Edges` that have been selected. In the row for `Type`, click on `Element Size` and an icon with a triangle will appear on the right side of the row to indicate a drop-down menu. Click on the triangle to open the drop-down menu and select `Number of Divisions`.

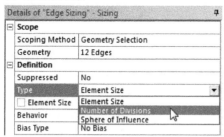

- A new row will appear labeled `Number of Divisions`; change this value to `20`. You should notice that all the edges of the rectangular box have dashed yellow lines with 20 divisions along each line. Change the row labeled `Behavior` from `Soft` to `Hard`, which will force the lines to have 20 divisions and cannot be altered. By leaving the option as `Soft` enables the meshing algorithm to modify the value if required to complete multiple meshing operations.

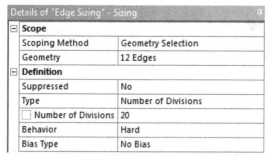

- Change the selection filter type to `Body` and click on the rectangular cavity so that it is highlighted in green. In the `Outline` window, right-click on the `Mesh` and select `Insert | Method` from the menu.

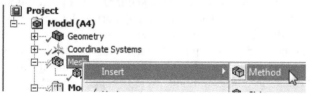

- In the window for Details of "Automatic Method" - Method, change the row labeled Method from Automatic to Sweep. Change the row Element Midside Nodes to Dropped. This will cause ANSYS to use FLUID30 acoustic elements. The analysis could also be done using FLUID220 acoustic elements, by selecting Kept. Change the row Free Face Mesh Type to All Quad.

Details of "Sweep Method" - Method	
Scope	
Scoping Method	Geometry Selection
Geometry	1 Body
Definition	
Suppressed	No
Method	Sweep
Element Midside Nodes	Dropped
Src/Trg Selection	Automatic
Source	Program Controlled
Target	Program Controlled
Free Face Mesh Type	All Quad
Type	Number of Divisions
Sweep Num Divs	Default
Sweep Bias Type	No Bias
Element Option	Solid

- In the mesh toolbar at the top of the window, click on the Update button which will mesh the solid model. Once the meshing has completed, in the Outline window click on the object Mesh to show the finite element model.

- By default the finite element model will use structural elements. The next step is to change the element type to acoustic elements. Change the selection method to Bodies in the Graphics Toolbar, by clicking on the green cube icon. Click on the rectangular cavity so it changes to a green color. In the Acoustics extensions menu bar, click on the Acoustic Body drop-down menu, and then select Acoustic Body. You should notice the appearance of a new branch in the Modal (A5) tree called Acoustic Body. By defining an Acoustic Body causes ANSYS to replace the default structural elements with appropriate acoustic elements. The type of acoustic element that is used depends on whether the mid-side nodes were Dropped or Kept.

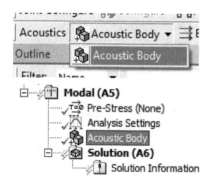

- A new window appears with the title Details of "Acoustic Body", where you can define the material properties for the fluid within the acoustic body. Change the values for Mass Density to 1.21 and Sound Speed to 343.

Details of "Acoustic Body"	
⊟ Scope	
Scoping Method	Geometry Selection
Geometry	1 Body
⊟ Definition	
Temperature Dependency	No
Mass Density	1.21 [kg m^-1 m^-1 m^-1]
Sound Speed	343 [m sec^-1]
Dynamic Viscosity	0 [Pa sec]
Thermal Conductivity	0 [W m^-1 C^-1]
Heat Coefficient Cp	0 [J kg^-1 C^-1]
Heat Coefficient Cv	0 [J kg^-1 C^-1]
Specific Heat C	0 [J kg^-1 C^-1]
Reference Pressure	2E-05 [Pa]
Reference Static Pressure	101325 [Pa]
Acoustic-Structural Coupled Body Options	Uncoupled

- In the Outline window, click on the Analysis Settings in the Modal (A5) branch. Change the Max Modes to Find to 40. Expand the Analysis Data Management tree and change Future Analysis to MSUP Analyses. Some of the options in the rows will change automatically. Note that at Release 14.5 of ANSYS Workbench it is not possible to conduct an acoustic harmonic analysis using the modal summation method and only the full harmonic analysis is supported. Hence this step of selecting MSUP Analyses is not necessary, but it is intended that ANSYS will implement this feature in a later release.

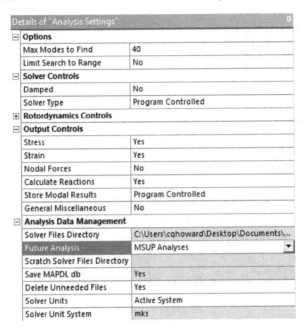

- That completes the setup of the analysis. Save the model by clicking on File
  | Save Project.

- Click the Solve button and wait until the computations complete.

- When the calculations have finished, click on the Solution (A6) branch in
  the Outline tree window. In the lower right corner of the screen should be
  a list of the 40 natural frequencies of the cavity.

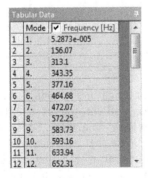

- We will now inspect one of the mode shapes of the room. Click on the
  Solution (A6) branch in the Outline tree window. In the Acoustics bar,
  click on the Results drop-down menu and select Acoustic Pressure.

- A new branch will appear under the Solution (A6) tree labeled Acoustic Pressure. In the window labeled Details of "Acoustic Pressure", change the value in row Mode to 18.

Details of "Acoustic Pressure"	
Scope	
Scoping Method	Geometry Selection
Geometry	All Bodies
Definition	
Type	User Defined Result
Expression	= PRES
Input Unit System	Metric (m, kg, N, s, V, A)
Output Unit	Pressure
Mode	18.
Coordinate System	Global Coordinate System
Identifier	
Suppressed	No
Results	
Minimum	-2418. Pa
Maximum	2418. Pa
Information	

- Right-click on the Acoustic Pressure branch under Solution (A6) and select Evaluate All Results from the menu. The 18th mode will be displayed, which is the $(1, 1, 2)$ mode. ANSYS does not have an in-built mechanism for determining the modal indices of acoustic (or structural) responses and therefore these have to be determined by viewing the response of the system and the user has to recognize the mode shape and appropriate modal index. The modal indices can be determined by counting the number of nodes along each axis where the pressure is zero. For this example, the color legend in ANSYS indicates that a green color is close to zero pressure (if reading this book where the images are grayscale, then you will have to inspect the legend to find the shade of gray between $+268$ Pa and -268 Pa) and there are two nodes along the long edge of the box (z axis) where the pressure is zero, hence the modal index is $n_z = 2$. By using this process of visual inspection you can determine that the modal indices are $n_x = 1, n_y = 1, n_z = 2$.

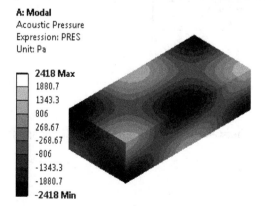

A: Modal
Acoustic Pressure
Expression: PRES
Unit: Pa

2418 Max
1880.7
1343.3
806
268.67
-268.67
-806
-1343.3
-1880.7
-2418 Min

- To export the list of natural frequencies, move the mouse into the Tabular Data window that contains the list of natural frequencies. Right-click with the mouse and select Export. Type in a filename such as ansys_workbench_res_freqs.txt and click the Save button.

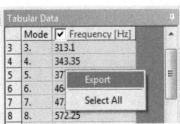

That completes the modal analysis of the rigid-walled cavity. The next step is to conduct a harmonic analysis.

Harmonic Response

Note that at Release 14.5 of ANSYS Workbench it is not possible to conduct an acoustic harmonic analysis using the modal summation method and only the full harmonic analysis is supported. It is possible to conduct a harmonic analysis using the modal summation method using ANSYS Mechanical APDL and is shown in Section 4.4.3.

- Return to the Workbench project window. In the left-hand column under Analysis Systems, click on the Harmonic Response analysis type, and with the mouse button held down, drag it on top of the row 4. Model in the Modal analysis object.

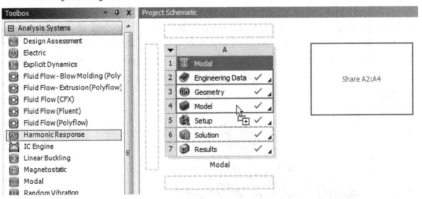

- Connection lines will be drawn between the Modal and Harmonic Response analyses.

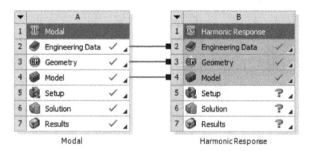

Modal   Harmonic Response

- Double-click on the row 5 `Setup` on the `Harmonic Response` analysis, which will start ANSYS Mechanical.

- Under the `Harmonic Response` (B5) branch, click on the `Analysis Settings` branch. In the window for `Details of "Analysis Settings"`, change the `Range Maximum` to 500 Hz. Change `Solution Intervals` to 500, which will give a frequency spacing of 1 Hz. Change the row `Solution Method` to `Full`. Change the row `Analysis Data Management | Save MAPDL db` to `Yes`, as the database is needed for post-processing of the acoustic results.

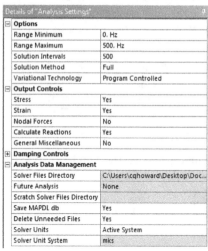

- The next step is to define two `Named Selection` objects that correspond to the nodes where the acoustic source and microphone are located as listed in Table 4.1. In the `Outline` window, right-click on `Model` (A4,B4) and left-click on `Insert | Named Selection`.

- In the window `Details of "Selection"`, change the row `Scoping Method` to `Worksheet`.

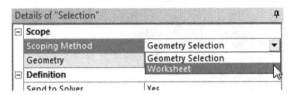

- In the Worksheet window, right-click and select Add Row.

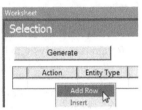

- The Worksheet can be used to select a node by initially selecting a group of nodes and then filtering the set. In this case the initial set of nodes will be along the $x = 0.15$ m, and this set will be filtered to keep only the nodes along the $y = 0.12$ m and lastly filtered to select the node along $z = 0$ m. After the filtering operations there should only be 1 node remaining in the selection set. In the Worksheet window, change the options in each of the cells as per the following table.

	Action	Entity Type	Criterion	Operator	Units	Value
☑	Add	Mesh Node	Location X	Equal	m	0.15
☑	Filter	Mesh Node	Location Y	Equal	m	0.12
☑	Filter	Mesh Node	Location Z	Equal	m	0

- When finished, click the Generate button and in the window Details of "Selection" the row Scope | Geometry should indicate 1 Node.

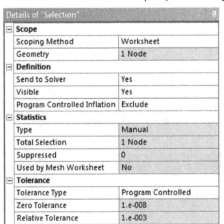

- Rename this named selection by right-clicking on `Named Selections |
  Selection` and left-click on `Rename` and type `nodesource`.

- Repeat these steps to define a named selection for the node at the location
  of the microphone at $(0.3, 0.105, 0.715)$, and rename the named selection as
  `nodereceiver`.

	Action	Entity Type	Criterion	Operator	Units	Value
☑	Add	Mesh Node	Location X	Equal	m	0.3
☑	Filter	Mesh Node	Location Y	Equal	m	0.105
☑	Filter	Mesh Node	Location Z	Equal	m	0.715

- Once the two named selections have been defined, right-click on `Model
  (A4,B4) | Named Selections` to open a menu, and left-click on `Generate
  Named Selections`.

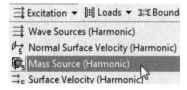

- Now that named selections have been defined for the nodes for the acoustic
  source and microphone, the acoustic source can be defined and the acoustic pressure can be calculated at the microphone. First, click on `Harmonic
  Response (B5)`, then in the ACT Acoustics extension toolbar click on
  `Excitation | Mass Source (Harmonic)`.

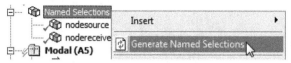

- In the window `Details of "Acoustic Mass Source"` change `Scope |
  Scoping Method` to `Named Selection`, `Scope | Named Selection` to
  `nodesource`, and `Definition | Amplitude of Mass Source` to 1.

Details of "Acoustic Mass Source"	
Scope	
Scoping Method	Named Selection
Named Selection	nodesource
Definition	
Amplitude Of Mass Source	1 [kg m^-1 m^-1 sec^-1]
Phase Angle	0 [°]

- The next step is to define an acoustic body. The model does not retain the
  definition of the acoustic body from the modal analysis, and it is necessary
  to redefine this object. In the Acoustics toolbar click on `Acoustic Body |
  Acoustic Body`.

- In the window Details of "Acoustic Body", define the Geometry as the rectangular block, change the Mass Density to 1.21, and the Sound Speed to 343.

- The next step is to request that the sound pressure level at the microphone location be calculated. Click on Solution (B6) and in the Acoustics toolbar select Results | Acoustic Time_Frequency Plot.

- In the window Details of Acoustic Time_Frequency Plot, change the Scoping Method to Named Selection, Named Selection to nodereceiver, and Definition | Result to SPL.

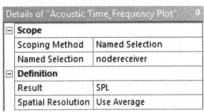

- That completes the setup of the analysis. Click on File | Save Project, and then click the Solve icon.

- If an error occurs, click on Solution (B6) Solution Information and scroll to the bottom of the printout to locate the cause of the error. There might be an error that indicates that the Component NODESOURCE could not be found, or that several nodes were not selected. If this is the case, then click on Harmonic Response (B5) | Acoustic Mass Source, and in the window Details of "Acoustic Mass Source" change the row Scope | Scoping Method to Geometry Selection. The row Scope | Geometry should still indicate 1 Node.

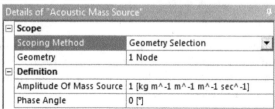

Details of "Acoustic Mass Source"	
⊟ **Scope**	
Scoping Method	Geometry Selection
Geometry	1 Node
⊟ **Definition**	
Amplitude Of Mass Source	1 [kg m^-1 m^-1 m^-1 sec^-1]
Phase Angle	0 [°]

- Click the Solve icon again.

- Once the calculations have completed, if there is a red lightning bolt next to Solution (B6) | Acoustic Time_Frequency Plot and an error listed in the Messages window, then double-click on the row with the error to inspect the message. The likely cause is that there was a conflict with the number of available ANSYS licenses. Right-click on Solution (B6) | Acoustic Time_Frequency Plot and left-click on Generate.

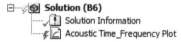

- Once there is a green tick next to Solution (B6) | Acoustic Time_ Frequency Plot, click on this object and a graph of the sound pressure level versus frequency will be displayed in the Worksheet tab.

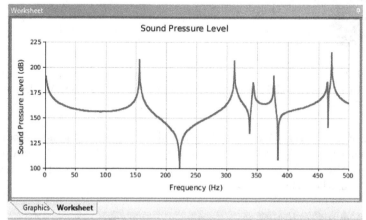

- The results can be exported by right-clicking on Solution (B6) | Acoustic Time_Frequency Plot and left-clicking on Export. Type an appropriate file-name such as ansys_wb_full_harm_spl_receiver.txt and click the Save button.

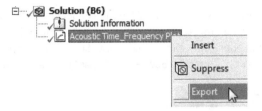

That completes the harmonic analysis of the rigid-walled cavity using ANSYS Workbench. Section 4.4.4 describes the comparison of results from theoretical predictions from the MATLAB model, and results from simulations using ANSYS Workbench and Mechanical APDL.

4.4.3 ANSYS Mechanical APDL

Modal Analysis

A modal analysis was conducted using ANSYS Mechanical APDL to calculate the natural frequencies of the acoustic cavity. Figure 4.2 shows the finite element model that was constructed.

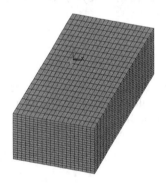

FIGURE 4.2
Finite element model of a rigid-walled cavity created using ANSYS Mechanical APDL.

The completed ANSYS Mechanical APDL file `rigid_cavity_modal_super. inp` is included with this book, and is used to conduct the modal analysis. The script is also used to conduct the harmonic analysis using the modal superposition method that is described later.

Harmonic Analysis: Full Method

A harmonic analysis (`ANTYPE,HARMIC`) was used to calculate the sound pressure level at the microphone location arising from an acoustic point source. The completed ANSYS Mechanical APDL script `rigid_cavity_full.inp` is available with this book.

The FULL method is used to solve the model, which is achieved by issuing the APDL command HROPT,FULL. The excitation was applied using an acoustic mass source using the APDL command BF,node,JS,mass_source. The script will generate the results file ansys_MAPDL_FULL_SPL.txt that contains the sound pressure level at the receiver microphone, and ansys_MAPDL_FULL_p_masssource_1.txt that contains the real and imaginary parts of the acoustic pressure at the receiver microphone.

The sound pressure and sound pressure level were evaluated and compared with the theoretical predictions that are discussed in Section 4.4.4.

Figure 4.3 shows the sound pressure level at the receiver location calculated and displayed using ANSYS Mechanical APDL for a constant mass flow rate, using the APDL command BF,node_a,JS,mass_source.

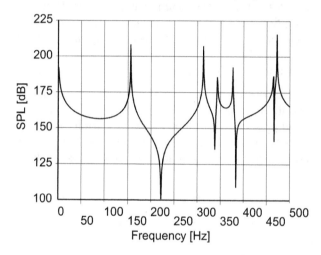

FIGURE 4.3
Sound pressure level at the receiver location calculated using ANSYS Mechanical APDL for a full harmonic analysis using a constant Mass Source as an acoustic source.

Harmonic Analysis: Modal Superposition Method

The sound pressure level at the microphone location can also be calculated using the modal superposition method and is implemented in the ANSYS Mechanical APDL script rigid_cavity_modal_super.inp .

The acoustic excitation is modeled as a point acoustic mass acceleration source using the APDL command F,node,FLOW,flowload, which is described in Section 2.9.2. Note that at Release 14.5 of ANSYS, an acoustic mass source excitation BF,node,JS,mass_source cannot be used for modal superposition analyses. The FLOW load is considered as an older style of applying acoustic loads, which has existed since Release 5.0.

An issue that occurs from the use of the FLOW load for modal superposition analyses is that ANSYS calculates the acoustic pressure, but not the sound pressure level, which is calculated when conducting a full harmonic analysis. The sound pressure level can be calculated from the real and imaginary parts of the acoustic pressure.

For most harmonic analyses using either the modal superposition or the full equations of motion, one normally defines the analysis frequency range with the HARFRQ command and the model is solved with the applied harmonic loads having constant amplitude, but varying excitation frequency. It is possible to alter the applied load at each analysis frequency by using load steps. This can be useful if one wishes to have ANSYS evaluate the response of a system to an applied acoustic volume velocity, where the applied acoustic excitation can be scaled appropriately for the desired acoustic volume velocity excitation. For example, if one wanted to apply a known volume velocity excitation Q_s using the APDL FLOW load, the equivalent FLOW load that should be applied is (refer to Section 2.9.2)

$$\texttt{flowload} = j\omega\rho_0 Q_s \,, \tag{4.6}$$

where $\omega = 2\pi f$ is the circular frequency in radians/s, f is the frequency of analysis in Hz, and ρ_0 is the fluid density.

Figure 4.4 shows the sound pressure level at the receiver location calculated and displayed using ANSYS Mechanical APDL for a constant flow source, using the APDL command F,node,FLOW,flowload. The sound pressure level

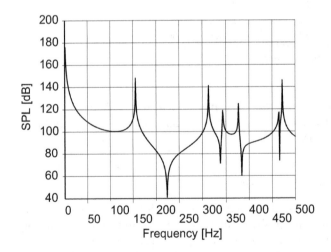

FIGURE 4.4

Sound pressure level at the receiver location calculated using ANSYS Mechanical APDL using modal superposition for a constant acoustic Mass Source.

is calculated as

$$\mathrm{SPL} = 20 \log_{10} \left[\frac{|p|}{\sqrt{2} \times 20 \times 10^{-6}} \right], \tag{4.7}$$

where $|p|$ is the magnitude of the complex nodal pressure.

The sound pressure level results can be converted into equivalent levels to simulate the application of a constant acoustic mass source rate, as was done for the examples using ANSYS Workbench and ANSYS Mechanical APDL for a full harmonic analysis. This can be achieved by multiplying the value of the absolute pressure at the receiver location by $j\omega = j2\pi f$. Figure 4.5 shows the result of multiplying the pressure by $j2\pi f$ and then calculating the sound pressure level. It can be seen that these results are the same as those presented in Figure 4.3.

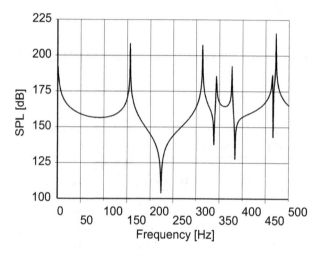

FIGURE 4.5
Sound pressure level at the receiver location calculated using ANSYS Mechanical APDL using modal superposition for a constant acoustic Mass Source and modified in post-processing to simulate the application of a constant acoustic Mass Source.

4.4.4 Results

Figure 4.6 shows the natural frequencies calculated using Equation (4.1) in MATLAB and by conducting a modal analysis using ANSYS. The frequencies and the order of the modal indices are in close agreement up to the 29th mode. Table 4.2 lists the natural frequencies and modal indices of the rigid-wall cavity calculated using MATLAB and ANSYS starting at the 29th mode. The results from the 30th mode and higher calculated using ANSYS show that the order of the modes starts to differ from the theoretical values calculated using MATLAB.

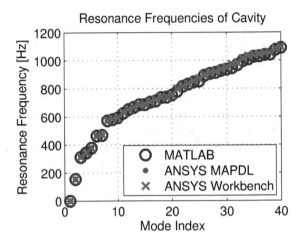

FIGURE 4.6
The natural frequencies of the rigid-walled cavity calculated using MATLAB and ANSYS.

TABLE 4.2
Table of Results Comparing the Natural Frequencies and Mode Indices using MATLAB and ANSYS with 20 Divisions along Each Side of the Cavity

Mode	MATLAB Frequency [Hz]	n_x	n_y	n_z	ANSYS Frequency [Hz]	n_x	n_y	n_z
29	929	2	0	4	938	2	0	4
30	938	0	0	6	951	2	1	2
31	948	2	1	2	973	0	0	6
32	969	0	1	5	986	0	1	5
33	999	1	0	6	1015	2	1	3
34	1011	2	1	3	1032	1	0	6

At 940 Hz, with 20 divisions along the z axis of the cavity, the number of elements per wavelength is

$$\mathrm{epw} = \frac{(c_0/f)}{(L_z/\text{number of divisions})}$$
$$= \frac{(343/940)}{(1.1/20)}$$
$$= 6.6.$$

It is recommended by ANSYS that when using the linear `FLUID30` elements, the mesh density should be at least 12 elements per wavelength.

TABLE 4.3
Table of Results Comparing the Natural Frequencies and Mode Indices Using
MATLAB and ANSYS with 40 Divisions along Each Side of the Cavity

Mode	MATLAB Frequency [Hz]	n_x	n_y	n_z	ANSYS Frequency [Hz]	n_x	n_y	n_z
29	929	2	0	4	932	2	0	4
30	938	0	0	6	947	0	0	6
31	948	2	1	2	949	2	1	2
32	969	0	1	5	974	0	1	5
33	999	1	0	6	1007	1	0	6
34	1011	2	1	3	1012	2	1	3

If the solid model is re-meshed with 40 divisions along each line (by changing the line `LESIZE,ALL,,,20` to `LESIZE,ALL,,,40`) and the finite element
modal analysis is recalculated, then the natural frequencies calculated using
ANSYS are closer to the theoretical values (as calculated using MATLAB),
and therefore the order of the modes is correct. Table 4.3 lists the natural
frequencies and modal indices calculated using MATLAB and ANSYS, when
there were 40 divisions along each line. In summary, it is important to consider
the required accuracy of results when selecting the mesh density for acoustic
analyses.

Figure 4.7 shows the sound pressure level calculated at the receiver location
using MATLAB, ANSYS Workbench, and ANSYS Mechanical APDL. The
simulations conducted using ANSYS were done using a full harmonic analysis

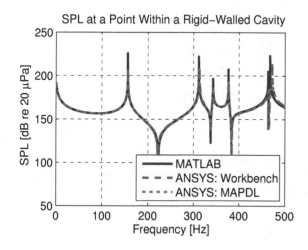

FIGURE 4.7
The sound pressure level at the receiver location in the rigid-walled cavity cal-
culated using MATLAB, ANSYS Workbench, and ANSYS Mechanical APDL.
All three results overlay each other up to about 450 Hz.

and applying an acoustic mass source (BF,node,JS,masssource) at the source location. It can be seen that the three sets of results overlay each other, and hence there is good agreement between all the methods. It can be seen that there are small variations between ANSYS predictions and the theoretical (MATLAB) results above 450 Hz, where there are small differences in the natural frequencies due to the marginally acceptable number of elements per wavelength used in the finite element model.

Figure 4.8 shows the sound pressure level in the cavity at the receiver location, calculated using MATLAB and ANSYS Mechanical APDL using the full and modal superposition methods for conducting a harmonic analysis. The ANSYS simulations were conducted with a full harmonic analysis where a mass source (BF,node,JS,masssource) was applied, and a modal superposition harmonic analysis was conducted where a FLOW load (F,node,FLOW,flowload) was applied. The sound pressure results from the modal superposition analysis were multiplied by $j\omega$ to simulate the mass source load, where a mass volume velocity was applied. It can be seen that the three sets of results overlay each other, and hence there is good agreement between all the methods.

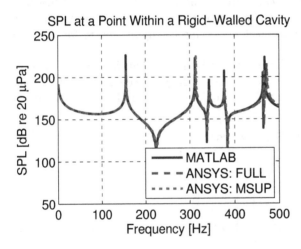

FIGURE 4.8
The sound pressure level at the receiver location in the rigid-walled cavity calculated using MATLAB and ANSYS Mechanical APDL for full and modal superposition harmonic analyses. The lines overlay each other up to about 450 Hz.

Comparison of Computation Times

Table 4.4 lists the comparison of the computation times to calculate the sound pressure level in the cavity using various methods. The computations were conducted on a laptop computer running Microsoft Windows 7 64-bit operating system, with an Intel Core i5 M540 2.53 GHz processor with 4 GB of RAM. Note that each method will calculate and store different sets of results. The

TABLE 4.4

Comparison of Computation Times of Sound Pressure Level in a
Rigid-Walled Cavity Calculated Using Full and Modal
Superposition Harmonic Analysis Methods

Method	CP Time [s]	Elapsed Time [s]
MAPDL: Full Method	1569	803
MAPDL: MSUP Method	34	28

full harmonic analysis will calculate the pressure response at every node and at every frequency, whereas the modal superposition harmonic analysis will calculate the natural frequencies and mode shapes of the model. If one wants to calculate the pressure throughout the model from a modal superposition analysis, another step is required to "expand" the modal solution at a single frequency or over a range of frequencies, using the APDL command EXPSOL or NUMEXP, respectively. For this example, only the pressure response at one node was evaluated. Hence it is not appropriate to directly compare the computation times for each analysis method. However, if one only wants to calculate the acoustic pressure at a few locations, then the modal superposition method is significantly faster than conducting a full harmonic analysis.

The use of the modal superposition method for calculating the vibro-acoustic response of a coupled structural and acoustic system is discussed in Section 9.3. In this technique the natural frequencies and mode shapes of the structure are evaluated without the presence of the fluid, and the natural frequencies and mode shapes of the cavity are calculated using rigid-wall conditions. The two solutions are combined to determine the coupled vibro-acoustic response.

Chapter 5 describes the analysis of a rigid-walled cavity where forms of acoustic damping are introduced into the system.

5

Introduction to Damped Acoustic Systems

5.1 Learning Outcomes

The learning outcomes from this chapter are:

- develop an understanding of the various ways in which damping can be implemented in ANSYS,

- understand how to measure the impedance, reflection coefficient, and sound absorption coefficient of a specimen in an impedance tube,

- understand the difference between a boundary impedance and an impedance sheet in ANSYS,

- develop an appreciation of classical absorption and how to model visco-thermal losses in ANSYS,

- develop an understanding of porous media and how to model these in ANSYS,

- develop an understanding of spectral (global) damping and the three forms suitable for acoustic analysis in ANSYS,

- understand the restrictions faced when applying damping to the 2D `FLUID29` acoustic element in ANSYS,

- understand how the "two-microphone" method is used to estimate impedance,

- develop an appreciation of the difference in ANSYS between a velocity estimate obtained from the pressure gradient, and a velocity estimate obtained from nodal displacements.

5.2 Introduction

Damping is present in all physical systems and is a result of processes that dissipate energy. An understanding of the loss mechanisms, both qualitative and quantitative, is an essential requirement for modeling vibro-acoustic systems since the peak response of such systems are often governed by damping. There are many types of loss mechanisms and only a few are demonstrated through examples in this chapter. As a way of an introduction to damping, four common damping models will be discussed in the following sections. There are two linear models of damping, namely viscous (also known as linear) damping and hysteretic (also known as structural) damping; and two non-linear models, namely air damping and finally Coulomb damping.

5.2.1 Viscous or Linear Damping

Consider the single-degree-of-freedom (DOF) spring-mass-damper system shown in Figure 5.1. It consists of a mass m, a linear spring with stiffness k, a viscous damper with damping constant b, and is excited by a force f acting on the mass.

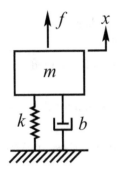

FIGURE 5.1
Single-degree-of-freedom spring-mass-damper system.

The differential equation that describes the dynamics of this system is given by [84, Eq. (2.26)]

$$m\ddot{x}(t) + b\dot{x}(t) + kx(t) = f(t) \qquad (5.1)$$

and may be found in any undergraduate textbook on vibrations or harmonic motion. The "inertial" force, $m\ddot{x}(t)$, and the spring force, $kx(t)$, are conservative forces and as such do not dissipate energy. The force due to the viscous damper, $b\dot{x}(t)$, on the other hand is non-conservative and is responsible for the removal of energy from the system. This type of damping is called *viscous damping* or *linear damping*.

Assuming a harmonic solution for the excitation force $f(t) = F_0 e^{j\omega t}$, where F_0 is the amplitude of the input force, and the response as $x(t) = X e^{j\omega t}$, where X is the complex valued displacement, then the velocity can be written as $\dot{x}(t) = j\omega x(t)$, the acceleration can be written as $\ddot{x}(t) = -\omega^2 x(t)$, and substituting these expressions into Equation (5.1) and re-arranging, the steady-state receptance for an excitation frequency ω is

$$\frac{X}{F_0} = \frac{1/m}{-\omega^2 + j2\zeta\omega_n\omega + \omega_n^2} ,$$ (5.2)

where the damping ratio is

$$\zeta = \frac{b}{2\sqrt{km}} ,$$ (5.3)

and the natural frequency is

$$\omega_n = \sqrt{\frac{k}{m}} .$$ (5.4)

Sometimes the term 2ζ in Equation (5.2) is written as $\eta = 2\zeta$, where η is the modal loss factor [84, Eq. (2.115)]. The quality factor, Q, which quantifies the ratio of the half-power bandwidth and the resonance frequency (see Equation (5.58)), is equal to the inverse of the modal loss factor, i.e., $Q = 1/\eta$. Resonance is defined to occur when the driving frequency is equal to the natural frequency, $\omega = \omega_n$, and thus the receptance at resonance is given by

$$\left.\frac{X}{F_0}\right|_{\omega = \omega_n} = \frac{1}{j2k\zeta} = \frac{-j}{2k\zeta} ,$$ (5.5)

which is purely imaginary. The $-j$ term in the numerator means that at resonance, for a single-degree-of-freedom system, a 90° phase lag is observed.

The damped natural frequency is the frequency that the *unforced* damped system will oscillate at and is given by [84, Eq. (1.37)]

$$\omega_d = \omega_n \sqrt{1 - \zeta^2} .$$ (5.6)

The frequency of the peak magnitude in the complex frequency response function in Equation (5.2) is given by [84, Eq. (2.41)]

$$\omega_p = \omega_n \sqrt{1 - 2\zeta^2} ,$$ (5.7)

with a corresponding receptance amplitude of [84, Eq. (2.42)]

$$\left|\frac{X}{F_0}\right|_{\omega = \omega_p} = \frac{1}{k\,2\zeta\sqrt{1 - \zeta^2}} .$$ (5.8)

The circular frequencies ω can be converted from radians per second to frequency f in Hertz using the relationship $\omega = 2\pi f$.

Looking at Equations (5.5) and (5.8) it becomes apparent that the magnitude of the response for frequencies around the natural frequency of the system are governed by the damping ratio ζ. It turns out that the viscous damping model is a very good approximation for the loss mechanisms in acoustic systems, and many of the damping mechanisms available in ANSYS are viscous-based models.

5.2.2   Hysteretic or Structural Damping

Many structures that undergo cyclic loading exhibit internal friction within the material itself or at non-welded joints. This type of energy dissipation is called *hysteretic damping*, *solid damping*, or *structural damping*. The approximate steady-state response of a single-degree-of-freedom system with hysteretic damping can be written as [84, Eq. (2.123)]

$$m\ddot{x}(t) + \frac{\beta_h k}{\omega}\dot{x}(t) + kx(t) = F_0 e^{j\omega t}, \tag{5.9}$$

where β_h is defined as the hysteretic damping constant and F_0 is the amplitude of the force. By comparing Equation (5.9) with Equation (5.1), an equivalent viscous damping constant can be defined as [84, Eq. (2.122)]

$$c_{\text{eq}} = \frac{k\beta_h}{\omega}. \tag{5.10}$$

This type of damping can be modeled in ANSYS using *alpha damping*, which is discussed further in Section 5.10.1 on the topic of Rayleigh damping.

Assuming a harmonic solution for the response as $x(t) = Xe^{j\omega t}$, then the velocity can be written as $\dot{x}(t) = j\omega\, x(t)$, and substituting this expression into Equation (5.9) the equation can be re-written as [84, Eq. (2.128)]

$$m\ddot{x}(t) + \frac{\beta_h k}{\omega}[\,j\omega\, x(t)\,] + kx(t) = F_0 e^{j\omega t}$$
$$m\ddot{x}(t) + k(1 + j\beta_h)x(t) = F_0 e^{j\omega t}. \tag{5.11}$$

This gives rise to the notion of a *complex stiffness* or *complex modulus*. The response of a system with hysteretic damping will peak with a magnitude of $X/F_0 = 1/(\beta_h k)$ at a frequency of $\omega = \omega_n$.

5.2.3   Air Damping

Another common damping model is *air damping*, *quadratic damping*, or *velocity-squared damping*. This type of damping is representative of the dissipative losses experienced when a structure vibrates in a fluid, where the force it experiences is proportional to the square of the velocity [84, Sec. (2.7)]. The equations of motion for such vibration is [84, Eq. (2.129)]

$$m\ddot{x}(t) + \alpha\frac{\dot{x}}{|\dot{x}|}\dot{x}^2 + kx(t) = F_0 e^{j\omega t}, \tag{5.12}$$

where α represents the air damping coefficient and the term $\dot{x}/|\dot{x}|$ is the sign of the velocity, such that the force resists the direction of motion. By comparing Equation (5.12) with Equation (5.1), an equivalent viscous damping constant can be calculated as [84, Eq. (2.131)]

$$c_{\text{eq}} = \frac{8\alpha\omega|X|}{3\pi}. \tag{5.13}$$

Note that this expression is proportional to the *magnitude* of the displacement, $|X|$, and thus is a form of non-linear damping. For a constant displacement, this type of damping can be modeled in ANSYS using *beta damping*, which is discussed further in Section 5.10.1 on the topic of Rayleigh damping.

5.2.4 Coulomb Damping

Coulomb damping or *dry friction damping* occurs between sliding surfaces and is governed by the following equation of motion [84, Eq. (2.97)]

$$m\ddot{x}(t) + F_c \frac{\dot{x}}{|\dot{x}|} + kx(t) = F_0 e^{j\omega t}, \tag{5.14}$$

where F_c represents the constant friction force and the term $\dot{x}/|\dot{x}|$ is the sign of the velocity, such that the force resists the direction of motion. By comparing Equation (5.14) with Equation (5.1) an equivalent viscous damping constant can be calculated as [84, Eq. (2.105)]

$$c_{eq} = \frac{4F_c}{\pi\omega|X|}. \tag{5.15}$$

Note that this expression is inversely proportional to the magnitude of the displacement and thus is a form of non-linear damping. For a constant displacement, this type of damping can be modeled in ANSYS using *alpha damping*, which is discussed further in Section 5.10.1 on the topic of Rayleigh damping.

5.3 General Discussion of Damping of Vibro-Acoustic Systems in ANSYS

The damping in a vibro-acoustic system, as with any dynamic system, is a critical factor in determining its response to an excitation. There are numerous ways in which damping can be applied to vibro-acoustic models in ANSYS [24, Section 1.4. Damping] and can be broadly classified as spectral damping, which is a systemwide approach (also known as global), and phenomenological damping, which attempts to accurately model the dissipation mechanisms. Phenomenological damping can be structural or purely acoustic. The latter can be further classified as locally reacting and bulk reacting. A summary of the available methods for incorporating damping in vibro-acoustic systems is presented in Tables 5.1, 5.2, and 5.3 for spectral, purely acoustic, and specialized structural elements, respectively.

Those familiar with modeling structural systems may have previously come across spectral damping techniques, which generally come in two forms: Rayleigh damping and modal damping. The former is comprised of a stiffness matrix multiplier (implemented using the APDL command BETAD or

MP,BETD) and a mass matrix multiplier (ALPHAD or MP,ALPD). The use of such damping comes with constraints on the type of analysis (ANTYPE,TRANS with TRNOPT,FULL; ANTYPE,MODAL with MODOPT,QRDAMP or MODOPT,DAMP; or ANTYPE,SUBSTR with SEOPT,,,3). The harmonic-type analysis allows additional types of damping to be added to the model, for example, a constant modal damping ratio (input using either the DMPRAT or MDAMP command). There are also a few specialized forms of structural damping element such as the Coriolis or gyroscopic damping matrix which are only applicable to structural systems and therefore will not be discussed here. Numerical damping present in transient analyses in ANSYS is discussed in detail in Chapter 7.

For acoustic systems on the other hand, there are specific features aimed at removing energy from the system, or in other words providing damping. These can be classified as purely acoustic, in which only pressure degrees of freedom are required, and structural-acoustic, where displacement degrees of freedom are necessary through the fluid–structure interaction (FSI) flag. Recent releases of ANSYS have seen a large number of both local- and bulk-reacting purely acoustic damping methods introduced as demonstrated in Table 5.2.

For coupled vibro-acoustic systems, any damping applied to the structure will generally lead to attenuation of the acoustic field. For example, the application of the COMBIN14 spring-damper element directly modifies the damping matrix, and were this to be coupled to the displacement of an acoustic node it will lead to power flow from the acoustic field into the spring-damper element, thus damping the system. There are too many such elements to be covered in this book. However, there are two elements that are suited to acoustic analyses: the SURF153 2D structural effect element and the SURF154 3D structural effect element, which can be used to apply surface impedances to acoustic systems within the acoustic domain.

TABLE 5.1
Summary of Ways in Which Spectral (Global) Damping May Be Applied to Vibro-acoustic Systems in ANSYS

Mechanism	APDL Command	Description and Restrictions	Example in This Book
Rayleigh Damping	ALPHAD and BETAD	Global damping for *mode superposition harmonic* and *full harmonic*, and *mode superposition transient* and *full transient* analysis.	Section 5.10.
	MP,BETD and MP,ALPD	Material damping for *full harmonic* and *full transient* analysis. Can also be used in *damped modal* analysis.	Section 5.10.
Constant Damping Ratio	DMPRAT	Sets a constant damping ratio for use in the *mode superposition harmonic* (ANTYPE,HARMIC) or *transient* (ANTYPE,TRANS) analysis and the *spectrum* (ANTYPE,SPECTR) analysis.	Section 5.10.
Mode Dependent Damping Ratio	MDAMP	Sets a constant damping ratio for use in the *mode superposition transient* (ANTYPE,TRANS) or *harmonic* (ANTYPE,HARMIC) analysis and the *spectrum* (ANTYPE,SPECTR) analysis. It should be noted that ANSYS 14.5 does not support the *modal superposition method* (HROPT,MSUP or TRNOPT,MSUP) for coupled vibro-acoustic systems.	Section 5.10.

TABLE 5.2
Summary of Ways in Which Damping May Be Applied to Purely Acoustic Systems in ANSYS

Mechanism	APDL Command	Description and Restrictions	Example in This Book
Bulk Reacting			
Viscosity	MP,VISC	The dissipative effect due to fluid viscosity can be included (input as MP,VISC).	Section 5.9.
		The 2D linear acoustic element FLUID29 is assumed to be inviscid. This option is only applicable to 3D acoustic elements.	Chapter 6.
		The dynamic viscosity defaults to $1.84 \times 10^{-5} \mathrm{N.s/m^2}$ for elements with TB,PERF defined (see entry for Equivalent Fluid below). For all other elements, the viscosity is assumed to be zero when undefined.	
		The symmetric matrix formulation (KEYOPT(2) = 2) is not allowed for coupled modal analysis with viscous material.	
Equivalent Fluid of Perforated Materials	The material coefficients used in the Johnson–Champoux–Allard model are input with the TBDATA command for the TB,PERF material as well as through MP commands.	Although ANSYS refers to this as an "Equivalent Fluid of Perforated Materials," it is more accurately described as a porous media model.	Chapter 6.
		Assuming that the skeleton of the porous material is rigid, then it may be approximated using the Johnson–Champoux–Allard equivalent fluid model, which uses the complex effective density and bulk modulus.	
		Restricted to 3D acoustic elements.	
		Suitable for *modal*, *harmonic*, and *transient* analysis.	
		Note that in ANSYS 14.5 and 15.0 this models the fluid phase in the pores and not the bulk equivalent fluid. This is expected to be rectified in Release 16.0.	

Locally Reacting

Description	Command	Notes	Reference
Surface impedance — real admittance only. For use with 2D elements, eg FLUID29.	SF,,IMPD,1 along with MP,MU,,Admittance	Only real impedances can be applied to 2D FLUID29 acoustic elements via the SF command. MU is the real specific acoustic admittance of the surface (and thus the specific acoustic conductance), $0 \leq MU = \rho_0 c_0/Z \leq 1$, where 0 represents no sound absorption and 1 represents full sound absorption. Suitable for *modal, harmonic,* and *transient* analyses. This technique also works for 3D acoustic elements, however it is a legacy feature and is undocumented.	Section 5.5.
Surface impedance — resistance and reactance. For use with 3D acoustic elements, eg. FLUID30, FLUID200 and FLUID221.	SF,,IMPD,REAL,IMAG	For 3D acoustic elements it is possible to have complex impedances for *harmonic* analysis. REAL is the resistance in Ns/m³ if >0 and is the conductance in m³/(Ns) (or mho) if <0. IMAG is the reactance in Ns/m³ if >0 and is the product of the susceptance and angular frequency if <0. See Table 2.22 for more information. For *transient* analyses the imaginary part is ignored. For *modal* analysis the SF,,IMPD command applies an admittance coefficient.	Section 5.8 in this chapter. Chapter 6.
Surface impedance — Robin boundary condition	SF,,INF	The impedance value of the Robin boundary condition can be defined by the sound impedance $Z_0 = \rho_0 c_0$ (input as INF using the SF command). May be used in *modal* and *harmonic* analyses.	Section 8.2.
Surface impedance — Attenuation coefficient	SF,,CONV,ALPHA	The attenuation coefficient is defined by ALPHA using the SF command. Suitable for *modal* and *harmonic* analyses.	Chapter 6, Chapter 7, Section 5.9.

Continued

TABLE 5.2 (Continued)
Summary of Ways in Which Damping May Be Applied to Purely Acoustic Systems in ANSYS

Mechanism	APDL Command	Description and Restrictions	Example in This Book
Infinite acoustic elements	FLUID129 or FLUID130	For external radiation problems, ANSYS provides two types of elements that allow the pressure wave to satisfy the Sommerfeld radiation condition (which states that the waves generated within the fluid are outgoing) at infinity. These are the FLUID129 for 2D problems and FLUID130 for the 3D problems. Acoustic elements adjacent to these cannot have viscosity or be an equivalent porous fluid. May be used in *modal, harmonic,* and *transient* analyses.	Section 8.2.
Perfectly Matched Layers (PML)	The PML material is defined using FLUID30, FLUID220, and FLUID221 elements with KEYOPT(4) = 1	Perfectly matched layers are artificial anisotropic materials that absorb all incoming waves without any reflections, except for the gazing wave that travel parallel to the PML interface. The edges of the PML region must be aligned to the local or global Cartesian system. Suitable for *modal* and *harmonic* analysis.	Section 8.2.
Impedance Sheet Approximation	BF,, IMPD, RESISTANCE, REACTANCE or BFA,,....	The sheet impedance is input as IMPD using the BF command. It can be on interior surfaces. The impedance values are [Pa.s/m]. Suitable for *modal, harmonic,* and *transient*. Note: The reactance term is not correctly implemented in ANSYS Release 14.5, therefore only resistive impedances can be modeled. To be rectified in 15.0	Not implemented in this book because of the issue with the reactance term.
Thermo-viscous Boundary Layer Impedance Model	SF,,BLI	This is a beta feature in ANSYS Release 14.5 and is not fully tested nor is it documented.	Section 5.9.

TABLE 5.3
A Locally Reacting Damping Mechanism for Vibro-acoustic Systems in ANSYS

Mechanism	Command	Description and Restrictions	Example in This Book
2D and 3D Structural Surface Effect	SURF153 or SURF154 along with MP, VISC RMODIF, TYPE, 4, EFS RMODIF, TYPE, 6, ADMSUA	The 2D SURF153 or the 3D SURF154 surface effect element allows the modeling of a surface impedance using structural coupling. The fluid elements in contact with the element require displacement DOFs and they require activation of the FSI flag, which leads to unsymmetric matrices. Another limitation is it can only be applied to the exterior surface of the acoustic domain.	SURF153 see Sections 5.6 and 5.7.

The following sections illustrate the various ways damping may be applied to vibro-acoustic systems using a number of examples. Sections 5.5, 5.6, and 5.7 contain examples that involve the use of 2D elements and are restricted to ANSYS APDL since ANSYS Mechanical and the ACT Acoustics extension do not readily support 2D elements. Despite this limitation, these three examples have been included as they provide a simple introduction to locally reacting damping, as well as describing how to add damping to 2D systems. Sometimes very large acoustic systems can be modeled as a 2D system, which reduces the number of elements, nodes, and computational requirements. For users unfamiliar with ANSYS Mechanical APDL, or without the need to use 2D acoustic elements, Sections 5.5 to 5.7 may be skipped.

Following the 2D examples, the analysis is extended to 3D models using ANSYS Workbench with examples of locally reacting surface impedances (Section 5.8), bulk reacting classical absorption (Section 5.9), and systemwide damping (Section 5.10).

Many of the examples illustrate the way in which anechoic terminations can be created. Later chapters also illustrate how damping may be applied, including Chapter 6 (which models local reacting liners as well as using a porous media model to model a bulk reacting liner) and Chapter 7 (locally reacting damping). Acoustic impedance sheets are not demonstrated in this book because in ANSYS Release 14.5 the imaginary (reactance) part of the impedance has not been implemented. All damping examples are solved using a harmonic analysis, with the exception of Chapter 7 where modal, harmonic, and transient analyses are undertaken.

5.4   Theory

As discussed in Section 1.3.2, there are two methods for solving harmonic vibro-acoustic problems in ANSYS, namely the full and modal superposition. The full method involves directly calculating the mass, stiffness, and damping matrices at each step solving the equations of motion for the force response (inhomogeneous differential equations). The modal superposition method involves initially conducting a modal analysis, then summing the response of the system across all modes. The advantage of the latter is the time to solve the model can be an order of magnitude faster than the full method. The damping matrices implemented for the full and modal superposition method are discussed below.

Harmonic Full Analyses

The equations of motion for a full harmonic analysis were introduced in Section 1.3.2 and given by Equation (1.7). From the *ANSYS Help* manual [25], the damping matrix, $\mathbf{C}$, used in harmonic analyses (ANTYPE,HARM with Method = FULL, AUTO, or VT on the HROPT command) includes the following components [25, Eq. (15–21)]:

$$
\begin{aligned}
\mathbf{C} &= \alpha\mathbf{M} + \left(\beta + \frac{2}{\Omega}g\right)\mathbf{K} + \sum_{i=1}^{N_{ma}} \alpha_i^m \mathbf{M}_i + \sum_{j=1}^{N_m}\left(\beta_j^m + \frac{2}{\Omega}g_j + \frac{1}{\Omega}g_j^E\right)\mathbf{K}_j \\
&\quad + \sum_{k=1}^{N_e}\mathbf{C}_k + \sum_{m=1}^{N_v}\frac{1}{\Omega}\mathbf{C}_m + \sum_{l=1}^{N_g}\mathbf{G}_l,
\end{aligned}
\tag{5.16}
$$

where

$\mathbf{C}$ is the structure damping matrix,

α is the mass matrix multiplier (input on ALPHAD command),

$\mathbf{M}$ is the structure mass matrix,

β is the stiffness matrix multiplier (input using the APDL command BETAD),

g is the constant structural damping ratio (input using the APDL command DMPRAT),

Ω is the excitation circular frequency,

$\mathbf{K}$ is the structure stiffness matrix,

N_{ma} is the number of materials with MP,ALPD input,

α_i^m is the mass matrix multiplier for material i (input as ALPD on the MP APDL command),

$\mathbf{M}_i$ is the portion of structure mass matrix based on material i,

N_m is the number of materials with MP,BETD, DMPR, or SDAMP input,

β_j^m is the stiffness matrix multiplier for material j (input as BETD with the MP APDL command),

g_j is the constant structural damping ratio for material j (input as DMPR with the MP APDL command),

g_j^E is the material damping coefficient (input as SDAMP with the TB APDL command),

$\mathbf{K}_j$ is the portion of the structure stiffness matrix based on material j,

N_e is the number of elements with specified damping,

$\mathbf{C}_k$ is the element damping matrix,

N_v are the number of elements with viscoelastic damping,

$\mathbf{C}_m$ is the element viscoelastic damping matrix,

N_g is the number of elements with Coriolis or gyroscopic damping,

and $\mathbf{G}_l$ is the element Coriolis or gyroscopic damping matrix.

All the terms that are preceded by $1/\Omega$ (namely g, g_j, g_j^E and $\mathbf{C}_m$) are the structural damping terms which represent an imaginary contribution to the stiffness matrix $\mathbf{K}$, and therefore represent a loss mechanism. Structural damping is independent of the forcing frequency, Ω, and produces a damping force proportional to displacement (or strain). The terms g, g_i, and g_i^E are damping ratios (i.e., the ratio between actual damping and critical damping, not to be confused with modal damping). Terms not preceded by $1/\Omega$ (namely α, β, α_i^m, β_j^m, $\mathbf{C}_k$, and $\mathbf{G}_l$) are the usual viscous damping terms and are linearly dependent on the forcing frequency, Ω, and produce damping forces proportional to velocity.

The types of elements that have element damping matrices available, and are commonly used in vibro-acoustic models, are the SURF153 and SURF154 structural surface effect elements and are demonstrated in Sections 5.6 and 5.7. Impedance boundaries and impedance sheets change the element damping matrix $\mathbf{C}_k$; the former is demonstrated in Section 5.8 in this chapter and Chapters 3, 6, and 7. Global modifications to the damping matrix such as α, β, and g are explored in Section 5.10.

Modal Superposition

Damped modal superposition is discussed in detail in the ANSYS Help manual [25, 15.3.3. Mode-Superposition Analysis]. For such harmonic analyses (ANTYPE,HARM with HROPT,MSUP), as well as transient (ANTYPE,TRANS with TRNOPT,MSUP) or PSD analysis (ANTYPE,SPECTRUM with Sptype = SPRS, MPRS, or PSD on the SPOPT command), the damping matrix is not explicitly computed, but rather the damping is defined directly in terms of a damping ratio ξ^d, which is the ratio between actual damping and critical damping. The damping ratio ξ_i^d for the ith mode is the combination of the following [25, Eq. (15–23)]

$$\xi_i^d = \xi + \xi_i^m + \frac{\alpha}{2\omega_i} + \frac{\beta\omega_i}{2}, \tag{5.17}$$

where ξ is the constant modal damping ratio (input using the DMPRAT APDL command), ξ_i^m is the modal damping ratio for the ith mode (see below), ω_i is the (angular) natural frequency associated with the ith mode, α is the mass matrix multiplier (input using the ALPHAD APDL command), and β is the stiffness matrix multiplier (input using the BETAD APDL command). The combination of the terms $\frac{\alpha}{2\omega_i}$ and $\frac{\beta\omega_i}{2}$ is commonly referred to a Rayleigh damping. The modal damping ratio ξ_i^m can be defined for each mode directly using the MDAMP APDL command (undamped modal analyses only).

It should be noted that as of ANSYS Release 14.5 the modal superposition method (HROPT,MSUP or TRNOPT,MSUP) does not support coupled vibro-acoustic or damped acoustic systems, but has been introduced in Release 15.0. The damping terms in Equation (5.17) are all functional for acoustic analysis.

5.5 Example: 2D Impedance Tube with a Real Admittance

Characterizing the absorptive properties of acoustic materials is critical for understanding their behavior when deployed in engineering applications. One bulk characteristic of such materials is the flow resistivity, R_1 (MKS rayls/m), which is a measure of the pressure drop across the material and the induced normal velocity per meter, and is discussed in more detail in Chapters 6 and 7, as well as Ref. [47]. The total flow resistance across a non-impervious media is given by the product of the flow resistivity of the media and the thickness of the media. Flow resistance (and the corresponding flow resistivity) is commonly measured in one of two ways. The first involves forcing a mean flow of gas with velocity, v_0, through a sample of length l, and measuring the pressure difference, ΔP, to give the flow resistivity $R_1 = \Delta P/(v_0 l)$ [41]. The second method uses an acoustic impedance tube [39], comprising a rigid closed-end tube, a sound source, and a microphone, and will be the focus of this section. Using the impedance tube it is possible to measure the normal incidence sound absorption coefficient, as well as estimate the normal incidence surface impedance and flow resistivity of the specimen. There is also a method for measuring random incidence absorption coefficients using a reverberation room [40], and this is explored in Chapter 7.

5.5.1 Description of the System

Impedance tubes are used to measure the acoustic impedance of a sound-absorbing material and are essentially one-dimensional wave guides, with a source at one end and the acoustic load (typically the test sample of the absorptive material) placed at the other end as depicted in Figure 5.2. The type of source does not matter but is typically a pressure or velocity source. There are two common methods by which the impedance of the material is measured in an impedance tube. The first involves a moveable microphone that traverses the length of the tube [87]. This method is the older and arguably simpler of the two methods but is slow. The second method [88] is known as the "two-microphone" or "transfer function" method [135, 57] and will be the focus of the method employed in this section.

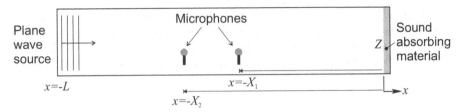

FIGURE 5.2
Impedance tube, with source at one end $(x = -L)$ and the impedance, Z, at the opposite $(x = 0)$. The two microphones are used to estimate the termination impedance using the two-microphone technique.

5.5.2 Theory

In this section the fundamental theory for application of the two-microphone method to determine the impedance, complex reflection coefficient, and the normal-incidence sound absorption coefficient is derived. Consider the one-dimensional wave guide shown in Figure 5.2, excited by some arbitrary plane wave source on the left-hand side $(x = -L)$ and with a complex impedance, Z, at the opposite end $(x = 0)$. Two pressure sensors (microphones) are located at $x = -X_1$ and $x = -X_2$ and are used to determine the magnitude of the forward and backward traveling waves, from which the impedance, flow resistivity, and sound absorption coefficient may be determined. It is assumed that there are no losses along the length of the tube and that only plane waves propagate (in other words, the frequency range of interest is below cut-on as calculated using Equation (3.17)). The origin of the system is at the right end of the tube and the wave incident on the specimen will be traveling in a positive x direction (left to right).

The sound absorption coefficient of a material, α, is defined as the ratio of sound power absorbed by a surface to the incident sound power [47, chapter 7]. For a plane wave at normal incidence, the sound absorption coefficient is given by [118, Eq. (2.62)][2, Eq. (2.23)]

$$\alpha = 1 - |r|^2 \,, \tag{5.18}$$

where r is the (complex) sound reflection coefficient and is defined as the ratio of reflected pressure, p_r, to incident pressure, p_i, at the absorbing surface and is given by [2, Eq. (2.20)]

$$r = \frac{p_r}{p_i} \,. \tag{5.19}$$

The incident and reflected waves may be written as [2, Eqs. (2.3) and (2.9)]

$$p_i = p_0 e^{j(\omega t - kx)} \text{ and } p_r = r p_0 e^{j(\omega t + kx)} \,, \tag{5.20}$$

where p_0 is the amplitude of the incident wave, $r p_0$ is the amplitude of the reflected wave, $\omega = 2\pi f$ is the angular frequency, and k is the wavenumber.

The total sound pressure, p_t, at a location x in the duct is the sum of the forward and backward traveling waves

$$p_t = p_0 e^{j(\omega t - kx)} + r p_0 e^{j(\omega t + kx)}, \tag{5.21}$$

and the total particle velocity, u_t, is given by

$$u_t = \frac{p_0}{\rho_0 c_0} \left(e^{j(\omega t - kx)} - r e^{j(\omega t + kx)} \right). \tag{5.22}$$

The complex transfer function between the two microphones located at $x = -X_1$ and $x = -X_2$ is

$$H_{12} = \frac{p_1}{p_2} = \frac{p_0 e^{j(\omega t + kX_1)} + r p_0 e^{j(\omega t - kX_1)}}{p_0 e^{j(\omega t + kX_2)} + r p_0 e^{j(\omega t - kX_2)}}, \tag{5.23}$$

which may be solved for r giving the reflection coefficient as a function of microphone positions and the transfer function between the two as [118, Section 2.7.1.2]

$$r = e^{2jkX_2} \frac{H_{12} - e^{-jk\Delta X}}{e^{jk\Delta X} - H_{12}}, \tag{5.24}$$

where $\Delta X = X_2 - X_1$. Hence, knowing the locations of the microphones, the complex sound reflection coefficient, and thus sound absorption coefficient, can be determined directly from the pressure transfer function between two microphones.

$$r = \frac{p_r}{p_i} = \frac{Z - Z_0}{Z + Z_0}, \tag{5.25}$$

where $Z_0 = \rho_0 c_0$ is the characteristic impedance of the fluid (typically air) and Z is the specific acoustic impedance of the absorbing surface.

The specific acoustic impedance at any point, x, is defined as the ratio of the total pressure and total particle velocity and is given by [47, Section 1.12.2]

$$Z(x) = \frac{p_t(x)}{u_t(x)} = \rho_0 c_0 \frac{1 + r e^{2jkx}}{1 - r e^{2jkx}}. \tag{5.26}$$

The specific acoustic impedance ratio (or normalized acoustic impedance) is defined as

$$\frac{Z}{Z_0} = \frac{p_t}{\rho_0 c_0 u_t} = \frac{1 + r e^{2jkx}}{1 - r e^{2jkx}}, \tag{5.27}$$

where $Z_0 = \rho_0 c_0$ is the characteristic impedance of the fluid. The impedance ratio at $x = 0$ is therefore

$$\frac{Z}{Z_0} = \frac{1 + r}{1 - r}, \tag{5.28}$$

and the complex reflection coefficient in terms of the impedance ratio is

$$r = \frac{Z/Z_0 - 1}{Z/Z_0 + 1}. \tag{5.29}$$

In terms of a real (resistive) part and an imaginary (reactive) part, the specific acoustic impedance ratio (also known as the normalized specific acoustic impedance) is

$$\frac{Z}{Z_0} = \underbrace{R}_{\text{Resistance}} + j \underbrace{X}_{\text{Reactance}}, \tag{5.30}$$

and the equivalent admittance is

$$\frac{Z_0}{Z} = \frac{1}{R + jX} = \underbrace{\left[\frac{R}{R^2 + X^2}\right]}_{\text{Conductance}} + j \underbrace{\left[\frac{-X}{R^2 + X^2}\right]}_{\text{Susceptance}}. \tag{5.31}$$

It can be shown that the sound absorption coefficient in terms of the real and imaginary parts of the impedance ratio is given by

$$\alpha = \frac{4R}{(R^2 + X^2) + 2R + 1} = 1 - |r|^2. \tag{5.32}$$

In ANSYS, when using the material property defined by the APDL command MP, MU, ..., then

$$MU = \text{Re}\left\{\frac{Z_0}{Z}\right\} = \frac{1}{R} \tag{5.33}$$

since it is assumed that the imaginary component is zero.

Table 5.4 lists typical configurations of end impedances, their impedances and the corresponding reflection coefficients.

TABLE 5.4
Some Typical Termination Impedances, the Corresponding Reflection
Coefficients and Value of MU

Description	Reflection Coeff.	Real Imped.	Imaginary Imped.	Absorp. Coeff.	MU
Rigid-wall	$r = 1$	$R = \infty$	$X = 0$	$\alpha = 0$	0
Soft-wall	$r = -1$	$R = 0$	$X = 0$	$\alpha = 0$	0
(pressure release)					
Total absorption	$r = 0$	$R = 1$	$X = 0$	$\alpha = 1$	1

5.5.3 Model

Now consider a specific example of an impedance tube terminated with a real impedance such that it is partially reflective. The parameters used in the example are listed in Table 5.5.

A harmonic analysis was conducted on the model. In the following sections a method is presented to model and analyze the system using MATLAB and ANSYS.

TABLE 5.5

Parameters Used in the Analysis of the Impedance Tube
System Using 2D Acoustic Elements

Description	Parameter	Value	Units
Duct length	L	1.0	m
Duct width	W	0.1	m
Speed of sound	c_0	343.24	m/s
Density	ρ_0	1.2041	kg/m^3
ANSYS acoustic flow	FLOW	1	kg/s^2
Real impedance ratio*	R	2	—
Imaginary impedance ratio*	X	0	—
Absorption coefficient*	α	0.8889	—
Reflection coefficient*	r	0.3333	—
ANSYS real admittance	MU	0.5	—
Mic 1 location	$-X_1$	−0.4	m
Mic 2 location	$-X_2$	−0.5	m
ANSYS mesh size	—	0.01	m

* Dependent parameters

5.5.4 MATLAB

The MATLAB code `impedance_tube.m` available with the book was used to analyze this system. The dimensions of the duct, the impedance of the specimen, and acoustic properties are defined in MATLAB, from which the dependent parameters in Table 5.5 were then calculated. Given the value of MU, the real specific impedance ratio R was determined using Equation (5.33), the imaginary specific impedance was set to zero, the reflection coefficient was obtained from Equation (5.29), and the sound absorption coefficient was obtained from Equation (5.32). The MATLAB code also imports the results from the ANSYS Mechanical APDL analysis and plots the results.

5.5.5 ANSYS Mechanical APDL

The ANSYS APDL code `code_ansys_impedance_tube.txt` supplied with this book was used to generate the FE model in Figure 5.3. The 1 m long impedance tube was modeled in ANSYS Mechanical APDL using the 2D acoustic element FLUID29, which is a linear element with optional displacement degrees of freedom (activated with KEYOPT(2)). The element size was 0.01 m, which even at the upper frequency limit of the analysis (1000 Hz) provides 34 elements per wavelength.

There are a variety of possible acoustic sources that could have been used to excite the system such as a pressure condition D,,PRES, a displacement condition D,,UX, or a flow condition F,,FLOW. Impedance tubes are designed to be used under plane wave conditions and therefore the acoustic source must generate acoustic plane waves from the source end. If a pressure condition is

FIGURE 5.3

Finite element model of the impedance tube meshed with linear 2D acoustic elements FLUID29. A FLOW source was applied to the left-hand side of the duct (indicated by the right-pointing arrows) and a boundary admittance MU was applied to the right (left-pointing arrows). The origin is at the lower right corner of the model.

applied equally to the end nodes in the duct, this will naturally create a plane wave. However, if the same is attempted with a FLOW source, then the acoustic near field is no longer characterized as an acoustic plane wave. This is because the nodes in the center have two elements to distribute the acoustic flow, whereas the corner nodes have only one element to distribute the flow. Therefore, to achieve plane wave conditions when applying the FLOW acoustic source directly to nodes, and assuming a uniform rectangular mesh, one must apply the desired flow on all nodes, with the exception of the corner nodes which must only have half the magnitude. This is demonstrated in Figure 5.4, where the left-hand images show the half unit FLOW applied to the corners, and the right images are for a unit FLOW.

A unit FLOW source was applied to the nodes on the left-hand side of the duct (with the exception of the corner nodes) of the model, as seen in the left-hand arrows of Figure 5.3, which created a right-traveling plane wave.

Real acoustic impedances may be added to FLUID29 2D elements using a surface load directly to the nodes using the APDL command SF,,IMPD,1. Alternatively, it may be applied to an element using SFE, a line using SFL, or an area using SFA. This activates the impedance and the magnitude of the admittance is defined using the APDL command MP, MU, , admittance. An admittance of 0.5 (as listed in Table 5.5) was used in this example.

In this example, the acoustic impedance was calculated in two ways: (a) from the ratio of pressure to particle velocity using Equation (5.27); and (b) indirectly using the transfer functions between two microphones using Equation (5.24), then followed by application of Equation (5.27). This provides a means of validating the two-microphone method using data obtained from ANSYS simulations.

It is worth noting that this is an entirely acoustic analysis and that there are no active displacement degrees of freedom. Hence it is not possible to calculate the particle velocity by differentiating the displacement with respect to time. In fact, plotting displacement results using PLNSOL,,U,X will show a plot with zero displacement. Instead particle velocities can be estimated from the pressure gradient across the elements using the APDL command

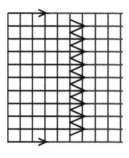
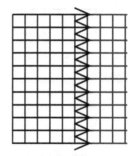

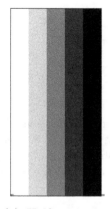

(a) Half `FLOW` applied to the corners.

(b) Unit `FLOW` applied to the corners.

FIGURE 5.4
Illustration of the effect of a `FLOW` acoustic source applied to the (left-hand) end of a duct. The upper images show the magnitude of the `FLOW`. The lower images are the resulting real pressure.

`ESOL,,ELEM_NUM,NODE_NUM,SMISC,3`, where `ELEM_NUM` is the number of the element attached to the node number `NODE_NUM`.

A full harmonic analysis (`ANTYPE,3` with the `HROPT,FULL` option) was performed in 100 Hz increments from 100 Hz to 1 kHz. This was used to calculate the acoustic pressures at the locations of the two microphones, from which the frequency response (transfer function) between the two microphones was calculated. Figure 5.5 shows the sound absorption coefficient calculated in ANSYS, which compares well against the value of 0.8889 listed in Table 5.5. The results were then exported to MATLAB for post-processing.

The frequency response between microphones 1 and 2 was calculated in MATLAB from 0 to 1000 Hz in 1 Hz increments and is presented in Figure 5.6 as a Bode diagram, where it is compared against the results generated in ANSYS. It can be seen that the two produce identical results.

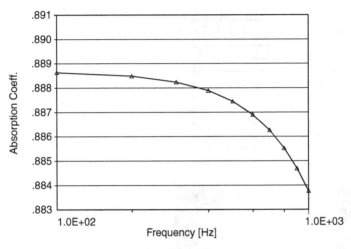

Absorption Coefficient vs. Frequency

FIGURE 5.5

Sound absorption coefficient, α, versus frequency in the impedance tube arising from an applied MU $= 0.5$ boundary condition specified in Table 5.5. Nominal absorption coefficient is 0.8889. The absorption coefficient was obtained using the particle velocity (estimated via the pressure gradient) at the termination boundary.

Using the pressure transfer function data shown in Figure 5.6, the reflection coefficient, sound absorption coefficient and normalized acoustic impedance (real and imaginary) were calculated using Equations (5.24), (5.32), and (5.28), respectively. The same parameters were also calculated using the ratio of surface pressure and the particle velocity at the termination end. These results from ANSYS are compared against theoretical results in Table 5.6. The percentage differences between the MATLAB and ANSYS results are also given. It can be seen that the results compare favorably, however, the estimates obtained by directly calculating the impedance from the velocity estimate is the less accurate of the two methods. This is because linear FLUID29 elements were used which provide poor estimates of the pressure gradients (compared to quadratic elements), and consequently the estimates of the acoustic particle velocities are also poor, especially in regions where the spatial second derivative of pressure (equivalent to curvature in solid mechanics) is high. For reactive sound fields, this is typically also the region of high sound pressure level, such as near walls.

The sound absorption coefficient calculated using the ANSYS results is compared against the theoretical value over the chosen frequency range in Figure 5.7, where it can be seen that the error in the ANSYS results is less than 1%.

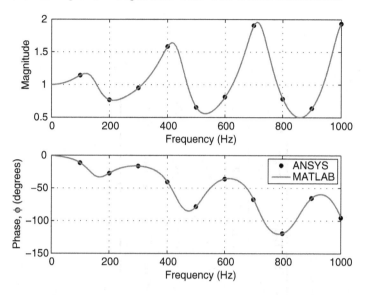

FIGURE 5.6

Amplitude and phase of the transfer function, $H_{12} = p_1/p_2$, between microphones 1 and 2 for a plane wave striking the real admittance defined in Table 5.5.

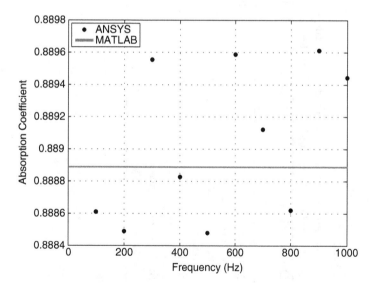

FIGURE 5.7

Sound absorption coefficient calculated using the results generated in ANSYS Mechanical APDL and in MATLAB for a plane wave striking the real admittance defined in Table 5.5. The ANSYS results were obtained using the two-microphone method, by applying Equations (5.24) and (5.28) to the data presented in Figure 5.6.

TABLE 5.6
Results from the Analysis at 100 Hz of the Impedance Tube When Using a Boundary Admittance MU. All Parameters Are Unit-less.

Description	Parameter	MATLAB	ANSYS 2 Mic	Diff.	ANSYS p/v	Diff.
Magnitude of reflection coefficient	$|r|$	0.3333	0.3338	0.13%	0.3303	0.93%
Sound absorption coefficient	α	0.8889	0.8886	0.03%	0.8909	0.23%
Surface impedance ratio - Real	R	2.000	2.002	0.09%	1.985	0.74%
Surface impedance ratio - Imaginary	X	0	−0.0116	N/A	−0.0363	N/A

5.6 Example: 2D Impedance Tube with a Complex Termination Impedance

In the previous section it was shown that it was possible to apply a real boundary admittance, $MU = \mathrm{Re}\{\rho_0 c_0 / Z\} = 1/R$, on the faces of FLUID29 elements. Unfortunately, this approach cannot be used to apply a complex impedance (both resistance and reactance). For 2D FLUID29 acoustic elements, the only way this can be achieved (as of ANSYS Release 14.5) is to use the 2D surface effect element SURF153. The same approach can also be used for 3D acoustic elements using the SURF154 element. Since these are structural elements, any fluid elements in contact with the SURF153/154 elements must have the displacement DOFs activated. This has a number of negative consequences: it increases the number of degrees of freedom, leads to unsymmetric matrices (or larger symmetric formulations), and the FSI flag must be activated. Another limitation is it can only be applied to the surface of acoustic elements, or in other words, on the exterior surface of the acoustic domain much like the APDL command SF,,IMPD. For a detailed discussion on the coupling of acoustic and structural elements see Section 2.4.

In this section it will be shown how to apply a complex impedance to a 2D model. The complex impedance will be applied using a SURF153 surface effect element. The real part of the impedance is defined using the command MP,VISC,,real_z, where real_z$= \mathrm{Re}\{Z\}$. The resistance must always be positive (i.e., boundary absorbs energy). When defining the imaginary part of the impedance, one of two commands must be used, depending on the sign of the reactance. If the imaginary part of the impedance is greater than zero, $\mathrm{Im}\{Z\} > 0$, then the "additional mass per unit area" admsua$= \mathrm{Im}\{Z\}/\Omega$ is issued by defining the 6th real parameter of the SURF element, e.g., RMODIF,TYPE,6,admsua, and Ω is the angular frequency. If the imaginary part of the impedance is less than zero, $\mathrm{Im}\{Z\} < 0$, then the "elastic foundation stiffness" efs$= -\Omega\,\mathrm{Im}\{Z\}$ is issued by defining the 4th real parameter of the SURF element, e.g., RMODIF,TYPE,4,efs, or alternatively R,,,,,efs. Note that both the admsua and the efs are normalized by area, so they represent the specific mass reactance and specific stiffness, respectively. It should be noted that it is possible to have both terms non-zero simultaneously, so it is important to delete one before setting the other (which can be easily done by using the R APDL command).

5.6.1 Description of the System

Consider the 2D model of the duct described previously in Section 5.5. The parameters used in this example are the same as detailed in Table 5.5 with the exception of the impedance, which is defined in Table 5.7.

A harmonic analysis was conducted on the model. In the following sections, methods to model and analyze the system are presented for both MATLAB and ANSYS Mechanical APDL.

TABLE 5.7

Parameters Used in the Analysis of the 2D Impedance Tube
System

Description	Parameter	Value	Units
Real impedance ratio	R	1	—
Imaginary impedance ratio	X	-1	—
Absorption coefficient*	α	0.8	—
Reflection coefficient*	r	$0.2 - j0.4$	—

* Dependent parameters

5.6.2 ANSYS Mechanical APDL

A finite element model of the duct shown in Figure 5.2 was developed in
ANSYS Mechanical APDL and is shown in Figure 5.8. The code that created
this, code_ansys_surf153.txt, is included with this book as is the MATLAB
code, impedance_surf153.m, used to post-process the results. The system was
modeled with linear FLUID29 2D acoustic elements and SURF153 2D surface
effect elements (with no mid-side nodes KEYOPT,,4,1) on the termination end
of the duct to provide the impedance. The layer of FLUID29 elements attached
to the SURF153 elements had the displacement DOFs activated (KEYOPT(2)=0).
All other FLUID29 elements had only the pressure DOFs (KEYOPT(2)=1). A
"unit" FLOW source was applied to the left-hand nodes of the model (with
the exception of the corner nodes) and was used to excite the cavity with a
right-traveling plane wave.

A harmonic analysis was undertaken from 100 Hz to 1 kHz inclusive in
100 Hz frequency increments. For a constant imaginary impedance term (as is
the case in this example) the mass (admsua) or stiffness (efs) is frequency dependent when using the SURF153/154 elements. This creates difficulties when
solving in ANSYS Release 14.5 since the standard method of solving the model
does not support varying the "real" set automatically with each frequency
step. This is because it is not possible to tabulate the real element constant

FIGURE 5.8

Finite element model of the 2D duct terminated with SURF153 elements on
the right-hand side. The layer of FLUID29 elements with displacement DOFs
are on the right-hand side of the model. The FLOW source excitation is shown
as arrows on the nodes on the left-hand side of the model.

for these elements, and using multiple load steps (LSWRITE/LSSOLVE) do not operate on real constants, so issuing RMODIF commands in multiple load steps are ignored. The "workaround" is to use the "multiple SOLVE method," which is done by placing the command defining the reactance (RMODIF) in a *DO loop along with the SOLVE command. This will vary the value of the impedance at each analysis frequency and has been employed in this example. Since it is necessary to discriminate between positive and negative imaginary impedances when using the SURF153 elements, the APDL code code_ansys_surf153.txt, employs the *IF command to switch between the mass (ADMSUA) or stiffness (EFS) definitions depending on the sign of the imaginary impedance.

The results for the analysis are presented in Table 5.8 (for 100 Hz only) which shows that at low frequencies the estimates of the parameters are very accurate. Figure 5.9 shows the frequency response (magnitude and phase) between the two microphones separated by 0.1 m. In Figure 5.10 the nominal complex termination impedance is compared against the calculated value using pressure frequency response results from ANSYS Mechanical APDL (Figure 5.9) along with Equations (5.24) and (5.28). The results at high frequencies begin to show a small error in the order of a few percent. This is due to a limitation of the linear FLUID29 element and the chosen element density at high frequencies. The use of a higher mesh density will cause this error to decrease. In Figure 5.11 the nominal sound absorption coefficient is compared against the calculated value using pressure frequency response data from ANSYS (Figure 5.9), directly using the velocity obtained from the pressure gradient across the acoustic element, as well as calculating the velocity by differentiating the nodal displacement data with respect to time.

Also displayed in Table 5.8 are the estimates of the parameters obtained from calculating the impedance using the ratio of the pressure to the particle velocity estimate (obtained from the pressure gradient), p/v. As was observed in Section 5.5, the impedance estimate obtained using the particle velocity estimate from the pressure gradient is less accurate than using the pressure estimates. Since the displacement DOFs were activated to couple the SURF153 elements to the FLUID29 acoustic elements, it is possible to extract the displacement of the nodes at this boundary. The acoustic impedance (and related parameters) was also calculated using the ratio of pressure to particle velocity estimate obtained from the surface displacement, $p/(j\omega u)$, and is shown in Table 5.8, where it is seen that these results are as accurate as those obtained using the pressure DOFs and the two-microphone method. This result is shown in Figure 5.11. This is an important learning outcome: using results directly obtained from the available DOFs almost always leads to better results than those obtained indirectly.

TABLE 5.8

Results from the Analysis at 100 Hz of the Impedance Tube Using the SURF153 Elements

Description	Parameter	MATLAB	ANSYS 2 Mic & $p/j\omega u$	Diff.	ANSYS p/v	Diff.
Complex reflection coefficient	r	$0.2 - j0.4$	$0.2000 - j0.4000$	—	$0.1978 - j0.4029$	—
Magnitude of reflection coefficient	$|r|$	0.4472	0.4472	Nil	0.4489	0.37%
Sound absorption coefficient	α	0.8	0.8000	Nil	0.7085	0.18%
Surface Impedance Ratio - Real	R	1	1.0000	Nil	0.9909	0.91%
Surface Impedance Ratio - Imaginary	X	-1	-1.0000	Nil	-1.0000	0.00%

Note: The column "2 Mic & $p/j\omega u$" represents the two-microphone method and directly calculating the results by differentiating the nodal displacement data. The column "p/v" represents the results obtained using the pressure-gradient-based velocity estimate.

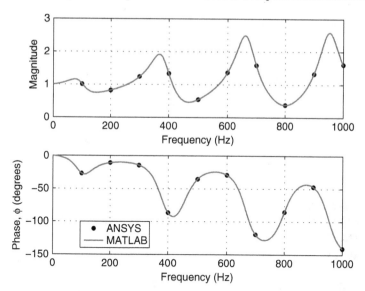

FIGURE 5.9

Amplitude and phase of the pressure transfer function, $H_{12} = p_1/p_2$, between microphones 1 and 2 for a plane wave striking a SURF153 surface effect element with the impedance defined in Table 5.7.

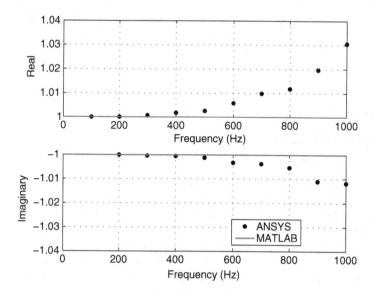

FIGURE 5.10

Calculated real and imaginary impedance ratios versus frequency for a plane wave striking a SURF153 surface effect element with the impedance defined in Table 5.7. The ANSYS results were obtained using the two-microphone method, by applying Equations (5.24) and (5.32) to the data presented in Figure 5.9.

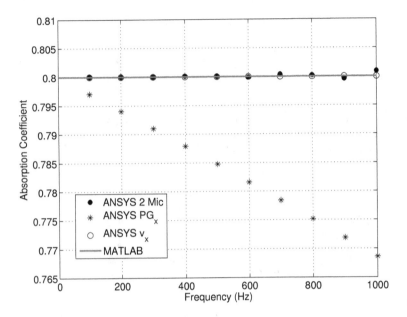

FIGURE 5.11

A comparison of sound absorption coefficient estimates in an impedance tube terminated with a SURF153 surface effect element with the impedance defined in Table 5.7. The nominal absorption coefficient was 0.8. The ANSYS results were calculated using three methods: the two-microphone method, using the element pressure gradient at the impedance surface, and the nodal displacement data at the impedance surface.

5.7 Example: 2D Impedance Tube with a Micro-Perforated Panel Absorber

Micro-Perforated Panels (or MPPs) are devices used to absorb sound and consist of a thin plate, shell, or membrane with many small holes in it. An MPP offers an alternative to traditional sound absorbers in that no porous material is required, and thus provide a clean and robust way to control sound. The absorption comes from the resistance offered by the viscous losses in the holes of the panel as the particles move back and forth under the influence of the sound. An MPP is normally 0.5–2 mm thick with holes typically covering 0.5% to 2% of the surface area. Typically an MPP will have a small backing cavity behind it to create a Micro-Perforated Panel Absorber (MPPA) as shown in Figure 5.12, the dimensions of which determine the attenuation and frequency range of performance.

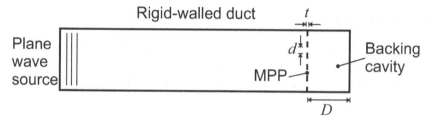

FIGURE 5.12
Schematic of a micro-perforated panel absorber (MPPA), comprised of an MPP and backing cavity.

5.7.1 Theory

A variety of expressions exist for the impedance of an MPP. The approach presented here will be to use the expressions derived by Maa [110], who showed that the normalized acoustic impedance of an MPP is approximately given by [110, Eq. (4)]

$$
Z_{\text{Maa}} = \frac{32\eta t}{d^2 \rho_0 c_0 P} \left(1 + \frac{K^2}{32}\right)^{\frac{1}{2}} + j \frac{\omega t}{c_0 P}\left(1 + \left(3^2 + \frac{K^2}{2}\right)^{-\frac{1}{2}}\right), \quad (5.34)
$$

where η is the dynamic viscosity of air, t is the thickness of the panel, d is the hole diameter, $K = \frac{d}{2}\sqrt{\omega \rho_0 \eta}$ is the perforate constant, P is the perforation ratio, c_0 is the speed of sound in the gaseous media, and ρ_0 is the density of the media.

Because the gaseous media is squeezed through the holes, the resistance should be increased by $\frac{1}{2}\sqrt{2\omega\rho_0\eta}$ and the reactance should be increased by $0.85d$ using the end corrections by Morse and Ingard [117]. Consequently, Maa's original equation for the normalized impedance was changed to [110, Eq. (5)]

$$
\begin{aligned}
Z_{\text{Maa,corrected}} \quad = \quad & \frac{4\sqrt{2\eta}K}{Pd\rho_0 c_0} + j0.85\frac{\omega d}{Pc_0} \\
& + \underbrace{\frac{32\eta t}{d^2 \rho_0 c_0 P}\left(1 + \frac{K^2}{32}\right)^{\frac{1}{2}} + j\frac{\omega t}{c_0 P}\left(1 + \left(3^2 + \frac{K^2}{2}\right)^{-\frac{1}{2}}\right)}_{Z_{\text{Maa}}}
\end{aligned}
$$

$$(5.35)$$

which may be separated into the real part of the impedance ratio (resistance), R, and the imaginary part (reactance), X, as shown in Equation (5.30).

The total normalized acoustic impedance of an MPPA comprising an MPP backed with a cavity of depth D is given by the sum of the impedance of the

MPP and the impedance of the backing cavity and is given by

$$Z_{MPPA} = Z_{Maa,corrected} - j \cot \frac{\omega D}{c_0}. \qquad (5.36)$$

The normal incidence absorption coefficient of the MPPA is [110, Eq. (9)]

$$\alpha = \frac{4R}{(1+R)^2 + (X - \cot(\omega D/c_0))^2}. \qquad (5.37)$$

Comparing Equation (5.37) with the equation for absorption arising from a surface impedance, Equation (5.32), the term $\cot(\omega D/c_0)$ is a contribution to the relative acoustic reactance from the backing cavity. It can be shown that the absorption coefficient has a maximum value of approximately [110, Eq. (10)]

$$\alpha_0 = \frac{4R}{(1+R)^2}, \qquad (5.38)$$

which occurs at the resonance frequency, ω_0, given by the solution to [110, Eq. (11)]

$$X - \cot\left(\frac{\omega_0 D}{c_0}\right) = 0. \qquad (5.39)$$

5.7.2   Example

Consider an impedance tube with an MPPA as shown in Figure 5.12 with properties as defined in Table 5.9. Using Equation (5.39), the resonance frequency of the MPPA is $\omega_0/2/\pi = 431$ Hz, and has a peak absorption coefficient of $\alpha_0 = 0.62$ given by Equation (5.38).

TABLE 5.9

Properties of the MPPA

Description	Parameter	Value	Units
Air properties:			
Speed of sound	c_0	343.24	m/s
Density	ρ_0	1.2041	kg/m^3
Viscosity	μ_0	1.84E-05	N.s/m^2
MPPA properties:			
Panel thickness	t	0.010	m
Hole diameter	d	0.001	m
Cavity depth	D	0.100	m
Perforation ratio	P	0.1	—
Peak absorption coeff. *	α_0	0.62	—
Resonance freq. *	f_0	431	Hz

* Dependent parameters

5.7.3 MATLAB

The MATLAB code `impedance_surf153_mpp.m` available with the book was used to analyze this system. The impedance and sound absorption co-efficients were calculated using Equations (5.36) and (5.37), respectively, along with the parameters listed in Table 5.9. The MATLAB code also reads in the ANSYS Mechanical APDL results and produces graphs of the results.

5.7.4 ANSYS Mechanical APDL

A finite element model of the micro-perforated array in an impedance tube was developed in ANSYS Mechanical APDL and is shown in Figure 5.13. The code that created this model, `code_ansys_surf153_mpp.txt`, is included with this book. The model is based on the 2D system presented in Section 5.6. The `SURF153` elements were used to model the complex impedance of the MPP given by Equation (5.35). Attached to the `SURF153` elements was the backing cavity meshed with 2D `FLUID29` acoustic elements. The elements of the backing cavity in contact with the `SURF153` elements had the displacement DOFs activated (as was done for the duct elements). The horizontal (Ux) and vertical

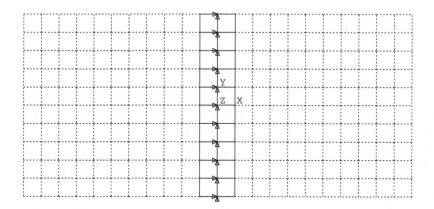

FIGURE 5.13
Finite element model of the impedance tube meshed with linear 2D acoustic elements `FLUID29` and terminated with a MPPA comprised of a `SURF153` and backing cavity. Only the region around the MPP is shown. A `FLOW` source was applied to the left-hand side of the duct (not shown). The `FLUID29` elements attached to the `SURF153` elements had the displacement DOFs activated. The vertical and horizontally aligned triangles represent coupling equations which couple the horizontal and vertical displacement DOFs of the two acoustic domains.

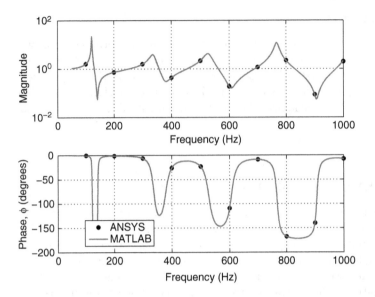

FIGURE 5.14
Amplitude and phase of the pressure transfer function, $H_{12} = p_1/p_2$, between microphones 1 and 2 for a plane wave striking the MPPA defined in Table 5.9.

(Uy) displacement DOFs of the coincident nodes along the boundary of the two contiguous acoustic domains were coupled using the ANSYS Mechanical APDL command CPINTF. This ensures that the two spaces are coupled by ensuring that the displacement of the nodes along the shared boundary are the same. An array of the complex values of the frequency-dependent MPP impedance was created within a *DO loop. A full harmonic analysis using the "multiple SOLVE method" was used to obtain the frequency response of two microphones (as previously discussed in Section 5.5), as well as the displacement and velocity estimate at the surface of the MPPA. The results were exported to a text file to be post-processed by the MATLAB script impedance_surf153_mpp.m. A comparison of the transfer function between the two microphones in the impedance tube is shown in Figure 5.14. The resulting complex impedance and sound absorption coefficient are shown in Figures 5.15 and 5.16, respectively. The * symbol in Figure 5.16 represents the frequency at which the surface impedance of the MPPA is entirely real, determined using Equations (5.38) and (5.39), and is very close to the frequency of the actual peak in the absorption.

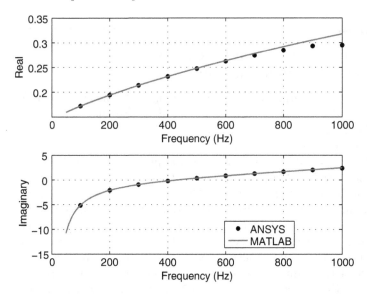

FIGURE 5.15
Calculated real and imaginary impedance ratios versus frequency for a plane wave striking the MPPA defined in Table 5.9. The ANSYS results were obtained using the two-microphone method, by applying Equations (5.24) and (5.28) to the data presented in Figure 5.14.

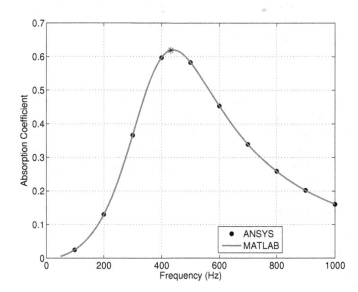

FIGURE 5.16
The normal incidence sound absorption coefficient versus frequency for a plane wave striking the MPPA defined in Table 5.9. Results were obtained using the two-microphone method. The * marker represents the approximate peak in the absorption coefficient determined using Equations (5.38) and (5.39).

5.8   Example: 3D Impedance Tube with a Complex Termination Impedance

ANSYS allows the application of a complex impedance to the exterior surface of a 3D acoustic domain, which avoids the need to use the surface effects elements SURF153/154 (demonstrated in Sections 5.6 and 5.7) and the issues that come with using structural elements.

The APDL command that is issued when applying a surface impedance to a node is SF, NODE, IMPD, VAL1, VAL2. There is an equivalent APDL command for elements (SFE) and areas (SFA). Table 2.22 describes the effect of positive and negative values of VAL1. When VAL1$\geq$ 0 the terms VAL1 and VAL2 correspond to the real (resistance) and imaginary (reactance) parts of the impedance, respectively. When VAL1$<$ 0 the two terms represent the negative real part of the admittance (conductance) and the product of the imaginary part of the admittance (susceptance) and the angular frequency, respectively. For the latter case, since VAL2 is frequency dependent even if the susceptance is constant, it is necessary to use the "multiple SOLVE method," which is done by placing the command defining the surface admittance (SF,ALL,IMPD,VAL1,VAL2) in a *DO loop along with the SOLVE command.

In this section it will be shown how to apply a constant (frequency-independent) complex impedance to a boundary. The use of positive and negative VAL1 will be demonstrated. Section 3.3.7.4 contains an example of an impedance varying with frequency, which was applied to the end of a duct to simulate it radiating into free space.

5.8.1   Model

The model used in this section to demonstrate the surface impedance capability is similar to that presented for 2D square duct models in Sections 5.5 to 5.7, and is also reused in Section 5.9.2. The square duct is 1 m long and is 0.01 m × 0.01 m in cross-section. The viscosity and thermal conductivity in the model were both set to zero. The complex termination impedance is the same as used in Section 5.6 and is listed in Table 5.10.

5.8.2   ANSYS Workbench

This section provides instructions on how to build a 3D duct and terminate it with a complex impedance boundary.

Constructing the Solid Model

The completed ANSYS Workbench project file Impedance_3D-SF-IMPD.wbpj is available with this book.

- Start ANSYS Workbench and start a new project.

TABLE 5.10

Parameters Used in the Analysis of the 3D Impedance Tube System

Description	Parameter	Value	Units
Air:			
Speed of sound	c_0	343.24	m/s
Density	ρ_0	1.2041	kg/m^3
Duct:			
Length	L	1.000	m
Width	W	0.010	m
Height	H	0.010	m
Termination:			
Real impedance ratio	R	1	—
Imaginary impedance ratio	X	-1	—
Absorption coefficient*	α	0.8	—
Reflection coefficient*	r	$0.2 - j0.4$	—

* Dependent parameters

- It is assumed that the ACT Acoustics extension is installed and is operating correctly. This can be checked in the Workbench project view by selecting the Extensions | Manage Extensions menu. The extension ExtAcoustics should be listed in the table and a tick present in the Load column.

Load	Extensions	Version
☑	ExtAcoustics	8

- Double-click on Harmonic Response under Analysis Systems in the Toolbox window, so that a new Harmonic Response cell appears in the Project Schematic window.

- Double-click on row 3 Geometry to start DESIGNMODELER.

- Select Meter as the desired length unit, and click the OK button.

- The first step is to create the solid model of the duct—a 1D waveguide. In the toolbar, left-click on Create | Primitives | Box.

An item called Box1 will appear in the Tree Outline. Rename the object to Duct by right-clicking over the Box object in the Tree Outline and selecting Rename in the context menu. An alternative way to rename is to left-click on the item Box1 and press <F2>.

- In the Details View window, change the row Box to Duct. Also ensure that the row Box Type is set to From One Point and Diagonal. Then proceed to define the geometry of the box as shown below. The dimensions of the box are $0.010 \times 0.010 \times 1.000\,\mathrm{m}$ as described previously and listed in Table 5.10. The coordinate origin will be in the center of the face of the source end of the duct.

Details View	
⊟ **Details of Duct**	
Box	Duct
Base Plane	XYPlane
Operation	Add Material
Box Type	From One Point and Diagonal
Point 1 Definition	Coordinates
☐ FD3, Point 1 X Coordinate	-0.005 m
☐ FD4, Point 1 Y Coordinate	-0.005 m
☐ FD5, Point 1 Z Coordinate	0 m
Diagonal Definition	Components
☐ FD6, Diagonal X Component	0.01 m
☐ FD7, Diagonal Y Component	0.01 m
☐ FD8, Diagonal Z Component	1 m
As Thin/Surface?	No

- To create the duct you need to generate it by either right-clicking over the Duct object in the Tree Outline and selecting Generate in the context menu or alternatively left-clicking the Generate button in the toolbar.

You will notice that the Tree Outline has 1 Part, 1 Body. Clicking on the + symbol next to 1 Part, 1 Body will show the Solid that was just created. A rendered solid model of the duct will have also appeared in the Graphics window. It will initially appear small because of the default scale. In the Toolbar, click on the Zoom to Fit icon.

The narrow duct should now be visible in the Graphics window and is shown in the following. The three text labels showing named selections will not be shown.

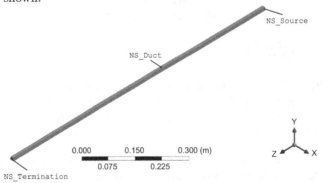

- Now is a good time to save your work. Click on `File | Save Project` and enter an appropriate filename such as `Impedance-3D-SF-IMPD.wbpj`.

- The final step is to define three named selections that will identify the source end of the duct, the termination end (which will be a complex impedance), and the duct itself. We will first define the duct. Either left-click the mouse on `Selection Filter: Bodies` in the toolbar or press `<Ctrl> b` on the keyboard. Then in the toolbar, click `Tools | Named Selection`. In the `Details View` window, in the row `Named Selection`, type `NS_Duct`.

  Click on the solid model of the duct in the `Graphics` window, then in the `Details View` window left-click on `Apply` in the row `Geometry`, which should now show `1 Body`. In the `Tree Outline`, right-click on `NS_Duct` and then click on `Generate` to create the named selection.

- Repeat the process for the source end of the duct (at $z = 0$). To do this you will need a view in which the end can be seen: start with an isometric view, then in the triad at the bottom of the screen move the mouse between the X and Y axes such that the negative Z axis is shown and click on it.

- Either click the mouse on `Selection Filter: Faces` in the toolbar or press `<Ctrl> f` to allow selection of faces. Create a new named selection (`Tools | Named Selection`). In the `Details View` window, change the name in the row `Named Selection` to `NS_Source`. Click on the source face in the `Graphics` window and then `Apply` in `Details View | Geometry`, which should now show `1 Face`. In the `Tree Outline` right-click on the `NS_Source` object and then left-click on `Generate` in the context menu to create the named selection.

- Repeat the process and define a new named selection for the other end of the duct and call it `NS_Impedance`. To select the correct face, select the positive Z axis as shown below.

- If you have done things correctly your `Tree Outline` should look like the following image.

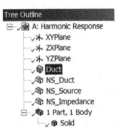

- The solid model is now complete. Click on File | Save Project. Exit the DESIGNMODELER.

Meshing

Prior to meshing we will define the material properties of some of the objects in the solid model.

- In the Workbench Project Schematic double-click on row 4 Model. This will start ANSYS Mechanical.

- We will now define a new acoustic body. In the ACT Acoustics extension toolbar click on Acoustic Body. This will insert an Acoustic Body entry in the Outline window under Harmonic Analysis (A5) | Acoustic Body. An alternative way to define this is to right-click on Harmonic Analysis (A5) and then in the context menu left-click on Insert | Acoustic Body. In the window Details of "Acoustic Body", change the row Scope | Scoping Method to Named Selection then choose NS_Duct. The values for the remaining rows beneath Definition should by default match the values defined in Table 5.10.

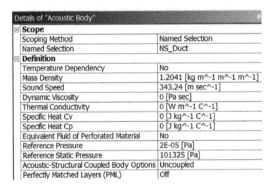

- Now define a new acoustic Mass Source that will generate the sound in the duct. In the ACT Acoustics extension toolbar, click on Excitation | Mass Source (Harmonic).

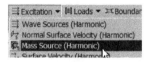

Click on the Details of "Acoustic Mass Source" entry and change the Scope | Scoping Method to Named Selection, then choose NS_Source. Under Definition set the Amplitude of Mass Source to 1.

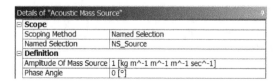

- Now define a complex termination impedance. In the ACT Acoustics extension toolbar click on Boundary Conditions | Impedance Boundary.

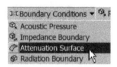

In the window Details of "Acoustic Impedance Boundary", change the row Scope | Scoping Method to Named Selection then choose NS_Impedance. Under Definition, ensure that the row Impedance or Admittance is set to Impedance. Set the Resistance to 413.3 in order to create a normalized resistance of 1. Set the Reactance to -413.3 to create a normalized reactance of −1.

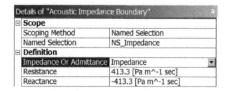

- Now it is time to mesh the solid model. Since the duct is such a simple geometry, it is unnecessary to explicitly define how the duct is to be meshed and the default settings are adequate. By default quadratic acoustic elements (FLUID220) will be used. In the Outline window, left-click on the Mesh object, then in the Details of "Mesh" window, under Sizing | Element Size type 0.01 (m). This will ensure that there are at least 33 elements per wavelength at 1 kHz, which is the highest frequency of interest.

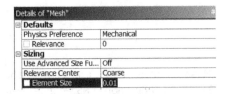

- Click on `File | Save Project` as sometimes the meshing crashes.

- Now mesh the model. This can be done by either clicking `Mesh | Generate Mesh` in the toolbar or alternatively right-click over `Mesh` in the `Outline` window and select `Generate Mesh`.

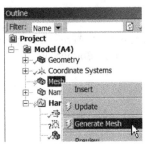

The duct will have been meshed and should look like the illustration shown below.

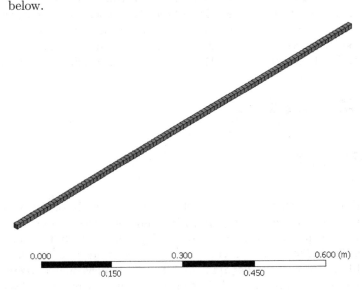

- The mesh is now complete. Click on `File | Save Project`.

Solution

With the model now meshed, and material properties of the elements, boundary conditions, and acoustic sources defined, it is possible to solve the model.

- In order to use the two-microphone method to estimate the impedance, it is necessary to know the sound pressures at two axial locations in the duct. We will select all nodes at two axial planes to determine the pressures. Create two named selections to represent microphones at locations $z = 0.4\,\mathrm{m}$ (to be referred to as Microphone 1) and $z = 0.5\,\mathrm{m}$ (Microphone 2). In the Outline window under Project | Model (A4,B4) | Named Selections right-click to Insert | Named Selection twice.

  In the two selections that are created under Named Selections, change the name of the first selection by right-clicking and selecting Rename in the context menu. Call the selection NS_Mic1. Rename the second named selection NS_Mic2.

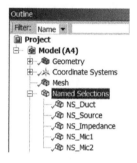

- In the Details of the "NS_Mic1" change Scope | Scoping Method to Worksheet.

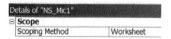

- In the new Worksheet that is created, right-click in the blank row and Add Row using the context menu. Then change each cell in the table as shown below.

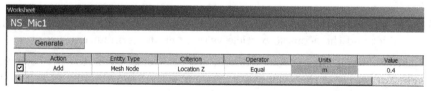

When complete, click on the Generate button in the Worksheet. This completes the selection of the nodes defining NS_Mic1 (Microphone 1). If successful, then 8 Nodes should be shown in the row Statistics | Total Selection.

- Repeat these steps to select 8 nodes for the named selection NS_Mic2 positioned at $z = 0.5\,\text{m}$.

- Set up Details of "Analysis Settings" by clicking on the Analysis Settings in the Outline. Under Options change the row Range Maximum to 1000Hz, the row Solution Intervals to 10, and the row Solution Method to Full. This will provide solutions from 100 Hz to 1 kHz in 100 Hz increments. Note that the first frequency to be solved is not the Range Minimum but rather (Range Maximum - Range Minimum)/Solution Intervals + Range Minimum. Under Analysis Data Management, change Save MAPDL db to Yes. This will allow you to post-process the results using the ACT ACOUSTICS extension option Acoustic Time Frequency Plot.

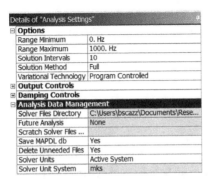

- Save the project by clicking on File | Save Project.

- Solve the harmonic analysis by clicking Solve in the toolbar or right-clicking over Harmonic Response (B5) and then selecting Solve in the context menu. The yellow lightning bolt next to Harmonic Response (B5) will turn green indicating the solver is working. Be patient, this may take several minutes to solve.

Results

We are interested in the attenuation of the sound pressure along the length of the duct and the acoustic particle velocity at the termination.

- Using the ACT Acoustics extension toolbar, select Results | Acoustic SPL. You will note that All Bodies are currently selected under Geometry | Scoping Method, which is what we wish to display.

- Using the ACT Acoustics extension toolbar, select Results | Acoustic Pressure. Rename the Acoustic Pressure object to Acoustic Pressure Termination by either right-clicking over the object and selecting Rename in the context menu or by pressing <F2> on the keyboard.

- Using the ACT Acoustics extension toolbar, left-click on Results |
  Acoustic Velocity Z. Rename the object to Acoustic Velocity Z
  Termination by either right-clicking over the object and selecting Rename
  in the context menu or by pressing <F2>. Change the row Geometry |
  Scoping Method to Named Selection, then change the row Named Selection
  to NS_Impedance.

- Also add to the Results two more Acoustic Pressure objects. Change the
  row Geometry | Scoping Method to Named Selection, and define the Named
  Selections as NS_Mic1 for the first object and NS_Mic2 for the second.

- Using the ACT Acoustics extension toolbar, add a new object Results |
  Acoustic Time Frequency Plot. Rename the object to Acoustic Time
  Frequency Plot NS_Mic1 Pres by either right-clicking over the object and
  selecting Rename in the context menu or by pressing <F2>. Change the row
  Geometry | Scoping Method to Named Selection, then change the row
  Named Selection to NS_Mic1. Change Display to Real and Imaginary.

- Repeat the above steps for named selection NS_Mic2.

- Save the project by clicking on File | Save Project.

- Right-click over any of the results objects under Solution (A6), and in the
  context menu left-click on Evaluate All Results.

- As a "sanity check" that the model was created properly, we can see if the
  impedance was applied correctly by dividing the pressure by the velocity
  at named selection NS_Impedance. Click on the object Acoustic Velocity
  Z Termination and in the window Details of "Acoustic Velocity Z
  Termination", change the row Frequency to 100 Hz and the row Phase
  Angle to 0 to get the real value of the velocity. Then change the Phase
  Angle to -90° to get the imaginary component. Repeat the exercise with
  the object Acoustic Pressure Termination to get the real and imaginary
  values of the termination pressure. This will give an impedance of

$$Z = \frac{p}{v} = \frac{-61.324 - j400.18}{0.40995 - j0.55833} = 413.3 - j413.3 \,, \qquad (5.40)$$

as defined using the Impedance Boundary.

- Now export the pressure data for named selections NS_Mic1 and NS_Mic2
  by right-clicking over the Acoustic Time Frequency Plot and selecting
  Export from the context menu. Save the data for later post-processing as the
  files Impedance-3D-SF-IMPD_Mic1_Pres.txt and Impedance-3D-SF-IMPD_
  Mic2_Pres.txt, respectively.

Figure 5.17 shows the frequency response (magnitude and phase) between
the two microphones separated by 0.1 m. In Figure 5.18 the nominal complex termination impedance is compared against the calculated value using

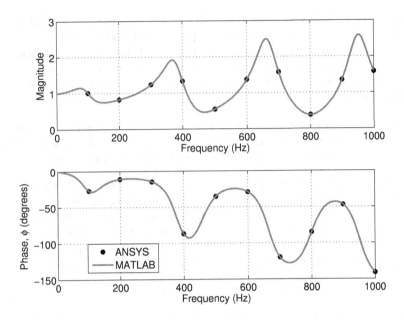

FIGURE 5.17
Amplitude and phase of the transfer function, $H_{12} = p_1/p_2$, between microphones 1 and 2 for a plane wave striking the impedance defined in Table 5.10
using a 3D model and an `Impedance Boundary`.

frequency response data from ANSYS Workbench (Figure 5.17) along with
Equations (5.24) and (5.28). It is interesting to note that the impedance estimates using the 3D quadratic elements (with mid-side nodes) shown in Figure
5.18 are over an order of magnitude more accurate than the 2D linear acoustic
elements used in Section 5.6 and shown in Figure 5.10, despite the mesh density being equivalent. This illustrates the benefits from using the quadratic
elements in preference to the older linear elements. To confirm this, repeat
this analysis using 3D linear acoustic `FLUID30` elements. Click on the `Mesh`
object in the `Outline` window, then in the `Details of "Mesh"` window, and
change the row `Advanced | Element Midside Nodes` to `Dropped`. Right-click
on the `Mesh` object and select `Generate Mesh` in the context menu. Save the
model, then click on the `Solve` icon.

5.8.3 Discussion

This exercise can be repeated using the *admittance* formulation of the
`Impedance Boundary` (instead the *impedance* formulation) as given in
Table 2.22. Since this is a frequency-dependent formulation it is necessary
to solve using multiple load steps using either the `LSWRITE/LSSOLVE` command

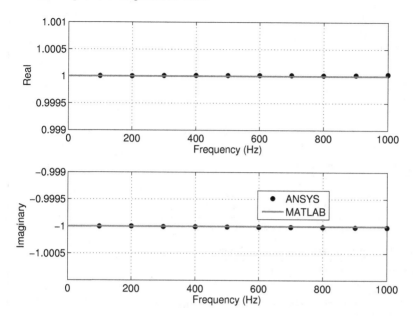

FIGURE 5.18
Calculated real and imaginary impedance ratios versus frequency for a plane
wave striking the impedance defined in Table 5.10 using a 3D model and an
Impedance Boundary.

or the "multiple SOLVE method," both of which require issuing APDL commands.

In the previous ANSYS Workbench model, right-click on Harmonic
Response (A5), then in the context menu left-click on Insert | Commands. In
the Commands (APDL) object created in the Outline window, press <F2> and
rename the object to Commands (APDL) - Multiple SOLVE method. In the
Commands window of the Commands (APDL) - Multiple SOLVE method object,
type the following.

```
1   ! This script is used to apply a frequency varying acoustic impedance
2   ! to the outlet of the duct and calculate the harmonic response.
3   ! These commands are to be inserted under the Harmonic (A5) branch.
4   !
5   ! The following Input Arguments need to be defined in the window
6   ! Details of "Commands (APDL)"
7   !
8   ! ARG1 = density (rho_0 = 1.2041 kg/ m  3 )
9   ! ARG2 = speed of sound (c_0 = 343.24 m/s)
10  ! ARG3 = start analysis frequency Hz (100 Hz)
11  ! ARG4 = end analysis frequency Hz (1000 Hz)
12  ! ARG5 = step increment of frequency Hz (100 Hz)
13  ! ARG6 = Real Z
14  ! ARG7 = Imag Z
15  !
16  ! Create new parameters from the input arguments
17  MY_DENS=ARG1
```

```
18   MY_SONC=ARG2
19   MY_STARTF=ARG3
20   MY_ENDF=ARG4
21   MY_INCF=ARG5
22   Real_Z=ARG6
23   Imag_Z=ARG7
24   ! Define a constant for PI
25   PI=4*ATAN(1)
26   ! Admittance
27   !
28   ! Conductance of the material [ m  3 /N.s]
29   Real_Y = Real_Z/(Real_Z*Real_Z + Imag_Z*Imag_Z)
30
31   ! Susceptance of the material [ m  3 /N.s]
32   Imag_Y = -Imag_Z/(Real_Z*Real_Z + Imag_Z*Imag_Z)
33
34   !-----------------------------------------------------
35   ! Specify harmonic analysis options
36   !-----------------------------------------------------
37   NSUBST,1 ! Single substep for each load step
38   !-----------------------------------------------------
39   ! Solving Loop Over Analysis Frequency Range
40   !-----------------------------------------------------
41   *DO,AR99,MY_STARTF,MY_ENDF,MY_INCF
42   MY_FREQ=AR99 ! Define the frequency to solve
43   HARFRQ,,MY_FREQ
44   ! Can define termination using an impedance or admittance.
45   ! Admittance Terms
46   VAL1 = -Real_Y ! Use conductance
47   VAL2 = Imag_Y*(2*PI*MY_FREQ) ! Use susceptance
48   ! Select all the nodes on the impedance face
49   ! Can do this with either named selection or directly by nodes
50   ! CMSEL,S,NS_Impedance
51   NSEL,S,LOC,Z,1,1
52   SF,ALL,IMPD,VAL1,VAL2 ! Apply the admittance to the selected nodes
53   ALLSEL
54   SOLVE
55   *ENDD
```

In order to resolve using the *admittance* model, right-click over the Acoustic Impedance Boundary object in the Outline window and left-click on Suppress in the context menu. Then solve by clicking on the Solve icon in the toolbar. The results should be identical to those obtained using the *impedance* model.

The APDL commands in the listing can be used to define any impedance that varies with frequency, for example, see Section 3.3.7.4. It should be noted that in ANSYS Release 14.5 there is an issue that prevents multiple load steps working on models with complex impedances, so it is necessary to use the "multiple-solve" method instead as was done above. It is expected that this will be rectified in ANSYS Release 15.0.

5.9   Example: 3D Waveguide with Visco-Thermal Losses

When sound propagates through acoustic media energy it is dissipated through a combination of viscous losses, thermal conductivity, and molecular relaxation [129, Section 10-8][117, Section 6.4]. The combination of viscous and thermal losses are often categorized as "classical absorption," which can be modeled in ANSYS. It should be noted that in many acoustic problems the intrinsic viscous and thermal losses are negligible compared to the losses associated molecular relaxation. This section demonstrates the intrinsic visco-thermal losses experienced by a plane wave propagating in a duct.

5.9.1 Theory

The linearized homogeneous wave equation with visco-thermal losses is given by [129, Eq. (10-3.13)]

$$\nabla^2 p - \frac{1}{c_0^2}\frac{\partial^2 p}{\partial t^2} + \frac{2\delta_{cl}}{c_0^4}\frac{\partial^3 p}{\partial t^3} = 0 , \tag{5.41}$$

where p is the acoustic pressure, c_0 is the speed of sound and [102, Eq. (8.5.5)]

$$\delta_{cl} = \frac{\nu}{2}\left(\frac{4}{3} + \frac{\gamma-1}{\mathrm{Pr}}\right) , \tag{5.42}$$

where the subscript cl is used to indicate classical (absorption), $\nu = \mu/\rho_0$ is the kinematic viscosity, μ is the dynamic viscosity, ρ_0 is the density of the fluid, $\gamma = c_p/c_v$ is the ratio of specific heats, c_p is the specific heat for constant pressure, c_v is the specific heat for constant volume (per unit mass), $\mathrm{Pr} = \mu c_p/\kappa$ is the Prandtl number, and κ is the thermal conductivity. The first term in the parentheses in Equation (5.42) is associated with viscous losses and the second term is associated with thermal conductivity. It should be noted that Equation (5.42) is strictly only valid for monatomic gases. For air it is suggested [129, Eq. (10-8.10c)] that the term $\frac{4}{3}$ should be replaced by $\frac{4}{3} + 0.6$, where the latter term is associated with the bulk viscosity.

The governing equation for a uniform density media that is solved by ANSYS [26, Eq. (8-1)] in the absence of a source is given by

$$\nabla^2 p - \frac{1}{c_0^2}\frac{\partial^2 p}{\partial t^2} + \frac{4\nu}{3c_0^2}\nabla^2\frac{\partial p}{\partial t} = 0. \tag{5.43}$$

Using the Helmholtz equation

$$\nabla^2 p - \frac{1}{c_0^2}\frac{\partial^2 p}{\partial t^2}, \tag{5.44}$$

Equation (5.41) can be written as

$$\nabla^2 p - \frac{1}{c_0^2}\frac{\partial^2 p}{\partial t^2} + \frac{\nu}{c_0^2}\left(\frac{4}{3} + \frac{\gamma-1}{\mathrm{Pr}}\right)\nabla^2\frac{\partial p}{\partial t} = 0. \tag{5.45}$$

Comparing Equations (5.43) and (5.45) it would appear that the effect of thermal conductivity $\frac{\nu}{c_0^2}\left(\frac{\gamma-1}{\mathrm{Pr}}\right)\nabla^2\frac{\partial p}{\partial t}$ has been neglected by ANSYS. However this is not the case as results from analyses in ANSYS show that thermal losses are also calculated and therefore the expression in the ANSYS Help manual [26, Eq. (8-1)] is incorrect.

If we assume that the solution to the wave equation in 1D is given by [117, Eq. (6.4.27)]

$$p(t) = P_0 e^{j(\omega t - kx)}, \tag{5.46}$$

where ω is the angular frequency, k is the complex wavenumber, and P_0 represents the pressure at $x = 0$, then it can be shown that

$$k = \pm \omega/c_0 \left[1 + j \frac{\omega \nu}{c_0^2} \left(\frac{4}{3} + \frac{\gamma - 1}{\mathrm{Pr}} \right) \right]^{-1/2} = \pm (\beta - j\alpha), \tag{5.47}$$

where the positive solution represents the propagation in the positive direction and vice versa. For propagation in the positive direction, $k = \beta - j\alpha$, then the solution to the pressure becomes

$$p(t) = P_0 e^{-\alpha x} e^{j\omega(t - x/(\omega/\beta))}. \tag{5.48}$$

Thus α represents how fast the acoustic field is attenuated by the effects of viscosity and thermal conductivity, and the propagation speed is given by $c_{ph} = \omega/\beta$. The attenuation of the sound pressure level in dB per meter is given by

$$-20 \log_{10}(e^{-\alpha}) = 8.69\alpha \; \mathrm{dB/m}. \tag{5.49}$$

At low frequencies, Equation (5.47) is almost entirely real and thus the visco-thermal affects are negligible. However at higher frequencies, when the dimensionless coefficient $\omega \nu / c_0^2$ approaches unity, attenuation can be significant.

For small ω the complex wavenumber given by Equation (5.47) may be approximated by

$$k = \frac{\omega/c_0}{\sqrt{1 + j \frac{\omega \nu}{c_0^2} \left(\frac{4}{3} + \frac{\gamma-1}{\mathrm{Pr}} \right)}} \approx \frac{\omega/c_0}{1 + j\frac{1}{2}\frac{\omega \nu}{c_0^2} \left(\frac{4}{3} + \frac{\gamma-1}{\mathrm{Pr}} \right)}$$

$$\approx \; \omega/c_0 \left(1 - j\frac{1}{2}\frac{\omega \nu}{c_0^2} \left(\frac{4}{3} + \frac{\gamma - 1}{\mathrm{Pr}} \right) \right) = \omega/c_0 \left(1 - j\frac{\omega \delta_{cl}}{c_0^2} \right), \tag{5.50}$$

and thus for low frequencies the attenuation coefficient may be approximated by

$$\alpha \approx \frac{\omega^2 \delta_{cl}}{c_0^3}, \tag{5.51}$$

which is consistent with the derivation in Pierce [129, Section 10-2, Eq. (10-2.12)], Kinsler et al. [102, Eq. (8.5.5)] and Morse and Ingard [117, Eq. (6.4.14)].

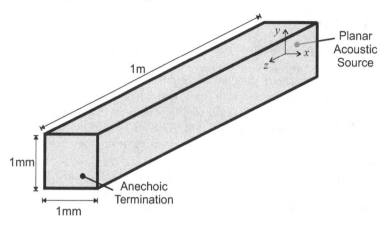

FIGURE 5.19
Duct geometry used in the visco-thermal model.

5.9.2 Model

Consider the 1D waveguide of a cross-section 1 mm × 1 mm and a length of 1 m shown in Figure 5.19. It will be excited by a 100 kHz tone. This high frequency has been chosen because very little attenuation is observed at frequencies below this unless significant lengths are used. The duct width and height of 1 mm has been chosen to ensure that no cross-modes are present in the duct and thus plane wave conditions exist. The physical parameters used in the model are listed in Table 5.11.

TABLE 5.11
Parameters Used in the Visco-thermal Example

Parameter	Symbol	Value	Units
Speed of Sound	c_0	343.24	m/s
Density	ρ_0	1.2041	kg/m^3
Dynamic Viscosity	μ	1.84×10^{-5}	Pa.s
Kinematic Viscosity*	ν	1.523×10^{-5}	m^2/s
Thermal Conductivity	κ	0.0257	W.m^{-1}K^{-1}
Specific Heat Capacity at Constant Pressure	c_p	1012	J.kg^{-1}K^{-1}
Specific Heat Capacity at Constant Volume (per unit mass)	c_v	722.9	J.kg^{-1}K^{-1}
Ratio of Specific Heats*	$\gamma = c_p/c_v$	1.4	—

* Dependent parameters

5.9.3 MATLAB

The MATLAB script plane_wave_viscous_losses.m included with this book is to be used with this example. The script is used to define the independent parameters and calculate the dependent parameters listed in Table 5.11, then using the expressions derived in Section 5.9.1, the attenuation per unit length is calculated using Equations (5.47) and (5.49). The script also analyzes the results from ANSYS Mechanical APDL. The ANSYS Workbench results are not analyzed in MATLAB as the attenuation arising from the classical absorption is directly calculated in Workbench.

Using the parameters in Table 5.11 and Equations (5.47) and (5.49), an attenuation of 1.22dB/m would be expected for a plane wave. This is comprised of 0.86dB/m from the viscous effects and 0.36dB/m from the thermal conductivity effects.

The system will now be modeled in ANSYS to demonstrate how the visco-thermal effects are incorporated into the model.

5.9.4 ANSYS Workbench

This section provides instructions on how to incorporate visco-thermal effects in acoustic models. The particular model used to illustrate the approach is the long narrow duct described previously in Section 5.9.2. A harmonic analysis will be conducted using ANSYS Workbench.

Constructing the Solid Model

The completed ANSYS Workbench project file Visco-thermal.wbpj is available with this book.

- Start ANSYS Workbench and start a new project.

- It is assumed that the ACT Acoustics extension is installed and is operating correctly. This can be checked in the Workbench project view by selecting the Extensions | Manage Extensions menu. The extension ExtAcoustics should be listed in the table and a tick present in the Load column.

Load	Extensions	Version
☑	ExtAcoustics	8

- Double-click on Harmonic Response under Analysis Systems in the Toolbox window, so that a new Harmonic Response cell appears in the Project Schematic window.

- Double-click on row 3 Geometry to start DESIGNMODELER.

- Select Meter as the desired length unit, and click the OK button.

- The first step is to create the solid model of the duct — a 1D waveguide. In the toolbar, left-click on Create | Primitives | Box.

Once complete, an item called Box1 will appear in the Tree Outline. Rename the object to Duct by right-clicking over the Box object in the Tree Outline and selecting Rename in the context menu. An alternative way to rename is to left-click on the item Box1 and press <F2>, or alternatively, under the Details View, change the Box row to Duct.

- In the Details View window, ensure that the row Box Type is set to From One Point and Diagonal. Then proceed to define the geometry of the box as shown below. The dimensions of the box are the same as described in Section 5.9.2. Note that the origin of the Cartesian coordinate system is located at the center of the face with the acoustic source.

Details View	
⊟ **Details of Duct**	
Box	Duct
Base Plane	XYPlane
Operation	Add Material
Box Type	From One Point and Diagonal
Point 1 Definition	Coordinates
☐ FD3, Point 1 X Coordinate	-0.0005 m
☐ FD4, Point 1 Y Coordinate	-0.0005 m
☐ FD5, Point 1 Z Coordinate	0 m
Diagonal Definition	Components
☐ FD6, Diagonal X Component	0.001 m
☐ FD7, Diagonal Y Component	0.001 m
☐ FD8, Diagonal Z Component	1 m
As Thin/Surface?	No

To create the duct, you need to generate it by either right-clicking over the Duct object in the Tree Outline and selecting Generate in the context menu or alternatively left-clicking the Generate button in the toolbar.

You will notice that the Tree Outline has 1 Part, 1 Body. Clicking on the + symbol next to 1 Part, 1 Body will show the Solid that was just created. A rendered solid model of the duct will have also appeared in the Graphics window. It will initially appear small because of the default scale. In the Toolbar click on the Zoom to Fit icon.

The very narrow duct should now be visible in the Graphics window.

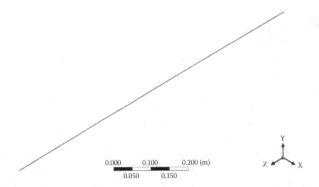

- Now is a good time to save your work. Click on File | Save Project and enter an appropriate filename such as Visco-thermal.wbpj.

- The final step is to define three named selections that will identify the source end of the duct, the termination end (which will be anechoic), and the duct itself. We will first define the duct. Either left-click the mouse on Selection Filter: Bodies in the toolbar or press <Ctrl> b on the keyboard. Then in the toolbar, click Tools | Named Selection. In the Details View window, in the row Named Selection, type NS_Duct.

Details View	
Details of NS_Duct	
Named Selection	NS_Duct
Geometry	1 Body
Propagate Selection	Yes
Export Selection	Yes

Click on the solid model of the duct in the Graphics window, then in the Details View window left-click on Apply in the row Geometry, which should now show 1 Body. In the Tree Outline right-click on NS_Duct and then click on Generate to create the named selection.

- Repeat the process for the source end of the duct (at $z = 0$). To do this you will need a view in which the end can be seen: start with an isometric view, then in the triad at the bottom of the screen move, the mouse between the X and Y axes such that the negative Z axis is shown and click on it.

- Either click the mouse on Selection Filter: Faces in the toolbar or press <Ctrl> f to allow selection of faces. Create a new named selection (Tools | Named Selection). In the Details View window, change the name in the row Named Selection to NS_Source. Click on the source face in the Graphics window then Apply in Details View | Geometry, which should now show

1 Face. In the Tree Outline right-click on the NS_Source object and then left-click on Generate in the context menu to create the named selection.

- Repeat the process and define a new named selection for the other end of the duct and call it NS_Anechoic. To select the correct face, select the positive Z axis as shown below.

- If you have done things correctly, your Tree Outline should look like the following image.

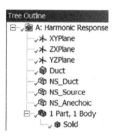

- The solid model is now complete. Click on File | Save Project. Exit the DESIGNMODELER.

Meshing

Prior to meshing, we will define the material properties of some of the objects in the solid model.

- In the Workbench Project Schematic, double-click on row 4 Model. This will start ANSYS Mechanical.

- We will now define a new acoustic body. In the ACT Acoustics extension toolbar, click on Acoustic Body. This will insert an Acoustic Body entry in the Outline window under Harmonic Analysis (A5) | Acoustic Body. An alternative way to define this is to right-click on Harmonic Analysis (A5) and then in the context menu left-click on Insert | Acoustic Body. In the window Details of "Acoustic Body", change the row Scope | Scoping Method to Named Selection then choose NS_Duct. Then change the rows beneath Definition to match the values defined in Table 5.11.

Details of "Acoustic Body"	
Scope	
Scoping Method	Named Selection
Named Selection	NS_Duct
Definition	
Temperature Dependency	No
Mass Density	1.2041 [kg m^-1 m^-1 ...
Sound Speed	343.24 [m sec^-1]
Dynamic Viscosity	1.84E-05 [Pa sec]
Thermal Conductivity	0.0257 [W m^-1 C^-1]
Specific Heat Cv	722.9 [J kg^-1 C^-1]
Specific Heat Cp	1012 [J kg^-1 C^-1]
Equivalent Fluid of Perforated Material	No
Reference Pressure	2E-05 [Pa]
Reference Static Pressure	101325 [Pa]
Acoustic-Structural Coupled Body Options	Uncoupled
Perfectly Matched Layers (PML)	Off

- Now define a new acoustic Mass Source to excite the duct. In the ACT Acoustics extension toolbar click on Excitation | Mass Source (Harmonic).

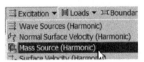

Click on the Details of "Acoustic Mass Source" entry and change the Scope | Scoping Method to Named Selection then choose NS_Source. Under Definition, set the Amplitude of Mass Source to 1e-07. The reason for this is that the attenuation from the visco-thermal effects is small (in the order of 1dB) and therefore in order to have the results presented with sufficient significant figures it is necessary to have a small source such that the resulting sound pressure levels are close to $0\,\mathrm{dB}$ re $20\,\mu\mathrm{Pa}$.

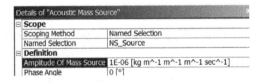

- We will now define a new surface to emulate the anechoic termination. In the ACT Acoustics extension toolbar, click on Boundary Conditions | Attenuation Surface.

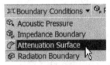

In the window Details of "Acoustic Attenuation Surface", change the row Scope | Scoping Method to Named Selection, then choose NS_Anechoic. Under Definition, set the Attenuation Coefficient to 1 in order to create an anechoic termination.

Details of "Acoustic Attenuation Surface"	
Scope	
Scoping Method	Named Selection
Named Selection	NS_Anechoic
Definition	
Attenuation Coefficient	1

Now it is time to mesh the solid model.

- Since the duct is such a simple geometry it is unnecessary to explicitly define how the duct is to be meshed and the default settings are adequate. By default, quadratic acoustic elements (FLUID220) will be used. Click on the Mesh object in the Outline window. Then in the Details of "Mesh" window, in the row Definition | Element Size type 0.001 (m). This will ensure that there are at least 3 elements per wavelength, which is below the recommended 6 elements per wavelength as described in Section 2.11, but is adequate for this example.

- Now mesh the model. This can be done by either clicking Mesh | Generate Mesh in the toolbar or alternatively right-clicking over Mesh in the Outline window and selecting Generate Mesh.

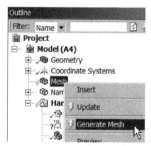

The duct will have been meshed, although it is difficult to tell since the mesh size is only 1 mm. Click on the Box Zoom icon in the Toolbar and zoom in around the anechoic end of the duct. The mesh should look like the illustration below.

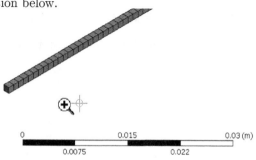

0		0.015		0.03 (m)
	0.0075		0.022	

- The mesh is now complete. Click on File | Save Project.

Solution

With the model now meshed, and material properties of the elements, boundary conditions, and sources defined, it is possible to solve the model.

- Set up `Details of "Analysis Settings"` by clicking on the `Analysis Settings` in the `Outline`. Under `Options`, change the `Range Maximum` to `100000Hz`, `Solution Intervals` to `1`, and `Solution Method` to `Full`.

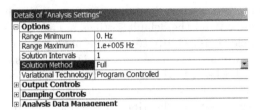

- Solve the harmonic analysis by clicking `Solve` in the toolbar or right-clicking over `Harmonic Response (B5)` and then selecting `Solve` in the context menu. The yellow lightning bolt next to `Harmonic Response (B5)` will turn green indicating the solver is working. Be patient, this may take several minutes to solve.

Results

We are interested in looking at the attenuation of the sound pressure level (SPL) along the length of the duct.

- Using the ACT Acoustics extension toolbar, select `Results | Acoustic SPL`. You will note that `All Bodies` are currently selected under `Geometry | Scoping Method`.

- Right-click over the `Acoustic SPL` object and select `Evaluate All Results`.

- The figure below shows the SPL along the length of the duct at 100 kHz.

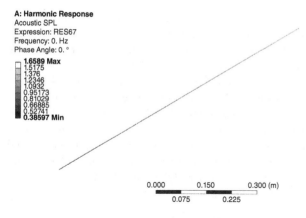

A: Harmonic Response
Acoustic SPL
Expression: RES67
Frequency: 0. Hz
Phase Angle: 0. °

1.6589 Max
1.5175
1.376
1.2346
1.0932
0.95173
0.81029
0.66885
0.52741
0.38597 Min

0.000   0.150   0.300 (m)
      0.075   0.225

- Since the termination of the duct is anechoic and only forward-traveling waves exist, the attenuation per meter due to visco-thermal losses is given by the difference between the maximum SPL and the minimum SPL, namely $1.659 - 0.386 = 1.273 \, \text{dB/m}$ compared to the theoretically predicted value of $1.222 \, \text{dB/m}$, which is a 4% error. This error is associated with the very low number of elements per wavelength. To rectify this, redo the analysis with twice the number of elements per unit length by clicking on Mesh in the Outline window, then change the Sizing | Element Size to 0.0005m, then click on Solve. Note that this will exceed the node limit of the ANSYS Academic Teaching license. With the higher mesh density the attenuation is $1.677 - 0.452 = 1.225 \, \text{dB/m}$, which is an error of less than 0.3%.

5.9.5 ANSYS Mechanical APDL

The ANSYS Mechanical APDL file code_ansys_visco_thermal.txt that is available with this book was used to repeat the analysis described in Section 5.9.4 for the Workbench model. When the simulation is run, a finite element model is generated, the appropriate material properties are defined, an acoustic Mass Source is applied to one end, and an anechoic termination is applied to the other end. A harmonic analysis is performed at the frequency of 100 kHz and the results are exported to a text file visco_thermal_scalars.txt to be read by the MATLAB script plane_wave_viscous_losses.m.

5.10 Application of Spectral Damping to a Rigid-Walled Cavity

As discussed in the introduction to this chapter, it is possible to apply damping to the entire system or sub-systems. This approach is often desirable as it does not require detailed knowledge of the source of damping and therefore effort spent on modeling the actual loss mechanisms can be avoided. It is simply sufficient to define the amount of damping, which is often determined experimentally. This section illustrates how three types of spectral damping can be applied to acoustic systems. The model used for this purpose is the rigid-walled cavity discussed in Chapter 4 and illustrated in Figure 4.1.

As of ANSYS Release 14.5, spectral damping of pressure-formulated acoustic elements using full analyses is not supported in Workbench (unlike structural elements). Furthermore, as mentioned at the beginning of this chapter, acoustic-based modal superposition analyses are not as yet fully supported in Workbench either. However, it is possible to use the latter under certain conditions using APDL code. Due to such restrictions, this section explores spectral damping using only ANSYS Mechanical APDL.

5.10.1   Spectral Damping Types

Three types of damping will be explored in this section. With reference to Equation (5.17), these are a model-wide constant structural damping ratio ξ, Rayleigh damping (comprised of α and β), and a mode dependent damping ratio ξ_i^m (which is only applicable for the modal superposition method).

Constant Structural Damping Ratio

The constant structural damping ratio ξ is the simplest way of specifying damping in a structure [24, Section 1.4. Damping]. It represents the ratio of actual damping to critical damping, and is specified as a decimal number with the DMPRAT command. It is available only for spectrum, harmonic, and mode-superposition transient dynamic analyses. It is possible to define material dependent damping ratios using the MP,DMPR command.

Mode-Dependent Damping Ratio

Mode-dependent damping ratio (implemented with the MDAMP command) provides the ability to specify different damping ratios for different modes. It is available only for the spectrum and mode-superposition method of solution (transient dynamic and harmonic analyses). Note that it cannot be used in conjunction with material-dependent damping (MP,DMPR). As of ANSYS Release 14.5, this is not supported directly in Workbench and can only be implemented using APDL commands.

Rayleigh Damping

Rayleigh damping uses a linear combination of the mass and stiffness matrices to create the damping matrix and has certain mathematical advantages over other forms of damping. It is also known as *proportional damping* since the damping matrix is proportional to the mass and stiffness matrices. When using Rayleigh damping, the damping matrix $\mathbf{C}$ is defined by scaling the mass matrix $\mathbf{M}$ and stiffness matrix $\mathbf{K}$ by constants α (*alpha damping*) and β (*beta damping*), respectively [152, Eq. (16.13)]

$$\mathbf{C} = \alpha\mathbf{M} + \beta\mathbf{K}. \qquad (5.52)$$

In ANSYS Mechanical APDL, the commands ALPHAD and BETAD are used to specify α and β, respectively. The values of α and β are not generally known directly, but are calculated from modal damping ratios, ξ_i, where ξ_i is the ratio of actual damping to critical damping for a particular mode, i. If ω_i is the (angular) natural frequency of mode i, then α and β are chosen to satisfy the relation [152, Eq. (16.58)]

$$\xi_i = \frac{\alpha}{2\omega_i} + \frac{\beta\omega_i}{2}. \qquad (5.53)$$

If alpha damping (or mass damping) is ignored ($\alpha = 0$), then β can be evaluated from known values of ξ_i and ω_i,

$$\beta = 2\xi_i/\omega_i. \qquad (5.54)$$

Therefore, as the frequency increases, the damping increases for stiffness damping. According to Park [127, page 16-7] "as such, this representation of system damping is often used in the modeling of structural damping due to joint effects, acoustic noise and internal material friction." Note that only one value of β can be input in a load step, so it is necessary to choose the most dominant frequency active in that load step to calculate β.

For the special case of mass proportional damping (when beta damping or stiffness damping is ignored, $\beta = 0$), α can be evaluated from known values of ξ_i and ω_i,

$$\alpha = 2\xi_i\omega_i. \qquad (5.55)$$

Park [127, page 16-7] states that "from the physical viewpoint, the case of mass-proportional damping introduces higher modal damping for lower frequency solution components and the degree of damping decreases as the frequency increases. This does not, however, necessarily mean that the response components of the high-frequency modes will decay slower than those of the low-frequency modes within a time period. As a matter of fact, the decay rate is uniform for all frequency components."

To specify both α and β for a given damping ratio ξ, it is commonly assumed that the sum of the α and β terms is nearly constant over a range of frequencies (see Figure 5.20). Therefore, given ξ and a frequency range ω_1 to ω_2, two simultaneous equations can be solved for α and β:

$$\alpha = 2\xi\frac{\omega_1\omega_2}{\omega_1 + \omega_2}, \qquad (5.56)$$

and

$$\beta = \frac{2\xi}{\omega_1 + \omega_2}. \qquad (5.57)$$

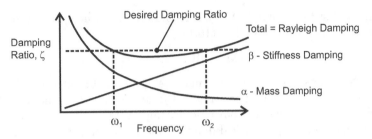

FIGURE 5.20
Schematic of Rayleigh damping and approximation of a constant damping ratio.

5.10.2   Example: Damping in a Rigid-Walled Cavity

This example is based on the model of the rigid-walled cavity described previously in Chapter 4. Three forms of spectral damping have been added to the system with a target damping ratio of $\xi = 0.01$. To validate that the desired damping has been achieved, the half-power (3 dB) bandwidth will be used to estimate the damping ratio achieved using [44, Eq. (2.27)]

$$\eta = 2\xi = \frac{\Delta f_{3\mathrm{dB}}}{f},\qquad\qquad (5.58)$$

where η is the modal loss factor, and $\Delta f_{3\mathrm{dB}}$ is the frequency bandwidth 3 dB down from the peak at the resonance frequency f. Note that this expression is strictly only valid for lightly damped modes.

5.10.3   MATLAB

The MATLAB script rigid_wall_cavity_damping.m included with this book is to be used with this example. The script is a modified version of the script rigid_wall_cavity.m presented previously in Chapter 4. The modifications to the file include code to read in and process the damped ANSYS results.

    The pressure at the receiver location [0.150, 0.120, 0.000] arising from a unit Mass Source at the source location [0.300, 0.105, 0.715] is plotted in Figure 5.21, where it is compared against the results from the three damped models.

5.10.4   ANSYS Mechanical APDL

This analysis is based on the model detailed in Chapter 4, and involves minor changes to the ANSYS Mechanical APDL file rigid_cavity_modal_super.inp included with this book. Open the file and increase the number of frequency steps to 1000 by modifying the scalar parameter numsteps using the APDL command numsteps=1000. Save the file as

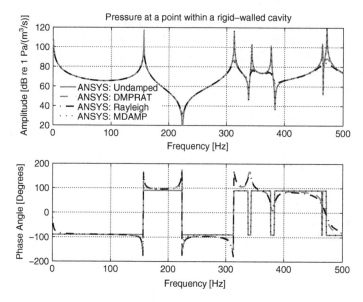

FIGURE 5.21

Pressure response of a receiver microphone due to a unit Mass Source in a rigid cavity. The three damped models have a nominal damping ratio of 1%. The model using DMPRAT has consistent damping over the entire frequency range. The model using MDAMP has damping only applied to the bulk compression mode and first dynamic mode (156 Hz). The model using Rayleigh damping has been optimized to produce a damping ratio of 1% in the frequency range 100 Hz to 200 Hz, after which the damping ratio exceeds the desired value.

rigid_cavity_modal_super_damped.inp, run the undamped case by typing \INPUT,rigid_cavity_modal_super_damped,inp in the command entry line, and rename the output file ansys_MSUP_p_receiver.txt to ansys_MSUP_p_receiver_undamped.txt.

5.10.4.1 Constant Damping Ratio

A constant damping ratio of $\xi = 0.01$ will be applied to the rigid cavity. To do this, edit the rigid_cavity_modal_super_damped.inp file and after the /SOLU command under the harmonic analysis, add the following command DMPRAT,0.01. Run the script in ANSYS Mechanical APDL. The resulting magnitude and phase of the pressure at the receiver is shown in Figure 5.21. Notice that the amplitude of the pressure levels at the receiving node have decreased substantially compared to the undamped case. Rename the output file ansys_MSUP_p_receiver.txt to ansys_MSUP_p_receiver_DMPRAT.txt.

The magnitude of the pressure at the receiver microphone near the resonance peak of the first (non-zero) mode is plotted in Figure 5.22, where the magnitude peaks at 90.4 dB at 156 Hz. Locating the frequencies 3 dB down from the peak (using linear interpolation) gives 154.4 Hz and 157.6 Hz. Using Equation (5.58) the damping ratio is

$$\xi = \frac{\Delta f_{3dB}}{2f} = \frac{157.6 - 154.4}{2 * 156} = 1.02\% ,\qquad (5.59)$$

which is close to the desired 1% and only differs because of the frequency resolution.

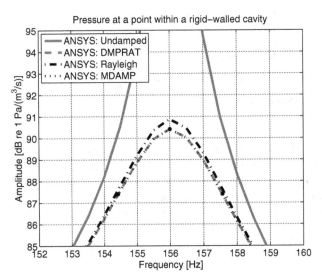

FIGURE 5.22
Pressure response of a receiver microphone due to a unit `Mass Source` in a rigid cavity. Three types of spectral damping have been used in an attempt to deliver a damping ratio of 1%. The solid circle markers indicate the peak frequency for the `DMPRAT` model, as well as the frequencies 3 dB down from the peak magnitude. The results using `DMPRAT` and `MDAMP` damping are coincident.

5.10.4.2 Rayleigh Damping

It is possible to approximate a constant damping ratio using Rayleigh damping. To illustrate this process we will attempt to create a damping ratio for the frequency range 100 Hz to 200 Hz, which spans the resonance of the first dynamic mode (with a natural frequency of 156 Hz). Using Equation (5.56) the `alpha` damping constant is

$$\alpha = 2 \times 0.01 \times (2\pi)\frac{100 \times 200}{100 + 200} = 8.378 ,\qquad (5.60)$$

and using Equation (5.57), the beta damping constant is

$$\beta = 2 \times 0.01 \times \frac{1}{2\pi(100 + 200)} = 1.061 \times 10^{-5}. \tag{5.61}$$

Edit the rigid_cavity_modal_super_damped.inp file and comment out the MDAMP command by inserting an exclamation mark ! at the start of the line. Immediately following this command, add the following two commands ALPHAD, 2*0.01*2*PI*(100*200)/(100+200) and BETAD, 2*0.01/(2*PI*(100+200)). Run the file by typing /INPUT,rigid_cavity_modal_super_damped,inp in the command entry line and rename the output file ansys_MSUP_p_receiver.txt to ansys_MSUP_p_receiver_Rayleigh.txt. The resulting complex pressure is shown in Figure 5.21 and the magnitude around the first resonance peak is shown in Figure 5.22. It can be seen from the figure, that around the octave over which α and β were solved, the pressure response for the Rayleigh damping is almost identical to the pressure response using a global damping ratio, with Figure 5.22 showing only a very minor deviation at the resonance peak, with the peak being slightly higher than desired. This is expected as illustrated in Figure 5.20, which shows that the effective damping in the center of the chosen frequency range is less than desired and more than desired at the limits of the range. Outside this frequency range, the pressure response begins to significantly deviate from desired as expected, with the deviation increasing as the frequency moves further from the lower or upper bounds of the range.

5.10.4.3 Mode-Dependent Damping

To illustrate the difference between a mode-dependent damping ratio and a global damping ratio, damping has been applied to a limited subset of the modes in this example. A damping ratio of 0.01 will be applied to only the first two modes of the cavity using the MDAMP command. Note that the first mode is the bulk compression mode (equivalent to a structural rigid body mode) and the second mode can be considered the fundamental dynamic mode. Edit the rigid_cavity_modal_super_damped.inp file and after the /SOLU command add the following command MDAMP, 1, 0.01, 0.01, , , , (ensuring all previous damping comments have been commented out). Run the file by typing /INPUT, rigid_cavity_modal_super_damped,inp in the command entry line and rename the output file ansys_MSUP_p_receiver.txt to ansys_MSUP_p_receiver_MDAMP.txt.

Looking at Figure 5.21 it is apparent that over the frequency range where the first two modes dominate (<200 Hz) the pressure response is almost identical to the pressure generated with the global damping ratio. At higher frequencies the pressure is almost identical to the undamped case, which is expected given that the higher-order modes are not damped. Figure 5.22 shows that at the resonance of the first non-zero mode the desired damping ratio is achieved.

6

Sound Absorption in a Lined Duct

6.1 Learning Outcomes

The learning outcomes for this chapter are:

- to learn how to model a rectangular duct lined with a sound-absorbing material,

- to understand how to use the FLUID220 quadratic acoustic elements in AN-SYS,

- to understand how to apply the acoustic Mass Source in ANSYS, implemented using BF,,JS in ANSYS Mechanical APDL,

- to understand how to use boundary impedances to model locally reactive surfaces with the SF,,IMPD surface load in ANSYS,

- to understand how to model porous materials using the Johnson–Champoux–Allard equivalent fluid model to simulate bulk reacting impedances and implemented with the TB,PERF command in ANSYS Mechanical APDL, and

- to understand the difference between insertion loss and transmission loss.

6.2 Definitions

In order to understand some of the concepts presented in this chapter it is necessary to define the following terms:

Acoustic impedance is the ratio of the pressure to the volume velocity (product of particle velocity and duct cross-sectional area) [47, Section 1.12].

Specific acoustic impedance is defined as the ratio of the pressure to the particle velocity [47, Section 1.12] and is given by the product of the effective complex density in the media, ρ_{eff}, and the complex speed of sound c_{eff} [45].

Characteristic acoustic impedance is the specific acoustic impedance in the free field and is given by $z_0 = \rho_0 c_0$.

Specific acoustic impedance ratio is a dimensionless quantity that normalizes the specific acoustic impedance to the characteristic acoustic impedance of air, $Z/(\rho_0 c_0)$ [45]. This is also sometimes referred to the normalized specific acoustic impedance.

Locally reacting liner is one where sound propagation may only occur normal to the surface of the liner, which implies it is characterized only by its local impedance and is completely independent of whatever occurs elsewhere in the liner. If it is assumed that a liner is locally reacting, this greatly simplifies an acoustic analysis.

Bulk reacting liner is one where sound can propagate in all directions, and therefore sound can propagate in the liner parallel to the axis of the duct. It is not easy to perform acoustic analyses of this type of liner using spreadsheets that are commonly used by acoustic practitioners, and instead must be solved using specialized software. Although a bulk reacting liner more accurately describes most absorptive silencers, most analysts model silencers as having the simpler locally reacting liners.

6.3   Description of the System

Dissipative lined ducts are used in a wide variety of applications including heating ventilation and air-conditioning (HVAC), industrial silencers, and aircraft jet engine nacelles. The earliest theory to describe the absorption of sound energy in such ducts was derived by Morse [114] and was based on the normal impedance for a locally reacting lining, where acoustic wave propagation within the lining parallel to the axis of the duct is prohibited. Scott [134] derived the attenuation for a homogeneous bulk-reacting liner. Later, Kurze and Vér [106] extended the model of Scott to include non-isotropic liners (aligned with the axis of the duct). A summary of the work of Morse [114], Scott [134], and Kurze and Vér [106] can be found in Wassilieff [147]. Further enhancements to the theory include mean flow in the duct; see Bies et al. [48].

This chapter contains examples of the attenuation of a plane wave in a rectangular duct, with a finite lined section as shown in Figure 6.1. Although we will be using 3D acoustic elements, the system is essentially 2D as the sound field will be uniform along the z-axis. The termination of the end of the duct is anechoic, which can be expressed mathematically as a real impedance ratio of unity. In this chapter both locally and bulk reacting liners will be modeled.

In the example used in this chapter, the Insertion Loss (IL) and Transmission Loss (TL) were calculated for both the locally reacting and bulk reacting liners. In the presence of the anechoic termination, the IL of the lined section in such a duct is given by the change in sound pressure level (SPL) downstream of the absorptive element with and without the lined section present. The TL is defined as the difference in the incident sound power level to the transmitted power level through the silencer.

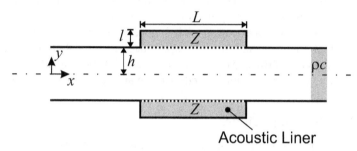

FIGURE 6.1
Schematic of a rectangular lined duct showing the acoustic liner and anechoic termination (indicated by ρc).

6.4 Theory

The following section describes the theories relevant to the calculation of IL and TL for both locally reactive and bulk reactive liners. The reader is referred to Section 3.2.3 for a more thorough discussion of IL and TL.

6.4.1 Insertion Loss (IL) and Transmission Loss (TL)

The Insertion Loss (IL) is typically used to classify the acoustic performance of silencing systems such as lined ducts. The IL of the silencing devices is defined [47, page 433] as the reduction in decibels in sound power transmitted through a duct with the silencer present, $L_{w,\text{with silencer}}$, compared to that transmitted through a rigid-walled duct, $L_{w,\text{w/o silencer}}$, and is given by

$$\text{IL} = L_{w,\text{w/o silencer}} - L_{w,\text{with silencer}} = \Delta L_w . \qquad (6.1)$$

Provided that there is an anechoic termination downstream of the silencing section, then the reduction in sound pressure level at a point (sufficiently far) downstream of the silencer is equal to the IL, i.e., [47, page 433]

$$\text{IL} = L_{p,\text{w/o silencer}} - L_{p,\text{with silencer}} = \Delta L_p . \qquad (6.2)$$

The Transmission Loss (TL) is defined as the difference in the incident sound power level to the transmitted sound power level (when the termination is anechoic) and is given by [47, page 433]

$$\text{TL} = L_{w,\text{ upstream}} - L_{w,\text{ downstream}} = \Delta L_w \ . \tag{6.3}$$

The IL provides a measure of the efficacy of the silencer as a function of the duct geometry, termination, and source impedances. For one-dimensional waveguides (when the wavelength is significantly greater than the characteristic dimensions) the TL is solely associated with attenuation across the silencer and is independent of the duct and source impedance or the distribution of energy within the modes of a duct. Note that the IL and TL become identical when both the source and termination impedance are anechoic [46, page 374].

6.4.2 Locally Reacting Liners

In this section a theoretical expression for the amount of absorption from a locally reacting liner is derived. Consider a section of a duct lined with non-rigid-walls comprising a locally reacting layer with normal impedance (defined as the ratio of the acoustic pressure to normal velocity at the absorbing surface) given by

$$Z_b = R + jX \ , \tag{6.4}$$

where R is the real (or resistive) part and X is the imaginary (or reactive) part. With reference to Figure 6.1, the boundary conditions at the locally reacting layer in a rectangular duct are [59, Chapter 8, Eq. (16)] [103, Eqs. (19) and (20)]

$$Z_b = \rho_0 c_0 \frac{k_0}{k_y} j \cot(k_y h) \ , \text{ for symmetrical modes} \tag{6.5}$$

$$= -\rho_0 c_0 \frac{k_0}{k_y} j \tan(k_y h) \ , \text{ for asymmetrical modes.} \tag{6.6}$$

where ρ_0 is the density of air, c_0 is the speed of sound in air, $k_0 = \omega/c_0$ is the wavenumber in free space, h is half the duct height, and k_y is the complex wavenumber in the y-axis (perpendicular to the axis of the duct).

Typically the plane wave (which is symmetrical) is the least attenuated mode and thus from Kurze and Allen [105, Eq. (1)] the cross distribution of waves in the duct (represented in terms of the wavenumber k_y) as a function of boundary conditions at the absorbing wall at $y = h$ (see Fig. 6.1) is

$$k_y h \tan(k_y h) = \frac{j k_0 h \rho_0 c_0}{Z_b} \ , \tag{6.7}$$

where $\rho_0 c_0$ is the characteristic acoustic impedance for plane waves. For a 2D system (zeroth order modes in the z-axis) the complex wavenumber in the x-axis (axis of the duct), k_x (also referred to as the axial wavenumber), is defined as

$$k_x^2 = k_0^2 - k_y^2 = (k_{x,R} + jk_{x,I})^2 \,. \tag{6.8}$$

Solving Equations (6.7) and (6.8) using a finite difference approximation, the complex axial propagation constant, $\gamma_x = jk_x$, (for the fundamental mode in the duct) in the presence of a locally reacting liner of impedance Z_b is given by Beranek [45, Eq. (15.14)] and Kurze and Allen [105, Eq. (5)]

$$\gamma_x = jk_x$$

$$\approx jk_0 \sqrt{1 - \left(\frac{2}{k_0 h}\right)^2 \left[1 + \cfrac{1}{1 + \cfrac{4Z_b}{2k_0 h \rho_0 c_0}} \pm \sqrt{1 + \cfrac{1}{\left(1 + \cfrac{4Z_b}{2k_0 h \rho_0 c_0}\right)^2}}\right]} \,. \tag{6.9}$$

The significance of the $\pm$ sign in the square root terms is that there are two solutions to the transcendental equation. We are interested in the result that produces the smallest value (least attenuation) and it is typically the negative sign which leads to the least attenuation [45, Page 510]. It should be noted that Equation (6.9) is only an approximation and tends to diverge from the true result for very high levels of attenuation and at high frequencies.

The attenuation per unit length of 2D duct is given by Beranek [45, Page 510], Mechel [112, Eq. (17)], and Kurze and Allen [105, Eq. (1)],

$$\frac{D_h}{h} = -8.69\mathrm{Im}\,\{k_x\} = 8.69\mathrm{Re}\,\{\gamma_x\} \ \ \mathrm{dB/m} \,, \tag{6.10}$$

where D_h/h is the attenuation per unit length of duct (and D_h is the attenuation along the x-axis for a length h of duct). The term $8.69 = 20\log_{10}(e) = 20/\ln(10)$ comes from converting nepers per unit length of duct to dB per unit length of duct. The total attenuation is thus given by the product of the above expression and the length of the lined section of duct. It can be shown that the optimal value of the impedance of the liner (to maximize the attenuation per unit length of duct) is [45, page 503],[112, Eq. (25)]

$$Z_{\mathrm{opt}} = \rho_0 c_0 \eta \,(0.92 - j0.77) \,, \tag{6.11}$$

where $\eta = 2h/\lambda$ is the duct height normalized by the wavelength of the sound. It is interesting to note that the optimal impedance is neither equal to the characteristic impedance of the duct, $\rho_0 c_0$, as is required to maximize attenuation at normal incidence; nor is it greater than $\rho_0 c_0$ in order to maximize oblique angles; nor is it entirely real. The reason is when maximum attenuation occurs at $l = \lambda/4$ some stiffness is required. From Mechel [112, Eq. (26)] the maximum attenuation is given by

$$\frac{D_{h,\mathrm{max}}}{h} = \frac{1}{h}8.69\mathrm{Re}\left(\sqrt{3.42 - (k_0 h)^2 + j5.24}\right) \ \ \mathrm{dB/m} \,, \tag{6.12}$$

where for low frequencies ($k_0 h \ll 3.42$) then $D_{h,\max} \simeq 19$ dB which represents the attenuation (per axial length h of duct) of the least attenuated mode [45, Section 15.3].

6.4.3 Darcy's Law, Flow Resistivity, and the Relationship with Impedance

Before commencing with a discussion on porous media, it is worthwhile briefly discussing the flow through such media and flow resistivity.

6.4.3.1 Darcy's Law

Darcy's law is the relationship between the instantaneous discharge rate, Q, in m^3/s through a porous medium, the dynamic viscosity of the fluid, μ_0, in Pa.s and the pressure drop, ΔP, in Pa over a given distance L, such that

$$Q = \frac{-k_D A}{\mu_0} \frac{\Delta P}{L} , \qquad (6.13)$$

where k_D is the permeability of the medium in m^2 and A is the cross-sectional area normal to the flow (m^2). Dividing both sides of Equation (6.13) by the area through which the flow is discharging leads to

$$q = \frac{-k_D}{\mu_0} \nabla P , \qquad (6.14)$$

where q is the mean (or Darcy) flux (with units of m/s) and $\nabla P = \Delta P / L$ is the pressure gradient vector (Pa/m). Note the Darcy flux is not the velocity at which the fluid travels through the pores. The pore velocity is related to the Darcy flux and the porosity, ϕ, by $v = q/\phi$. The flux is divided by porosity to account for the fact that only a fraction of the total formation volume is available for flow.

6.4.3.2 Flow Resistivity

The flow resistivity in acoustics is defined as the ratio of the pressure drop per unit length across a sample to the velocity through the sample, and is given by Bies and Hansen [47]

$$R_1 = \frac{\Delta P A}{Q L} = \frac{\Delta P}{q L} , \qquad (6.15)$$

and therefore the flow resistivity is related to the permeability by

$$R_1 = \frac{-\mu_0}{k_D} \quad \text{or} \quad k_D = \frac{-\mu_0}{R_1} . \qquad (6.16)$$

There are standard tests for measuring the flow resistivity. This may be done directly by applying a constant flow rate of air across a test sample of the porous material, measuring the pressure drop across the surface, and then applying Equation (6.15) [41]. Alternatively, it may be estimated from the acoustic properties by conducting impedance tube measurements [87, 88, 39] or indirectly from tests using reverberation rooms to measure the sound absorption coefficient [40, 85] (discussed further in Chapter 7).

6.4.3.3 Delany and Bazley

Delany and Bazley ([2, Section 2.5.3]) measured the characteristic impedance of many fibrous materials over a large range of frequencies and found the following relationship as a function of frequency, f, and flow resistivity, R_1, [2, Eqs. (2.28) and (2.29)][100, Eqs. (1) to (3)]

$$Z = \rho_0 c_0 \left(1 + 0.0570 \left(\frac{\rho_0 f}{R_1} \right)^{-0.754} - j0.0870 \left(\frac{\rho_0 f}{R_1} \right)^{-0.732} \right), \qquad (6.17)$$

with the associated complex characteristic wavenumber given by

$$k = \frac{\omega}{c_0} \left(1 + 0.0978 \left(\frac{\rho_0 f}{R_1} \right)^{-0.700} - j0.1890 \left(\frac{\rho_0 f}{R_1} \right)^{-0.595} \right), \qquad (6.18)$$

where $\omega = 2\pi f$. The fit used to derive the above expressions was found to be acceptable within the bounds $0.01 < \left(\frac{\rho_0 f}{R_1} \right) < 1$. Mechel [65, Eq. (7.14)] presented the more accurate expressions for $\rho_0 f / R_1 \leq 0.025$,

$$Z = \rho_0 c_0 \left(1 + 0.081 \left(\frac{\rho_0 f}{R_1} \right)^{-0.699} - j0.191 \left(\frac{\rho_0 f}{R_1} \right)^{-0.556} \right), \qquad (6.19)$$

$$k = \frac{\omega}{c_0} \left(1 + 0.136 \left(\frac{\rho_0 f}{R_1} \right)^{-0.641} - j0.322 \left(\frac{\rho_0 f}{R_1} \right)^{-0.502} \right), \qquad (6.20)$$

and for $\rho_0 f / R_1 > 0.025$

$$Z = \rho_0 c_0 \left(1 + 0.0563 \left(\frac{\rho_0 f}{R_1} \right)^{-0.725} - j0.127 \left(\frac{\rho_0 f}{R_1} \right)^{-0.655} \right), \qquad (6.21)$$

$$k = \frac{\omega}{c_0} \left(1 + 0.103 \left(\frac{\rho_0 f}{R_1} \right)^{-0.716} - j0.179 \left(\frac{\rho_0 f}{R_1} \right)^{-0.663} \right). \qquad (6.22)$$

The above models, whilst commonly used to model porous media, are not available in ANSYS Release 14.5, however, the Delany and Bazley model has been implemented in ANSYS Release 15, along with a Miki model. A more elaborate phenomenological model available in ANSYS is discussed in Section 6.4.4.3.

6.4.3.4 The Effect of Temperature on Impedance

One final point worth noting is the effect of temperature on impedance. It is well known that temperature affects the speed of sound and the density of a gas, [83, page 287, Eqs. (8.6) and (8.7)]

$$c_T = c_0 \left(\frac{T}{T_0} \right)^{0.5} , \qquad (6.23)$$

$$\rho_T = \rho_0 \left(\frac{T_0}{T} \right) , \qquad (6.24)$$

where T is the actual absolute temperature (in Kelvin) and T_0 is the ambient absolute temperature (in Kelvin). What is sometimes neglected is the effect of temperature on viscosity, which thus influences flow resistivity. The viscosity of a gas (derived using kinetic theory) shows that the viscosity is proportional to $\sqrt{T}$, and therefore the flow resistance is also proportional to the square root of the absolute temperature [83, pages 252, 287],

$$R_{1,T} = R_1 \left(\frac{T}{T_0} \right)^{0.5} , \qquad (6.25)$$

where $R_{1,T}$ is the flow resistivity at the absolute temperature T. Therefore the normalized flow resistance is [83, Eq. (8.12)]

$$\frac{R_{1,T}}{\rho_T c_T} = \frac{R_1}{\rho_0 c_0} \left(\frac{T}{T_0} \right) , \qquad (6.26)$$

which is proportional to the temperature.

Since ANSYS Release 14.5, temperature dependence is possible in its acoustic elements as discussed in Section 3.6, where the density and the speed of sound are automatically compensated for temperature. For the 2D FLUID29 element, the sound absorption coefficient, MU, is also adjusted for temperature (inversely proportional to the characteristic impedance which is proportional to the root of temperature). For the 3D acoustic elements FLUID30, FLUID220, and FLUID221, ANSYS evaluates the dynamic viscosity, VISC, at the average nodal temperature. However, it is still up to the user to modify the impedances of surfaces (using the ANSYS Mechanical APDL command SF,,IMPD) and sheets (using the command BF,,IMPD) to accommodate temperature changes of the fluid local to the impedance. A final note on temperature dependence, when using the surface absorption coefficient SF,,CONV in ANSYS it is automatically adjusted for temperature.

6.4.4 Bulk Reacting Liners

In bulk reacting liners, sound can propagate within the liner along the axis of the duct. The theory which describes their acoustic behavior is considerably more involved than the case for the locally reacting liner presented in Section 6.4.2, especially for non-isotropic materials. Here we will only present the theory for isotropic materials since this is the model that is supported in the current version of ANSYS (Release 14.5).

6.4.4.1 Isotropic Media with No Mean Flow

There is considerable literature on the attenuation achieved by isotropic lined ducts. The approach used to derive the transcendental equation for the locally reacting liner, Equation (6.7), can be extended for a bulk reacting liner. The theory presented below has been drawn from Scott [134] and Wassilieff [147]. From Wassilieff [147, Eq. (1)] the transcendental equation may be written as

$$
\sqrt{(k_0 h)^2 + (\gamma_x h)^2} \tan \sqrt{(k_0 h)^2 + (\gamma_x h)^2} =
$$
$$
\frac{-j\rho_0 c_0 k_0}{Z_l \gamma_l} \sqrt{(\gamma_x h)^2 - (\gamma_l h)^2} \tan \left(\frac{l}{h} \sqrt{(\gamma_x h)^2 - (\gamma_l h)^2} \right), \quad (6.27)
$$

where l is the liner thickness, h is half the duct height, Z_l is the liner characteristic impedance, and γ_l is the bulk propagation constant in the liner. The left-hand term is the same as the locally reacting transcendental equation, Equation (6.7) after the substitution of Equation (6.8) is made and noting that $\gamma_x = jk_x$.

In order to solve Equation (6.27) it is necessary to rewrite it as

$$
f(\gamma_x) = \frac{-j\rho_0 c_0 k_0}{Z_l \gamma_l} \sqrt{(\gamma_x h)^2 - (\gamma_l h)^2} \tan \left(\frac{l}{h} \sqrt{(\gamma_x h)^2 - (\gamma_l h)^2} \right)
$$
$$
- \sqrt{(k_0 h)^2 + (\gamma_x h)^2} \tan \sqrt{(k_0 h)^2 + (\gamma_x h)^2} \quad (6.28)
$$

and solve the roots for this iteratively using numerical methods such as a Newton–Raphson method (for example, the nonlinear solver `fsolve` in MATLAB). The attenuation along the silencer may then be calculated using Equation (6.10).

6.4.4.2 Perforated and Limp Surface Facings

In many real industrial silencers the sound-absorbing material is faced with a perforated sheet and a limp liner which improves durability. The effect of this is to add an additional impedance (mass reactance). For a limp liner with surface density ρ_m and perforated sheet with hole diameter D_s the transcendental equation becomes [48, Eq. (53)]

$$\sqrt{(k_0 h)^2 + (\gamma_x h)^2}\, \tan\sqrt{(k_0 h)^2 + (\gamma_x h)^2} =$$
$$\frac{-j\rho_0 c_0 k_0}{Z_l \gamma_l}\sqrt{(\gamma_x h)^2 - (\gamma_l h)^2}\, \tan\left(\frac{l}{h}\sqrt{(\gamma_x h)^2 - (\gamma_l h)^2}\right)$$
$$\times \left(1 - \frac{\rho_t}{\rho_0 h}\frac{j\rho_0 c_0 k_0}{Z_l \gamma_l}\sqrt{(\gamma_x h)^2 - (\gamma_l h)^2}\, \tan\left(\frac{l}{h}\sqrt{(\gamma_x h)^2 - (\gamma_l h)^2}\right)\right)^{-1}$$

$$(6.29)$$

where $\rho_t = \rho_m + \rho_0(h_s + a D_s)/\alpha$ is the total surface density, α and h_s are the fraction of open area and the thickness of the perforated sheet, respectively, and $a = 0.8$ is a constant.

6.4.4.3 Porous Media

The infill for bulk reacting silencers is generally some form of porous media. These may be broadly classified as either having a rigid frame (where the frame does not move under the acoustic field) or an elastic frame (which does move). The discussion below will be restricted to rigid-framed models. Consider Figure 6.2, which shows a plane wave propagating through two contiguous spaces; an isotropic fluid and an isotropic porous media. The porous media comprises a rigid (sometimes referred to as a motionless) frame and a fluid which fills the pores. On a microscopic scale, the modeling of sound propagation is generally very difficult because of the complicated geometries of the frame. However, as engineers we are interested in the macroscopic properties which are obtained by averaging over a "homogenization volume" with dimensions sufficiently large to be statistically accurate, while at the same time being smaller than the acoustic wavelength [145]. The following discussion will explore the relationship between the microscopic properties of the porous media and bulk acoustic characteristics.

With reference to Figure 6.2, the continuity of mass flow and pressure at the interface between the two layers implies [2, Eqs. (4.130) and (4.131)]

$$p_2 = p_1 \tag{6.30}$$

$$v_2 = \phi v_1, \tag{6.31}$$

where $\phi = V_{pores}/V_{total}$ is the porosity (or ratio of open volume) of the media, and v_1 and v_2 are the particle velocities in the porous media and isotropic fluid, respectively. The porosity represents the ratio of the volume occupied by the fluid phase in the pores, V_{pores}, to the total volume of the material, V_{total}, and is therefore less than unity. From Equation (6.31) it can be seen that for a constant particle velocity, v_2, in the isotropic fluid, as the porosity decreases the velocity through the pores v_1 must increase in order to maintain conservation of volume velocity—for example, if half of the space was filled with rigid material, then one would expect that the velocity in the

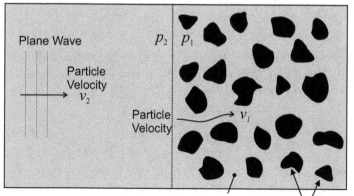

FIGURE 6.2
Schematic of an isotropic fluid in contact with a rigid (motionless) frame isotropic porous media at a microscopic scale.

porous media region would be double the velocity in the isotropic fluid region. The two impedances at the surface are [145][2, Eqs. (4.132) and (4.133)]

$$Z_2 = \frac{p_2}{v_2} \tag{6.32}$$

$$Z_1 = \frac{p_1}{v_1} = \phi Z_2 . \tag{6.33}$$

The homogeneous wave equation for the fluid phase was established by Zwikker and Kosten [92] and is given by Allard and Atalla [2, Eq. (4.134)]

$$\nabla^2 p + \omega^2 \frac{\tilde{\rho}}{\tilde{K}} p = 0 , \tag{6.34}$$

where $\tilde{\rho}$ and $\tilde{K}$ are the dynamic density and dynamic bulk modulus of the fluid phase in the pores and are complex functions of the frequency and pore geometry of the media. The over-tilde indicates that the associated variable is frequency dependent and complex valued. Thus, the complex speed of sound in the pore is given by

$$\tilde{c} = \sqrt{\frac{\tilde{K}}{\tilde{\rho}}} . \tag{6.35}$$

The specific acoustic impedance in the pore is given by Allard and Atalla [2, Eq. (4.135)]

$$Z_c = \sqrt{\tilde{\rho}\tilde{K}} = \tilde{\rho}\tilde{c}. \tag{6.36}$$

The complex wavenumber of the fluid in the pore is [2, Eq. (4.135)]

$$\tilde{k} = \omega\sqrt{\tilde{\rho}/\tilde{K}} = \omega/\tilde{c}. \tag{6.37}$$

The dynamic density of the fluid phase, $\tilde{\rho}$, is given by Panneton and Olny [126, Eqs. (1)–(4)], Allard and Atalla [2, Eqs. (5.50) and (5.52)], Jaouen [92, Visco-inertial effects], and Jaouen and Becot [93, Eq. (14)]

$$\tilde{\rho} = \frac{\sigma\phi}{j\omega}\sqrt{1 + \frac{j4\alpha_\infty^2\,\mu_0\omega\rho_0}{\phi^2\Lambda^2\sigma^2}} + \rho_0\alpha_\infty \,, \tag{6.38}$$

where ρ_0 is the density of the ambient fluid (kg.m^{-3}), σ is the flow resistivity (N.s.m^{-4}), α_∞ is the tortuosity, Λ is the viscous characteristic length (m), ω is the circular frequency (rad/s), and μ_0 is the dynamic viscosity of the ambient fluid (kg.m^{-1}.s^{-1}). The tortuosity is defined by the ratio of the average aperture length to the thickness of the material, such that as the structure of the porous media becomes more complicated, the tortuosity increases, and therefore is greater than unity. The viscous characteristic length is associated with the viscous effects at mid to high frequencies.

The dynamic bulk modulus of the fluid phase is given by Olny and Panneton [123, Eqs. (1)–(3)], Allard and Atalla [2, Eqs. (5.51) and (5.52)], Jaouen and Becot [93, Eq. (20)], and Jaouen [92, Thermal effects]

$$\tilde{K}(\omega) = \frac{P_0\gamma}{\gamma - (\gamma - 1)\left[\dfrac{8\mu_0}{j\omega\rho_0 P_{rt}\Lambda'^2}\sqrt{1 + \dfrac{\Lambda'^2}{16}\dfrac{j\omega\rho_0 P_{rt}}{\mu_0}} + 1\right]^{-1}} \,, \tag{6.39}$$

where γ is the specific heat ratio ($= 1.4$ for air), P_0 is the static reference pressure (Pa), P_{rt} is the Prandtl number, and Λ' is the thermal characteristic length (m).

Other useful relationships are [60, Eqs. (7), (8), (9), and (10), respectively]

$$\sigma = -\frac{1}{\phi}\lim_{\omega\to 0}\left(\text{Im}\,\{\omega\tilde{\rho}\}\right) \tag{6.40}$$

$$\alpha_\infty = \frac{1}{\rho_0}\left(\text{Re}\,\{\tilde{\rho}\} - \sqrt{\text{Im}\,\{\tilde{\rho}\}^2 - \left(\frac{\sigma\phi}{\omega}\right)^2}\right) \tag{6.41}$$

$$\Lambda = \alpha_\infty\sqrt{\frac{2\rho_0\mu_0}{\omega\text{Im}\,\{\tilde{\rho}\}\,(\rho_0\alpha_\infty - \text{Re}\,\{\tilde{\rho}\})}} \tag{6.42}$$

$$\Lambda' = \delta_t\sqrt{2}\left(-\text{Im}\left\{\left(\frac{1 - \tilde{K}/K_0}{1 - \gamma\tilde{K}/K_0}\right)^2\right\}\right)^{-\frac{1}{2}} , \tag{6.43}$$

where $K_0 = \rho_0 c_0^2$ is the adiabatic bulk modulus (and for an ideal gas $K_0 = \gamma P_0$) of the ambient air and $\delta_t = \sqrt{2\mu_0/(\rho_0 \omega P_{rt})}$ is the "thermal skin depth."

When testing such porous media, we are not so much interested in the behavior of the fluid particles in the pores, but rather the bulk or average properties of the media. This leads to the concept of an equivalent fluid [2, Sections 4.8.2, 5.7]. The homogeneous wave equation for such an equivalent fluid media was established by Zwikker and Kosten [92] and is given by [125, Eq. (4)]

$$\nabla^2 p + \omega^2 \frac{\tilde{\rho}_{eq}}{\tilde{K}_{eq}} p = 0 , \qquad (6.44)$$

where the subscript eq indicates the equivalent fluid. The relationship between the dynamic density of the fluid phase of the medium and the dynamic density of the rigid-frame equivalent fluid (bulk property) is given by Panneton and Olny [126, Eq. (6)], Allard and Atalla [2, Eq. (5.43)], and Panneton [125, Eq. (4)]

$$\tilde{\rho}_{eq} = \frac{\tilde{\rho}}{\phi} . \qquad (6.45)$$

The relationship between the bulk modulus of the fluid phase of the medium, $\tilde{K}$, and the equivalent bulk modulus (bulk property) is given by Allard and Atalla [2, Eq. (5.44)] and Panneton [125, Eq. (4)]

$$\tilde{K}_{eq} = \frac{\tilde{K}}{\phi} . \qquad (6.46)$$

The complex velocity of the equivalent fluid is given by

$$\tilde{c}_{eq} = \sqrt{\frac{\tilde{K}_{eq}}{\tilde{\rho}_{eq}}} . \qquad (6.47)$$

The specific acoustic impedance of the equivalent fluid is given by [93, Eq. (5)]

$$Z_{c,eq} = \sqrt{\tilde{\rho}_{eq} \tilde{K}_{eq}} = \tilde{\rho}_{eq} \tilde{c}_{eq} . \qquad (6.48)$$

The complex wavenumber of the equivalent fluid [2, Eq. (5.41)][93, Eq. (6)]

$$\tilde{k}_{eq} = \omega \sqrt{\tilde{\rho}_{eq}/\tilde{K}_{eq}} = \omega/\tilde{c}_{eq} . \qquad (6.49)$$

Therefore, the relationship between the impedance of the fluid phase in the porous media and the bulk properties of an equivalent fluid is

$$Z_{c,eq} = \frac{Z_c}{\phi}, \qquad (6.50)$$

and the other characteristics are related by $\tilde{c}_{eq} = \tilde{c}$ and $\tilde{k}_{eq} = \tilde{k}$. Therefore, the surface impedance at normal incidence of a layer of isotropic porous medium of thickness l is given by $Z_s = -jZ_c \cot(kl)/\phi$, which is identical to the surface impedance of a layer of isotropic fluid of the same thickness given by $Z_{s,\text{eq}} = -jZ_{c,\text{eq}} \cot(k_{\text{eq}}l)$. Hence it is possible to replace a porous medium by a homogenous fluid layer, with equivalent density, $\tilde{\rho}_{\text{eq}}$, and bulk modulus, $\tilde{K}_{\text{eq}}$, without modifying the reflected field in the external medium [2, Section 5.7][145].

Implementation of "Equivalent Fluid of Perforated Materials" in ANSYS: The Johnson–Champoux–Allard Equivalent Fluid Model

As discussed in Table 5.2 in Chapter 5, a bulk reacting material may be modeled in ANSYS by using the Johnson - Champoux - Allard Equivalent Fluid Model of Perforated Material [27], which is a rigid-frame porous media model. Note that ANSYS refers to this model as a *perforated material* rather than the more conventional term *porous media*. In this book, the more common "porous media" terminology will be used unless explicitly referring to actual ANSYS commands, and "perforated material" will be used to describe perforated sheets.

In ANSYS the Johnson–Champoux–Allard model is activated by issuing the TB,PERF command in an acoustic full harmonic analysis to define an equivalent fluid model of a porous medium. The physical constants are input using the TBDATA command, and are presented in Table 6.1. Note that the Johnson–Champoux–Allard model is limited to 3D acoustic elements.

TABLE 6.1

Physical Constants for the Johnson–Champoux–Allard Model in ANSYS Entered Using the APDL Command TBDATA

Constant	Symbol	Description	Units
C1	σ	Flow resistivity	N.s/m^4
C2	ϕ	Porosity (defaults to 1)	—
C3	α_∞	Tortuosity (defaults to 1)	—
C4	Λ	Viscous characteristic length	m
C5	Λ'	Thermal characteristic length	m

It should be noted that the flow resistivity, σ, the viscous characteristic length, Λ, and the thermal characteristic length, Λ', are all temperature dependent. Since ANSYS (Release 14.5) does not provide a tabular form of Johnson–Champoux–Allard model data, it is not clear how one would model the effects of varying temperature in an absorbent except to define each element to have different material properties.

An implicit assumption in the Johnson–Champoux–Allard model is that the "frame" of the porous medium is rigid. From Panneton and Olny [126, p. 2028]:

In this case, the porous medium is seen as an equivalent fluid characterized by an equivalent dynamic density $\tilde{\rho}_{\text{eq}}$ and a dynamic bulk modulus $\tilde{K}_{\text{eq}}$ —

the over-tilde indicates that the associated variable is frequency dependent and complex valued.

The ANSYS theory manual [27, Eq. (8-80)] defines the wave equation for a "perforated material" as

$$\nabla \cdot \left(\frac{1}{\rho_{\text{eff}}} \nabla p_a \right) + \frac{\omega^2}{\rho_{\text{eff}} c_{\text{eff}}} p_a = 0 , \qquad (6.51)$$

where p_a is the acoustic pressure, and ρ_{eff} is the effective density and can be shown to be equal to the dynamic density of the fluid phase given by Equation (6.38) [27, Eq. (8-81)]. The effective bulk modulus, K_{eff} can be shown to be equal to the dynamic bulk modulus of the fluid phase and is given by Equation (6.39) [27, Eq. (8-82)]. The complex effective velocity, c_{eff}, is given by Equation (6.35) [27, Eq. (8-83)]. The specific acoustic impedance of the material is given by Equation (6.36) [27, Eq. (8-84)].

It can be shown that the "effective" terms employed in ANSYS are the same as the terms used to describe the fluid phase in the pores of the porous media. Unfortunately the implementation in ANSYS (Release 14.5) is not what a user would expect, as one would normally solve for the bulk equivalent not just the fluid phase as discussed in Panneton and Olny [126] and Allard and Atalla [2, Section 5.7]. This functionality will be changed in Release 16.0 of ANSYS so that the JCA model will use the equivalent fluid properties. Until then, the solutions from ANSYS (Release 14.5) will lead to an incorrect estimate for the impedance, i.e.,

$$Z_c = \sqrt{\tilde{\rho}_{\text{eq}} \tilde{K}_{\text{eq}}} = \frac{1}{\phi} \sqrt{\rho_{\text{eff}} K_{\text{eff}}} = \frac{1}{\phi} Z_{c,\text{Ansys}}. \qquad (6.52)$$

Furthermore, the pressure and particle velocity will be incorrect by a factor of ϕ. This also implies that velocity sources and volume velocity sources (such as the acoustic Mass Source and FLOW acoustic source) will also be incorrect by a factor of ϕ. It should be noted that the porosity ϕ for many sound-absorbing materials are close to unity, and therefore the solutions delivered by ANSYS (Release 14.5) using Johnson–Champoux–Allard model are likely to be very close to the correct solution. Note that despite the characteristic impedance differing by a factor of ϕ, the surface impedance of a finite thickness layer is the same [2, Section 5.7 Fluid layer equivalent to a porous layer] and therefore "the porous medium can be replaced by the homogeneous fluid layer without modifying the reflected field in the external medium."

6.5   Example: Locally Reacting Liner

We will now consider a specific example of a 5.0 m lined duct, with a plane wave source at one end and an anechoic termination at the other (as shown in

Figure 6.1). A 1.0 m section of lined duct spans from 2.0 m to 3.0 m. The liner was initially modeled as a locally reacting liner with an optimal impedance at 250 Hz determined using Equation (6.11). The optimal impedance was determined by first calculating the wavelength at the frequency of $f = 250$ Hz using $\lambda = c_0/f$, which may then be inserted into Equation (6.11) along with the substitution $\eta = 2h/\lambda$. This value of impedance was chosen to provide an extreme value of attenuation and highlight the numerical limits of the calculation within ANSYS. In Section 6.6 a bulk reacting liner with a more realistic attenuation is modeled. The parameters used in the example are listed in Table 6.2.

In the following sections a method to model and analyze the system is presented for both MATLAB and ANSYS. A finite element model was built in ANSYS using the parameters detailed in Table 6.2. After which a harmonic analysis was conducted on the model from 25 Hz to 1 kHz. The attenuation in dB/m for the locally reacting liner was calculated in MATLAB using Equations (6.9) and (6.10).

The IL from the ANSYS results was calculated using Equation (6.2) by calculating the SPL at the anechoic termination for the rigid-walled case and then later for the same location in the presence of the silencer elements. The IL is equal to the change in the SPL.

The time-averaged sound power from a plane wave passing through the duct of a cross-sectional area, A, is $W = \bar{I}A$, with corresponding sound power level (given by Equation (2.31)),

$$L_w = 10\log_{10}\left(\frac{\bar{I}A}{W_{\text{ref}}}\right) \text{ dB re } 10^{-12}\text{W}, \qquad (6.53)$$

where $\bar{I}$ is the time-averaged sound intensity, and $W_{\text{ref}} = 10^{-12}\text{W}$ is the reference sound power. For a duct with an anechoic termination, the time-averaged sound intensity of a plane traveling through the duct is given by,

$$\bar{I} = \frac{\overline{p^2}}{\rho_0 c_0}, \qquad (6.54)$$

where $\overline{p^2}$ is the time-averaged squared pressure. Therefore the sound power level as a function of the sound pressure level in the anechoic duct is [47, Eq. (1.87)]

$$L_w = L_p + 10\log_{10}\left(\frac{A}{\rho_0 c_0}\right) + 26\,\text{dB re } 10^{-12}\text{W}, \qquad (6.55)$$

where $L_p = 10\log_{10}\left(\overline{p^2}/p_{\text{ref}}^2\right)$ is the sound pressure level, $p_{\text{ref}} = 20\mu\text{Pa}$, and the term $26\,\text{dB} = 10\log_{10}\left(W_{\text{ref}}/p_{\text{ref}}^2\right)$. Hence the TL from the ANSYS results were calculated using a modified version of Equation (6.3) as follows

$$\begin{aligned} \text{TL} &= L_{w,\text{ upstream}} - L_{w,\text{ downstream}} \\ &= L_{w,\text{ upstream}} - \left(L_{p,\text{ downstream}} + 10\log_{10}\left(\frac{A}{\rho_0 c_0}\right) + 26\right) \end{aligned} \qquad (6.56)$$

TABLE 6.2
Parameters Used in the Analysis of a Lined Duct System

Description	Parameter	Value	Units
Air:			
Speed of sound	c_0	344	m/s
Density of air	ρ_0	1.21	kg/m^3
Ratio of specific heats	γ	1.4	
Adiabatic bulk modulus of air*	$K_0 = \rho_0 c_0^2$	1.4319e+5	Pa
Static Pressure (ideal gas)*	$P_0 = K_0/\gamma$	1.0228e+5	Pa
Prandtl number (at 20°C)	P_{rt}	0.713	—
Duct:			
Duct length	L_d	5.0	m
Duct height	$W_d = 2h$	0.25	m
Duct depth - arbitrary	D_d	0.25	m
ANSYS Mass Source	Mass Source	1	kg/s/m^2
Termination absorption coefficient	α	1	—
Silencer:			
Silencer length	L_s	1.0	m
Silencer thickness	l	0.1	m
Locally Reacting:			
Locally reacting impedance	Z_{opt}	69.575 $-j58.231$	kg/s/m^2
Max. attenuation*	$D_{h,max}/h$	153	dB/m
Freq. at max. attenuation*	f_{max}	250	Hz
Absorbent:			
Flow resistivity	σ	10800	MKS Rayls/m
Material porosity	ϕ	0.98	—
Material tortuosity	α_∞	1.04	—
Viscous characteristic length	Λ	129e-6	m
Thermal characteristic length	Λ'	198e-6	m
Effective density*	$\rho_{eq}(\omega \to 0)$	1.53	kg/m^3
Effective bulk modulus*	$K_{eq}(\omega \to 0)$	0.707	kg/m/s^2

* Dependent parameters

where the subscripts downstream and upstream are in reference to the silencer section calculated at the termination and source end of the duct, respectively. The area term can be removed from Equation (6.56). Since the TL across the rigid-walled duct is zero, $\text{TL}_{\text{rigid}} = 0$, then Equation (6.56) can be written as

$$
\begin{aligned}
\text{TL} \;=\;& \text{TL}_{\text{silencer}} - \text{TL}_{\text{rigid}} \\[4pt]
=\;& L_{w,\text{ upstream, silencer}} \\[4pt]
& - \left(L_{p,\text{ downstream, silencer}} + 10\log_{10}\left(\frac{A}{\rho_0 c_0}\right) + 26 \right) \\[4pt]
& - L_{w,\text{ upstream, rigid}} \\[4pt]
& + \left(L_{p,\text{ downstream, rigid}} + 10\log_{10}\left(\frac{A}{\rho_0 c_0}\right) + 26 \right) \\[4pt]
=\;& \left(L_{w,\text{ upstream, silencer}} - L_{w,\text{ upstream, rigid}} \right) \\[4pt]
& - \left(L_{p,\text{ downstream, silencer}} - L_{p,\text{ downstream, rigid}} \right) \qquad (6.57)
\end{aligned}
$$

where the subscripts "silencer" and "rigid" refer to the duct with the silencer present and rigid-walled duct, respectively.

6.5.1   MATLAB

The MATLAB script lined_duct.m included with this book is to be used with this example. The script is used to define the key parameters defined in Table 6.2, then using the expressions derived in Section 6.4.2, the attenuation per unit length is calculated using Equations (6.9), (6.10) and (6.11). The script also analyzes the results from ANSYS Workbench and ANSYS Mechanical APDL.

6.5.2   ANSYS Workbench

Note that throughout this example, the Scoping Method will switch between Geometry Selection and Named Selection to illustrate both methods. In practice a user would tend to use to one method.

6.5.2.1   Rigid-Walled Duct

In this section we will first build the model and evaluate the results for the rigid-walled duct, since this forms the baseline result and is simple to validate. In Section 6.5.2.2 the locally reacting liner will be simulated.

Constructing the Solid Model

The completed ANSYS Workbench project file Lined_Duct.wbpj is available with this book.

- Start ANSYS Workbench and start a new project.

- It is assumed that the ACT Acoustics extension is installed and is operating correctly. This can be checked in the Workbench project view by selecting the Extensions | Manage Extensions menu. The extension ExtAcoustics should be listed in the table and a tick should be present in the Load column.

- Double-click on Harmonic Response under Analysis Systems in the Toolbox window, so that a new Harmonic Response cell appears in the Project Schematic window.

- Double-click on row 3 Geometry to start DESIGNMODELER. Select the desired length unit as Meter and click the OK button.

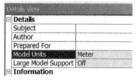

- In the Tree Outline window, left-click on A: Harmonic Response. In the Details View window, ensure that the row Model Units shows Meter.

Details View	
Details	
Subject	
Author	
Prepared For	
Model Units	Meter
Large Model Support	Off
Information	

- The first step in the creation of the solid model of the acoustic system is to create a rectangular duct. In the toolbar, click on Create | Primitives | Box.

- In the Details View window, ensure that the Box Type is set to From One Point and Diagonal. Then proceed to define the geometry of the box as shown below, where the coordinates for the Point 1 Definition are $[0, 0, 0]$ and the coordinates for the Diagonal Definition are $[5.000, 0.250, 0.250]$. Change the row Box from Box1 to Duct.

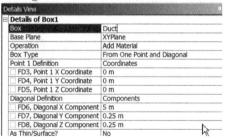

Inspect the model by either clicking on the lone point in the triad in the Graphics window (Model View tab) or click the ISO button in the toolbar and you will see an isometric view of the box you have just created.

- Left-click the mouse button in the Graphics window and then use the middle mouse wheel to scroll the view until the dimension scale at the bottom of the screen shows 5.000 m. This is to make it easier to construct the model.

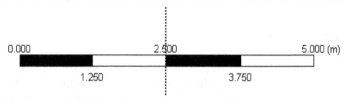

- We will now define the dimensions of the box as parameters. In the Details View window, click on the square check box to the left of the FD6, Diagonal X Component, and in the dialog box Parameter Name: type Duct.Length.

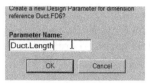

Repeat the process for the FD7, Diagonal Y Component and FD8, Diagonal Z Component naming these items Duct.Height and Duct.Width, respectively. If this was successful FD6, FD7, and FD8 will have the letter D next to them.

- In the toolbar, click on Tools | Parameters. This will open the Parameter Window. Click on the Design Parameters tab and you should see the three parameters that have just been defined.

- Now generate the box by left-clicking the Generate button in the toolbar, or alternatively in the Tree Outline right-click on the row Duct and in the context menu left-click on Generate.

In the Graphics window a rectangular duct will be shown as per the following figure. You will also notice that the Tree Outline has 1 Part, 1 Body. Clicking on the + symbol next to 1 Part, 1 Body will show the Solid that was just created.

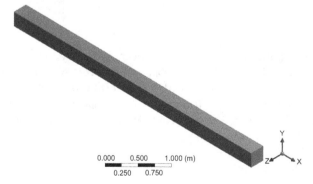

- Now is a good time to save your work. Click on File | Save Project and enter an appropriate filename such as Lined_Duct.wbpj.

- Although not necessary for this analysis of the rigid-walled duct or the locally reacting liner, we are going to add two more volumes which will form the upper and lower sides of the bulk reacting liner to be used in Section 6.6.

Repeat the previous process to define two additional boxes, and name them Upper_Silencer and Lower_Silencer with parameter definitions as below. When creating the boxes, use Operation | Add Frozen to create the new bodies.

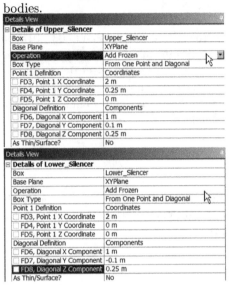

Details View	
□ **Details of Upper_Silencer**	
Box	Upper_Silencer
Base Plane	XYPlane
Operation	Add Frozen
Box Type	From One Point and Diagonal
Point 1 Definition	Coordinates
□ FD3, Point 1 X Coordinate	2 m
□ FD4, Point 1 Y Coordinate	0.25 m
□ FD5, Point 1 Z Coordinate	0 m
Diagonal Definition	Components
□ FD6, Diagonal X Component	1 m
□ FD7, Diagonal Y Component	0.1 m
□ FD8, Diagonal Z Component	0.25 m
As Thin/Surface?	No

Details View	
□ **Details of Lower_Silencer**	
Box	Lower_Silencer
Base Plane	XYPlane
Operation	Add Frozen
Box Type	From One Point and Diagonal
Point 1 Definition	Coordinates
□ FD3, Point 1 X Coordinate	2 m
□ FD4, Point 1 Y Coordinate	0 m
□ FD5, Point 1 Z Coordinate	0 m
Diagonal Definition	Components
□ FD6, Diagonal X Component	1 m
□ FD7, Diagonal Y Component	-0.1 m
■ FD8, Diagonal Z Component	0.25 m
As Thin/Surface?	No

- Click on the Generate icon to create the two remaining bodies. In the Outline window the Tree branch shows 3 Parts, 3 Bodies. Rename these 3 bodies with more descriptive names such as Duct, Upper_Silencer, and Lower_Silencer, by clicking on the objects in the Tree Outline, then in the Details View window in the row Details of Body | Body type the desired names as shown below.

- We will now define parameters for the silencer geometry; specifically the offset, height, and length. For the Upper_Silencer object in the Details View window in the row Point 1 Definition, define FD3, Point 1 X Coordinate, FD7, Diagonal Y Component, and FD6, Diagonal X Component as Upper_Silencer.Offset, Upper_Silencer.Height, and Upper_Silencer.Length, respectively.

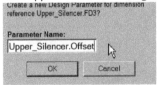

- Now left-click on Tools | Parameters to open the Parameter Manager and in the Design Parameters tab you should see the six parameters that have just been defined.

```
Parameter Manager
Duct.Length = 5
Duct.Height = 0.25
Duct.Width = 0.25
Upper_Silencer.Offset = 2
Upper_Silencer.Height = 0.1
Upper_Silencer.Length = 1
```

Rather than repeating the same process as described previously to dimension the Lower_Silencer, we can define the dimensions using the Parameter/Dimension Assignments tab. In this tab, directly type the additional 6 assignments shown below. Click on the Generate icon.

```
Parameter Manager
Upper_Silencer.FD6 = @Upper_Silencer.Length
Upper_Silencer.FD7 = @Upper_Silencer.Height
Upper_Silencer.FD3 = @Upper_Silencer.Offset
Duct.FD8 = @Duct.Width
Duct.FD7 = @Duct.Height
Duct.FD6 = @Duct.Length

Lower_Silencer.FD3 = @Upper_Silencer.Offset
Lower_Silencer.FD6 = @Upper_Silencer.Length
Lower_Silencer.FD7 = -@Upper_Silencer.Height
Lower_Silencer.FD8 = @Duct.Width
Upper_Silencer.FD4 = @Duct.Height
Upper_Silencer.FD8 = @Duct.Width
```

| Design Parameters | Parameter/Dimension Assignments | Check | Close |

- At this stage there are three separate parts that are not coupled. To join them into a single body, select all three parts by either holding down the <Shift> or <Ctrl> key on the keyboard whilst selecting the parts by left-clicking on the mouse. Then right-click on the selected objects and from the context menu left-click on Form New Part.

In the Tree Outline you should see 1 Part, 3 Bodies.

- We will now Imprint the silencer sections onto the duct in order to form a common face between the duct and silencer sections. In the toolbar, left-click on Create | Boolean. In the Details View window of the Boolean1 object, change the row Operation to Imprint Faces. Set the selection filter to bodies by left-clicking on the Selection Filter: Bodies icon in the toolbar, or alternatively press the hot key <Ctrl> b on the keyboard, then select the duct body (by moving the mouse over the duct in the Graphics window) and left-click. Then in the Details View window of the Boolean1 object, in the row Target Bodies, click Apply.

Details View	
⊟ **Details of Boolean1**	
Boolean	Boolean1
Operation	Imprint Faces
Target Bodies	Apply

Repeat these steps for the Tool Bodies but this time select both the upper and lower silencer elements. If done correctly, the row Tool Bodies should indicate 2 Bodies. Finally, in the row Preserve Tool Bodies? select Yes, Imprinted.

Details View	
⊟ **Details of Boolean1**	
Boolean	Boolean1
Operation	Imprint Faces
Target Bodies	1 Body
Tool Bodies	2 Bodies
Preserve Tool Bodies?	Yes, Imprinted

- Click on the Generate icon in the toolbar to create the bodies.

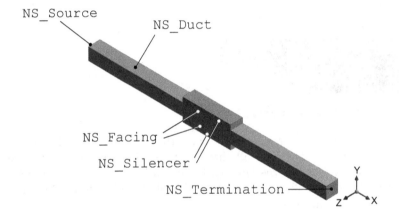

- The final step will be to generate five Named Selections, which will be used to identify various parts in the model (and are shown in the previous figure). In the toolbar left-click on Tools | Named Selection. Repeat this step another four times to create five Named Selections. To change the name of these objects, click on the field to the right of Named Selection and type the names:

  1. NS_Duct, which will be the duct body. Left-click on the first named selection in the Tree Outline window. With the body selection filter active (left-click on the Selection Filter: Bodies icon or press <Ctrl> b on the keyboard to) select the duct body in the Graphics window, then in the Details View window click on Apply in the Geometry row. In the row Named Selection type NS_Duct. Click on the Generate icon.

  2. NS_Silencer, which forms the two silencer bodies. Repeat the

above process, using the selection filter to select the upper and lower bodies.

3. NS_Termination, which is the duct face at $x = 5$m. Make sure that you have the faces selection filter active (in the toolbar left-click on the Selection Filter: Faces icon or press <Ctrl> f on the keyboard).

4. NS_Source, which is the duct face at $x = 0$ m. To see this you can either rotate the model or simply select the top face shown, right-click, then Hide Face(s).

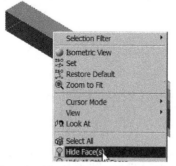

You can then select the desired face (the source end of the duct), then click on Geometry | Apply.

To show the hidden face, select any face, right-click, and left-click on Show All Faces.

5. NS_Facing is the contact between the silencer elements and the duct. With the body selection filter active, select the duct body, then right-click Hide All Other Bodies. You will see the Imprint left by the silencer sections. Make the face selection filter active and in the Graphics window left-click on both of these imprints, then click on Geometry | Apply in the Details View window.

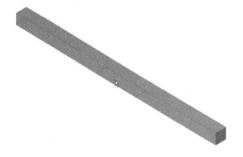

- The model is now complete.

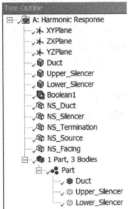

- In the toolbar, click on File | Save Project and exit the DESIGNMODELER.

Meshing

- The next step is to mesh the solid model. In the Workbench Project Schematic double-click on row 4 Model. This will start ANSYS Mechanical.

- In the Details of "Mesh" window, in the row Sizing | Element Size, type 0.05 (m). This will mean we have elements of length 5 cm giving us 5 (25/5) elements across the duct width and 2 (10/5) elements across the silencer height.

- We are now going to define how the system is to be meshed. In the Outline window, right-click in Project | Model (A4) | Mesh, then in the context menu left-click on Insert | Method. With the body selection filter active (by clicking on the Selection Filter: Bodies icon or pressing <Ctrl> b on the keyboard) right-click over any body and left-click on Select All.

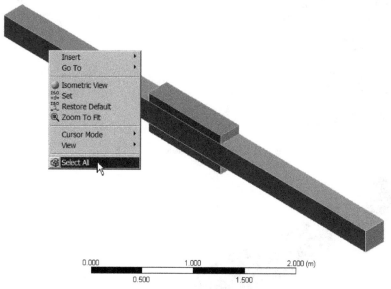

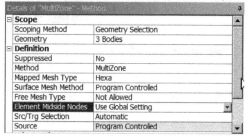

In the Details of "Automatic Method" Method window, in the row Scoping Method | Geometry Selection, click Apply. The Geometry row should now indicate 3 Bodies. In the row Definition | Method, select MultiZone. This will be used to create a hexahedral mapped mesh. In the row Element Midside Nodes choose Use Global Setting since we are using the 20-noded quadratic acoustic elements (FLUID220).

Details of "MultiZone" - Method		
Scope		
Scoping Method	Geometry Selection	
Geometry	3 Bodies	
Definition		
Suppressed	No	
Method	MultiZone	
Mapped Mesh Type	Hexa	
Surface Mesh Method	Program Controlled	
Free Mesh Type	Not Allowed	
Element Midside Nodes	Use Global Setting	
Src/Trg Selection	Automatic	
Source	Program Controlled	

- The next step is to mesh the model. This can be done by either clicking Mesh | Generate Mesh in the toolbar, or alternatively, by right-clicking over Mesh in the Outline window and left-clicking on Generate Mesh.

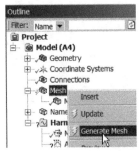

The resulting `MultiZone` mesh is shown below.

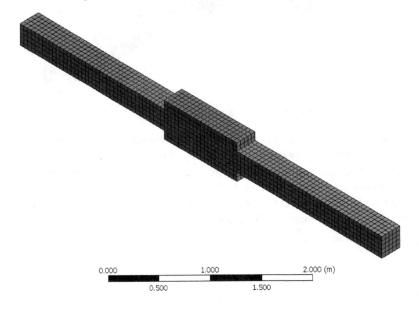

```
0.000              1.000              2.000 (m)
      0.500              1.500
```

- In the toolbar, click on `File | Save Project`.

Part Definitions

In the following paragraphs the properties of the acoustic components will be defined.

- Define a new acoustic body for the duct by either left-clicking the mouse on the `Selection Filter: Bodies` icon in the toolbar or press `<Ctrl>` b on the keyboard. In the `Geometry` window, click on the duct body. In the ACT Acoustics extension toolbar, click on `Acoustic Body`. This will insert an `Acoustic Body` entry in the `Outline` window under the object `Harmonic Analysis (A5)`. An alternative way to define this is to right-click in `Harmonic Analysis (A5)` and then in the context menu left-click on `Insert | Acoustic Body`.

An alternative way to select the body would be to change the Scope | Scoping Method to Named Selection, then choose NS_Duct. Rename the Acoustic Body by right-clicking over the entry in the Outline, then selecting Rename, and naming it Acoustic Body - Duct.

Click on this entry to see the window Details of "Acoustic Body - Duct".

Details of "Acoustic Body - Duct"	
Scope	
Scoping Method	Named Selection
Named Selection	NS_Duct
Definition	
Temperature Dependency	No
Mass Density	1.21 [kg m^-1 m^-1 m^-1]
Sound Speed	344 [m sec^-1]
Dynamic Viscosity	0 [Pa sec]
Thermal Conductivity	0 [W m^-1 C^-1]
Specific Heat Cv	0 [J kg^-1 C^-1]
Specific Heat Cp	0 [J kg^-1 C^-1]
Equivalent Fluid of Perforated Material	No
Reference Pressure	2E-05 [Pa]
Reference Static Pressure	102280 [Pa]
Acoustic-Structural Coupled Body Options	Uncoupled
Perfectly Matched Layers (PML)	Off

- In the window Details of "Acoustic Body - Duct", change the row Definition | Mass Density to 1.21 and the row Definition | Sound Speed to 344. Change the Reference Static Pressure to 102280 [Pa].

- Repeat these steps for the sound-absorbing material in the silencer by adding a second Acoustic Body and renaming it to Acoustic Body - Silencer. A quick way to copy the previous properties is to right-click over the Acoustic Body - Duct, then left-click on Duplicate in the context menu. In this body we will use the Johnson–Champoux–Allard Equivalent Fluid Model, which can be activated by changing the row Equivalent Fluid of Perforated Material to Yes.

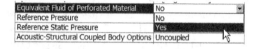

Change the rows for the following five properties to reflect those listed in Table 6.2: Fluid Resistivity as 10800, Porosity as 0.98, Tortuosity as 1.04, Viscous Characteristic Length as 0.000129, and Thermal Characteristic Length as 0.000198.

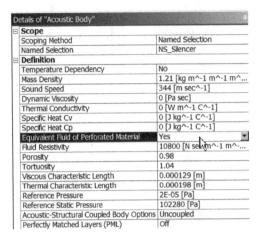

Details of "Acoustic Body"	
⊟ **Scope**	
Scoping Method	Named Selection
Named Selection	NS_Silencer
⊟ **Definition**	
Temperature Dependency	No
Mass Density	1.21 [kg m^-1 m^-1 m^...
Sound Speed	344 [m sec^-1]
Dynamic Viscosity	0 [Pa sec]
Thermal Conductivity	0 [W m^-1 C^-1]
Specific Heat Cv	0 [J kg^-1 C^-1]
Specific Heat Cp	0 [J kg^-1 C^-1]
Equivalent Fluid of Perforated Material	Yes
Fluid Resistivity	10800 [N sec m^-1 m^-...
Porosity	0.98
Tortuosity	1.04
Viscous Characteristic Length	0.000129 [m]
Thermal Characteristic Length	0.000198 [m]
Reference Pressure	2E-05 [Pa]
Reference Static Pressure	102280 [Pa]
Acoustic-Structural Coupled Body Options	Uncoupled
Perfectly Matched Layers (PML)	Off

It should be noted that this equivalent fluid definition is not needed for the locally reacting liner but is defined now for convenience for use in Section 6.4.4.

- We will now define a new surface to create an absorbing termination. In the Outline window, right-click in Harmonic Analysis (A5) and in the context menu left-click on Insert | Acoustic Attenuation Surface.

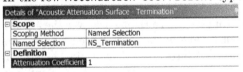
Acoustic Attenuation Surface

This will create a new entry. Rename this to Acoustic Attenuation Surface - Termination (by either right-clicking on the object and selecting Rename or pressing <F2> on the keyboard. In the row Scope | Scoping Method select Named Selection. Then in the row Named Selection select NS_Termination. In the row Attenuation Coefficient type 1.

Details of "Acoustic Attenuation Surface - Termination"	
⊟ **Scope**	
Scoping Method	Named Selection
Named Selection	NS_Termination
⊟ **Definition**	
Attenuation Coefficient	1

- Now add an Acoustic Mass Source by right-clicking over Harmonic Response (A5) then left-click on Insert | Acoustic Mass Source, or alternatively left-click on the ACT Acoustics extension toolbar Excitation | Mass Source (Harmonic). In the window Details of "Acoustic Mass Source", change the row Scope | Scoping Method to Named Selection. Then in the row Named Selection, select NS_Source. In the row Amplitude Of Mass Source type 1. Keep the Phase Angle as 0.

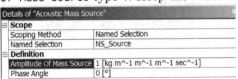

Details of "Acoustic Mass Source"	
⊟ **Scope**	
Scoping Method	Named Selection
Named Selection	NS_Source
⊟ **Definition**	
Amplitude Of Mass Source	1 [kg m^-1 m^-1 m^-1 sec^-1]
Phase Angle	0 [°]

- Now define an `Acoustic Impedance Boundary` by right-clicking over `Harmonic Response` (A5), then left-clicking on `Insert | Acoustic Impedance Boundary`, or alternatively, left-click on `Boundary Conditions | Impedance Boundary` in the ACT Acoustics extension toolbar. Rename this `Acoustic Impedance Boundary - Local` since this will be our locally reacting liner. The impedance will be equal to the values given in Table 6.2. Select the `Named Selection NS_Facing` and define the `Resistance` as `69.575` and the `Reactance` as `-58.2313`, as shown below.

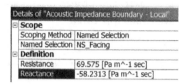

- The definitions are now complete. In the toolbar click on `File | Save Project`.

Analysis

This section configures the analysis settings.

- In the `Outline` window left-click on the `Analysis Settings`. In the window `Details of "Analysis Settings"`, change the row `Options | Range Minimum` to `0 Hz`, `Options | Range Maximum` to `1000 Hz`, `Options | Solution Intervals` to `40`, and `Options | Solution Method` to `Full`. This will conduct a full harmonic analysis from 25 Hz to 1000 Hz in 25 Hz increments.

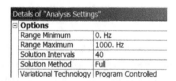

- In the window `Details of "Analysis Settings"`, expand the `Analysis Data Management` tree and change the row `Save MAPDL db` to `Yes`. This will allow you to post-process the results using the ACT Acoustics extension `Acoustic Time_Frequency Plot`.

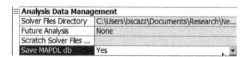

- Now `Suppress` (by right-clicking over) the objects not required to solve the rigid-walled duct, namely `Acoustic Body - Silencer` and `Acoustic Impedance Boundary - Local`.

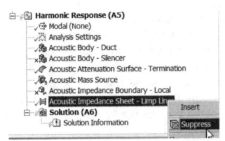

When you suppress an object, the green tick next to the object in the tree changes to a blue cross to indicate that the item is suppressed.

- Also `Suppress` the `Upper_Silencer` and `Lower_Silencer Part` in the tree `Model (A4) | Geometry` as these are not required for the locally reacting liner.

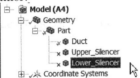

- In the toolbar, click on `File | Save Project`.

- Solve the harmonic analysis by clicking `Solve` in the toolbar, or alternatively right-clicking over `Harmonic Response (A5)` then left-clicking on `Solve` in the context menu. The yellow lightning bolt next to `Harmonic Response (A5)` will turn green indicating the solver is working. Be patient, this may take several minutes to solve.

Reviewing the Results

- Using the ACT Acoustics extension toolbar, select `Results | Acoustic Pressure` which will add a plot showing contours of equal pressure across the surface of the solid body (once the result has been "evaluated"). Note that in the window `Details of "Acoustic Pressure"`, `All Bodies` are currently selected under `Geometry | Scoping Method`.

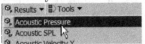

Right-click on the object and then left-click on `Rename`, or alternatively press `<F2>` on the keyboard, and rename the object to `Acoustic Pressure - Real All Bodies`.

- In the ACT Acoustics extension toolbar, click on `Results | Acoustic SPL` to create a SPL contour plot. Again ensure `All Bodies` are currently selected. Rename the object to `Acoustic SPL - All Bodies`. In the row `Definition | Frequency`, type `250` which will display the results at 250 Hz (the frequency where the attenuation peaks).

- Using the ACT Acoustics extension toolbar, click on `Results | Acoustic Time_Frequency Plot`. This will generate a graph versus frequency (for harmonic analyses).

Rename the object to `Acoustic Time_Frequency Plot - Source Pres`. In the window `Details of "Acoustic Time_Frequency Plot - Source Pres"`, change the row `Scope | Scoping Method` to `Named Selection`, then change the `Named Selection` to `NS_Source`. In the row `Definition | Result`, select `Pressure` and change the row `Display` to `Real and Imaginary` which will display the real and imaginary parts of the complex pressure.

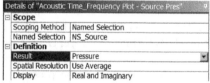

- Create another time-frequency plot for the source SPL. Rename the object to `Acoustic Time_Frequency Plot - Source SPL`. In the window `Details of "Acoustic Time_Frequency Plot - Source SPL"`, change the row `Named Selection` to `NS_Source` and the row `Definition | Result` to `SPL`.

- Create another time-frequency plot for the termination SPL. Rename the object to `Acoustic Time_Frequency Plot - Termination SPL`. In the window `Details of "Acoustic Time_Frequency Plot - Termination SPL"`, change the row `Geometry | Scoping Method` to `Named Selection`, then change the row `Named Selection` to `NS_Termination`, and then change the row `Definition | Result` to `SPL`.

- Finally, right-click over any of the objects and select `Evaluate All Results`.

- The figure below shows the SPL along the length of the (surface of the) rigid duct at 250 Hz. This was obtained by selecting the `Acoustic SPL - All Bodies` object.

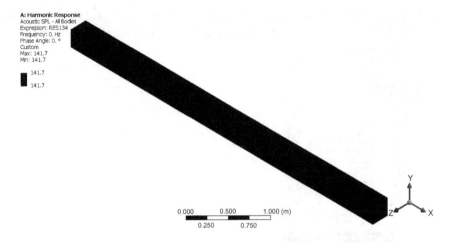

Note that the SPL is uniform along the length of the duct as would be expected from the anechoic termination. The expected sound pressure level can be determined from the fact that we have excited the duct with a unit Mass Source of $Q\rho_0 = 1\text{kg.s}^{-1}.\text{m}^{-2}$, where Q is the volume velocity of the source. Since the duct termination is anechoic, the specific impedance of the duct at the source is $Z_c = \rho_0 c_0 = p/Q$, where p is the acoustic pressure. Using these two expressions, the expected acoustic pressure amplitude is $|p| = Q\rho_0 c_0 = c_0$. Therefore the SPL is $20\log_{10}(344) + 94 - 3\,\text{dB} = 141.7\,\text{dB}$, where the 344 is the speed of sound, the 94 dB term comes from the reference pressure $(-20\log_{10}(20E - 6))$ and the $-3\,\text{dB}$ comes from converting peak to RMS $(-20\log_{10}\sqrt{(2)})$. If you look at the real part of the pressure (by clicking on the Acoustic Pressure - Real All Bodies object), then you will note that the amplitude of the pressure is approximately $344\,\text{Pa}$. For a more detailed discussion on the relationship between the Mass Source and the radiated power, see Section 7.3.2.

- In the Outline window, left-click on the Acoustic Time_Frequency Plot - Termination SPL. You will note that the SPL versus frequency at the termination remains nearly constant at $141.7\,\text{dB}$.

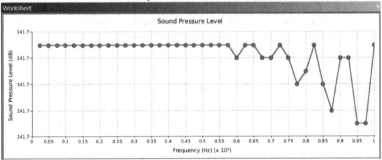

Exporting the Results

- To export the results for later post-processing in MATLAB, right-click on the Acoustic Time_Frequency Plot - Source Pres object in the Outline window, then left-click on Export in the context menu. Save the file as Lined_Duct_Rigid_Source_Pres.txt .

- Repeat these steps to export the results for Acoustic Time_Frequency Plot - Termination SPL and save the file as Lined_Duct_Rigid_Termination_SPL.txt.

- This concludes the analysis of the rigid-walled duct. In the toolbar, click on File | Save Project.

6.5.2.2 Local Reacting Liner

In this section the locally reacting liner will be analyzed to show:

- how sound is attenuated by the locally reacting impedance, and

- how to calculate insertion loss.

The following are the ANSYS Workbench instructions.

- The first step is to Unsuppress the object Acoustic Impedance Boundary - Local, which will activate the impedance that represents the locally liner.

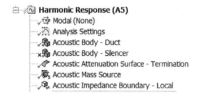

- In the toolbar, click on File | Save Project.

- Solve the analysis by pressing the Solve icon in the toolbar, or alternatively right-click over Harmonic Response (A5) and left-click on Solve in the context menu.

- The figure below shows the SPL along the length of the (surface of the) duct at 250 Hz, by clicking the Solution (A6) | Acoustic SPL - All Bodies object. Ensure that the row Definition | Frequency still shows 250.

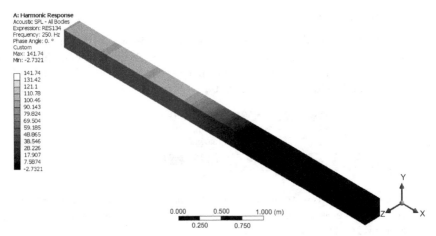

The contour plot shows that the SPL upstream of the silencer (left-hand side of the figure) is similar in magnitude to the SPL of the rigid-walled duct. There is a small variation in the SPL in this region due to weak reflection of sound from the impedance change caused by the addition of the locally reacting liner. The SPL of the termination has been reduced significantly. Also note that the cut-on of the higher-order modes occurs when the excitation frequency is greater than $f_{\text{cut on}} = c_0/(2h) = 344/(2 \times 0.25) = 688\,\text{Hz}$ (Eq. 3.17), and therefore one could expect that the pressure distribution across the height of the duct in the silencer section would be uniform at 250 Hz; however, it is not.

- In the `Outline` window, left-click on the `Acoustic Time_Frequency Plot - Termination SPL`. Notice that the SPL at the termination shows a minimum at 250 Hz as we expected because the impedance was defined to provide maximum attenuation at this frequency.

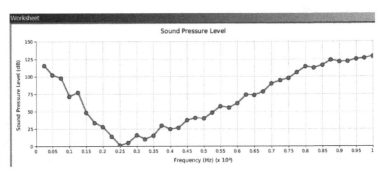

- Due to the anechoic termination, the insertion loss is equal to the SPL at the termination for the rigid-walled duct minus the SPL at the termination in the presence of the locally reacting liner. Therefore, the previous plot represents an affine version of the negative insertion loss. Note that IL is not a smoothly varying function with respect to frequency but instead shows small "wiggles" around a general curve that dips down (at 250 Hz) only to rise again. These small "wiggles" are caused by the reactive energy present in the upstream part of the duct as a result of the hard-walled termination. If the termination was made anechoic, then the wiggles will disappear. The TL, on the other hand, will be independent of the duct impedance and will not exhibit the frequency-dependent fluctuations. Although it is possible to calculate the TL in ANSYS Workbench using APDL commands, it was not done here but rather post-processed in MATLAB for simplicity.

Exporting the Results

- To export the results for later post-processing in MATLAB, right-click on the `Acoustic Time_Frequency Plot - Source Pres` object in the `Outline` window, then left-click on `Export` in the context menu. Save the file as `Lined_Duct_Local_Source_Pres.txt`.

- Repeat the process for the `Acoustic Time_Frequency Plot - Termination SPL` and save the file as `Lined_Duct_Local_Termination_SPL.txt`.

6.5.3 ANSYS Mechanical APDL

The ANSYS Mechanical APDL file `code_ansys_lined_duct.txt` that is available with this book was used to generate the FE model shown in Figure 6.3, as well as conduct the harmonic that is described later in this section. The file also includes additional code to write the ANSYS results to a text file, as well as the code for generating the plots shown in this section.

The duct and silencer was modeled in ANSYS using the 3D quadratic acoustic element `FLUID220` with the option of displacement degrees of freedom deactivated (with `KEYOPT(2)`). Hence, plotting these DOFs using `PLNSOL,,U,X` will result in a plot showing zero displacement. The particle velocities, however, can be estimated from the pressure gradients if necessary.

The material properties for the gas in the duct are defined by the speed of sound and density of the fluid, whereas the acoustic elements in the silencer section use the Johnson–Champoux–Allard Equivalent Fluid Model parameters defined in Table 6.1 and quantified in Table 6.2. The duct and silencer were mapped mesh with hexahedral elements. The element length was 0.05 m which gives the number of elements per wavelength at the highest frequency (1000 Hz) as 6.9 which is sufficient to accurately model the system.

The termination was made anechoic using the command `SF,,CONV,Alpha`, where `Alpha` is the sound absorption coefficient, which was set to unity to completely absorb all incident sound.

The locally reacting liner was modeled as an impedance boundary, and was added using a surface load directly applied to the nodes between the duct and silencer bodies via the surface command `SF,,IMPD,RESISTANCE,REACTANCE`, where `RESISTANCE` is the resistance and `REACTANCE` is the reactance of the impedance. Alternatively, it may be applied to an element using `SFE`, a line using `SFL` or an area using `SFA`.

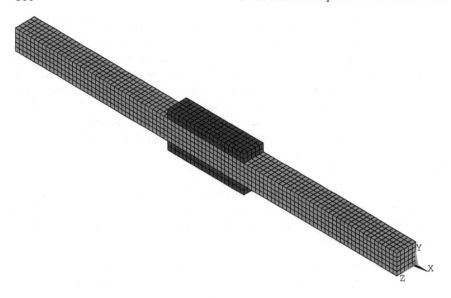

FIGURE 6.3

Finite element model of a lined duct created using ANSYS Mechanical APDL. This model shows the duct elements (light shading) as well as the silencer elements (dark shading) used for the bulk reacting liner. The silencer elements and associated nodes were not selected when the locally reacting liner model was analyzed, and instead an impedance boundary was placed at the interface between the duct and silencer bodies.

For the locally reacting example, the acoustic elements and associated nodes that form the silencer section of the system were unselected, so that when the model is solved they do not participate in the calculations. This is similar to the practice in ANSYS Workbench where unwanted parts are Suppressed. This differs from hidden objects in ANSYS Workbench which still participate in the analysis.

There are a variety of possible acoustic sources that could have been used to excite the system such as a pressure condition (implemented using the APDL code D,,PRES), a velocity condition (D,,UX), a plane wave definition (AWAVE), a flow source (F,,FLOW), or a mass source (BF,,JS). The mass source with a unit amplitude was chosen for three reasons: it does not require that the acoustic elements have their displacement DOFs activated; it is directly supported in the ACT Acoustics extension in ANSYS Workbench; and finally, the power from the source can be directly calculated (see Section 7.3.2).

A harmonic analysis (ANTYPE,3) with the (HROPT,FULL) option was used to calculate the frequency response of the system. These results were then exported to MATLAB where the IL and TL were calculated. Figure 6.4 shows the resulting sound pressure level at 250 Hz arising from the Mass Source excitation on the left-hand side of the duct. The attenuation is very low at

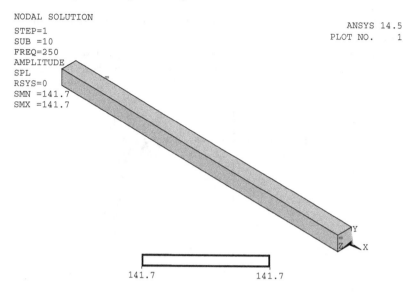

(a) Rigid-walled duct.

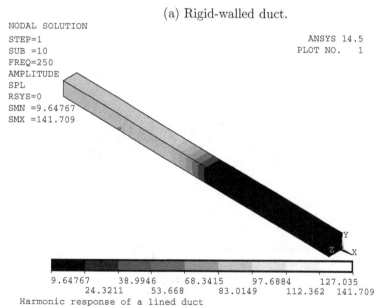

(b) Locally reacting liner.

FIGURE 6.4

Comparison of the sound pressure level in the rigid-walled duct with a locally reacting liner at 250 Hz arising from the applied unit amplitude Mass Source calculated using ANSYS Mechanical.

this frequency as the silencer was designed to attenuate sound at a much higher frequency.

Figure 6.5 shows frequency response plots of the SPL upstream and down-stream of the silencer element. The figure shows that there is a decrease at

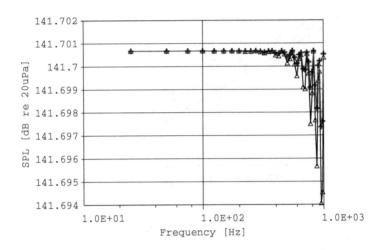

(a) Rigid-walled duct.

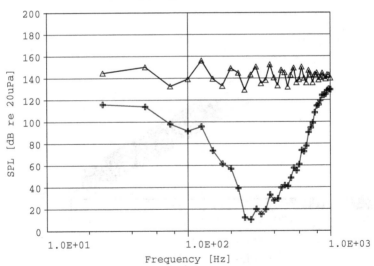

(b) Locally reacting silencer.

FIGURE 6.5
SPL at the source (triangle markers) and termination (cross markers) end of the duct, for the rigid-walled duct and with the locally reacting silencer element calculated using ANSYS Mechanical.

the SPL at the exit of the duct at 250 Hz which is to be expected because the value of the impedance was selected to maximize attenuation at 250 Hz. Note for the rigid-walled duct the source SPL and termination SPL are essentially identical, which is to be expected since there are no impedance changes or absorption mechanisms along the length of the duct. Once the locally reactive impedance is activated, it can be seen that the SPL near the source fluctuates with respect to frequency due to reflections from the silencer. The IL is equal to the difference between the termination SPL for the rigid-walled duct and the termination SPL with the silencer activated.

6.5.4 Results

The IL calculated in MATLAB is presented in Figure 6.6, where it is compared against the IL and TL loss results from ANSYS. The ANSYS IL results in Figure 6.6 shows some local variability with respect to frequency, unlike the TL. This is to be expected since the upstream termination where the source is located is not anechoic. The power radiated by the source is affected by the impedance of the silencer (since it is not purely absorptive), as well as being affected by the reactive impedance of the finite length of the rigid-walled duct

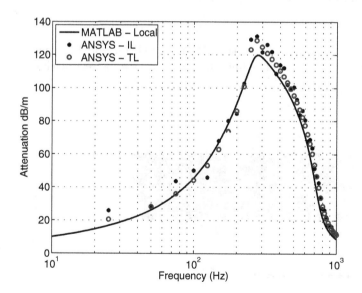

FIGURE 6.6
Attenuation per unit length provided by a locally reacting liner optimized for 250 Hz. The theoretical results were calculated in MATLAB using the expressions presented in Section 6.4.2. The ANSYS results represent the insertion loss and transmission loss calculated over a 1 m length of silencer.

upstream of the silencer section. Since the radiated power is affected by this reactance, it subsequently impacts on the SPL downstream of the silencer and thus impacts the IL. The TL on the other hand is not influenced by the source type or the backward traveling waves in the upstream part of the duct.

In Figure 6.6 the IL and the TL results obtained in ANSYS and the MATLAB results all peak at approximately 250 Hz as expected (see Table 6.2). The TL results from ANSYS diverge slightly from the MATLAB results which is not surprising given that the theoretical attenuation predicted by Equation (6.9) is only an approximation and diverges at high levels of attenuation and at high frequencies. The peak attenuation predicted by Equation (6.12) gives $D_{max}/h = 153$ dB/m, whereas the peak attenuation obtained from AN-SYS is $D_h/h = 132$ dB/m and the finite difference approximation given by Equations (6.9) and (6.10) is $D_h/h = 120$ dB/m. The error in the latter is expected due to its inherent approximation. It is possible that the ANSYS results do not achieve the peak theoretical attenuation rate due to the following:

- finite length effects associated with the change of impedance at the inlet and outlet of the silencer section,

- excitation of evanescent higher order modes, as well as

- the very large dynamic range and an issue with numerical precision.

Figure 6.7 shows a contour plot of the SPL in the vicinity of the silencer section at a frequency of 250 Hz. The finite length effects can be seen with the

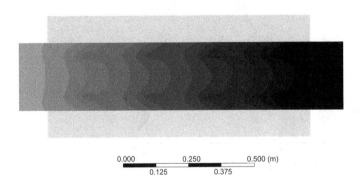

FIGURE 6.7
SPL at a frequency of 250 Hz around the locally reacting liner. Dynamic range is from -3 dB (dark) to 141 dB (light) re 20 μPa, with contours approximately every 10 dB. The upper and lower silencer sections have been shown to mark where the impedance boundary is located but were not used in the analysis.

clear axial modulation of the SPL superimposed on the steady axial decay. The non-planar nature of the sound field is also evident with SPL gradients normal to the axis of the duct.

Muffler Transmission Loss from the ACT Acoustics Extension

Note, the TL of the silencer may also be calculated using the ACT Acoustics extension feature from the toolbar Results | Muffler Transmission Loss. See Section 2.8.5.8 for how to use this feature, and Section 3.4.2.3 for an example of the use of calculation of the TL of a duct with a quarter wavelength side-branch resonator. This approach requires that both the source end and the termination end are anechoic, which would mean that the TL and IL are equal. The current arrangement was deliberately chosen to illustrate that the IL and TL differ because of the effect that the source, termination, and silencer impedance has on the IL.

To use the Muffler Transmission Loss object in this example, first insert a new Acoustic Attenuation Surface from the ACT Acoustics extension. Set the Scoping Method to Named Selection, and the row Named Selection to NS_Source. Set the Attenuation Coefficient to 1. Then in the ACT Acoustics extension toolbar, left-click on Results | Muffler Transmission Loss. Use named selections for the scoping method, and set the row Outlet | Named Selection to NS_Termination, and Inlet | Named Selection to NS_Source. Change the row Inlet Source to Acoustic Mass Source. Ensure that Mass Density of Environment Media and Sound Speed in Environment Media are 1.21 and 344, respectively. Re-solve to obtain the results, which will match the TL shown in Figure 6.6. Note that the IL will have changed and is now the same as the TL.

6.6   Example: Bulk Reacting Liner

The previous example was repeated, this time for a bulk reacting liner, with acoustic properties of the equivalent fluid based on Doutres et al. [60]. The parameters used in the example are listed in Table 6.2. Based on the chosen geometry and material properties, this bulk reacting silencer will provide a peak attenuation around approximately 750 Hz with more realistic levels of attenuation compared to the highly optimized locally reacting liner used in the previous section.

6.6.1   MATLAB

The MATLAB script lined_duct.m available with this book was used to calculate the theoretical attenuation per unit length using the expressions derived in Section 6.4.4. The script calls the function scott.m, also provided with this book, which contains Equation (6.28) and was solved using the non-linear solver, fsolve, in MATLAB. The script lined_duct.m also post-processes the ANSYS Mechanical and ANSYS Workbench results.

6.6.2   ANSYS Workbench

In this section the duct with the bulk reacting liner will be analyzed using ANSYS Workbench.

- Start up ANSYS Workbench and load the previously saved lined duct model in the project file Lined Duct.wbpj.

- In the Outline window, right-click over the object Acoustic Body - Silencer and left-click on Unsuppress in the context menu. Also Unsuppress the Upper_Silencer and Lower_Silencer objects in Modal (A4) | Geometry | Part. Suppress the object Acoustic Impedance Boundary - Local.

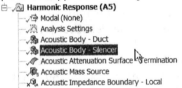

- In the toolbar, click on File | Save Project.

- Left-click on the Solve icon in the toolbar, or alternatively in the Outline window right-click on any object in the Harmonic Response (A5) tree and then left-click on Solve in the context menu.

- The figure below shows the SPL along the length of the (surface of the) duct and silencer at 750 Hz. This was obtained by selecting the Acoustic SPL - All Bodies object, then in the row Definition | Frequency, typing 750. You may need to Evaluate All Results by right-clicking over the object.

As was observed with the locally reacting liner, the SPL upstream of the silencer (on the left side in the figure above) shows a slight increase due to reflection of sound from the impedance change. The SPL at the termination has been reduced compared to the rigid-walled duct.

- In the `Outline` window, left-click on the `Acoustic Time_Frequency Plot - Termination SPL`. The SPL at the termination shows a minimum at approximately 750 Hz.

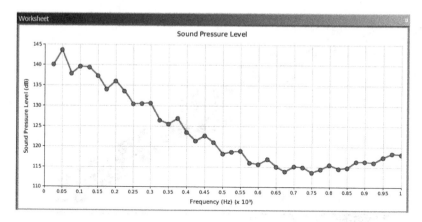

Exporting the Results

- To export the results for post-processing in MATLAB, in the `Outline` window left-click on the `Acoustic Time_Frequency Plot - Source Pres` plot to select it, then right-click on it and left-click on `Export` in the context menu. Save the file as `Lined_Duct_Bulk_Source_Pres.txt`.

- Repeat the process for `Acoustic Time_Frequency Plot - Termination SPL` and save the file as `Lined_Duct_Bulk_Termination_SPL.txt`.

6.6.3 ANSYS Mechanical APDL

The ANSYS Mechanical APDL `code_ansys_lined_duct.txt` that is available with this book was used to model the bulk reacting liner. Figure 6.8 shows the resulting sound pressure level arising from the `Mass Source` excitation on the left-hand side of the duct. Figure 6.9 shows frequency response plots of the SPL upstream and downstream of the silencer, with the maximum attenuation occurring at approximately 750 Hz, which is consistent for a silencer with the dimensions and flow resistivity of the current design [47].

6.6.4 Results

Figure 6.10 compares the IL and TL across the 1 m silencer generated in ANSYS against the theoretical expressions given by Equations (6.28) and (6.10). The TL and the theoretical expression for attenuation per unit length match very well and any minor differences are likely to be associated with higher-order modes which are not accounted for in the theoretical expression. The scatter in the IL is expected and is a result of the reactance of the silencer element.

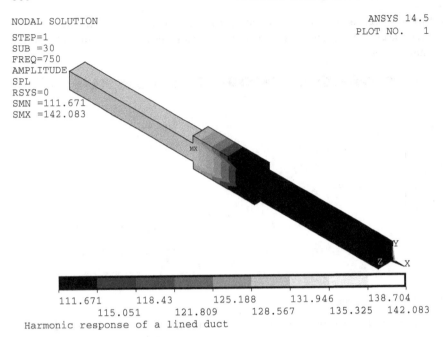

Harmonic response of a lined duct

FIGURE 6.8
Sound pressure level in the lined duct at 750 Hz with bulk reacting elements arising from the applied `Mass Source` calculated using ANSYS Mechanical.

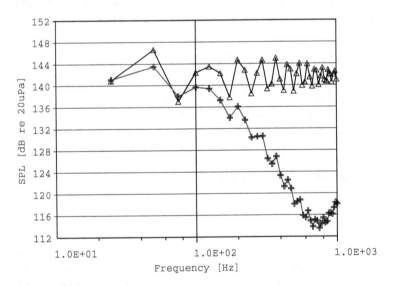

FIGURE 6.9
SPL at the source (triangle markers) and termination (cross markers) ends for the bulk reacting silencer calculated using ANSYS Mechanical.

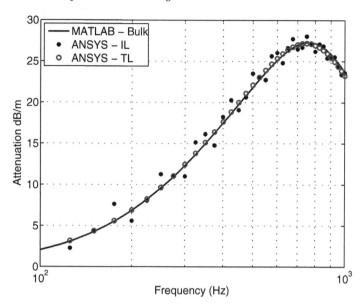

FIGURE 6.10

Attenuation per unit length provided by a bulk reacting liner. The theoretical results were calculated in MATLAB using the expressions presented in Section 6.4.4. The ANSYS results represent the insertion loss and transmission loss calculated over a 1 m length of silencer.

Bulk Reacting Liner with Limp Liner and Perforated Sheet

As previously mentioned, in practice, the sound-absorbing material in silencers is faced with a limp liner and perforated sheet to improve the operational life of the material. This requires modeling a thin impedance layer between the fluid in the duct and the equivalent fluid that represents the porous media. This is not a simple exercise in the current release of ANSYS (14.5). The surface impedance (applied using the APDL command `SF,,IMPD`) is not suitable as this can only be used on the outer layer of acoustic elements. Neither is the impedance sheet (applied using the APDL command `BF,,IMPD`) since this applies an acoustic side branch, where pressure is maintained on both sides of the sheet, whereas the limp liner and perforated sheet have continuity of velocity. The only current way to solve this problem is to use the `SURF153/154` elements to model the impedance and then couple the two acoustic domains via these surface elements as was done in Section 5.7. This is left as an exercise for the reader. In ANSYS 15.0 it will be possible to model such an impedance using the `Trim` element.

7

Room Acoustics

7.1 Learning Outcomes

The learning outcomes for this chapter are:

- how to model a reverberation room and how to estimate the absorption coefficient of an absorptive surface in a diffuse field,

- how to conduct an undamped and damped modal analysis in ANSYS,

- how to perform a harmonic analysis in ANSYS in the presence of acoustic damping from an absorptive surface,

- how to undertake a transient analysis in ANSYS and an awareness of the importance of numerical damping,

- how to use the acoustic flow source and mass source in ANSYS and how it is possible to calculate the acoustic power radiated from these sources,

- an understanding of the capabilities of the FLUID30 acoustic elements in ANSYS, and

- how to apply an absorption coefficient to a surface in ANSYS.

7.2 Description of the System

As discussed previously in Chapters 5 and 6, the characterization of the absorption properties of materials is essential for acousticians. In room acoustics there are two absorption coefficients that are commonly used: the Sabine absorption coefficient and the random incidence absorption coefficient. The latter is obtained by measuring the absorption coefficient of a specimen at angles ranging from normal to grazing, then integrating to obtain a single averaged value. The method for measuring the Sabine absorption coefficient of a specimen using a reverberation room is given in ASTM C423 [40] and ISO 35-2003 [85]. The process involves exciting the room using either an impulsive source

369

or steady broadband source at several locations not immediately adjacent to any walls, absorbing surface, or the receiver microphones. The sound source is abruptly terminated (for a steady source) and the sound in the room is left to decay. Several microphones are used to measure the sound pressure level, from which the reverberation time is determined. The reverberation time is defined as the time it takes for the sound pressure level to decay by 60 dB. The Sabine absorption coefficient is proportional to the reverberation time, the details of which are presented in Section 7.3. For a more detailed explanation of measurement of the sound absorption coefficient in diffuse fields the reader is referred to Beranek [45, Chapter 9, Sound in large rooms], Bies and Hansen [47, Chapter 7, Sound in enclosed spaces] or the standards ASTM C423 [40] and ISO 35-2003 [85].

Reverberation rooms are also often used to determine the sound power level of a source [86]. This is a steady-state approach, as opposed to the transient decay approach used to determine the reverberation time, and involves exciting the room with the acoustic source, measuring the resulting mean spatially averaged sound pressure level (at locations some distance from the source and walls), and if the (constant percentage bandwidth) reverberation times are known, then the source sound power level can be calculated [86].

The reverse logic can also be applied in that if one knows the sound power level of the source (for example, an ILG reference sound source), and the resulting spatially averaged sound pressure levels are measured, then the sound absorption coefficient can be determined. For more information on sound power measurements in diffuse fields, see [47, Section 6.6.2] or [45, Section 6.6].

7.3 Theory

This section includes theory relevant to room acoustics as well as how to determine the sound power radiated from an acoustic source in ANSYS.

7.3.1 Room Acoustics

The Sabine absorption coefficient is typically determined by first measuring the reverberation time of an "empty" reverberation room without the absorptive test material. The material to be tested is then installed in the reverberation room and the reverberation time is remeasured, which will typically be less than the measurement in the empty room. The change in reverberation time can be used to calculate the absorption coefficient of the test material. Implicit in the approach is that the sound field can be described as diffuse, which implies [107] that

- the local average energy density in the room is uniform,

- the energy is uniformly incident onto a surface from all directions, and

- the total sound absorption in the room is the sum of the absorptions of individual surfaces.

Sabine's formula for reverberation time is defined as the time it takes for the sound pressure level to decrease by $60\,\mathrm{dB}$ and is given by Bies and Hansen [47, Eq. (7.52)]

$$T_{60} = \frac{55.25V}{c_0 S_\alpha \bar{\alpha}}, \qquad (7.1)$$

where c_0 is the speed of sound in the air, $V = L_x L_y L_z$ is the volume of the (rectangular) room, S_α is the surface area of the absorbing material, $\bar{\alpha}$ is the Sabine absorption coefficient, and L_x, L_y, L_z are the dimensions of the (rectangular) room in the x, y, and z axes, respectively.

Before proceeding with the analysis of the acoustics in a room, it is important to understand the definition of a reverberant field and how that is related to the number of modes over the frequency range of interest. As discussed previously in Section 4.3.1, the natural frequencies of an undamped rectangular room are given by Bies and Hansen [47, Eq. (7.17), page 294]

$$f(n_x, n_y, n_z) = \frac{c_0}{2} \sqrt{\left(\frac{n_x}{L_x}\right)^2 + \left(\frac{n_y}{L_y}\right)^2 + \left(\frac{n_z}{L_z}\right)^2}, \qquad (7.2)$$

where n_x, n_y, n_z are the modal indices in the x, y, and z axes, respectively.

The number of modes, N, which may be excited in the frequency range from 0 to f Hz for a rectangular room, is given by Bies and Hansen [47, Eq. (7.21)]

$$N(f) = \frac{4\pi f^3 V}{3 c_0^3} + \frac{\pi f^2 S}{4 c_0^2} + \frac{fL}{8 c_0}, \qquad (7.3)$$

where $S = 2(L_x L_y + L_x L_z + L_y L_z)$ is the total surface area of the room, and $L = 4(L_x + L_y + L_z)$ is the total perimeter of the room, which is the sum of lengths of all the edges. The modal density, which is the number of modes per unit frequency, can be obtained by differentiating the above expression with respect to the frequency, giving [47, Eq. (7.22)]

$$\frac{dN}{df} = \frac{4\pi f^2 V}{c_0^3} + \frac{\pi f S}{2 c_0^2} + \frac{L}{8 c_0}. \qquad (7.4)$$

The modal density in a rectangular room for a finite bandwidth is given by

$$\frac{\Delta N}{\Delta f} = \frac{N(f_{\mathrm{upper}}) - N(f_{\mathrm{lower}})}{f_{\mathrm{upper}} - f_{\mathrm{lower}}}, \qquad (7.5)$$

where f_{upper} and f_{lower} are the upper and lower frequencies of the band, and $N(f)$ is calculated using Equation (7.3).

The modal overlap, M, is a measure of the number of modes in a bandwidth and is defined as [47, Eq. (7.25)]

$$M = \Delta f \frac{dN}{df}, \tag{7.6}$$

where $\Delta f = 2.20/T_{60}$ is the average bandwidth as a function of the reverberation time.

For a purely statistical representation of a broadband sound field there needs to be a minimum of between 3 and 6 modes in the frequency band of interest [47, Section 7.3.4]. The number of modes required increases as the damping decreases or higher accuracy is required. In the case of a pure tone excitation, the modal overlap should be $M \geq 3$ in order to approximate diffuse conditions.

The relationship between the sound power level of a source, L_w, and the resulting mean spatially averaged sound pressure level, L_p, in a reverberant room is [47, Eq. (6.13)],

$$L_w = L_p + 10 \log_{10}(V) - 10 \log_{10}(T_{60})$$
$$+ 10 \log_{10}\left(1 + \frac{S\lambda}{8V}\right) - 13.9 \ \text{dB re } 10^{-12}\text{W}, \tag{7.7}$$

where λ is the wavelength of sound at the band center frequency. The constant 13.9 dB has been calculated for a pressure of one atmosphere and a temperature of 20°C. Rewriting Equation (7.7), one can obtain an expression for the reverberation time as a function of the mean sound pressure level in the room and the sound power from the source

$$T_{60} = 10^{\left[L_p + 10 \log_{10}(V) - L_w + 10 \log_{10}\left(1 + \frac{S\lambda}{8V}\right) - 13.9\right]/10}. \tag{7.8}$$

For the above approach to be valid, there should be at least 20 acoustic modes in the frequency band of interest [47, Section 6.6.2], i.e., $M \geq 20$, which implies that a minimum room volume of $V_{min} = 1.3\lambda^3$ is required if measuring in octave bands.

The standards [40, 85] for measuring reverberation times require that multiple reverberation time measurements are made in order to obtain a statistically accurate estimate. When averaging N reverberation times, it is necessary to use the expression [47, Eq. (7.75)]

$$\overline{T}_{60} = \frac{N}{\sum_{n=1}^{N} \frac{1}{T_{60,n}}}, \tag{7.9}$$

where $\overline{T}_{60}$ represents the average and $T_{60,n}$ is the nth reverberation time measurement.

It is possible to estimate the power that will be radiated from a harmonic monopole source in a diffuse field using [90, Eq. (3.28)]

$$E\{W\} = \frac{\rho_0 c_0 k^2 |Q|^2}{8\pi}, \tag{7.10}$$

where E represents the expectation operator, $k = \omega/c$ is the acoustic wavenumber, and $|Q|$ is the amplitude of the volume velocity of the source. It must be noted that this expectation represents the most likely value and that the actual radiated sound power will vary throughout the field. The relative standard deviation of the expected sound power (accounting for coherent back-scattering) is given by [90, Eq. (3.30)]

$$\varepsilon\{W\} \approx \sqrt{2}\frac{1}{k}\sqrt{\frac{8\pi}{A}} = \sqrt{2}\sqrt{\frac{1}{M\pi}}, \tag{7.11}$$

where A is the total absorption in the room (which is the sum for all surfaces, the product of absorption coefficient, and the corresponding area) and M is the modal overlap. The sound power emitted by a source will therefore vary substantially with position unless the room is very large and heavily damped. For a modal overlap of $M = 3$, the relative standard deviation is equal to 1 [90, Fig. 3.8], which implies that for 68% of all source locations the sound power will range from zero to double the expected value, whilst the remaining 32% will radiate power outside this range.

7.3.2 Sound Power from Harmonic Sources

The sound power radiated from an acoustic source will vary depending on the impedance it is presented with and therefore the power from an acoustic cannot be calculated based on the source properties alone, but information about the response of the system it is driving is also required. In the harmonic analysis example to follow, it is shown how to determine the sound power radiated from two types of acoustic sources in ANSYS. In order to calculate the radiated acoustic power it is necessary to know the amplitude and phase of both the acoustic pressure and particle velocity surrounding the source. Consequently, when using acoustic elements that only have pressure DOFs it is not possible to use either a Pressure or Velocity condition to excite a system and easily determine how much acoustic power is supplied by the source. Whilst ANSYS does allow indirect calculation of the particle velocity from the pressure gradient, these estimates suffer from accuracy issues in regions of high spatial gradients. By adding displacement degrees of freedom to the acoustic elements, it is possible to directly calculate the particle velocity (via the nodal displacement) and hence estimate sound power from an acoustic source; however, this results in more degrees of freedom and longer solution times. ANSYS provides two types of acoustic sources which allow for the direct calculation of sound power for acoustic elements with only pressure

DOFs; namely the FLOW source and Mass Source. These are described in more detail below.

7.3.2.1 Determination of Sound Power from a Flow Acoustic Source

The FLOW source (actually a load in ANSYS) allows one to calculate the sound power of the source directly. The FLOW source, specified by F,,FLOW using ANSYS APDL, was described in Section 2.9.2 and (for nodes interior to a boundary) is defined as the product of the local density of the fluid, ρ_0, and the volume acceleration of the source, $\dot{Q}$, i.e., [28, 1.5.4.3 Load Types][26, Eq. 8-1][99, Eq. (6)]

$$\text{FLOW} = \rho_0 \dot{Q} \ . \tag{7.12}$$

For nodes that are on the boundary of a fluid mesh, then "a FLOW fluid load is equal to the negative of the fluid particle acceleration normal to the mesh boundary (+ outward), times an effective surface area associated with the node, times the mean fluid density"[149, p. 3-4]. Consequently, the FLOW source on a boundary creates an inward acceleration (normal to the surface).

For a harmonic volume velocity source $Q = Q_0 e^{j\omega t}$ of angular frequency ω, the volume acceleration may be written as $\dot{Q} = j\omega Q$, and therefore the FLOW in terms of the volume velocity of the source is

$$\text{FLOW} = j\omega \rho_0 Q \ . \tag{7.13}$$

The volume velocity of the FLOW source is defined as $Q = Av$, where A is a representative area associated with the source, and $v = V_0 e^{j\omega t}$ is the outward normal velocity of the fluid particle at the node with amplitude V_0. Therefore, the particle velocity close to the source is given by

$$v = \frac{\text{FLOW}}{jA\omega\rho_0} \ , \tag{7.14}$$

where ω is the angular frequency of the source.

The time-averaged active (real) sound power generated by an infinitesimally small acoustic source is defined as [47, Eqs. (1.36), (1.37), (1.72b), and (1.80)]

$$W = \frac{1}{2}\text{Re}\left\{Apv^*\right\} \ , \tag{7.15}$$

where $*$ denotes the complex conjugate and the $1/2$ term arises from the temporal integration of instantaneous power to obtain the average power over the period. The time-averaged sound power in terms of the complex FLOW source and the resulting pressure is therefore

$$W = \frac{1}{2}\text{Re}\left\{A\left(P_R + jP_I\right)\left(\frac{\text{FLOW}_R + j\text{FLOW}_I}{jA\omega\rho_0}\right)^*\right\}$$

$$= \frac{1}{2}\text{Re}\left\{A\left(P_R + jP_I\right)\left(\frac{j\text{FLOW}_R + \text{FLOW}_I}{A\omega\rho_0}\right)\right\}$$

$$= \frac{1}{2\omega\rho_0}\left(P_R\text{FLOW}_I - P_I\text{FLOW}_R\right), \tag{7.16}$$

where the subscripts R and I are the real and imaginary parts of the complex parameters. Thus FLOW_R and FLOW_I represent the real and imaginary amplitudes of the FLOW source, respectively, and likewise for the pressure amplitudes. The time-averaged sound power level is therefore given by

$$L_w = 10\log_{10}\left(\frac{P_R\text{FLOW}_I - P_I\text{FLOW}_R}{2\omega\rho_0}\right) + 120 \text{ dB re } 10^{-12} \text{ W}. \tag{7.17}$$

7.3.2.2 Determination of Sound Power from an Acoustic Mass Source

As mentioned in Section 2.9.2, the FLOW acoustic source has been deprecated for use with 3D acoustic elements. It still remains feasible to use the FLOW load by using APDL in ANSYS Workbench and for 2D analyses it still remains the only load option of its type. The more recent Mass Source, issued with the APDL command BF,,JS, applies a mass flow rate (in harmonic analyses) to a point, line, surface, or volume, and is defined as a mass flow rate per unit volume (units of $kg.s^{-1}.m^{-3}$). Note that if one applies the Mass Source to a single node, ANSYS uses the Dirac delta function to handle degenerate dimensions, so the "volume" effectively becomes a unit volume. Therefore, a Mass Source applied to a node has units $kg.s^{-1}$ which is a mass flow rate. For a Mass Source applied to a single node the following holds [26, Eq. 8-1]

$$\text{Mass Source} = Q\rho_0, \tag{7.18}$$

and the particle velocity close to the Mass Source is given by

$$v = \frac{\text{Mass Source}}{A\rho_0}. \tag{7.19}$$

It is simple to show that the Mass Source and FLOW source are related to each other by $\text{FLOW} = j\omega\,[\text{Mass Source}]$. Using Equation (7.15), the sound power arising from a Mass Source on a node can be obtained from the product of the complex conjugate of the source velocity and the resulting time-averaged pressure,

$$W = \frac{1}{2}\text{Re}\left\{A\left(P_R + jP_I\right)\left(\frac{[\text{Mass Source}]_R - j[\text{Mass Source}]_I}{A\rho_0}\right)\right\}$$

$$= \frac{1}{2\rho_0}\left[P_R\left([\text{Mass Source}]_R\right) + P_I\left([\text{Mass Source}]_I\right)\right], \tag{7.20}$$

where the subscripts R and I represent the real and imaginary terms of the complex parameters, with the corresponding time-averaged sound power level

$$L_w = 10 \log_{10} \left[\frac{P_R \left([\text{Mass Source}]_R\right) + P_I \left([\text{Mass Source}]_I\right)}{2\rho_0} \right] \tag{7.21}$$
$$+ \, 120 \text{ dB re } 10^{-12} \text{ W} \, .$$

The above two expressions are consistent with Equations (7.16) and (7.17) when substituting

$$\text{FLOW}_R + j\text{FLOW}_I = -\omega [\text{Mass Source}]_I + j\omega [\text{Mass Source}]_R \, . \tag{7.22}$$

7.4   Example: Reverberation Room

This section describes an example of a reverberation room with its walls partially lined with an sound-absorbing material of known absorption coefficient. The system is modeled and the absorption coefficient is then calculated from the pressure response in the room as an engineer would in practice. The learning outcomes of this example are:

- how to calculate the sound power radiated from a point acoustic source when conducting harmonic analyses,

- how to calculate the natural frequencies of a room that has no damping and also when damping is added to the surface, and

- how to conduct a transient analysis on a lightly damped acoustic system.

Figure 7.1 shows a sketch of a $5\,\text{m} \times 6\,\text{m} \times 7\,\text{m}$ rectangular reverberation room, with a single monopole source located near a corner, and 6 microphones to measure the pressure response. The dimensions of the room were chosen to reflect the reverberation chamber in the School of Mechanical Engineering at the University of Adelaide, Australia. Neither the source nor the microphones are within $0.75\,\text{m}$ of the walls or each other. An attempt has also been made to keep the separation between these points greater than the desired half a wavelength, but this is difficult for such a low frequency. A $9\,\text{m}^2$ square area of sound-absorbing material is located centrally in the room on the floor, with edges parallel to the walls. Note that in practice it is not recommended to have the arrangement of the specimen to be tested aligned to the walls but is convenient in numerical methods for mesh alignment. The parameters used in the example are listed in Table 7.1. Using these parameters, the expected reverberation time using Equation (7.1) is

$$T_{60} = \frac{55.25V}{c_0 S_\alpha \bar{\alpha}} = 3.75 \text{ s} \, . \tag{7.23}$$

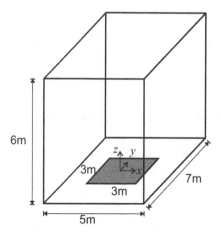

FIGURE 7.1
Schematic of the reverberation room with the sound-absorbing material on the floor.

TABLE 7.1
Parameters Used in Room Acoustics Example

Description	Parameter	Value	Units
Room depth	L_x	5.0	m
Room width	L_y	7.0	m
Room height	L_z	6.0	m
Surface area of sound-absorbing material	S_α	9.0	m^2
Surface area of room*	S	214	m^2
Volume of room*	V	210	m^3
Absorption coefficient	$\bar{\alpha}$	1	—
Speed of sound	c_0	344	m/s
Density of the air	ρ_0	1.21	kg/m^3
Sample rate for the transient analysis	f_s	500	Hz
Octave band center frequency	f_c	125	Hz
Octave band lower frequency for harmonic analysis*	f_{lower}	89	Hz
Octave band upper frequency for harmonic analysis*	f_{upper}	176	Hz
Corresponding wavelength at center frequency*	λ	2.75	m
Harmonic acoustic source. ANSYS	Mass Source	$0+j$	kg/s
Harmonic acoustic source. ANSYS	FLOW Source	$1+0j$	kg/s^2

* Dependent parameters

The analysis of the system will be conducted over four sections:

1. First, the solid and finite models of the reverberation room will be developed; followed by

2. a modal analysis of the room, both without and with the sound-absorbing material;

3. a harmonic analysis of the room subjected to a point acoustic source; and

4. finally, a transient analysis which simulates a standard sound absorption coefficient test in a reverberation room.

> Note that in a real reverberation room it is necessary to quantify the amount of absorption that is inherent in the room (without the test specimen) since this affects the sound pressure levels (both steady state and transient). In the current model we have assumed that there is no damping on the walls of the room and have neglected air absorption. This has been done to simplify the example and is sufficient to illustrate the important learning outcomes. If a small amount of damping is added to the walls, then two entire separate analyses are required, thereby effectively doubling the size of the problem and the effort from the user. This is left as an exercise for the reader. Try adding a wall absorption coefficient of 0.02, which is typical of hard plaster or concrete walls [47].

The 125 Hz octave band, which covers the frequency range 88.4-176.8 Hz [89], has been selected for the analysis. The harmonic analysis was truncated to the integer frequencies in this range for convenience and has little impact on the accuracy of the results.

There are a number of constraints associated with the physics of this example that restrict the bandwidth over which the transient analysis is valid. There is a lower bound on the frequency limit which is due to the diffuse field requirement and an upper bound arising from the time step used in the transient analysis.

In order for Equation (7.8) to be valid, the wavelength, λ, must not be less than $\sqrt[3]{V/1.3} = 5.4$ m. Consequently, the lower frequency limit of the diffuse field assumption for the chosen room dimensions is approximately 63 Hz, which is sufficiently below the lower frequency limit ($125/\sqrt{2} = 88.4$ Hz) of the 125 Hz octave band.

The transient analysis will be conducted with a time step of 2 ms, which is equivalent to a sample rate of 500 Hz (as shown in Table 7.1). Therefore, the Nyquist sampling theorem implies that the maximum (or Nyquist) frequency that can be modeled is 250 Hz, which is greater than the upper bound of the 125 Hz octave band (namely $125\sqrt{2} = 176.8$ Hz). In all one-third octave bands below the Nyquist frequency of 250 Hz, the modal overlap is too low for the diffuse field assumption to hold, and therefore the 125 Hz octave band is the only ISO standard [89] bandwidth suitable for the chosen sample rate.

The locations of the source and receiver microphones are listed in Table 7.2. In practice, more microphones and source locations are often necessary to obtain an accurate estimate of the absorption coefficient [85] but the six used here are sufficient for the exercise of illustrating the analysis technique. Furthermore, in this example the microphone locations are a rational fraction of the room dimensions and hence will sit on nodal lines/planes of some of the acoustic modes of the room, which is not ideal, but again has been used for simplicity.

TABLE 7.2

Location of Source and Receiver Microphones, and the Span of the Absorbent and Room Used in the Example

Description	Location	
Source	[1.750, −2.750, 5.250]	
Microphone 1	[1.250, 2.250, 4.750]	
Microphone 2	[0.750, 1.750, 4.250]	
Microphone 3	[−0.250, 0.750, 3.250]	
Microphone 4	[−1.250, −0.250, 2.250]	
Microphone 5	[0.250, 0.250, 3.250]	
Microphone 6	[0.250, 0.250, 4.750]	
Corners of room	[−2.500, −3.500, 0.000] to	[2.500, 3.500, 5.000]
Span of absorbent	[−1.500, −1.500, 0.000] to	[1.500, 1.500, 0.000]

7.4.1 Model

Models were built in MATLAB and ANSYS using the parameters listed in Tables 7.1 and 7.2.

7.4.1.1 Model: MATLAB

The MATLAB script Sabine.m included with this book is to be used with this example. The MATLAB script can be used to calculate the modal density, modal overlap, frequency bounds, and reverberation time using the equations in Section 7.3 as well as post-process the ANSYS Workbench and ANSYS Mechanical APDL results.

7.4.1.2 Model: ANSYS Workbench

This section describes the instructions to create the model of the reverberation room in ANSYS Workbench. This example will use the FLUID30 3D linear acoustic element type with the displacement degrees of freedom deactivated since they are unnecessary. The element size will be 0.25 m, which means that there are almost 11 elements per wavelength for the 125 Hz octave band center frequency and more than 7.5 at the upper bound of the octave band for the harmonic analysis, thereby satisfying the number of elements per wavelength constraint for the linear element.

The transient analysis will be conducted at 500 Hz and therefore the highest frequency that can be modeled is the Nyquist frequency of 250 Hz. At this upper frequency limit there are only 5.4 elements per wavelength, which although not ideal, is still is adequate for this exercise. Since the pressure response is filtered by an ANSI standard 125 Hz octave band filter, the number of elements per wavelength in the frequency range of interest is the same as the harmonic analysis.

The use of linear elements and a course mesh density have been used for two reasons: primarily to keep the analysis time down as much as possible, but equally important, this improves accuracy. This might seem counter to standard logic, in that having a low mesh density will improve the accuracy, but the statistical room acoustics works on having spatial averages of the pressure field. By employing too fine a mesh, the local spatial variance increases, which results in a worse estimate of the mean field in the vicinity of the node. The user can try repeating this example but halving the element size to 0.125 m and using quadratic elements, and apart from taking considerably longer to solve, the estimates of reverberation times and sound absorption coefficient are made slightly worse.

Note that throughout this example, the Scoping Method option for selecting objects will alter between Geometry Selection and Named Selection to illustrate both methods. In practice, a user would tend to use only one method.

Start of ANSYS Workbench Instructions

The completed ANSYS Workbench project file Sabine.wbpj is available with this book. It is assumed that the ACT Acoustics extension is installed and is operating correctly. This can be checked in the Workbench project view by selecting the Extensions | Manage Extensions menu. The extension ExtAcoustics should be listed in the table and a tick present in the Load column.

Load	Extensions	Version
☑	ExtAcoustics	8

Constructing the Solid Model

- Start ANSYS Workbench and start a new project.

- Double-click on Modal under Analysis Systems in the Toolbox window, so that a new Modal cell appears in the Project Schematic window.

- Double-click on row 3 `Geometry` to start DESIGNMODELER.

- Under `Tree Outline`, click on `A: Modal`. Then under `Details View | Model Units` ensure that the desired length unit is `Meter`.

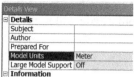

- Left-click the mouse button in the `Graphics` window and then use the middle mouse wheel to scroll the view until the dimension scale at the bottom of the screen shows 5.000 m. This is to make it easier to sketch the model.

- In the toolbar at the top of the screen, left-click on `Create | Primitives | Box` to generate the "room."

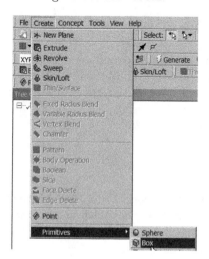

- In the window Details View, ensure that the Box Type is set to From One Point and Diagonal. Then proceed to define the geometry of the box by changing the parameters as shown below.

Details View	
Details of Box1	
Box	Box1
Base Plane	XYPlane
Operation	Add Material
Box Type	From One Point and Diagonal
Point 1 Definition	Coordinates
☐ FD3, Point 1 X Coordinate	-2.5 m
☐ FD4, Point 1 Y Coordinate	-3.5 m
☐ FD5, Point 1 Z Coordinate	0 m
Diagonal Definition	Components
☐ FD6, Diagonal X Component	5 m
☐ FD7, Diagonal Y Component	7 m
☐ FD8, Diagonal Z Component	6 m
As Thin/Surface?	No

- If you choose an isometric view, by either clicking on the lone point in the triad in the screen, or on the ISO button in the toolbar, you will see the box you have just created.

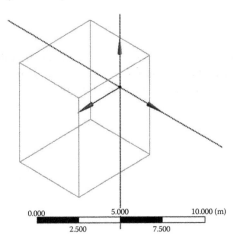

- We will now define the dimensions of the box as parameters. Under the Details View, click on the square check box to the left of the FD6, Diagonal X Component. Define the parameter LEN, which will be the length of the room.

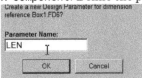

- Repeat the process for the FD7, Diagonal Y Component and FD8, Diagonal Z Component naming these items WIDTH and HGT (the width and height of the room), respectively. If this was successful, FD6, FD7, and FD8 will have the letter D next to them.

Diagonal Definition	
D	FD6, Diagonal X Component
D	FD7, Diagonal Y Component
D	FD8, Diagonal Z Component

- Click on Tools | Parameters and in the Design Parameters tab within the Parameter Manager window you should see the three parameters that have just been defined.

Parameter Manager

```
LEN = 5
WIDTH = 7
HGT = 6
```

- The box is to have a coordinate offset. We will now define parameters for the offset in the Details View window under the row Point 1 Definition using entries FD3, Point 1 X Coordinate and FD4, Point 1 Y Coordinate. Rather than repeating the same process we used to define FD6, FD7, and FD8 above we will assign dimensions to these directly using the Parameter/Dimension Assignments tab in the Parameter Manager window. In this tab, type Box1.FD3 = -@LEN/2 and Box1.FD4 = -@WIDTH/2. This will ensure that changes in the LEN or WIDTH parameters will cause changes in the values calculated for these offset definitions.

Parameter Manager

```
Box1.FD8 = @HGT
Box1.FD7 = @WIDTH
Box1.FD6 = @LEN
Box1.FD3 = -@LEN/2
Box1.FD4 = -@WIDTH/2
```

Design Parameters   Parameter/Dimension Assignments

- Now generate the box by either left-clicking the Generate button in the toolbar or right-clicking on Box | Generate in the Tree Outline.

If all has gone well, you will see a rendered solid.

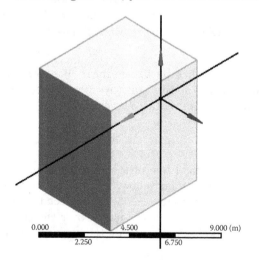

You will also notice that the `Tree Outline` has `1 Part, 1 Body`. Clicking on the + symbol next to `1 Part, 1 Body` will show the `Solid` that was just created.

⊟⋯🐌 1 Part, 1 Body
   ⌞⋯🐌 Solid

- Now is a good time to save your work. Click on `File | Save Project` and enter an appropriate filename such as `sabine_absorption`.

- We will now create a rectangle which will form the sound-absorbing material on the floor of the room. Move the mouse cursor over the Z axis on the triad in the lower right corner of the screen and click on the +Z axis so that the X-Y Plane is shown. Alternatively, you can right-click on the `XYPlane` in the tree outline and left-click on `Look at` in the context menu.

- Click on the `Sketching` tab to open the `Sketching Toolboxes` window. Make sure that the point and edges selection filters are both active, which is indicated by the icons with a sunken appearance. If they are not active, you can click on the icons or type with the keyboard `<Ctrl> p` for points and `<Ctrl> e` for edges.

- In the `Sketching Toolboxes` window click on `Draw | Rectangle`.

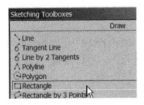

Then move the mouse to approximately $[-1.5\,\text{m}, -1.5\,\text{m}]$ as shown in the status window at the bottom of the screen in the X-Y Plane, left-click, and hold the button down while dragging the cursor to approximately $[1.5\,\text{m}, 1.5\,\text{m}]$.

- In the `Sketching Toolboxes` window, click on `Dimensions| Horizontal`. You will notice that in the lower left of the screen it says "`Horizontal -- Select first point or 2D Edge for Horizontal dimension.`"

   🔵 Horizontal -- Select first point or 2D Edge for Horizontal dimension

Move the mouse over the left-hand vertical line of the rectangle you just created and left-click on it; the line will become yellow. You will notice that you must then select the second line "`Horizontal -- Select second point or 2D Edge for Horizontal dimension.`" Move the mouse over the right-hand vertical line of the rectangle and click on it, release the mouse button, and then move the cursor upward until it is above the rectangle to place the dimension.

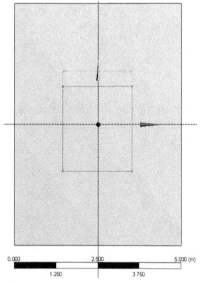

A new horizontal dimension has been created and can be seen in the Details View window and the row Dimensions: 1 | H1. To the right of H1, click on the number and set this to 3 m.

- Repeat these steps for the vertical dimension by clicking on Dimensions| Vertical to create a dimension V2.

- We will now define the parameters for the dimensions of the rectangle. Click on the Dimensions: 2 | H1 check-box

and create a new parameter called SPAN, which will define the horizontal span of the sound-absorbing material.

- We will now position the rectangle relative to the global coordinate system. In the Sketching Toolboxes window, click on Dimensions | Horizontal. Move the mouse over the left-hand line of the rectangle, which defines the sound-absorbing material, and click on it to define the first edge of the dimension. Now move the mouse over the y-axis in the middle of the rectangle and click on this to define the second edge of the dimension. This will create the dimension H3.

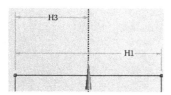

- Repeat this process to add a new constraint for the location of the sound-absorbing material in the vertical dimension by clicking on Dimensions | Vertical in the Sketching Toolboxes window.

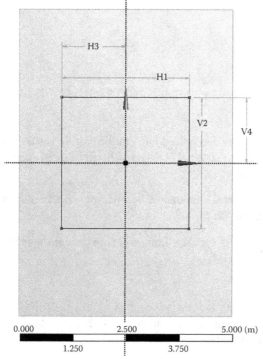

0.000   2.500   5.000 (m)

1.250   3.750

- You will notice that in the Details View window in the row Dimensions: 4 there are four dimensions, two of which are the ones just defined (H3 and V4). You may define the magnitude of these two new dimensions as both 1.5 m.

Dimensions: 4	
D H1	3 m
H3	1.5 m
V2	3 m
V4	1.5 m

- Left-click in Tools | Parameters. Select the Parameter/Dimension Assignments tab and type XYPlane.H3 = @SPAN/2, XYPlane.V2 = @SPAN, and `XYPlane.V4 = @SPAN/2`. This will ensure that the offset of the sound-absorbing material on the floor will be automatically updated based on the room dimensions given by the `LEN` and `WIDTH` parameters.

```
Parameter Manager
XYPlane.H1 = @SPAN
XYPlane.H3 = @SPAN/2
XYPlane.V2 = @SPAN
XYPlane.V4 = @SPAN/2
Box1.FD8 = @HGT
Box1.FD7 = @WIDTH
Box1.FD6 = @LEN
Box1.FD3 = -@LEN/2
Box1.FD4 = -@WIDTH/2
```

You will notice that all 4 items in the `Details View` window, row `Dimensions: 4`, are now dimensioned with parameters, which is indicated by the presence of the `D`.

Dimensions: 4	
D H1	3 m
D H3	1.5 m
D V2	3 m
D V4	1.5 m

- The next step is to attach the rectangle (sound-absorbing material) to the box (room). In the toolbar click `Create | Extrude`. In the `Details View` window, change the `Operation` to `Imprint Faces`.

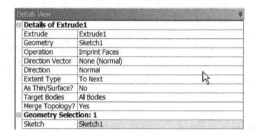

Details of Extrude1	
Extrude	Extrude1
Geometry	Sketch1
Operation	Imprint Faces
Direction Vector	None (Normal)
Direction	Normal
Extent Type	To Next
As Thin/Surface?	No
Target Bodies	All Bodies
Merge Topology?	Yes
Geometry Selection: 1	
Sketch	Sketch1

Once this is done, either left-click the `Generate` button in the toolbar or right-click on `A: Modal | Extrude1 | Generate` in the `Tree Outline` window. You will notice that `Extrude1` now has a green tick next to it. The final model should look like the following figure (once the `ISO` view and `Display Plane` are selected).

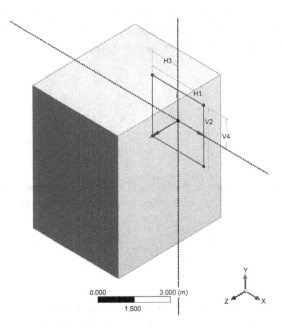

- The next step is to define two named selections that will identify the body associated with the room and the area associated with the sound-absorbing material. We will first define the named selection for the room. Either left-click the mouse on Selection Filter: Bodies in the toolbar or press <Ctrl> b on the keyboard. Then in the toolbar, click Tools | Named Selection. In the window Details View in the row Named Selection, type NS_Room. Click on the box (of the reverberation room) in the Model View window, then in the window Details View, click on the Apply button in the Geometry row, which should now show 1 Body. In the Tree Outline window, right-click on NS_Room and in the context menu left-click on Generate. You should see the following:

Details View	
Details of NS_Room	
Named Selection	NS_Room
Geometry	1 Body
Propagate Selection	Yes
Export Selection	Yes

- Repeat these steps for the rectangle imprinted onto the surface of the box. To do this you will need a view in which the imprint can be seen. Start with an isometric view, then in the triad at the bottom of the screen, move the mouse between the X and Y axes such that the negative Z axis is shown and click on it.

- In the toolbar, left-click the mouse on Selection Filter: Faces or alternatively press <Ctrl> f on the keyboard to activate the filter to select faces. Create a new named selection by selecting from the toolbar Tools | Named Selection. In the window Details View, type NS_Absorbent in the row Named Selection. Left-click on the rectangle (which represents the sound-absorbing material) in the Model View window. In the window Details View click on the Apply button in the Geometry row, which should now show 1 Face. The face will be highlighted in a different color.

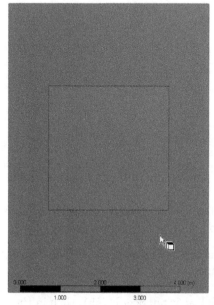

In the window Tree Outline, right-click on NS_Absorbent and in the context menu, right-click on Generate, which will generate the named selection.

It is worth noting that although Workbench will accept having Named Selections with spaces, when solving the model, which involves the creation of a script of APDL commands that is "hidden" from the user, the spaces are stripped (and characters are capitalized) so if one had "NS Absorbent" then it would become "NSABSORBENT" in Mechanical APDL. Therefore, be wary when using Named Selections for loads, boundary conditions, etc., if one intends to use APDL commands.

- The model is now complete; in the toolbar click on File | Save Project. Exit the DESIGNMODELER.

Meshing and Part Definitions

- The next step is to mesh the solid model. In the Workbench Project Schematic double-click on row 4 Model. This will start ANSYS Mechanical.

- We will now define the box representing the reverberation room as an acoustic body. Either click the mouse on `Selection Filter: Bodies` in the toolbar or press `<Ctrl> b` on the keyboard. In the `Geometry` window, click on the room body. In the ACT ACOUSTICS EXTENSION toolbar, left-click on `Acoustic Body` and select `Acoustic Body`. This will insert an `Acoustic Body` entry in the `Outline` window under `Modal (A5) | Acoustic Body`. An alternative way to define this is to right-click in `Modal (A5)`, then in the context menu left-click on `Insert | Acoustic Body`. Click on this entry to see the following window of `Details of "Acoustic Body"`.

Details of "Acoustic Body"	
Scope	
Scoping Method	Geometry Selection
Geometry	1 Body
Definition	
Temperature Dependency	No
Mass Density	1.2041 [kg m^-1 m^-1 ...
Sound Speed	343.24 [m sec^-1]
Dynamic Viscosity	0 [Pa sec]
Thermal Conductivity	0 [W m^-1 C^-1]
Specific Heat Cv	0 [J kg^-1 C^-1]
Specific Heat Cp	0 [J kg^-1 C^-1]
Reference Pressure	2E-05 [Pa]
Reference Static Pressure	101325 [Pa]
Acoustic-Structural Coupled Body Options	Uncoupled

In the window `Details of "Acoustic Body"`, change the row `Definition | Mass Density` to 1.21 and the row `Definition | Sound Speed` to 344.

An alternative way to select the body would be to change the `Scope | Scoping Method` to `Named Selection`, and then choose NS_Room.

- The next step is to define the acoustic absorption surface. Either click the mouse on `Selection Filter: Faces` in the toolbar or press `<Ctrl> f` on the keyboard. Change the view so you can see the face with the sound-absorbing material. In the window `Outline`, right-click in `Modal (A5)`, then in the context menu, left-click on `Insert | Acoustic Attenuation Surface`.

This will create a new entry in `Modal (A5)` called `Acoustic Attenuation Surface`. In the window `Details of "Acoustic Attenuation Surface"`, change the row `Scope | Scoping Method` to `Named Selection`. Then change the row `Named Selection` to NS_Absorbent. In the row `Attenuation Coefficient`, type 1 as per the property of the sound-absorbing material listed in Table 7.1. It should be noted that the coefficient many be any value from 0 to 1, which ranges from completely reflective to completely absorptive, respectively.

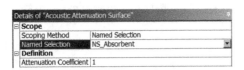

Details of "Acoustic Attenuation Surface"	
Scope	
Scoping Method	Named Selection
Named Selection	NS_Absorbent
Definition	
Attenuation Coefficient	1

- The next step is to define how the room will be meshed. In the `Outline` window, right-click in `Project | Model (A4) | Mesh` then left-click on

Insert | Method. There are two ways to select the body to which this method will be applied. First, in the row Scope | Scoping Method select Named Selection. Then in the row Named Selection choose NS_Room, which was defined previously. Alternatively, the selection process can be achieved with Scoping Method | Geometry Selection and using the mouse to select the body of the reverberation room in the Geometry window.

Change the row Definition | Method to MultiZone. This will be used to create a hexahedral mapped mesh. Change the row Element Midside Nodes to Dropped, since we will be using the 8-noded FLUID30 elements. If this is not done, then quadratic elements (FLUID220) will be used.

Details of "MultiZone" - Method	
Scope	
Scoping Method	Named Selection
Named Selection	NS_Room
Definition	
Suppressed	No
Method	MultiZone
Mapped Mesh Type	Hexa
Surface Mesh Method	Program Controlled
Free Mesh Type	Not Allowed
Element Midside Nodes	Dropped
Src/Trg Selection	Automatic
Source	Program Controlled
Advanced	
Mesh Based Defeaturing	Off
Minimum Edge Length	3. m
Write ICEM CFD Files	No

- Now we will define the default element size. In the Outline window, right-click in Project | Model (A4) | Mesh, then left-click on Insert | Sizing. The next step will illustrate how to manually select an object. In the window Details of "Body Sizing" - Sizing, change the row Scoping Method to Geometry Selection. Click in the cell next to the row Geometry. Ensure that the Body Selection filter is activated in the selection toolbar, and left-click on the body of the reverberation room in the Geometry window. The status bar at the bottom of the screen should indicate 1 Bodies selected. Click on the Apply button next to Geometry.

Details of "Body Sizing" - Sizing	
Scope	
Scoping Method	Geometry Selection
Geometry	Apply · Cancel
Definition	
Suppressed	No
Type	Element Size
Element Size	Default
Behavior	Soft

If you have done this correctly, the MultiZone object in the Outline window will have a small green tick next to it.

- In the window Details of "Body Sizing" - Sizing, in the row Definition | Element Size, type 0.25. The element size will now be defined as a parameter that can be altered. Left-click in the small box to the left of the cell Element Size. The letter P will appear inside the box indicating a parameter.

P Element Size

- Now mesh the model. This can be done by either clicking Mesh | Generate

Mesh in the toolbar, or alternatively, right-click over Mesh in the Outline window and in the context menu left-click on Generate Mesh.

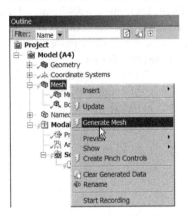

The resulting mapped mesh is shown in the following figure.

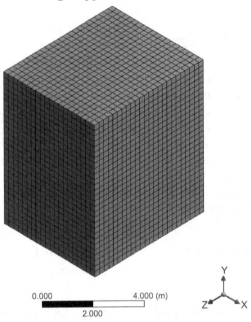

0.000   4.000 (m)

2.000

- Rotate the view of the model to show the face with the sound-absorbing material by clicking on the -Z axis on the triad. You will see that the surface also has a mapped meshed.

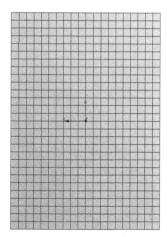

- The mesh is now complete. In the menu bar, left-click on File | Save Project.

7.4.1.3 Model: ANSYS Mechanical APDL

The ANSYS Mechanical APDL file code_ansys_sabine.txt that is available with this book was used generate the FE model shown in Figure 7.2, as well as conduct the modal, harmonic, and transient analyses that are described later in this chapter.

The sound-absorbing material was modeled using a surface load applied directly to the nodes using the APDL command SF,,CONV,ALPHA. This directly applies an entirely real, locally reacting impedance to the node, where ALPHA represents the absorption coefficient which must be between 0 and 1, where the latter represents complete absorption. Alternatively, it may be applied to an element, line, or area using the APDL command SFE, SFL, or SFA, respectively.

As was discussed in Section 7.3.2, using simple post-processing it is possible to directly calculate the power radiated from either the acoustic FLOW source or Mass Source. For the harmonic analysis presented in Section 7.4.3, the ANSYS APDL code can be used with either acoustic source type.

For the transient analysis described in Section 7.4.4 there are a variety of possible purely acoustic sources that could have been used to excite the system such as a pressure condition D,,PRES, a flow condition F,,FLOW, or a Mass Source BF,,JS. However, in ANSYS Release 14.5 the Mass Source does not support tabular data for multiple load steps, which is necessary for this example, and consequently the FLOW acoustic source was chosen. This is consistent with the Workbench analysis presented in Section 7.4.4.3.

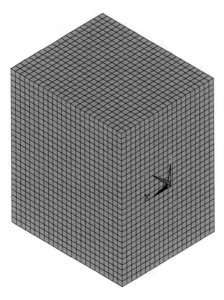

FIGURE 7.2
FE model of the reverberation room created in ANSYS Mechanical APDL.
The image has been rotated from the standard isometric view in order to see
the face with the sound-absorbing material (right-hand side face in the image
where the triad is located).

7.4.2 Modal Analysis

A modal analysis can be conducted to determine the number of modes of
the reverberation room in the frequency range of interest. This is not necessary for the purposes of determining the sound absorption coefficient of a
material. However, it is good practice to ensure that the constraints necessary
to approximate a diffuse field are met. MATLAB was used to determine the
natural frequencies and modal density of the room based on the theory presented in Section 7.3. ANSYS has been used to conduct modal analyses for
both an undamped and damped models, to determine the natural frequencies,
to visualize the acoustic modes of the room, and to provide validation of the
theoretical results.

7.4.2.1 Modal Analysis: MATLAB

Using the MATLAB file Sabine.m and Equation (7.2), all natural frequencies
below the upper limit (176.8 Hz) of the 125 Hz octave band were calculated.
There are 168 modes in total, the lowest 10 of which are presented in Table
7.3. Within the 125 Hz octave band there are 140 modes, which means that the
room is sufficiently diffuse to meet both constraints discussed in Section 7.3.

Using Equation (7.5) to calculate the modal density of the entire 125 Hz octave band we get $\dfrac{\Delta N}{\Delta f} = \dfrac{140}{176.8 - 88.4} = 1.6$. This is consistent with the expected modal density at the 125 Hz center frequency using Equation (7.4), which equates to $\dfrac{dN}{df} = 1.4$. Note that the modal density increases with frequency, and therefore the modal density over a band is always greater than that at the center frequency of the band.

TABLE 7.3

First 10 Natural Frequencies (Hz) of the Reverberation Room

Mode	Indices	Theory Undamped	ANSYS Undamped	ANSYS Damped
1	[0,0,0]	0	8.8E-06	0.00
2	[0,1,0]	24.6	24.6	24.7
3	[0,0,1]	28.7	28.7	29.6
4	[1,0,0]	34.4	34.4	34.7
5	[0,1,1]	37.6	37.8	38.0
6	[1,1,0]	42.3	42.3	42.4
7	[1,0,1]	44.8	44.8	45.3
8	[0,2,0]	49.1	49.2	50.0
9	[1,1,1]	51.1	51.1	51.2
10	[0,2,1]	56.9	57.0	57.3

Also presented in Table 7.3 are the results from the modal analysis performed in ANSYS. It includes both the undamped and damped analyses. The fundamentals of undamped modal analysis were introduced previously in Chapter 4 for a rigid-walled rectangular room. In a damped system the eigenvalues are complex and are given by $\sigma \pm j\omega_d$, where σ is the real part of the eigenvalue, and the imaginary part of the eigenvalue, ω_d, is the damped natural frequency. The (undamped) natural frequency is given by the magnitude of the eigenvalue

$$\omega_n = \sqrt{\sigma^2 + \omega_d^2} \qquad (7.24)$$

and the damping ratio is given by $\zeta = \sigma/\omega_n$. When performing a damped modal analysis, ANSYS Mechanical APDL returns the real and imaginary eigenvectors, whereas ANSYS Workbench returns the damped natural frequency and the damping ratio (amongst other things). The natural frequency may be obtained from the latter using

$$\omega_n = \omega_d/\sqrt{1 - \zeta^2}. \qquad (7.25)$$

To illustrate the effect that damping has on the mode shapes of the modeled room, the mode shape of the 5th mode (first tangential mode) is displayed in Figure 7.3 for both an undamped model (without sound-absorbing material) and a damped model (with sound-absorbing material). The effect of the

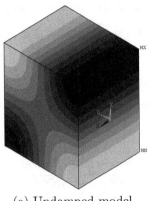
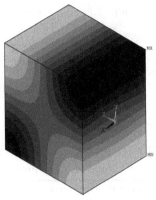

(a) Undamped model   (b) Damped model

FIGURE 7.3
Mode shape of the 5th acoustic mode (calculated using ANSYS Mechanical
APDL) for the (a) undamped and (b) damped model. The effect of the local
damping can be seen by the curved contours on the right-hand face near the
triad.

sound-absorbing material on the wall of the room can be seen by the curvature
in the pressure contours in the vicinity of the triad in Figure 7.3(b).

7.4.2.2   Modal Analysis: ANSYS Workbench

The following instructions describe how to conduct a modal analysis of the
room in ANSYS Workbench.

Setting up the Modal Solver

- The properties of the modal analysis will be defined. In the `Outline` win-
  dow, left-click on the row `Project | Modal (A5) | Analysis Settings`. In
  the `Details of "Analysis Settings"` window, locate the row `Options |
  Max Modes to Find` and type 170. Change the row `Limit Search to Range`
  to `Yes`, and set the `Range Maximum` to `250. Hz`, which is twice the center
  frequency of the 125 Hz octave band. Note that it is not always necessary
  to set the search range, but for the settings we have chosen, the solver will
  fail and return an error without constraining the range.

  Initially, an undamped modal analysis will be conducted. Ensure that the
  `Solver Controls | Damped` is set to `No`.

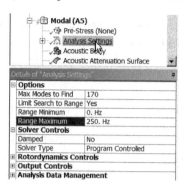

- That completes the setup of the analysis and it can now be solved. In the Outline window, right-click on Project | Modal (A5) and left-click on Solve, or alternatively in the toolbar click the Solve icon. This may take some minutes to solve depending on the computer you are using.

- Note that only pressure and acoustic particle velocity post-processing is available for modal analyses. However, the velocities are not stored in the .rst results file by default — to do this you need to specify that you want to store stresses in the analysis results files. In the Outline window, left-click on Project | Modal (A5) | Analysis Settings, then in the window Details of "Analysis Settings" locate the row Output Controls | Stress and select Yes.

Note that since the size of the results file is increased, this can slow the solution process, so if you do not require the particle velocities, then you may leave this option as No.

Reviewing and Exporting the Results

- You can review the results by clicking on Project | Modal (A5) | Solution (A6). This will show the Tabular Data which contains the natural frequencies of the system.

- Now is a good time to Export the results. In the Tabular Data window, right-click, then in the context menu, left-click on Export. Save the file as Sabine_Modal_Undamped.txt which will be used by the MATLAB script for post-processing.

- To view the calculated acoustic pressure response, in the ACT Acoustics Extension toolbar click on Results | Acoustic Pressure.

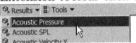

- In the Outline window, in the object Project | Model (A4) | Modal (A5) | Solution (A6) left-click on the newly created entry Acoustic Pressure. You can select a particular mode to display in the row Definition | Mode. Select the fifth mode by typing 5 opposite the row Mode. Note that the eigen-frequency associated with the model is reported in the Details of "Acoustic Pressure" window in the row Information | Reported Frequency.

You must "evaluate" the results to display the mode shape of interest. To do this, right-click on the Solution (A6) entry, or alternatively the Acoustic Pressure entry, then left-click on Evaluate All Results,

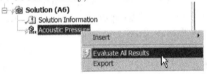

which will show an image like the following.

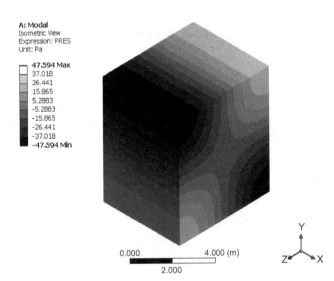

Damped Modal Analysis

- To conduct a damped modal analysis, which incorporates the effects of the sound-absorbing material, click on Modal (A5) | Analysis Settings, then

in the Details of "Analysis Settings" window, change the row Solver
Controls | Damped to Yes. In the row Solver Controls | Solver Type
select Full Damped.

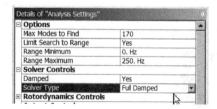

Make sure that the row Limit Search to Range is set to Yes, and set
the Range Maximum to 250. Often, ANSYS will reset fields to their default
when certain analysis settings are changed.

- Click the Solve button in the toolbar or alternatively right-click on Modal
  (A5) and left-click on Solve in the context menu to obtain the damped
  modal solution.

- As shown previously, the results may be reviewed by clicking on Project
  | Modal (A5) | Solution (A6). This will show the Tabular Data which
  contains the damped natural frequencies of the system.

	Mode	☑ Damped Frequency [Hz]	☐ Stability [Hz]	☐ Modal Damping Ratio	☐ Logarithmic Decrement
1	1.	0.	-8.3272e-012	N/A	N/A
2	2.	24.664	-0.30328	1.2296e-002	-7.7263e-002
3	3.	29.47	-2.18	7.3772e-002	-0.46479
4	4.	34.647	-0.49331	1.4237e-002	-8.9461e-002
5	5.	37.979	-0.58114	1.53e-002	-9.6143e-002
6	6.	42.358	-0.14392	3.3976e-003	-2.1348e-002
7	7.	45.24	-1.0004	2.2107e-002	-0.13894
8	8.	49.961	-0.62907	1.259e-002	-7.9113e-002
9	9.	51.22	-0.28307	5.5265e-003	-3.4724e-002
10	10.	57.295	-0.12077	2.1079e-003	-1.3244e-002

The Damped Frequency [Hz] column represents the damped natural frequency, ω_d, and the Modal Damping Ratio column is the damping ratio, ζ.
The undamped natural frequency can be calculated using Equation (7.24).

- In the Tabular Data window, right-click then left-click Export in the context
  menu. Save the file as Sabine_Modal_Damped.txt.

- Left-click on Solution | Acoustic Pressure. This will show the damped
  mode shape of the 5th acoustic mode which occurs at 37.9 Hz and is shown
  below.

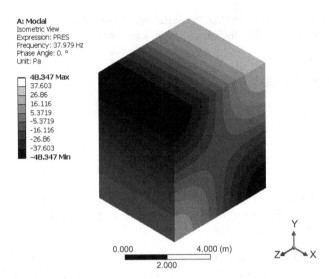

A: Modal
Isometric View
Expression: PRES
Frequency: 37.979 Hz
Phase Angle: 0. °
Unit: Pa

48.347 Max
37.603
26.86
16.116
5.3719
-5.3719
-16.116
-26.86
-37.603
-48.347 Min

0.000   4.000 (m)
        2.000

- On first appearances the results look identical to the undamped case. However, there are subtle differences in the legend — the damped case displays the Frequency and Phase Angle of the displayed mode shape, whereas the undamped case does not. Since it is not immediately obvious what affect the sound-absorbing material has had on the modes, we will create three views of the mode shapes. To do this, in the Outline window left-click on the Acoustic Pressure object to select it, then in the toolbar click on the New Figure or Image icon, then select Figure.

  Repeat these steps two more times to create three figures.

- Left-click on the first figure object in the Outline window. In the Geometry window, left-click on the triad to create an isometric view, or alternatively in the toolbar left-click on the ISO icon. In the Details of "Figure" window type Isometric View in the row Caption | Text. Rename the figure to Isometric View by right-clicking on the figure object in the Outline window and selecting Rename in the context window, or alternatively press the <F2> button on the keyboard.

- Edit the remaining two figures to create a Top View by clicking on the +Z axis on the triad in the Geometry window, and a Bottom View by clicking on the −Z axis on the triad in the Geometry window. The following figure shows the three renamed figures and the Caption for the Bottom View.

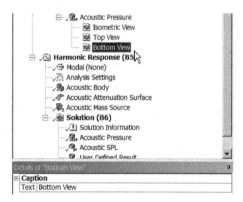

- The acoustic pressure of the 5th acoustic mode is shown for the Top View and Bottom View below, where the left image is the top plane and the right image is the bottom plane where the sound-absorbing material is positioned. If we compare the sound pressure on the top and bottom walls of the room, then the effect of the sound-absorbing material becomes apparent. As expected, the sound pressures are slightly lower adjacent to the sound-absorbing material.

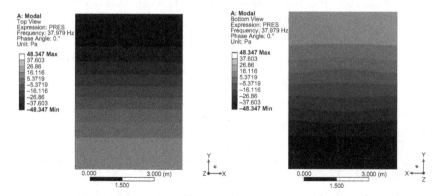

- The modal analysis is now complete. In the menu bar, click on File | Save Project.

- Close ANSYS Mechanical.

7.4.2.3 Modal Analysis: ANSYS Mechanical APDL

The ANSYS Mechanical APDL file code_ansys_sabine.txt may be used to conduct the modal analysis. Both undamped and damped analyses can be conducted by changing the variable called ModalType from 'Undamped' to 'Damped'. The eigenvalues (natural frequencies) are exported to a text file for post-processing in MATLAB using the file Sabine.m included with this book.

7.4.3   Harmonic Analysis

A harmonic analysis can be conducted to calculate the steady-state sound pressure levels in the room due to an acoustic source, which may then be used to determine the sound absorption coefficient of the sound-absorbing material in the room using Equation (7.8). This section provides details on how to achieve this in both ANSYS Workbench and ANSYS Mechanical APDL. The results from a harmonic analysis conducted in ANSYS will be compared against the theoretical results obtained in MATLAB. These will also be compared against the results obtained from the transient analysis presented in Section 7.4.4.

7.4.3.1   Harmonic Analysis: MATLAB

Using the file Sabine.m included with this book, the key parameters (reverberation time and sound-absorption coefficient) in Table 7.1 were calculated in MATLAB from the equations in Section 7.3. The MATLAB file was used to post-process both the text files containing the results from ANSYS Mechanical APDL and Workbench, which are displayed in Figure 7.4 and in Table 7.4. The frequency response graphs in Figure 7.4 show some modal behavior, in particular the mean sound pressure level. The expected sound power level from the single unit-amplitude Mass Source used in this example is given by Equation (7.15), which is equal to

$$E\{L_w\} = 10\log_{10}(\rho_0 c_0 k^2 |Q|^2 /(8\pi)) + 120\,\mathrm{dB}\ \mathrm{re}\ 10^{-12}\mathrm{W}$$
$$= 137.7\,\mathrm{dB}\ \mathrm{re}\ 10^{-12}\mathrm{W}$$

at the 125 Hz octave band center frequency. This value is consistent with the values in Figure 7.4 given the variance expected because of the low modal density. The total sound power level shown in Figure 7.4 is 152.0 dB re 10^{-12}W, which is close to 157.8 dB re 10^{-12}W obtained by evaluating Equation (7.15) over discrete frequencies across the 125 Hz octave band.

The results in Table 7.4 from the harmonic analyses obtained in ANSYS were calculated using Equation (7.8). The results from the transient analysis, discussed in Section 7.4.4, were calculated using Equation (7.9). The errors in the harmonic analysis, whilst on first impressions might appear very large, are purely as a result of the sensitivity of Equation (7.8) to small errors in the magnitude of the sound pressure level and sound power level estimates. It turns out that an error of less than 1 dB in the difference between the mean sound pressure level of the microphones and the sound power level from the source will lead to errors of greater than 20% in the T_{60} estimates. The errors in the estimates obtained from the transient analysis are half that of the harmonic analysis. A 10% error is not surprising given the expected variance in the sound power and pressure levels in the chosen frequency range. Furthermore, the test standards [40, 85] require the reverberation times to be calculated for multiple source locations, which has not been undertaken here.

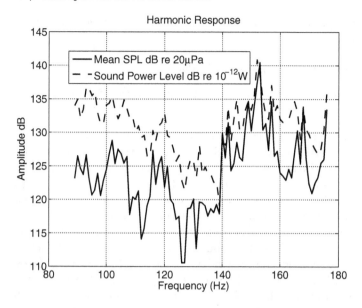

FIGURE 7.4

Sound power from a unit amplitude harmonic Mass Source and resulting mean SPL (from 6 locations) versus frequency for the 125 Hz octave band.

TABLE 7.4

Results from the Analysis of the Reverberation Room

Descript.	Param.	Units	Theory	HARM	Diff	TRANS	Diff
Reverb. time	T_{60}	s	3.748	4.514	20.4%	3.354	10.5%
Sound absorption coefficient	$\bar{\alpha}$	—	1.000	0.830	17.0%	1.117	11.7%

Note: The theoretical results calculated in MATLAB are compared against the harmonic (HARM) and transient (TRANS) analysis in ANSYS Mechanical APDL.

7.4.3.2 Harmonic Analysis: ANSYS Workbench

The following ANSYS Workbench instructions describe how to conduct a harmonic analysis on the model Sabine.wbpj developed in the previous section. This involves inserting a single acoustic source into the finite element model as defined in Table 7.2 and measuring the pressure response at six locations. The response is calculated across the entire bandwidth of the 125 Hz octave band (89 Hz to 176 Hz) in 1 Hz increments, and then the results are averaged to enable calculation of the reverberation time using Equation (7.8). The harmonic analysis was solved using the FULL option (see Chapter 4) since ANSYS

Release 14.5 does not support the modal superposition approach for damped acoustic systems.

Setting Up the Model for the Harmonic Analysis

The next few steps are to set up the model in preparation for the harmonic analysis.

- The harmonic analysis will share many of the features used in the previously conducted modal analysis including the geometry, mesh, and parameters. Rather than starting a new ANSYS Workbench model, it is possible to re-use appropriate information from the previous analysis. In the ANSYS Workbench project within the Toolbox window, left-click and hold on to the Harmonic Analysis branch in the Analysis Systems tree. With the mouse button still pressed, drag the Harmonic Analysis into the Project Schematic window and release the mouse button over row 4 Model in the Modal cell. The Project Schematic window should look like the image below, where connecting lines are drawn between the Engineering Data, Geometry and Model cells, indicating that information from the Modal analysis will be used in the Harmonic Response analysis.

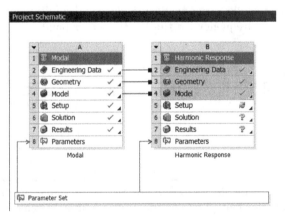

- Double-click on row 5 Setup in the Harmonic Response cell, which will start ANSYS Mechanical.

- We first have to define additional points that will be used as locations for the source and six microphones. In the Outline window, right-click on Project | Model (A4,B4) | Named Selections and in the context menu left-click on Insert | Named Selection. An alternative way to rename the object is to press <F2> on the keyboard.

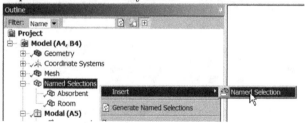

- Rename the object that was created under Named Selections by right-clicking on the object and in the context menu left-click on Rename and type Source Location.

- In the window Details of the "Source Location" change the row Scope | Scoping Method to Worksheet.

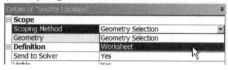

- In the Worksheet window that is created, right-click in the blank row beneath the cell Action and then left-click on Add Row.

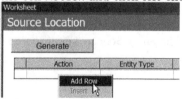

Add another two rows using the same method so that there are a total of three blank rows.

- We will now set up a filter to select the desired nodes. The first point of interest is at [1.750, −2.750, 5.250] in the global Cartesian coordinate system, which is where the acoustic source will be placed.

    In the first row of the Source Location table, under the column Action, select the option Add, change Entity Type to Mesh Node, Criterion to Location X, Operator to Equal, and Value to 1.75. In the second row of the table, change the column Action to Filter, Entity Type to Mesh Node, Criterion to Location Y, Operator to Equal, and Value to -2.75. In the third row of the table, change the column Action to Filter, change Entity Type to Mesh Node, Criterion to Location Z, Operator to Equal, and Value to 5.25. This completes the definition of the coordinates filter that will be used to locate a node at the desired location. The completed table should appear as shown below.

Worksheet

NS_Source_Location

Generate

	Action	Entity Type	Criterion	Operator	Units	Value	Lower Bo...	Upper Bo...	Coordinat...
☑	Add	Mesh Node	Location X	Equal	m	1.75	N/A	N/A	Global Co...
☑	Filter	Mesh Node	Location Y	Equal	m	-2.75	N/A	N/A	Global Co...
☑	Filter	Mesh Node	Location Z	Equal	m	5.25	N/A	N/A	Global Co...

Note that *tolerance settings* are used when the Criterion is defined as an Equal comparison. For more information on the tolerance settings see ANSYS Help [29, Adjusting Tolerance Settings for Named Selections by Worksheet Criteria]. When complete, click on the Generate button in the Worksheet window. If the filtering of the nodal coordinates was able to identify a node, then in the window Details of "Source Location", the row Statistics | Total Selection should show 1 Node.

Details of "Source Location"

⊟ **Scope**	
Scoping Method	Worksheet
Geometry	1 Node
⊟ **Definition**	
Send to Solver	Yes
Visible	Yes
Program Controlled Inflation	Exclude
⊟ **Statistics**	
Type	Manual
Total Selection	1 Node
Suppressed	0
Used by Mesh Worksheet	No

- The same process must be repeated for the six microphone locations. The fastest way to do this is to right-click on the entry Source Location and in the context menu left-click on Duplicate.

- Edit the worksheet for each of the entries and enter the coordinates for each microphone location as listed in Table 7.2. Once the six named selections have been created, in the Outline window right-click on Model (A4, B4) | Named Selection and left-click on Generate Named Selections.

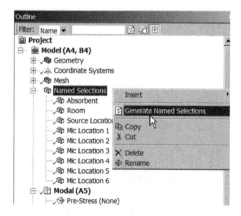

- Since we are not sharing the analysis information between the modal and harmonic analyses, it is necessary to define an Acoustic Body and Acoustic Attenuation Surface in the Harmonic Response (B5). This can be done by repeating the same process described for the modal analysis.

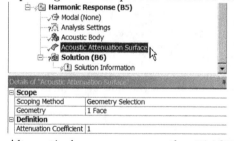

Alternatively, one can press the <Shift> key on the keyboard, left-click on the Acoustic Body and Acoustic Attenuation Surface objects in Modal (A5) branch in the Outline window, right-click over the highlighted objects, and select Copy in the context menu. Then left-click on Harmonic Response (B5), right-click on the object, and left-click on Paste in the context menu.

- The next step is to create a Mass Source at the source location. This can be done by either right-clicking on Harmonic Response (B5) and then in the context menu, left-click on Insert | Acoustic Mass Source, or alternatively, use the ACT Acoustics extension toolbar and select Excitation | Mass Source (Harmonic). Click on the Acoustic Mass Source in the Outline window. In the window Details of "Acoustic Mass Source" select Scope | Scoping Method | Named Selection. Assign the Named Selection to NS_Source_Location. Give the source a unit amplitude by changing the Amplitude of Mass Source to 1 and the Phase Angle set to 90°, which defines it as an imaginary value (i.e., $[0 + j1]$). The reason for making the Mass Source entirely imaginary is so that the results may be checked against Equation (7.20) discussed in Section 7.3.2. It also means it will be in phase with an entirely real FLOW source.

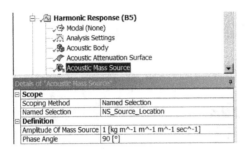

- Now add a FLOW acoustic source which we will use to compare against the Mass Source defined above. The ACT Acoustics extension does not support the FLOW source, so it is necessary to define this using ANSYS APDL code. In the Outline window, right-click in the row Harmonic Response (B5) and then in the context menu left-click on Insert | Commands.

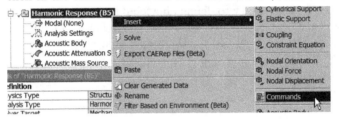

- In the Commands window, click on the Commands tab and type the following APDL code:

```
1  ! Attach these commands to the Harmonic analysis.
2  ! This will add a FLOW acoustic source.
3  ! --------------------------------------------------
4  ! Units are m, kg, s
5  ! --------------------------------------------------
6  ! Apply a FLOW Source to the source node
7  ! Find the appropriate node
8  CMSEL,S,NS_Source_Location
9  *GET, Flow_Node, NODE, 0, Num, Max
10 ALLSEL
11 ! Apply the FLOW condition. Last two entries are the Real and Imag terms
12 F,Flow_Node,FLOW,1,0
```

- In the Outline window, right-click over the new Commands (APDL) object and left-click on Rename in the context menu, and type Commands (APDL) - FLOW source, to rename the object, or alternatively press <F2> on the keyboard.

  Commands (APDL) - FLOW source

- We do not need both sources active at the same time; therefore, in the Outline window, right-click on the object Commands (APDL) - FLOW source and in the context menu click on Suppress.

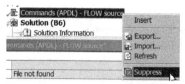

If successful, then a small blue cross will appear to the left of the object.

Harmonic Settings

- The next step is to define the parameters for the harmonic analysis such
  as the analysis frequency range, intervals, and method for calculating har-
  monic analysis. In the Outline window click on Harmonic Response (B5) |
  Analysis Settings to reveal the Details of "Analysis Settings" window.
  In the row Options | Range Minimum type 88, Options | Range Maximum
  type 176, Options | Solution Intervals type 88, and change the row
  Options | Solution Method to Full since ANSYS Release 14.5 does not
  support damped modal superposition with acoustic elements.

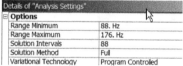

- In the Details of "Analysis Settings" window, change the row Analysis
  Data Management | Save MAPDL db to Yes. This will allow you to post-
  process the results later using the ACT Acoustics extension object Acoustic
  Time_Frequency Plot.

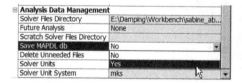

- Solve the harmonic analysis by clicking Solve in the toolbar or right-clicking
  on Harmonic Response (B5) | Solve. The yellow lightning bolt next to
  Harmonic Response (B5) will turn green indicating the solver is working.
  Be patient, this may take several minutes to solve.

Reviewing the Results

It is possible to request the types of result parameters derived in post-
processing before commencing the computations to solve the finite element
model. However, this example will show how the desired results may be defined after the computations of the harmonic response have completed.

- With the results available we will now inspect the acoustic pressure and sound pressure level at the six microphone locations. Initially we will plot the acoustic pressure across the entire outer face of the acoustic body. To do this using the ACT Acoustics toolbar, select Results | Acoustic Pressure.

You will note in the window Details of "Acoustic Pressure", in the row Geometry | Scoping Method, that 1 Body is selected.

- To evaluate the results from the previously solved analysis, in the results branch Solution (B6) in the Outline window, right-click over the Acoustic Pressure object and in the context menu left-click on Evaluate All Results.

- The frequency at which the object is evaluated can be selected by typing the desired frequency in the row Definition | Frequency in the Details of "Acoustic Pressure" window. The figure below shows the acoustic pressure across the face of the room at 152 Hz (which was selected as it is a frequency in the vicinity at which the maximum SPL was observed in Figure 7.4).

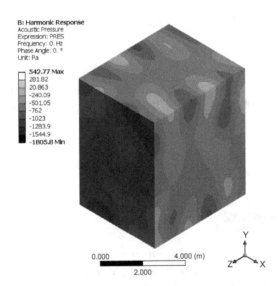

- The sound pressure level (SPL) is another commonly used acoustic quantity. To display the SPL on the faces of the body, in the ACT Acoustics toolbar select Results | Acoustic SPL. Click on the Acoustic SPL object in the Outline window and in the Details of "Acoustic SPL" window ensure that Geometry | Scoping Method has All Bodies selected. The figure below shows the SPL across the faces of the room at 152 Hz.

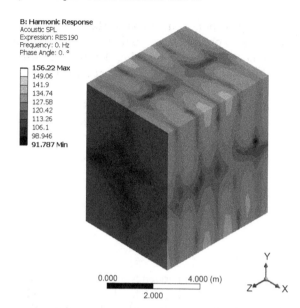

- The SPL response shown in the previous figure resembles the mode shape of the [1,1,5] acoustic mode, which has an undamped natural frequency of 151.91 Hz. To visualize the mode shape of this mode, in the Outline window left-click on the Solution (A6) | Acoustic Pressure object and in the Details of "Acoustic Pressure" window, change the row Definition Mode to 109, then right-click on Acoustic Pressure and select Evaluate All Results. The mode shape of the 109th acoustic mode is shown below.

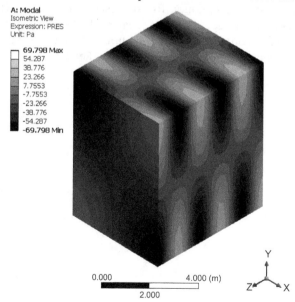

- Also of interest are the frequency response functions of the pressures and sound pressure levels at both the acoustic source and the six microphone locations. It is possible to extract this information using the ACT Acoustics toolbar by selecting Results then Acoustic Time_Frequency Plot.

Use this method create an Acoustic Time_Frequency Plot and rename the object to Source_Location Pres by left-clicking on the object in the Outline window and then pressing the <F2> button on the keyboard, or alternatively, right-click over the object and left-click on Rename in the context menu. In the window Details of "Source_Location Pres", change the row Geometry | Scoping Method to Named Selection, then change the Named Selection to NS_Source_Location. Ensure that the Result field in Definition is selected as Pressure.

- Create another six Acoustic Time_Frequency Plots to display the sound pressure level (SPL) of the six microphones. Rename each object as shown below and in the Details window for each plot, change the Result cell to SPL. Right-click over any of the objects in the results branch below Solution (B6) and in the context menu left-click on Evaluate All Results.

Warning

Note that although it is possible to create a named selection in Workbench that contains all six microphones, and that an "average" can be created in an Acoustic Time_Frequency Plot, this average is a linear average and not a logarithmic average, and therefore cannot be used to obtain the spatial mean sound pressure level of all the microphones. In other words, for N measurements this process would return $\bar{L}_p = \frac{1}{N}\sum_{i=1}^{N} L_{p,i}$, where $L_{p,i}$ is the sound pressure level at the ith location, which is not correct. To correctly average the SPLs for the microphones, the following expression is required $\bar{L}_p = 10\log_{10}\left(\frac{1}{N}\sum_{i=1}^{N} 10^{L_{p,i}/10}\right)$.

Exporting the Results

- The final step is to export the results for later post-processing in MATLAB using the file Sabine.m provided with this book. In the Outline window right-click on the entry Solution (B6) | Source_Location Pres then left-click on Export. Save the file as Sabine_Harm_Source_Pres.txt .

- Repeat the process for the six microphones saving the files as Sabine_Harm_Mic1_SPL.txt to Sabine_Harm_Mic6_SPL.txt, respectively.

- The post-processing by the file Sabine.m involves reading in the above text files, then using the source pressure to estimate the radiated sound power level using Equation (7.21) by logarithmically summing the individual frequency results to obtain the total radiated power level over the entire octave band. The six microphone sound pressure levels are logarithmically averaged, and then the individual frequency results are logarithmically summed to obtain the space average total sound pressure levels. Using these two results, the reverberation times are calculated using Equation (7.8).

7.4.3.3 Harmonic Analysis: ANSYS Mechanical APDL

The previous harmonic analysis was also conducted in ANSYS Mechanical APDL using (ANTYPE,3) with the (HROPT,FULL) option. This was used to calculate the frequency response between the source and pressures at the location of the six microphones. The frequency response was calculated in 1 Hz frequency increments across the entire 125 Hz octave band (i.e., from 89 to 176 Hz inclusive). These results were then exported to MATLAB for post-processing.

Note that FLUID30 linear elements were used in this example to keep the number of nodes in the model less than 32,000, which is a limitation of the teaching license of ANSYS. If you are not constrained by this node limit, then as an exercise you could try repeating the analysis with the acoustic element FLUID220 which has a quadratic shape function.

The results are very sensitive to changes in the model. Try changing the speed of sound slightly (for example, use the ANSYS Workbench default value of $c_0 = 343.24$ m/s) and you will see a small change source and microphone pressures which will impact on the reverberation time estimates. If you choose different locations for the source or microphones you will also find the estimates for the reverberation time change.

The results presented for the ANSYS Mechanical APDL and Workbench were obtained using a Mass Source, defined in APDL using the command BF,,JS. As an exercise you could repeat the ANSYS Mechanical APDL analysis using a FLOW source. This can be done by commenting out the line that defines the Mass Source and uncommenting the line where the FLOW source is defined in the file code_ansys_sabine.txt provided with this book. If repeating the analysis with the FLOW source in ANSYS Workbench, then it is necessary to Suppress the Acoustic Mass Source and Unsuppress the Commands (APDL) - FLOW source object, then click on Solve again.

Users familiar with ANSYS would be aware that it is possible to conduct random vibration analysis of structural systems over a finite frequency range. This is called a *Power Spectral Density (PSD) analysis*. This type of analysis does not calculate the results at specific frequencies as in a harmonic analysis, or amplitudes at specific moments in time as in a transient analysis, but rather statistically calculates the response of a structure to a given random vibration excitation. Given the nature of the steady-state analysis used in the example above, it would be desirable to repeat the harmonic analysis with a PSD analysis over the entire octave band. However, PSD analysis of coupled vibro-acoustic systems are not supported in ANSYS Release 14.5 as according to the developers it is not clear how one combines the complex eigenvectors, arising from the unsymmetric matrices with the FSI (see Chapter 9), in the PSD analysis.

7.4.4 Transient Analysis

The standards Ref. [40, 85] describe the process for measuring the absorption coefficient in a reverberation room, where either a steady-state broadband noise source or an impulse excitation may be used. The impulsive source has been used here for its simplicity. The impulse, or balloon burst method as it is sometimes referred to, requires generation of a sharp impulsive noise with a temporal duration no longer than the inverse of the highest frequency range of interest. In this simulation a step size of 2 ms was used, which is equivalent to a sample rate of 500 Hz. Therefore, activating a source for one step (2 ms), then deactivating the very next step would create a broadband excitation up to approximately 500 Hz, and well above the upper bound (176 Hz) of the 125 Hz octave band. There is an issue with this approach when using ANSYS, regardless of the source type used (i.e., PRES, FLOW or Mass Source), in that since the room is enclosed, an impulse leads to an increase in the static pressure because more media has been injected into the (constant volume) room. This would also happen in practice if a pressure line was suddenly opened then closed in an airtight room. This increase can cause difficulties in the numerical analysis as the static pressure can be many orders of magnitude higher than the transient sound pressures needed to determine the decay times. If the static pressures are too high, then ANSYS lacks the numerical precision to accurately calculate the much smaller-magnitude sound pressures. A way around this is to use a doublet: a unit positive excitation immediately followed by a unit negative excitation as shown in Figure 7.5. For a FLOW source, the net sum of volumetric flow is zero and therefore the static pressure does not increase. In this analysis the following time steps were used:

Step 0 the initial condition of the FLOW was set to zero,

Step 1 a unit positive real FLOW was applied for 1 time increment,

Step 2 at the next time increment a unit negative real FLOW was applied for 1 time increment,

Step 3 at the next time increment the FLOW was set to zero for 1 time increment,

Step 4 the solution was allowed to run to the desired finish time.

An acoustic FLOW source was used in preference to an acoustic Mass source since the FLOW source supports tabular data in ANSYS Release 14.5, whereas the Mass Source do not.

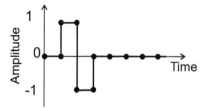

FIGURE 7.5
Illustration of amplitude profile of the doublet used for the FLOW source in the transient analysis.

The transient analysis in ANSYS was conducted over a 5 second duration with a time step of 2 ms (sample rate of 500 Hz). The pressure responses at the six microphone locations were then calculated. Although it may be possible to post-process the pressure results using ANSYS APDL commands, some of the analysis required, such as the octave band filtering, would be extremely difficult (see Section 7.4.4.1). Hence the pressure responses of the six microphones were saved to a text file for post-processing in MATLAB.

7.4.4.1   Transient Analysis: MATLAB

In order to calculate the reverberation time, it is necessary to filter the pressure response of the microphones in the desired octave (or third-octave) band. An ANSI standard [3] 125 Hz octave band filter (see Figure 7.6) was implemented in the MATLAB script Sabine.m. The instantaneous sound pressures calculated using ANSYS were read into MATLAB and then filtered using the octave band filter. The filtered sound pressures were then converted to instantaneous sound pressure levels, after which a 10-point moving average was applied to smooth the amplitude of the sound pressure levels. The first second of the response was discarded until the decay rate stabilized as is normal practice [47]. A linear regression (line of best fit) was conducted on the final four seconds of the filtered and smoothed sound pressure levels to establish the

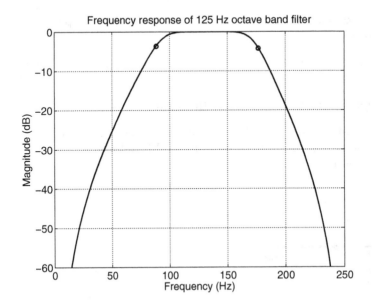

FIGURE 7.6
Frequency response of the 125 Hz octave band filter implemented in MATLAB for a 500 Hz sample rate. The circles indicate the bounds of the octave band.

rate of decay of the sound, with the coefficient of determination (R^2) typically around 95%.

7.4.4.2 Discussion of Transient Solvers in ANSYS

The choice of the options used in transient analyses determines the numerical accuracy of the results. For example, the appropriate value for the Time Step has to be selected so that the dynamics are accurately simulated. ANSYS [30, Guidelines for Integration Step Size] recommends that there should be at least 20 time steps for the maximum frequency of interest. However, the absolute minimum value is the Nyquist frequency (two time steps for the maximum frequency). The choice of Solver Type and Numerical Damping also have a significant effect on the results as discussed below.

Solver Type

In the damped *implicit* acoustic transient analysis of the current problem, only the Full solver may be used, for which there is a choice of two methods: the Newmark method and (an improved algorithm called) the HHT method. The ANSYS manual [31, 17.2. Transient Analysis] has written

in low frequency modes the Newmark method fails to retain the second-order accuracy. To circumvent the drawbacks of the Newmark family of methods, the ANSYS program implements the generalized HHT — a method which sufficiently damps out spurious high-frequency response via introducing controllable numerical dissipation in higher frequency modes, while maintaining the second-order accuracy.

In ANSYS Mechanical APDL the solver type is set using the command TRNOPT, FULL, , , , , TINTOPT, where the TINTOPT is the time integration method for the transient analysis: NMK or 0 which is the Newmark algorithm (default for ANSYS Mechanical APDL); HHT or 1 which is the HHT algorithm (valid only for the full transient method and default in ANSYS Workbench).
    The use of numerical dissipation is discussed in detail below.

Numerical Damping:

Numerical Damping is employed to improve the stability of the numerical integration. The ANSYS manual [31, 17.2.2.1. Time Integration Scheme for Linear Systems] describes in how numerical damping is applied. The amount of numerical damping required depends on the system being analyzed. Despite the implicit transient solver being unconditionally stable, in the absence of any damping (numerical or otherwise), the higher natural frequencies of the system can produce unacceptable levels of numerical noise, and therefore in most analyses a certain level of numerical damping is added. The higher the damping the less numerical noise, however, this will artificially increase the damping in the system and consequently the response of the system will have higher decay rates. Hence care must be exercised when selecting the magnitude and type of Numerical Damping.

In ANSYS Mechanical APDL, the amount of Numerical Damping is controlled using the command TINTP, GAMMA, ALPHA, DELTA, THETA, OSLM, TOL, -, -, AVSMOOTH, ALPHAF, ALPHAM. This allows direct and indirect control over the damping values. The simplest approach is to set GAMMA ($>$ 0) only, which is the Amplitude Decay Factor for second-order transient integration and in ANSYS Mechanical APDL the default value is 0.005. However, in ANSYS Workbench the default value for GAMMA is 0.1, which can significantly alter the response of lightly damped vibro-acoustic systems.

In conclusion, since we are trying to determine the absorption coefficient in the 125 Hz octave band, the very low frequency modes (those below, say, 35 Hz) are not of concern, hence the restrictions on the Newmark method are not of concern. Furthermore, the system is relatively heavily damped by the sound-absorbing material so any spurious high-frequency response will be damped. Therefore either the Newmark or HHT method is suitable for this example.

Since the objective is to determine the level of damping in the room, it is essential that the value used for the Numerical Damping does not affect the total damping. A value of GAMMA of 0.005 (default value in ANSYS Mechanical APDL) is acceptable. However, if the analysis is conducted using ANSYS Workbench, which uses as default value of 0.1, the calculated reverberation time of the room is significantly reduced and the absorption coefficient estimates are incorrect.

7.4.4.3 Transient Analysis: ANSYS Workbench

The transient analysis will share many of the features used in the previous two analyses including the geometry, the mesh, and the parameters.

Setting up the model for the transient analysis

- In the Workbench Project window the available analysis types are listed in the Toolbox window. In the Analysis Systems tree, click and hold over Transient Structural and drag the mouse into the Project Schematic and release over row 4 Model of the Harmonic Response object. Do not be concerned about the title Transient Structural as this is the default label in ANSYS. If successful, then the Project Schematic window should resemble the following figure.

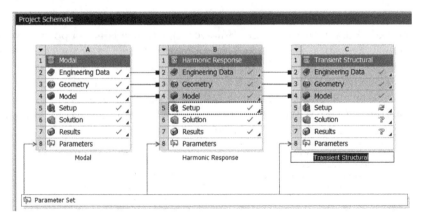

- Double-click on row 5 `Setup` in the `Transient Structural` object which will start ANSYS Mechanical.

Transient Settings

- The next step is to define the parameters for the transient analysis. Since we are not sharing the analysis information between the modal and harmonic analyses, it is necessary to re-create the `Acoustic Body` and the `Acoustic Attenuation Surface` in the `Transient (C5)` branch by repeating the same process used in the `Modal (A5)` or `Harmonic Response (B6)` tree. The most efficient way to do this is to click on all the objects that you wish to duplicate whilst holding down the `<Ctrl>` or `<Shift>` key on the keyboard. Select only the `Acoustic Body` and the `Acoustic Attenuation Surface`, and while holding the left mouse button down, drag these objects to the `Transient (C5)` branch. A small + symbol in a box will appear just below the cursor (not indicated in the figure below) and then release the left mouse button to copy the objects.

This will duplicate the two objects necessary for the transient analysis.

- Regarding the acoustic source, the ACT Acoustics extension has two versions of the `Acoustic Mass Source` (found in the toolbar under `Excitation`): the `Mass Source (Harmonic)` and `Mass Source Rate (Transient)`, with the latter being appropriate for transient analysis. Note that in ANSYS Release 14.5 neither the harmonic nor the transient `Mass Source` support tabular data, and therefore it is not possible in ANSYS Workbench to create multiple load steps for this source by defining a table of values. Instead, we will use a FLOW acoustic source. However, this source is not supported with the ACT Acoustics extension and therefore we will need to create a command object that contains APDL commands to define the acoustic excitation.

In the window Outline, left-click on the Transient (C5) branch, and either left-click on the Insert Commands button in the toolbar or right-click over Transient (C5) and then select Insert | Commands from the context menu. A Command window will be displayed with following text.

```
1   ! Commands inserted into this file will be executed just prior to the
2         ANSYS SOLVE command.
3   ! These commands may supersede command settings set by Workbench.
4   ! Active UNIT system in Workbench when this object was created:
5         Metric (m, kg, N, s, V, A) with temperature units of C !
6   NOTE: Any data that requires units (such as mass) is assumed to be in the
7         consistent solver unit system.
8   ! See Solving Units in the help system for more information.
```

At the end of this text you need to insert the following APDL code. The file flow_code_snippet.txt supplied with this book contains the required code.

```
1   ! Attach these commands to the Transient analysis.
2   ! This will add a Flow acoustic source.
3   ! The source is a doublet, that is the source is positive, then
4   ! negative, then zero for the remainder of the analysis.
5
6   ! ----------------------------------------------------
7   ! Units are m, kg, s
8   ! ----------------------------------------------------
9
10  ! Apply a FLOW Source to the source node
11  ! Find the appropriate node
12  CMSEL,S,NS_Source_Location
13  *GET, Flow_Node, NODE, 0, Num, Max
14  ALLSEL
15
16  ! Dimension the FLOW tabular data
17  *DIM,FLOWID,TABLE,4,1,1
18  FLOWID(1,0,1) = 0.0           ! TIME VALUES
19  FLOWID(2,0,1) = 0.0025
20  FLOWID(3,0,1) = 0.005
21  FLOWID(4,0,1) = 0.0075
22  FLOWID(1,1,1) = 0.0           ! FLOW IMPULSE LOAD VALUES
23  FLOWID(2,1,1) = 1.0
24  FLOWID(3,1,1) = -1.0
25  FLOWID(4,1,1) = 0.0
26
27  ! Apply the FLOW condition
28  F,Flow_Node,FLOW,%FLOWID%    ! APPLY TABULAR LOADS
```

- In the Outline window, click on Transient (C5)| Analysis Settings then select Details of "Analysis Settings".
  The row Step Controls | Number of Steps should be set to 1.
  The row Step Controls | Step End Time should be set to 5 which will set the analysis time from 0 to 5 seconds.
  Change the row Step Controls | Auto Time Stepping to Off as we wish

to control the step size.

The row `Step Controls | Define By` should be set to `Time`.

Set the row `Step Controls | Time Step` to `0.002`. This will force the solver to calculate the results every 2 ms.

Ensure that the row `Step Controls | Time Integration` is `On` as this is necessary for transient analyses. Selecting `Off` is used for initial static analyses, for example pre-stressing a model.

Details of "Analysis Settings"	
Step Controls	
Number Of Steps	1.
Current Step Number	1.
Step End Time	5. s
Auto Time Stepping	Off
Define By	Time
Time Step	2.e-003 s
Time Integration	On
Solver Controls	
Solver Type	Program Controlled
Weak Springs	Off
Large Deflection	Off
Restart Controls	
Nonlinear Controls	
Output Controls	
Stress	No
Strain	No
Nodal Forces	No
Contact Miscellaneous	No
General Miscellaneous	No
Store Results At	All Time Points
Cache Results in Memory ...	Never
Max Number of Result Sets	Program Controlled
Damping Controls	
Stiffness Coefficient Define...	Direct Input
Stiffness Coefficient	0.
Mass Coefficient	0.
Numerical Damping	Manual
Numerical Damping Value	0.005
Analysis Data Management	
Solver Files Directory	C:\Users\bscazz\Documents\...
Future Analysis	None
Scratch Solver Files Direct...	
Save MAPDL db	Yes
Delete Unneeded Files	Yes
Nonlinear Solution	No
Solver Units	Active System
Solver Unit System	mks

- In the window `Details of "Analysis Settings"` under `Solver Controls`, turn `Off` both `Weak Springs` and `Large Deflection` (as shown in the image above) as these are not relevant to this model.

  In `Output Controls` select the first 5 fields as `No` (as shown in the image above) as we only want the pressures.

  Change the row `Damping Controls | Numerical Damping` to `Manual`, then set the `Damping Controls | Numerical Damping Value` to `0.005`. If the default of 0.1 is retained, then the system will be overdamped leading to erroneous results.

- In the `Details of "Analysis Settings"` window, change the `Analysis Data Management | Save MAPDL db` to `Yes`. This will allow you to post-process the results using the `Acoustic Time_Frequency Plot` in the ACT Acoustics extension.

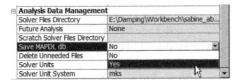

- Save the project by pressing <Ctrl> and s on the keyboard.

- Solve the transient analysis by clicking the Solve icon in the toolbar or right-click over Transient (C5), and in the context menu left-click on Solve. The yellow lightning bolt next to Transient (C5) will turn green indicating the solver is working. Be patient, this may take over an hour to solve.

Results

The results of interest are the instantaneous sound pressures and the sound pressure levels. We will now define objects that will allow us to view these results. As discussed previously, it is also possible to define these result objects prior to solving, however, these are defined post solving to illustrate how this can be done.

- Create an Acoustic Pressure object by clicking on the ACT Acoustics toolbar Results | Acoustic Pressure. right-click on the object and select Rename or press <F2> on the keyboard and define the name of the object as Acoustic Pressure - All nodes. Note that the row Geometry has All Bodies selected.

- Repeat this process of defining and renaming new Acoustic Pressure objects for the following entities:

  · Acoustic Pressure - Source: Choose Scoping Method | Named Selection and set Named Selection to NS_Source_Location.

  · Acoustic Pressure - Mic 1: Choose Scoping Method | Named Selection and set Named Selection to NS_Mic_1_Location. Repeat this for microphones 2 to 6.

- Right-click on Solution (C6), then left-click on Insert | User Defined Result. Right-click on the object and left-click on Rename or press <F2> on the keyboard and define the name of the object as SPL - All nodes. In the window Details of "SPL - All nodes", locate the row Definition | Expression and type 10*log10(PRES*PRES).

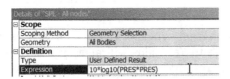

This will calculate the instantaneous SPL of the selected nodes. Note that the Definition | Output Unit is Pressure and that the labels in the Tabular Data and Graph indicate Pa, which is not correct, but cannot be altered as there are no user-defined units.

- Repeat this process of defining and renaming new User Defined Result objects for the following:
  SPL - Mic 1: Choose Scoping Method | Named Selection and set the Named Selection to NS_Mic_1_Location. Under Details of "SPL - Mic 1", locate the row Definition | Expression and type 10*log10 (PRES*PRES). Repeat this for microphones 2 to 6.

- The result we are trying to calculate is the mean decay rate of the sound pressure level at all the microphones. In the window Outline, right-click over Model (A4, B4, C4) | Named Selections and left-click on Insert — Named Selection. Rename this entry to NS_All_Mics. Under Scope | Scoping Method choose Worksheet. Add six entries, one for each microphone. In the first row of the NS_ALL_Mics table, under the column Action select the option Add, change Entity Type to Mesh Node, Criterion to Named Selection, Operator to Equal, and Value to NS_Mic_1_Location. Modify the remaining five rows for the other five microphone locations. The completed Worksheet table is shown below.

	Action	Entity Type	Criterion	Operator	Units	Value	Lower Bound	Upper Bound	Coordinate System
☑	Add	Mesh Node	Named Selection	Equal	N/A	NS_Mic_1_Location	N/A	N/A	N/A
☑	Add	Mesh Node	Named Selection	Equal	N/A	NS_Mic_2_Location	N/A	N/A	N/A
☑	Add	Mesh Node	Named Selection	Equal	N/A	NS_Mic_3_Location	N/A	N/A	N/A
☑	Add	Mesh Node	Named Selection	Equal	N/A	NS_Mic_4_Location	N/A	N/A	N/A
☑	Add	Mesh Node	Named Selection	Equal	N/A	NS_Mic_5_Location	N/A	N/A	N/A
☑	Add	Mesh Node	Named Selection	Equal	N/A	NS_Mic_6_Location	N/A	N/A	N/A

NS_All_Mics — Generate

Once complete, press Generate. You will notice in the Details of "NS_All_Mics" window that the row Statistics | Total Selection indicates 6 Nodes.

- Right-click on Solution (C6), then left-click Insert | User Defined Result. Right-click on the object and select Rename (or press <F2> on the keyboard) and define the name of the object SPL - All Mics. In the window Details of "SPL - All Mics", locate the row Definition | Expression and type 10*log10(PRES*PRES). Change Scoping Method to Named Selection and set the row Named Selection to NS_All_Mics. There should be a total of 16 User Defined Result objects.

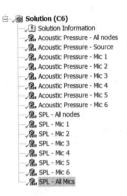

- Right-click on Solution (C6), then left-click on Evaluate All Results. Select the object Acoustic Pressure - All nodes. In the Tabular Data window, select the tenth time step (2.0e-002) then right-click and select Retrieve This Result.

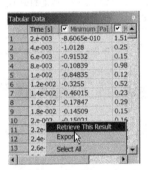

- In the Geometry window, right-click over the legend and change the Color Scheme to Reverse Grayscale. Also select Adjust to Visible.

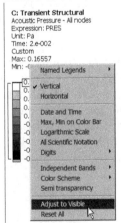

You can see the initial impulse doublet arising from the acoustic source propagating outward, indicated by the positive pressure followed by the negative pressure. The reflections off the wall can be seen as small regions of high pressure in the corner and at the edges near the boundary of the wavefront.

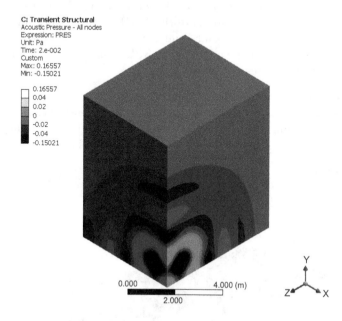

• In the Outline window, select the object Solution (C6) | SPL - Mic 1. In the window Graph, the SPL versus time is displayed and exhibits a linear decay as expected. Note that the units on the graph are Pa, which is incorrect, and should be dB re 20 μPa as described previously.

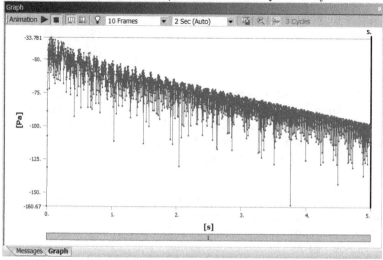

By looking at the graph, the time taken for the SPL to decay by 60 dB is approximately 5 seconds, which is greater than the 3.75 seconds that theory predicts for the 125 Hz octave band. This is because the sound pressure results from the transient analysis contains energy from 0 Hz to 250 Hz (the Nyquist frequency). The lower frequency modes decay more slowly than the higher frequency modes (which dominate the 125 Hz octave band) and therefore the reverberation time across all frequencies is larger than the reverberation time for the 125 Hz octave band. The SPL results for the other five microphones show similar responses.

- In the Outline window, select the object Solution (C6) | SPL - All Mics. This will show the instantaneous response of all six microphones, plotting the maximum and minimum SPLs. The decay rates for both the minimum and maximum SPLs exhibit similar decay rates to the individual microphone SPLs.

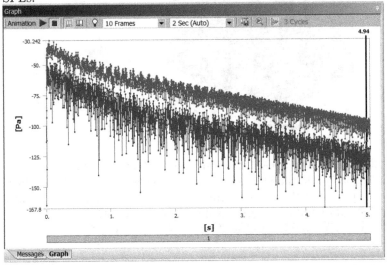

Exporting Data

- To export the data, in the window Outline, click on the Acoustic Pressure - Mic 1 object. Then, in the right hand side of the page in the Tabular Data window, right-click and left-click on Export. Save the file as Sabine_Trans_Mic1_Pres.txt. Repeat the process for the other five microphones, replacing the 1 in the filename with the corresponding microphone number.

- The MATLAB file Sabine.m may be used to post-process these results to determine the reverberation times and absorption coefficients for the 125 Hz octave band. This process is illustrated below using the results from the ANSYS Mechanical APDL.

7.4.4.4 Transient Analysis: ANSYS Mechanical APDL

The ANSYS Mechanical APDL file code_ansys_sabine.txt was used to conduct the transient analysis of the reverberation room. When executed, the file will create the model and mesh as previously described, and the transient analysis is set up and then solved. The file also post-processes the results, creating graphs of the pressure versus time and exports the results for post-processing in MATLAB.

Figure 7.7 shows the absolute value of the unfiltered transient pressure measured at the node where the acoustic source was located. It can be seen that there is an initial transient and after approximately 0.5 seconds the slope remains constant. This initial transient response may be used to provide an estimate of the early decay time (EDT), defined as the reverberation time computed by the slope of the decay of SPL in the range between 0 and −10 dB. The time it takes the sound pressure shown in Figure 7.7 to decay by a factor of 1000 (equivalent to a reduction in the sound pressure level by 60 dB) is approximately 2 to 2.5 seconds depending on what part of the curve is used. Note that this estimate was obtained from unfiltered data and therefore contains all frequencies from 0 Hz to 250 Hz (the Nyquist frequency).

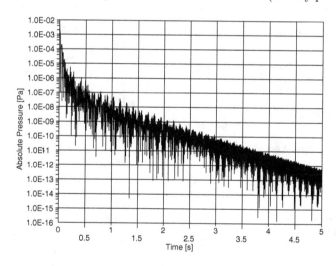

FIGURE 7.7
Instantaneous absolute pressure versus time measured at the source node.

A more accurate estimate of the decay time may be obtained by exporting the instantaneous pressures from the six microphones generated from the transient analysis in ANSYS, importing these into MATLAB then filtering with a 125 Hz octave band filter, then fitting a linear regression as discussed earlier. The result of this process is shown in Figure 7.8. The reverberation time obtained from this analysis was calculated using the mean slope of the slope of the SPL from 1 to 5 seconds. The six reverberation times were av-

TABLE 7.5

Results from the Transient Analysis of the Reverberation Room

Descript.	Param.	Units	Theory	APDL	Diff	WB	Diff
Reverb. time	T_{60}	s	3.748	3.354	10.5%	3.364	10.2%
Sound absorption coefficient	$\bar{\alpha}$	—	1.000	1.117	11.7%	1.114	11.4%

Note: The theoretical results calculated in MATLAB are compared against the Mechanical APDL and Workbench (WB) analysis.

eraged using Equation (7.9) and tabulated in Table 7.5. The results for the ANSYS Mechanical APDL and ANSYS Workbench differ slightly as a result of the different options used in the analysis, the most important being the solver type which was discussed on page 417.

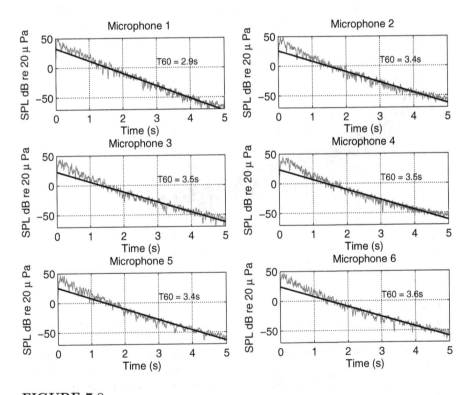

FIGURE 7.8

Plots of the 125 Hz octave band SPLs versus time, lines of best fit and corresponding reverberation times (T_{60}) for the six microphones.

The results obtained from the transient analysis are quite close to what theory would predict. The difference is typical of the statistical accuracy observed when conducting actual experiments on real samples. The accuracy can be improved by decreasing the `Time Step`, moving the microphones to more asymmetric locations, increasing the number of microphones and sources, all of which, however, will increase the solution times.

8

Radiation and Scattering

8.1 Learning Outcomes

The learning outcomes of this chapter are to:

- understand the ways to define wave-absorbing boundaries in ANSYS and their limitations,

- understand the options for examining the scattered pressure wave field when using Perfectly Matched Layer elements,

- learn how to conduct a 2D axi-symmetric analysis using FLUID29 in ANSYS, and

- learn how to determine the nodal areas using 2 methods.

8.2 Wave-Absorbing Conditions

In the previous chapters many simulations in ANSYS were described where the acoustic domain was finite and bounded either by a rigid wall or there was a vibrating structure such as a plate. In Section 3.3.6 an example was shown of an anechoic termination on a duct that required a wave-absorbing condition. In this chapter, the application of wave-absorbing conditions is further illustrated by the creation of infinite or semi-infinite acoustic domains to investigate the radiation of sound from acoustic sources, an oscillating piston, and to investigate the scattering of an incident acoustic plane wave by an object. There are three recommended ways that a wave-absorbing boundary can be simulated using ANSYS for 3D analyses: Perfectly Matched Layers (PMLs), radiation boundary conditions, and infinite fluid elements. These wave-absorbing boundary conditions can only be used for certain types of analyses as listed in Table 8.1.

For 2D models created in ANSYS using FLUID29 acoustic elements, there are a couple of ways that wave-absorbing boundaries can be simulated. In early releases of ANSYS, the main way that an absorbing boundary was created was

TABLE 8.1

Acoustic Wave-Absorbing Conditions and the Applicable Analysis
Types

Wave-Absorbing Condition	Modal	Harmonic	Transient
Perfectly Matched Layers	no	yes	no
Radiation boundary conditions	yes	yes	no
Infinite fluid elements	yes	yes	yes

through the use of the material property MU, which is the "boundary admittance associated with a fluid-structure interface, and has a value between 0 and 1.0 that is equal to the ratio of the fluid's characteristic impedance ($\rho_0 c_0$) to the real component of the specific acoustic impedance (resistance term) associated with the sound-absorbing material" [150, p. 3-3]. Sound-absorbing conditions can be created on the boundary of FLUID29 elements that have their displacement DOFs turned off, by setting the material property of MU=1 and applying a surface load to the appropriate nodes using the APDL command SF, Nlist, IMPD, 1. Another method of simulating wave-absorbing boundaries is to use the FLUID129 infinite acoustic elements, as described in Section 2.7.3. The FLUID29 elements there have no capability for using Perfectly Matched Layers.

The following sections describe the wave-absorbing conditions listed in Table 8.1.

8.2.1 Perfectly Matched Layers

Perfectly Matched Layers (PMLs) are used to absorb incident acoustic waves and do not reflect waves except those traveling tangentially to the layer. Figure 2.18 (on page 62) shows the typical configuration for the use of a PML layer in ANSYS where an acoustic source is surrounded by 3D acoustic elements (FLUID30 or FLUID220). An equivalent source surface is defined that encases the acoustic source, and a buffer layer of acoustic elements separates it from the PML region. There are some guidelines for the size of each of these regions. It is recommended that there should be at least a half a wavelength or greater separation between the radiator or scatter and the equivalent source surface. There should be some buffer elements (at least 3 layers of elements) between the equivalent source surface and the PML region. The PML region should be three or four elements thick and may need to be greater than a quarter-wavelength thick to provide adequate attenuation of outgoing acoustic waves. However, if the PML region is excessively thick, the computational requirements will increase and take longer to solve. The recommended number of PML elements through the thickness of the PML region to obtain acceptable numerical accuracy can be determined using the APDL command PMLSIZE, FREQB, FREQE, DMIN, DMAX, THICK, ANGLE where

FREQB is the minimum analysis frequency,

FREQE is the maximum analysis frequency,

DMIN is the minimum distance from a radiation source to the PML interface,

DMAX is the maximum distance from radiation source to the PML interface,

THICK is the thickness of PML region and has a default value of 0, and

ANGLE is the incident angle of an outgoing acoustic wave to the PML interface and defaults to 0.

If the thickness of the PML region is known, then the value of THICK can be defined when using the PMLSIZE command, which will return the recommended element size h and issue the APDL command ESIZE, h, which will define the size for the PML elements. If the thickness of the PML region has not been defined, then the value for THICK can be set to 0 or left empty, and the number of divisions in the thickness direction of the PML region n will be determined and will issue the APDL command ESIZE,, n.

The exterior of the PML region should have a pressure constraint of 0, which can be applied using the ACT Acoustics extension Boundary Conditions | Acoustic Pressure. An important point about PML regions is that the edges of the bodies should be aligned with the Cartesian coordinate system; if this is not done, then an error message will be generated.

Also note that excitation sources are not permitted in the PML region.

It was shown in Section 3.3.5 that an anechoic termination can be simulated in ANSYS Workbench using the ACT Acoustics extension feature Boundary Conditions | Radiation Boundary. An anechoic termination on a duct can also be created using PML elements. Figure 8.1 shows a square duct comprising three acoustic bodies. The central body is a standard acoustic domain and the faces on each end are attached to acoustic bodies with PML conditions that absorb incident acoustic waves, thereby providing an anechoic termination.

A harmonic response analysis was conducted where an acoustic plane wave, that was defined using the ACT Acoustics extension option Excitation | Wave Sources, propagated along the axis of the duct. Figure 8.2 shows the sound pressure in the duct where it can be seen that the pressure contours are uniformly spaced and are not bent. This indicates that the pressure wave is propagating uniformly down the duct.

Now consider a situation where the downstream PML region (acoustic body on the righthand end) is removed, thereby creating a semi-infinite duct, as shown in Figure 8.3, and the harmonic analysis was repeated. Figure 8.4 shows the sound pressure results where it can be seen that the pressure contours are different from those shown in Figure 8.2, as the incident pressure wave has been reflected off the end-wall of the duct.

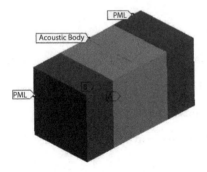

FIGURE 8.1
Image of an infinite acoustic duct bounded by two PML acoustic bodies.

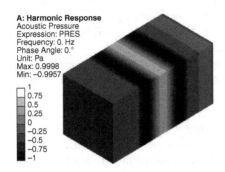

FIGURE 8.2
Real part of the sound pressure inside infinite acoustic duct bounded by two
PML acoustic bodies.

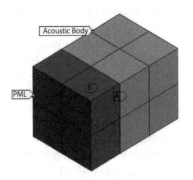

FIGURE 8.3
Image of a semi-infinite acoustic duct bounded by a PML acoustic body on
the upstream end.

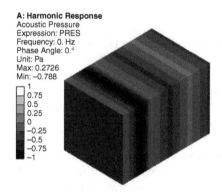

A: Harmonic Response
Acoustic Pressure
Expression: PRES
Frequency: 0. Hz
Phase Angle: 0.°
Unit: Pa
Max: 0.2726
Min: −0.788

FIGURE 8.4

Real part of the sound pressure inside a semi-infinite acoustic duct bounded by a PML acoustic body on the upstream end.

8.2.2 Radiation Boundary

Another method of simulating a wave-absorbing region is to use a "radiation boundary condition" which can be applied to a surface ("face" in ANSYS terminology) and defines that it has a surface impedance of

$$Z = \frac{p}{v_n} = \rho_0 c_0 \,, \tag{8.1}$$

which is mathematically called a "Robin boundary condition" [17, Eq. (8-16)], which defines that on the boundary of the acoustic domain there is a relationship between the pressure and the acoustic particle velocity.

The acoustic absorption for this type of boundary condition works best where the incident waves are normal to the surface and plane wave conditions exist. When an incident wave strikes the radiation boundary at non-normal (oblique) angles of incidence, or there are non-plane wave conditions, there can be some slight (unwanted) reflection of sound back into the acoustic domain. For these more complicated sound fields, the PML regions work well.

One of the advantages of using the radiation boundary compared to the PML region is that no additional elements and nodes are required to define the wave-absorbing boundary, whereas when using PML regions, several layers of elements must be used that are only used to absorb outgoing waves. If the acoustic domain of interest is large, which can be the case for 3D scattering simulations, the additional layers of PML elements and nodes can increase the computational requirements and increase the time taken to solve a model.

Section 2.8.4.5 describes how to specify a radiation boundary in ANSYS Workbench using the ACT Acoustics extension, which is available from the toolbar under Boundary Conditions | Radiation Boundary that implements the APDL command SF,nodes,INF.

8.2.3   Infinite Acoustic Elements

The third type of wave-absorbing boundary is through the use of "infinite fluid elements." An example that made use of these elements was described in Section 3.3.7.2, where the end of a duct was mounted in a planar infinite baffle and radiated sound into a free-field, which was simulated using a 3D hemispherical acoustic domain with FLUID130 elements on the surface of the hemisphere to absorb the outgoing waves.

These infinite fluid elements are applied to the outside surface of a spherical shaped region (hemispheres and other curved are also acceptable) that have a constant radius of curvature. For 2D simulations using FLUID29 elements, an infinite domain is created by defining a circular acoustic domain (semi- or quarter-circular areas are also acceptable), and overlaying FLUID129 infinite acoustic elements on the outer circular perimeter.

In order to absorb outgoing acoustic waves, the Sommerfeld radiation condition must be satisfied, which simply means that acoustic waves generated in the acoustic domain continue to propagate outward and do not propagate inward. It is written mathematically as [32, Eq. (8.23)] [129, Eq. (4-5.5), p. 178]

$$\lim_{r \to \infty} \left[r \left(\frac{\partial p}{\partial r} + \frac{1}{c} \frac{\partial p}{\partial t} \right) \right] = 0 , \tag{8.2}$$

where r is the distance from the origin, p is the acoustic pressure, c_0 is the speed of sound of the acoustic medium, and t is the time variable. The partial differential of pressure with respect to time $\partial p/\partial t$, can be written as $-i\omega p$. Hence Equation (8.2) can be written as

$$\lim_{r \to \infty} \left[r \left(\frac{\partial \hat{p}}{\partial r} - ik\hat{p} \right) \right] = 0 , \tag{8.3}$$

where $\hat{p}$ indicates that the pressure is harmonic. It is assumed that the sound generation in the acoustic domain, such as from an acoustic source or vibrating structure that is generating sound, or from scattering of sound, is in a small region that is near the origin. At large distances r from the origin (i.e., this applies to the term $\lim_{r \to \infty}$), the acoustic response varies more in the radial direction than in directions that are perpendicular to the radial direction, which is characteristic of spherical spreading. An equivalent statement of the Sommerfeld radiation condition is [129, Eq. (4-5.6)]

$$\lim_{r \to \infty} \left[r(p - \rho_0 c_0 v_r) \right] = 0 , \tag{8.4}$$

where v_r is the acoustic particle velocity in the radial direction, and ρ_0 is the density of the acoustic medium. Equation (8.4) suggests that at large distances, $\lim_{r \to \infty}$, the acoustic field resembles an outward-traveling plane wave. Rearranging Equation (8.4) leads to the suggestion that at large distances r, the specific acoustic impedance is $Z = \hat{p}/\hat{v}_r$.

The previous discussion concerned what happens at *infinite* distances from the origin, and needs to be translated into something that can be applied to a *finite* finite element model, where the external surface of the acoustic domain has an appropriate boundary condition to satisfy Equation (8.4). What is required to implement this into an element are appropriate mass, stiffness, and damping matrices that satisfy the Sommerfeld radiation condition, which has been translated into an appropriate expression for a finite radius r (i.e., not the abstract r approaches infinity). The derivation of how that is accomplished is in the ANSYS online help manual starting at [32, Eq. (8.24)].

ANSYS recommends that there should be at least 0.2 times the acoustic wavelength (i.e., 0.2λ) separation between an acoustic source or vibrating structure and where the infinite acoustic elements are defined on the surface that has a constant radius of curvature. However, remember that it is assumed that in order to satisfy the Sommerfeld radiation boundary condition, the outgoing acoustic wave should be spherically spreading; if it isn't spherically spreading and propagating radially once the wavefront has reached the outer boundary of the model, then one should expect that the outgoing waves will not be absorbed completely and some will reflect back into the acoustic domain. In summary, take the separation recommendation of 0.2λ as guidance only and perform your own assessment by inspecting the sound pressure results from a simulation for the existence of reflected waves at the boundary. If it is apparent that waves have reflected at the boundary, then increase the size of the acoustic domain (and radius r of the infinite elements) and conduct the simulation again.

The decision of whether to use PML regions, a radiation boundary condition, or infinite acoustic elements is a "trade-off":

- Using PML regions usually provides excellent absorption but may have an additional cost of requiring more elements, nodes, computational resources and therefore takes longer to solve than the other two methods of simulating an infinite boundary.

- Using radiation boundary conditions gives excellent results when there are plane waves that are normally incident to the surface on which the boundary condition has been applied. The use of radiation boundary conditions is well suited for simulating ducts with anechoic terminations where plane wave conditions exist at the terminations. For other situations where the sound field is complicated, then it is likely the outgoing waves will not be absorbed completely and will reflect back into the acoustic domain. In which case, using PML regions will improve the results from simulations.

- Using infinite acoustic elements can also provide excellent attenuation of outward propagating acoustic waves, provided that the sound field that approaches the outer boundary of the model has the right conditions as described in this section. The restriction that the acoustic domain must be spherically shaped for 3D simulations, or circular shaped for 2D simulations, means that the size of the acoustic domain may be larger than necessary, and could easily be replaced with a suitable PML region that encases the region of interest.

Section 2.8.4.6 describes how to specify infinite acoustic elements on a boundary in ANSYS Workbench using the ACT Acoustics extension, which is available from the toolbar under Boundary Conditions | Absorbing Elements (Exterior To Enclosure), which uses the APDL command ESURF to overlay a mesh of FLUID130 elements on the outer boundary.

Note that an error message sometimes occurs in ANSYS (and the analysis will halt) when using the FLUID129 and FLUID130 elements indicating that the nodes of the element are not precisely on the curved surface of the defined radius, even though the mesh was created using regular meshing techniques in ANSYS. This error is discussed further in Appendix D.2.4 as well as how to fix the issue.

8.3 Example: Directivity of Acoustic Wave Sources

The ANSYS Workbench archive file called acoustic_sources_PML.wbpz, which contains the .wbpj project file, is available with this book can be used to explore the behavior, including the directivity, of four types of Acoustic Wave Sources. The model used to investigate these sources, shown in Figure 8.5, is a cube sliced into several smaller bricks. All the bodies were meshed with quadratic FLUID220 acoustic elements. A harmonic response analysis will be conducted at a single frequency of 500 Hz. It is recommended that for FLUID220 elements, at least 6 elements per wavelength should be used. For this example we will use epw = 8. The element size for all bodies can be calculated using Equation (3.22) and was set to esize=343/(500 × 8) = 0.08575 m. An inner acoustic body in the shape of a cube with 1 m edge lengths (labeled acoustic_interior) has the acoustic source defined at the center. Surrounding this is an acoustic buffer layer (labeled acoustic_buffer) that is 0.1 m thick. Surrounding this is an outer layer that is 0.4 m thick of Perfectly Matched Layer elements (labeled acoustic_PML). The PML layer is defined as an Acoustic Body with the row Perfectly Matched Layers (PML) turned On.

An acoustic Equivalent Source Surface is defined on the faces that are on the exterior of the central acoustic body acoustic_interior, as shown in Figure 8.6. This is necessary to enable the calculation of the directivity and other acoustic results that can be obtained by inserting Acoustic Far Field from the Results menu in the ACT Acoustics extension menu bar. Note that ANSYS will automatically create an equivalent source surface on a PML-acoustic medium interface or the exterior surface of an acoustic radiation boundary.

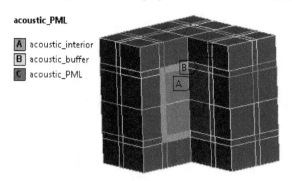

FIGURE 8.5

Model of an acoustic domain bounded by buffer and PML regions, used to plot the directivity of Acoustic Wave Sources.

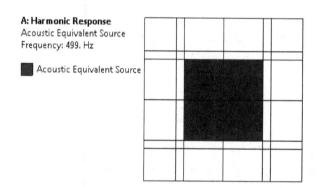

FIGURE 8.6

Acoustic Equivalent Source faces on the exterior of the acoustic_interior acoustic bodies of the model shown in Figure 8.5.

The model was used to calculate the sound pressure level versus angle for four types of acoustic wave sources: monopole, dipole, back-enclosed loudspeaker, and bare (unbaffled) loudspeaker, for a harmonic analysis at 500 Hz. The input parameters used for the acoustic wave sources are listed in Table 8.2. Note that the last row in the table for the Pure Scattering Options is set to On (Output Scattered Pressure), which will calculate the scattered pressure field. However for these examples, as there is only an acoustic source inside the equivalent source surface, and there is no object that can cause scattering, the scattered pressure field will be the same as the total pressure field. The row in the table "Radius of Pulsating Sphere" applies to a monopole and dipole, but for the examples of a bare and back-enclosed loudspeaker, it should really describe the radius of the loudspeaker cone.

TABLE 8.2

Input Parameters for Four Acoustic Wave Sources Used in the ANSYS Workbench Example in Section 8.3

Input Parameter	Units	Monopole	Dipole	Wave Type Back-Enclosed Loudspeaker	Bare Loudspeaker
Excitation Type		Velocity	Velocity	Velocity	Velocity
Source Location		Inside	Inside	Inside	Inside
Velocity Amplitude	m/s	1.0	1.0	1.0	1.0
Phase Angle	°	0	0	0	0
Global X Coordinate at Source Origin	m	0	0	0	0
Global Y Coordinate at Source Origin	m	0	0	0	0
Global Z Coordinate at Source Origin	m	0	0	0	0
Mass Density of Environment Media	kg/m^3	1.2041	1.2041	1.2041	1.2041
Sound Speed in Environment Media	m/s	343.24	343.24	343.24	343.24
Radius of Pulsating Sphere	m	0.01	0.01	0.01	0.01
Dipole Length	m	n/a	0.1	n/a	0.1
X Component of Unit Dipole Vector	m	n/a	0	n/a	0
Y Component of Unit Dipole Vector	m	n/a	0	n/a	0
Z Component of Unit Dipole Vector	m	n/a	1	n/a	1
Pure Scattering Options		On	On	On	On

In the window Details of "Acoustic Far Field", the row Properties | Result was set to SPL In Cartesian Plot, which can be used to plot the sound pressure level versus angle. The starting and ending angles for Phi was 0°, and Theta were 0° to 360°, with the number of divisions as 72, so that the results are calculated in increments of 5°. The row Properties | Sphere Radius was set to 4 m, which is the radius at which the sound pressure level results will be calculated. The value of the radius is not particularly important in this case; 4 m was selected so that the location is in the acoustic far-field.

Details of "Acoustic Far Field"	
Properties	
Result Set	1
Boundary Condition On Model Symmetric Plane	No
Result	SPL In Cartesian Plot
Starting Angle Phi (From X Axis Toward Y Axis)	0 [°]
Ending Angle Phi	0 [°]
Number Of Divisions Phi	0
Starting Angle Theta (From Z Axis Toward X Axis)	0 [°]
Ending Angle Theta	360 [°]
Number Of Divisions Theta	72
Sphere Radius	4 [m]
Reference RMS Sound Pressure	2E-05 [Pa]
Model Thickness in Z Direction (2D extension)	0 [m]
Spatial Radiation Angle	Full Space

8.3.1 Comparison of Monopole Acoustic Sources Calculated Theoretically and Using ANSYS Workbench

This section describes the calculation of the sound pressure level versus angle for a monopole acoustic source using ANSYS Workbench, using the project file acoustic_sources_PML.wbpj. The results predicted using ANSYS will be compared with theoretical predictions using MATLAB code.

As described in Section 8.3, the ANSYS Workbench model has been defined with an acoustic domain as a cube with 1 m edge lengths, and is surrounded by an acoustic body that acts as a buffer before reaching the perfectly matched layer acoustic body that absorbs outgoing sound waves. In this example, the acoustic source that will be used is a monopole, and the sound pressure level versus angle will be calculated.

Instructions

- Start ANSYS Workbench.

- Load the project file acoustic_sources_PML.wbpj that is supplied with this book.

- In the Project Schematic window, double-click on row 5 Setup to start Mechanical.

- The next step is to define the acoustic wave source type as a monopole. In the Outline window, click on Harmonic Response (A5) | Acoustic Wave Sources. In the window Details of "Acoustic Wave Sources", change the row Definition | Wave Type to Monopole. Make sure that the remaining parameters are defined as per the column labeled Monopole in Table 8.2 and shown below.

Details of "Acoustic Wave Sources"	무
⊟ **Definition**	
Wave Number	1
Wave Type	Monopole ▾
Excitation Type	Velocity
Source Location	Inside The Model
Velocity Amplitude	1 [m sec^-1]
Phase Angle	0 [°]
Global X Coordinate At Source Origin	0 [m]
Global Y Coordinate At Source Origin	0 [m]
Global Z Coordinate At Source Origin	0 [m]
Mass Density Of Environment Media	1.2041 [kg m^-1 m^-1 m^-1]
Sound Speed In Environment Media	343.24 [m sec^-1]
Radius Of Pulsating Sphere	0.01 [m]
Pure Scattering Options	On (Output Scattered Pressure)

- In the menu bar, click on File | Save Project.

- Click on the Solve icon.

- When the calculations have completed, in the Outline window click on Solution (A6) | Acoustic Far Field, which will display a graph of sound pressure level versus angle in the Worksheet window, and a table of results in the Data View window. The results appear to be a sine wave, however the range of sound pressure levels varies between 70.4 dB to 70.7 dB, which is shown below, compares favorably with the theoretical value of 70.48 dB.

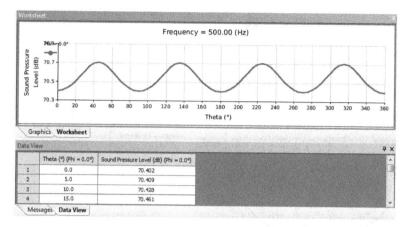

- To export the numerical results, right-click on `Acoustic Far Field` and in the context menu that opens, left-click on `Export`. In the `Export` window that opens, select the appropriate disk folder to save the results, and type a file-name such as `monopole_spl_vs_angle.txt`. The results file contains columns for angles `PHI`, `THETA`, and `SPL (dB)` at the analysis frequency.

   The MATLAB code `monopole_spl_vs_angle.m` included with this book can be used to calculate the sound pressure level versus angle for a monopole source radiating into a free-field.

   Figure 8.7 shows the comparison of the sound pressure level versus angle for a monopole source calculated theoretically using the MATLAB code and using ANSYS Workbench. The monopole source radiates omni-directionally and has a theoretical sound pressure level at 4 m, which can be calculated using Equations (2.13) and (2.30), of 70.48 dB. The predictions using ANSYS were between 70.45 dB and 70.7 dB, which has excellent agreement with theoretical predictions.

   It is possible to calculate other acoustic results using the `Acoustic Far Field` object, such as directivity and sound power.

   The directivity of a sound source is calculated in ANSYS as the sound intensity normalized (divided) by the radiated sound power as [23, Eq. (8-141)]

$$G_D(\phi, \theta) = \frac{\Omega U(\phi, \theta)}{W_{\text{rad}}}, \qquad (8.5)$$

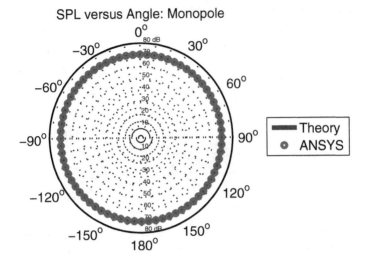

FIGURE 8.7

Sound pressure level versus angle for a monopole calculated theoretically and using ANSYS Workbench for $a = 0.01$ m, $u = 1.0$ m/s, and $r = 4$ m.

where $G_D(\phi, \theta)$ is the directivity at angles ϕ, θ, the solid angle Ω is given by

$$\Omega = \iint \sin \theta \, d\theta d\phi \,, \tag{8.6}$$

$U(\phi, \theta)$ is the sound radiation intensity given by

$$U(\phi, \theta) = \frac{1}{2} \mathrm{Re} \, (pv^*) \cdot \hat{r} r^2 \,, \tag{8.7}$$

p is the pressure, v^* is the conjugate of the acoustic particle velocity, $\hat{r}$ is the unit vector in spherical coordinates, r is the distance from the source, and W_{rad} is the radiated sound power given by [23, Eq. (8-140)]

$$W_{\text{rad}} = \frac{1}{2} \mathrm{Re} \iint (pv^*) \cdot \hat{r} r^2 \sin \theta \, d\theta d\phi \,. \tag{8.8}$$

The sound power is calculated as [23, Eq. (8-139)]

$$L_w = 10 \log 10 \left(\frac{W_{\text{rad}}}{W_{\text{ref}}} \right) \tag{8.9}$$

where W_{ref} is the reference sound power and has a default value of 10^{-12} W.

The instructions below describe how to calculate the directivity and radiated sound power for this example of a monopole source using ANSYS Workbench and the ACT Acoustics extension.

- In the Outline window, click on Solution (A6) | Acoustic Far Field, and change the row Properties | Result to Directivity in Cartesian Plot.

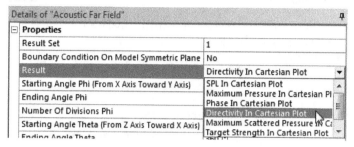

- Right-click on the object Acoustic Far Field and left-click on Generate, which will calculate the results.

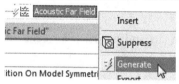

The directivity of the monopole source will appear to have a sinusoidal profile with a range between $-0.22 \cdots 0.03$ dB, which is nearly omni-directional—if the result was exactly 0 dB then it would be omni-directional. The log mean average of these results is calculated as

$$10 \times \log_{10} \left(\frac{1}{N} \sum_i^N 10^{\mathrm{DI}_i/10} \right) = -0.07 \text{ dB} , \qquad (8.10)$$

where $N = 72$ is the number of directivity measurements, and DI_i is the ith directivity measurement in dB, which indicates that the monopole source is nearly omni-directional.

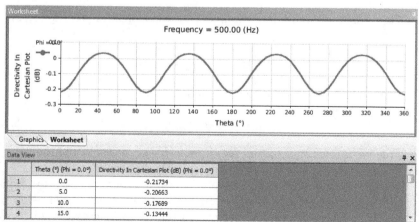

- Now we will calculate the radiated sound power. In the Outline window click on Solution (A6) | Acoustic Far Field, and change the row Properties | Result to Sound Power Level.

- In the window Details of "Acoustic Far Field", make sure that the row Reference Sound Power is set to 1E-12.

Details of "Acoustic Far Field"	
□ **Properties**	
Result Set	1
Boundary Condition On Model Symmetric Plane	No
Result	Sound Power Level
Reference Sound Power	1E-12 [W]
Model Thickness in Z Direction (2D extension)	0 [m]
Spatial Radiation Angle	Full Space

- Right-click on the object Acoustic Far Field and left-click on Generate, which will calculate the results.

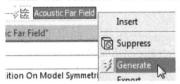

- The window Data View contains a table of the results, and shows that the sound power level is 93.475 dB re 10^{-12} W.

Data View	Frequency (Hz)	Sound Power Level (dB)
1	500.0	93.475

The theoretical sound power radiated by a monopole is [47, Eq. (5.12)]

$$W_{\text{monopole}} = \frac{Q_{\text{RMS}}^2 k^2 \rho_0 c_0}{4\pi \left(1 + k^2 a^2\right)}, \tag{8.11}$$

where the RMS volume velocity of the monopole is [47, Eq. (5.7)]

$$Q_{\text{RMS}} = 4\pi a^2 \, U_{0,\text{RMS}}, \tag{8.12}$$

U_0 is the peak velocity amplitude of the velocity of the surface of the sphere, and $U_{0,\text{RMS}} = U_0/\sqrt{2}$ is the corresponding RMS amplitude.

For the example here, the theoretical sound power given by Equations (8.11) and (8.9) is 93.37 dB re 10^{-12} W, which is calculated using the MATLAB script monopole_spl_vs_angle.m that is included with this book.

Hence the result calculated using ANSYS Workbench of 93.47 dB re 10^{-12} W compares favorably.

Another useful feature of ANSYS Workbench and the ACT Acoustics extension is the ability to calculate results in the Acoustic Near Field of the equivalent source surface. The results should only be requested at locations that are *exterior* of the equivalent source surface. As an example, a Results | Acoustic Near Field object was selected from the ACT Acoustics toolbar, which added the object to the Solution (A6) tree. In the window Details of "Acoustic Near Field", the row Near Field Position was changed to Along Path, the row Result was changed to Sound Pressure Level, and the location of the path was from the origin at $(0,0,0)$ to a point in the far field $(4,0,0)$, which was achieved by leaving the position 1 coordinates for x, y, z as 0, and changing Position 2: X Coordinate to 4, and leaving the remaining y and z coordinates as 0, as shown in the image below.

Details of "Acoustic Near Field"	
Properties	
Result Set	1
Boundary Condition On Model Symmetric Plane	No
Near Field Position	Along Path
Result	Sound Pressure Level
Reference RMS Sound Pressure	2E-05 [Pa]
Position 1: X Coordinate	0 [m]
Position 1: Y Coordinate	0 [m]
Position 1: Z Coordinate	0 [m]
Position 2: X Coordinate	4 [m]
Position 2: Y Coordinate	0 [m]
Position 2: Z Coordinate	0 [m]
Model Thickness in Z Direction (2D extension)	0 [m]

The Geometry window should show the model and the path where the results will be calculated along the X axis from $r = x = 0 \cdots 4$ m.

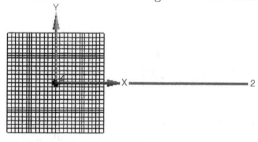

To request the results, in the Outline window right-click on Solution (A6) | Acoustic Near Field and in the context menu, left-click on Generate.

This will start a /POST1 post-processing task in ANSYS to calculate the near-field pressure results. Once the calculations have completed, a graph should be displayed. To export the results, right-click on the Acoustic Near Field object, left-click on Export, and type an appropriate filename such as monopole_spl_radius_near_field.txt. These ANSYS results can be compared with theoretical predictions.

Figure 8.8 shows the sound pressure level versus distance from the monopole source, from $r = 0.01 \cdots 4$ m, calculated theoretically using Equations (2.13), (2.14), and (2.30), and using ANSYS Workbench where the sound pressure level at nodes and the feature Acoustic Near Field along a path was used. The results show that the sound pressure levels calculated using ANSYS at the nodes within the model between $r = x = 0 \cdots 1.0$ m are in close agreement with theoretical predictions. However, when the sound pressure level is calculated using the Acoustic Near Field along a path feature, it shows that the results are inaccurate when the radius is $r \leq 0.5$ m, which is inside the equivalent source surface, and the results agree with theory at distances $r > 0.5$ m, which is outside the equivalent source surface. It was mentioned in Section 2.8.5.6 that results requested using the Acoustic Near Field feature should only be obtained *exterior* to the equivalent source surface, and the results shown in Figure 8.8 demonstrate what can occur if the feature is used incorrectly.

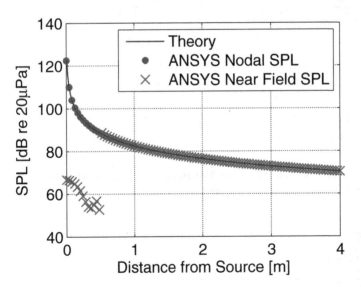

FIGURE 8.8
Sound pressure level versus distance from monopole source calculated theoretically for $a = 0.01$ m, $u = 1.0$ m/s, $r = 0.01 \cdots 4$ m, and using ANSYS Workbench to calculate the SPL at nodes and the Acoustic Near Field along a path feature.

Figure 8.9 shows the real and imaginary parts of the complex pressure versus distance from $r = 0.01 \cdots 4$ m, calculated theoretically using Equations (2.13), (2.14), and using ANSYS Workbench at nodes and the feature Acoustic Near Field along a path. In the window Details of "Acoustic Near Field" the row Result was set to Maximum Complex Pressure and then Pressure Phase. The results were exported and converted into complex pressure and then compared with the theoretical and nodal pressure results. Similar to the findings for the comparison of the sound pressure levels, the nodal pressure results agree with theoretical predictions, and for distances where $r \leq 0.5$ m, which is on the interior of the equivalent source surface, the Acoustic Near Field results do not match theoretical predictions, and do match at distances $r > 0.5$ m, which is on the exterior of the equivalent source surface.

8.3.2 Comparison of Monopole Acoustic Wave Source and Acoustic Mass Source

In the previous discussion, an acoustic monopole was simulated using an Acoustic Wave Source where the Wavetype was set to Monopole. It is also possible to simulate an acoustic monopole source by using an Acoustic Mass Source excitation. The following instructions describe how to replace the Acoustic Wave Source with an equivalent Acoustic Mass Source, and shows that the radiated sound pressure levels are identical.

- In the Outline window, right-click on Harmonic Response (A5) | Acoustic Wave Sources and left-click on Suppress to turn off the previous acoustic source.

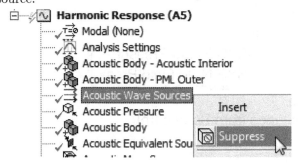

- The next step is to insert an acoustic mass source onto a vertex to simulate a monopole. In the ACT Acoustics extension toolbar, click on Excitation | Mass Source (Harmonic).

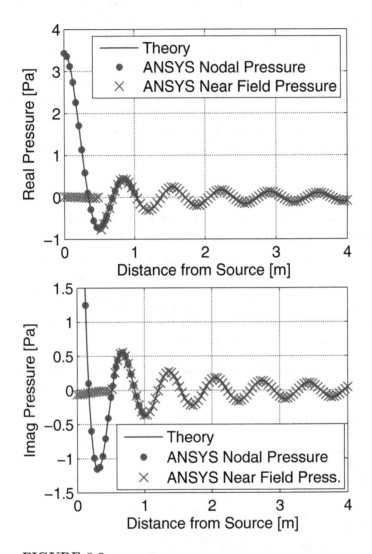

FIGURE 8.9
Real and imaginary parts of the complex pressure versus distance from a
monopole source calculated theoretically for $a = 0.01$ m, $u = 1.0$ m/s, $r = 0.01 \cdots 4$ m, and using ANSYS Workbench to calculate the pressure at nodes
and the Acoustic Near Field along a path feature.

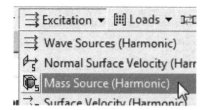

- It is necessary to define the location where the acoustic mass source will be applied in finite element model, which we want at the center of the cube. We also need to define the amplitude of the acoustic mass source, which we want to be the equivalent amplitude of the monopole source simulated in Section 8.3.1. Change the view type to wireframe by clicking on View | Wireframe in the menu bar. In the window Details of "Acoustic Mass Source", click in the row Scope | Geometry. Change the selection filter to Vertex. Rotate the view of the model as necessary until you can identify the vertex at the center of the cube and then select it. In the row Scope | Geometry click the Apply button. At the start of the solution phase, this acoustic load applied to the vertex will be transferred to the node at this location.

A: Harmonic Response
Acoustic Mass Source - monopole
Frequency: 499. Hz

■ Acoustic Mass Source - monopole

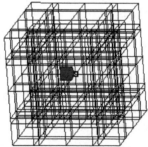

- Now we will define the amplitude of the acoustic mass source, which we want to be the equivalent of the monopole source, and is equal to the density of the fluid ρ_0 times the volume velocity $Q = S_{\mathrm{mono}}\, u$, hence

$$\text{Amplitude of Mass Source} = \rho_0 S_{\mathrm{mono}}\, u \tag{8.13}$$

$$= 1.2041 \times (4\pi \times 0.01^2) \times 1 \tag{8.14}$$

$$= 0.0015131\,\mathrm{kg/s}. \tag{8.15}$$

In the window Details of "Acoustic Mass Source", in the row Amplitude of Mass Source, enter the value 0.0015131.

Details of "Acoustic Mass Source - monopole"	
⊟ **Scope**	
Scoping Method	Geometry Selection
Geometry	1 Vertex
⊟ **Definition**	
Amplitude Of Mass Source	0.0015131 [kg m^-1 m^-1 m^-1 sec^-1]
Phase Angle	0 [°]

As described in Section 2.8.2.3, there is an inconsistency in the units that are displayed in the row Amplitude of Mass Source. In this example we are applying a mass source at a vertex and we are entering a value for the amplitude with units of kg/s. However, the units in the row Amplitude of Mass Source are $kg/m^3/s$, which is applicable only if the mass source were applied to a volume.

- In the menu bar, click on File | Save Project.

- Click on the Solve icon.

- When the calculations have completed, in the Outline window click on Solution (A6) | Acoustic Far Field, which will display a graph of sound pressure level versus angle in the Worksheet window, and a table of results in the Data View window. The results appear to be a sine wave, as was shown in the example of the monopole source using the Acoustic Wave Sources feature. The range of sound pressure levels varies between 70.4 dB to 70.7 dB, which compares favorably with the theoretical value calculated using Equations (2.13) and (2.30) of 70.48 dB. Hence, using an acoustic mass source gives nearly identical results as compared to using the monopole source.

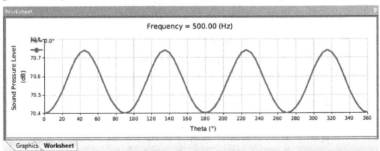

8.3.3 Comparison of Monopole and Back-Enclosed Loudspeaker Acoustic Sources

This section describes the comparison of the sound pressure level radiated from a monopole and back-enclosed loudspeaker acoustic wave sources calculated using ANSYS Workbench. It is assumed that the results are available from the previous analysis of the monopole sound source using the Acoustic Wave

Sources set to Monopole, which was described in Section 8.3.1. The instructions for simulating a back-enclosed loudspeaker using ANSYS Workbench are below.

Instructions

- Start ANSYS Workbench.

- Load the project file acoustic_sources_PML.wbpj that is supplied with this book.

- In the Outline window, make sure that the object Harmonic Response (A5) | Acoustic Wave Sources is Unsuppressed, which is indicated by a green tick next to the object.

- Click on the object Harmonic Response (A5) | Acoustic Wave Sources and in the window Details of "Acoustic Wave Sources", change the row Wave Type to back-enclosed loudspeaker, Velocity Amplitude to 1, and Radius of Pulsating Sphere to 0.01.

Details of "Acoustic Wave Sources"	
Definition	
Wave Number	1
Wave Type	Back Enclosed Loudspeaker
Excitation Type	Velocity
Source Location	Inside The Model
Velocity Amplitude	1 [m sec^-1]
Phase Angle	0 [°]
Global X Coordinate At Source Origin	0 [m]
Global Y Coordinate At Source Origin	0 [m]
Global Z Coordinate At Source Origin	0 [m]
Mass Density Of Environment Media	1.2041 [kg m^-1 m^-1 m^-1]
Sound Speed In Environment Media	343.24 [m sec^-1]
Radius Of Pulsating Sphere	0.01 [m]
Pure Scattering Options	On (Output Total Pressure)

- In the Outline window, make sure that the object Harmonic Response (A5) | Acoustic Mass Source - monopole is Suppressed, which is indicated by a cross next to the object. If it has a green tick next to it, then suppress it by right-clicking on the object and in the context menu that opens, left-click on Suppress. The Harmonic Response (A5) tree in the Outline window should look like the following image.

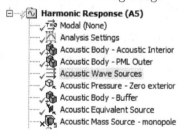

- The next step is to request that the results for sound pressure level versus angle are calculated. In the `Outline` window under `Solution (A6) |` `Acoustic Far Field`, in the window `Details of "Acoustic Far Field"`, change the row `Result` to `SPL in Cartesian Plot`. The remaining fields should be the same settings used previously and are shown in the following image.

Details of "Acoustic Far Field"	
⊟ **Properties**	
Result Set	1
Boundary Condition On Model Symmetric Plane	No
Result	SPL In Cartesian Plot ▼
Starting Angle Phi (From X Axis Toward Y Axis)	0 [°]
Ending Angle Phi	0 [°]
Number Of Divisions Phi	0
Starting Angle Theta (From Z Axis Toward X Axis)	0 [°]
Ending Angle Theta	360 [°]
Number Of Divisions Theta	72
Sphere Radius	4 [m]
Reference RMS Sound Pressure	2E-05 [Pa]
Model Thickness in Z Direction (2D extension)	0 [m]
Spatial Radiation Angle	Full Space

- That completes the setup of the analysis. Click on `File | Save Project` and then click the `Solve` icon.

The sound pressure level versus angle was calculated for a monopole source in Section 8.3.1. These results are compared with the sound pressure level from a back-enclosed loudspeaker acoustic wave source, and are shown in Figure 8.10 for the angles $\phi = 0°$ and $\theta = 0° \cdots 360°$ in the spherical coordinate system shown in Figure 2.14 (page 41). The results show that both sources radiate omni-directionally, and that the sound pressure level of the back-enclosed loudspeaker is 12 dB less than the monopole source. This is because the radiating surface area of the monopole source is a sphere of radius $a = 0.01$ m and hence has a surface area of $S_{\text{mono}} = 4\pi a^2$, whereas the back-enclosed loudspeaker has a radiating surface area from a circular piston of the same radius and hence has a surface area of $S_{\text{back}} = \pi a^2$, a factor of 4 less. Therefore, one would expect that the sound pressure level from the back-enclosed loudspeaker would be $20 \times \log_{10}(4) = 12$ dB less than the sound pressure levels of the monopole source. Monopole sound sources radiate omni-directionally when $ka \ll 1$. In other words, when the size of the sphere is very much less than the wavelength of sound it generates, the source will radiate equally in all directions. For this example $ka = (2\pi \times 500/343) \times 0.01 = 0.09 \ll 1$, and hence one would expect that the monopole would radiate omni-directionally. The radiation pattern will develop "lobes" when the $ka \approx 1$.

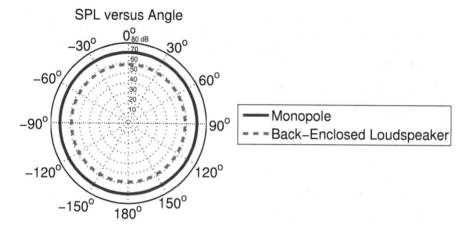

FIGURE 8.10
Sound pressure level versus angle for a monopole and a back-enclosed loudspeaker calculated using ANSYS Workbench for $a = 0.01$ m, $u = 1.0$ m/s, $r = 4$ m.

8.3.4 Comparison of Dipole Acoustic Source Calculated Theoretically and Using ANSYS Workbench

This section describes the calculation of the sound pressure level versus angle for a dipole acoustic source using ANSYS Workbench, using the project file acoustic_sources_PML.wbpj. The results predicted using ANSYS will be compared with theoretical predictions using MATLAB code. A dipole source was described in Section 2.8.2.1 that comprises two identical monopoles separated by a distance d, only the monopoles are oscillating $180°$ out of phase from each other.

The first step is to modify the ANSYS Workbench model to change the acoustic source from a monopole to a dipole Acoustic Wave Sources.

Instructions

• Start ANSYS Workbench.

• Load the project file acoustic_sources_PML.wbpj.

• In the Project Schematic window, double-click on row 5 Setup to start Mechanical.

• The next step is to alter the acoustic wave source type from a monopole to a dipole. In the Outline window, click on Harmonic Response (A5) | Acoustic Wave Sources. In the window Details of "Acoustic Wave Sources", change the row Definition | Wave Type to Dipole. Change the row Dipole Length to 0.1. Change the row Z Component of Unit Dipole

Vector to 1, and leave the X and Y components as 0, which defines a vector $(0, 0, 1)$ so that the orientation of the dipole axis is aligned with the Z axis. Note that if the vector is defined to have a length that is longer than 1.0, ANSYS will scale the vector to unit length.

Details of "Acoustic Wave Sources"	
Definition	
Wave Number	1
Wave Type	Dipole
Excitation Type	Velocity
Source Location	Inside The Model
Velocity Amplitude	1 [m sec^-1]
Phase Angle	0 [°]
Global X Coordinate At Source Origin	0 [m]
Global Y Coordinate At Source Origin	0 [m]
Global Z Coordinate At Source Origin	0 [m]
Mass Density Of Environment Media	1.2041 [kg m^-1 m^-1 m^-1]
Sound Speed In Environment Media	343.24 [m sec^-1]
Radius Of Pulsating Sphere	0.01 [m]
Dipole Length	0.1 [m]
X Component Of Unit Dipole Vector	0 [m]
Y Component Of Unit Dipole Vector	0 [m]
Z Component Of Unit Dipole Vector	1 [m]
Pure Scattering Options	On (Output Scattered Pressure)

- Make sure that the object `Acoustic Wave Sources` is not suppressed and has a green tick next to it. If it doesn't, then right-click on `Acoustic Wave Sources` and left-click on `Unsuppress`. Also check that the object `Acoustic Mass Source - monopole` is suppressed and has a blue cross next to it.

- In the menu bar, click on `File | Save Project`.

- Click the `Solve` icon.

- When the calculations have completed, in the `Outline` window click on `Solution (A6) | Acoustic Far Field`, which will display a graph of sound pressure level versus angle in the `Worksheet` window, and a table of results in the `Data View` window.

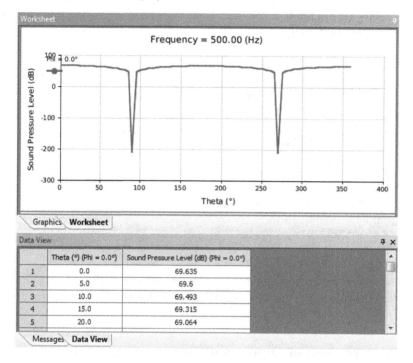

- The next step is to export the results so that they can be post-processed using MATLAB. In the Outline window, right-click on Solution (A6) | Acoustic Far Field, then left-click on Export. Type in an appropriate filename such as dipole_spl_vs_angle.txt.

That completes the analysis for the acoustic dipole using ANSYS Workbench.

The MATLAB code dipole_spl_vs_angle.m included with this book can be used to calculate the sound pressure level versus angle for a dipole source radiating into a free-field.

Figure 8.11 shows the comparison of the sound pressure level versus angle for a dipole source calculated theoretically using the MATLAB code and using ANSYS Workbench. The radiation pattern of a dipole source has a pressure null along the line $\theta = \pm 90$ where the out-of-phase monopole source are equidistant from each other and the pressures cancel. The predictions using ANSYS have excellent agreement with theoretical predictions.

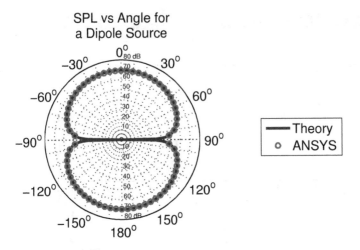

FIGURE 8.11
Sound pressure level versus angle for a dipole calculated theoretically and
using ANSYS Workbench for for $a = 0.01$ m, $u = 1.0$ m/s, $r = 4$ m, $d = 0.1$
m.

8.3.5 Comparison of Dipole and Bare Loudspeaker

This section describes the comparison of the sound pressure level radiated from
a dipole and bare (unbaffled) loudspeaker acoustic wave sources calculated
using ANSYS Workbench.

The same steps can be followed to calculate the sound pressure level versus
angle for a bare (unbaffled) loudspeaker as described for the analysis of the
dipole. The only difference is that in the Outline window, click on Harmonic
Response (A5) | Acoustic Wave Sources, and in the window Details of
"Acoustic Wave Sources", change the row Definition | Wave Type to Bare
Loudspeaker, and then follow the remaining steps as described in the last
section.

Details of "Acoustic Wave Sources"	
⊟ **Definition**	
Wave Number	1
Wave Type	Dipole
Excitation Type	Planar Wave
Source Location	Monopole
	Dipole
Velocity Amplitude	Back Enclosed Loudspeaker
Phase Angle	Bare Loudspeaker

The sound pressure level versus angle was calculated using ANSYS Workbench
for a dipole and bare (unbaffled) loudspeaker acoustic wave sources. The results are shown in Figure 8.12 for the angles $\phi = 0°$ and $\theta = 0° \cdots 360°$ in

the spherical coordinate system shown in Figure 2.14 (page 41). The results show that both sources have similar radiation patterns, only the sound pressure level of the bare loudspeaker is 12 dB less than the dipole source. This is because of the same reason as described previously for the comparison of the monopole and back-enclosed loudspeaker example.

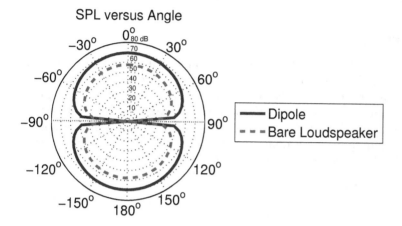

FIGURE 8.12
Sound pressure level versus angle ($\phi = 0°$ and $\theta = 0° \cdots 360°$) for a dipole and a bare loudspeaker calculated using ANSYS Workbench for $a = 0.01$ m, $u = 1.0$ m/s, $r = 4$ m, $d = 0.1$ m.

8.4 Example: Radiation of a Baffled Piston

8.4.1 Learning Outcomes

The learning outcomes from this example are:

- examination of the variation of the acoustic radiation pattern with changes in the ka normalized wavenumber frequency parameter,

- nulls in the acoustic pressure magnitude (and hence sound pressure level) occur on the axis of the piston,

- the acoustic source from an oscillating piston can be modeled without the use of structural elements,

- demonstration of how to conduct an axi-symmetric 2D acoustic analysis using ANSYS Workbench and ANSYS Mechanical APDL,

- methods of determining the area associated with nodes of an axi-symmetric model, and

- how to extract specialized results from an analysis by using the ANSYS help manual to examine the details of acoustic elements and the results that are produced.

8.4.2   Theory

This example shows the prediction of sound radiated from an oscillating circular piston in an infinite plane baffle, as shown in Figure 8.13. The theory can be found in most acoustic textbooks and is one of the fundamental examples of acoustic radiation. The results of this academic example are applicable in the radiation of sound from a cone loudspeaker oscillating in a baffled enclosure, an exhaust pipe protruding from a wall, underwater acoustic sonar emitters mounted to the surface of a structure, and others.

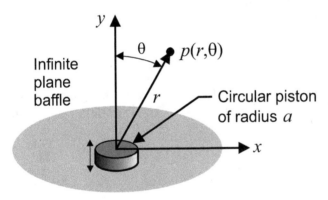

FIGURE 8.13
Schematic of an oscillating circular piston in an infinite plane baffle radiating sound.

The sound radiation behavior versus the ka parameter is one of the main learning outcomes from the analysis. There are three main regions of interest based on the frequency parameter:

$ka < 1$ is sometimes termed the low-frequency region, where sound will radiated nearly equally in all directions (omni-directionally).

$ka > 1$ is the mid-frequency region where the radiated sound is channeled into a single beam.

$ka \gg 1$ is the high-frequency region where the radiated sound is channeled into multiple beams.

The complex on-axis pressure radiated from an oscillating circular piston in an infinite plane baffle is given by [102, Eq. (7.4.4), p. 180]

$$p(r,0,t) = 2\rho_0 c_0 U_0 \left\{ 1 - \exp\left[-jk\left(\sqrt{r^2+a^2} - r\right)\right]\right\} e^{j(\omega t - kr)}, \quad (8.16)$$

where r is the distance from the piston, a is the radius of the piston, U_0 is the amplitude of the velocity of the piston, $k = \omega/c_0$ is the wavenumber, c_0 is the speed of sound of the fluid, and ρ_0 is the density of the fluid.

Figure 8.14 shows the normalized sound pressure along the axis of symmetry of the piston versus the normalized axial distance from the piston. The figure shows that there are 4 axial locations where the pressure is zero. This occurs when the exponential function in Equation (8.16) equals an integer multiple of $n\pi$. By solving for r when

$$k(\sqrt{r^2+a^2} - r) = n\pi, \quad (8.17)$$

results in

$$r(n) = \frac{a^2 f^2 - n^2 c_0^2}{2n c_0 f}, \quad (8.18)$$

where n is an integer. The valid range for n is where the radius is positive, and hence the numerator in Equation (8.18) must be greater than 0. Hence the valid range for n is 1 to $n < (af/c_0)$.

The radiation impedance from an oscillating circular piston in a plane infinite baffle is given by [102, Eq. (7.5.11)]

$$Z_r = \rho_0 c_0 \pi a^2 \left[R_1(2ka) + jX_1(2ka)\right], \quad (8.19)$$

where the piston resistance function R_1 and the piston reactance function X_1 are given by [47, p. 213]

$$R_1(x) = 1 - \frac{2J_1(x)}{x} = \frac{x^2}{2 \times 4} - \frac{x^4}{2 \times 4^2 \times 6} + \frac{x^6}{2 \times 4^2 \times 6^2 \times 8} - \cdots, \quad (8.20)$$

$$X_1(x) = \frac{2H_1(x)}{x} = \frac{4}{\pi}\left(\frac{x}{3} - \frac{x^3}{3^2 \times 5} + \frac{x^5}{3^2 \times 5^2 \times 7} - \cdots\right), \quad (8.21)$$

where J_1 is the Bessel function of the first kind and first order, and H_1 is the first-order *Struve* function.

The time-averaged radiated power from the piston is given by [47, Eq. (5.111b)]

$$W = \frac{1}{2}\rho_0 c_0 R_1 \pi a^2 |U_0|^2. \quad (8.22)$$

The following paragraphs describe the theory for calculating the radiation pattern of a baffled circular piston.

The pressure radiated by a baffled circular piston is given by [102, Eq. (7.4.17)]

$$p(r, \theta, t) = \frac{j}{2} \rho_0 c_0 U_0 \frac{a}{r} ka \left[\frac{2J_1(ka \sin \theta)}{ka \sin \theta} \right] e^{j(\omega t - kr)} . \qquad (8.23)$$

Note that all the angular variations occur within the terms contained in the square brackets in Equation (8.23). The other point to highlight about these bracketed terms is that when evaluating the on-axis response when the angle $\theta = 0$, the on-axis sound pressure response is described by Equation (8.16). However, the value of the Bessel function $J_1(0) = 0$, and the value of the denominator $ka \sin(0) = 0$, which causes numerical difficulties. This is resolved by the mathematical peculiarity that $0/0 = 1$.

When the beam pattern is plotted, which will be shown in the MATLAB section and Figure 8.16, it can be seen that there are pressure nodes that occur at several angles. These occur when the terms in the square brackets evaluate to zero

$$\left[\frac{2J_1(v(\theta))}{v(\theta)} \right] = 0 , \qquad (8.24)$$

$$\text{where } v(\theta) = ka \sin \theta . \qquad (8.25)$$

Hence, the pressure is zero at a particular angle when the numerator $J_1(ka \sin \theta_m) = 0$. The zeros of the Bessel function $J_1(v) = 0$ can be solved numerically, and the corresponding angles θ calculated using Equation (8.25). These calculations are further described in the following section.

8.4.3   MATLAB

The MATLAB code `pressure_on_axis.m` included with this book can be used to calculate the magnitude of the pressure on the axis of symmetry radiated from a baffled circular piston. The MATLAB code generates Figure 8.14 that reproduces Figure 7.4.2, p. 181 in Kinsler et al. [102].

Some readers might find this result surprising that the magnitude of the sound pressure is zero at a few locations directly in front of an oscillating piston. The reason this occurs is because of destructive interference, where sound generated by parts of the piston cancel out with sound generated with other parts of the piston.

The MATLAB code `baffled_piston.m` included with this book can be used to calculate the normalized impedance of an oscillating baffled circular piston. The MATLAB code generates Figure 8.15 that reproduces Figure 7.5.2, p. 187 in Kinsler et al. [102].

The calculation of the radiation impedance involves the use of the Struve function. However, MATLAB does not have an in-built Struve function and it is necessary to obtain an external MATLAB function, which is included with this book and is called `struve.m`.

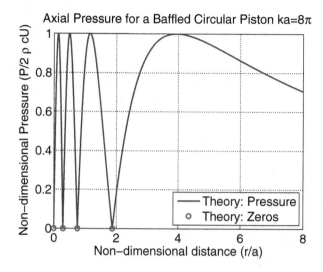

FIGURE 8.14
Sound pressure on axis for an oscillating circular piston in an infinite plane
baffle, at a frequency corresponding to $ka = 8\pi$.

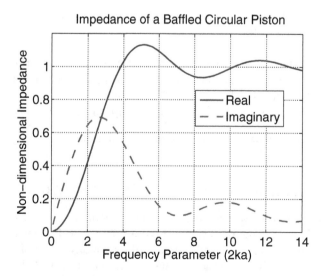

FIGURE 8.15
Real and imaginary normalized impedance of an oscillating piston installed in
a rigid plane baffle.

Figure 8.15 of the normalized impedance can be compared with results published in textbooks such as Morse and Ingard [117, p. 385, Fig. 7.9], Kinsler et al. [102, p. 187, Fig. 7.5.2], Fahy [65, p.130, Fig. 6.19], and Bies and Hansen [47, p. 214, Fig. 5.9].

The MATLAB code `radiation_pattern_baffled_piston.m` included with this book can be used to calculate the beam pattern or directivity of an oscillating baffled circular piston. The MATLAB code generates Figure 8.16 that reproduces Figure 7.4.5, p. 183 in Kinsler et al. [102].

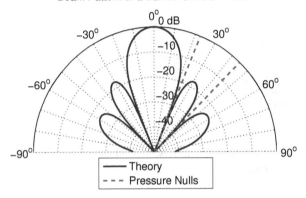

FIGURE 8.16
Radiation pattern of an oscillating piston installed in a rigid plane baffle at a frequency corresponding to $ka = 10$.

Figure 8.16 has two dashed lines drawn at 22.5° and 44.6° which are the angles where pressure zeros occur, and were calculated using the MATLAB code by solving for θ in Equations (8.24) and (8.25).

The MATLAB code evaluates the directivity at 3 groups of angles:

- $-\pi/2$ to -0.001 in increments of 0.001 radians, using Equation (8.23)

- 0, using Equation (8.16)

- $+0.001$ to $+\pi/2$ in increments of 0.001 radians, using Equation (8.23)

The reader might also like to modify the constants in the code to reproduce Figure 7.8, p. 382 in Morse and Ingard [117], or Figure 5.7, p. 226 in Pierce [129], which illustrate that a piston will radiate omni-directionally when the wavelength is longer than the circumference of the piston $2\pi a$ [115, p. 328], and for short wavelengths (high frequencies) will have a narrow beam radiation pattern.

Sound Power

Figure 8.17 shows the sound power radiated from a piston with a radius of $a = 0.1$ m calculated using MATLAB and ANSYS (to be shown in the following sections) where the peak displacement of the piston was 1 micron (i.e., 1×10^{-6} m). The sound power is calculated using Equation (8.22).

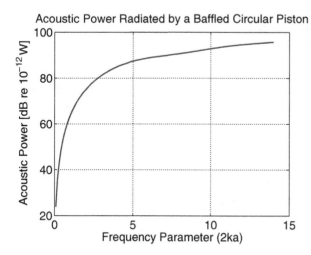

FIGURE 8.17
Sound power radiated from a piston of radius $a = 0.1$ m and peak displacement 1×10^{-6} m, calculated theoretically and using ANSYS.

8.4.4 ANSYS Workbench

The finite element model that will be created of the circular piston in a plane infinite baffle uses 2D axi-symmetric elements. As this problem is rotationally "symmetric," only a slice of the acoustic domain needs to be modeled as shown in Figure 8.18.

This system will be modeled in ANSYS as shown in Figure 8.19. The acoustic domain is modeled using 2D FLUID29 elements that have 4 nodes per element. This element has the capability of modeling an axi-symmetric system by changing various options using the KEYOPT command. The solid moving piston is not actually modeled, but the acoustic fluid around the piston face is modeled using FLUID29 elements that have pressure and displacement degrees of freedom. The acoustic free-field is simulated using 2D FLUID129 "infinite" acoustic elements that have 2 nodes per element.

ANSYS Workbench will be used to calculate the following results and will be compared with theoretical predictions:

- acoustic pressure versus distance on the axis of the piston

- radiation impedance

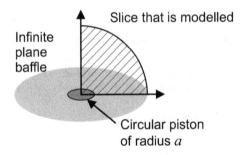

FIGURE 8.18
Schematic of the circular piston in an infinite plane baffle. Only a slice of the acoustic domain needs to modeled as the problem is axi-symmetric.

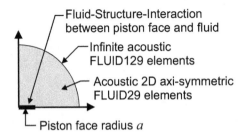

FIGURE 8.19
Schematic of the finite element model of the circular piston of radius a in an infinite plane baffle.

- directivity or beam pattern

- sound power

This analysis will be conducted up to a value of $2ka = 10$ where $k = \omega/c_0$ is the wavenumber, $c_0 = 343$ m/s is the speed of sound of air, $a = 0.1$ m is the piston radius, and $\rho_0 = 1.21$ kg/m^3. Hence the maximum frequency of analysis ($f_{\max}$) is

$$2ka = 2\frac{\omega}{c_0}a = \frac{4\pi f}{c_0}a = 10 \tag{8.26}$$

$$f_{\max} = \frac{10c_0}{(4\pi a)} = \frac{10 \times 343}{4\pi \times 0.1} = 2729 \text{ Hz} . \tag{8.27}$$

The minimum wavelength will be

$$\lambda_{\min} = \frac{c_0}{f_{\max}} = \frac{(4\pi a)}{10} = 0.126 \text{ m} . \tag{8.28}$$

The number of elements per wavelength for this analysis will be set at

EPW $= 20$. For linear acoustic elements, such as the FLUID29 elements, it is recommended that at least 12 elements per wavelength are used. We will use EPW $= 20$ to further improve the accuracy. The reader can experiment with changing the value of EPW and see the effect on results. The required element size is

$$\text{element size} = \frac{\lambda_{\min}}{\text{EPW}} = \frac{(4\pi \times 0.1)}{10} \frac{1}{20} = 6.3 \times 10^{-3} \text{m} . \tag{8.29}$$

It is possible to parameterize many of these expressions in ANSYS Workbench so that parametric studies can be conducted, which is beneficial for reproducing the results in Section 8.4.3.

Workbench Instructions

The completed ANSYS Workbench archive file of the system piston_baffle_ axisym.wbpz, which contains the .wbpj project file, is included with this book. Note that the mesh generated for this model has around 190,000 nodes, and cannot be solved using a teaching license of ANSYS, which is restricted to 32,000 nodes.

Although the model that will be created is to simulate a 2D axi-symmetric system, as shown in Figure 8.18, it is recommended that ANSYS Workbench is set to a 3D analysis, because ANSYS Workbench does not support Line Bodies in 2D mode, which are used for the infinite elements (FLUID129) used on the outer boundary of the quarter circle. This can be done by changing Properties of Schematic A3: Geometry, Advanced Geometry Options, row 16 Analysis Type to 3D.

Instructions

- Start a new ANSYS Workbench project.
- Double-click on Harmonic Response under Analysis Systems in the Toolbox window, so that a new Harmonic Response cell appears in the Project Schematic window.

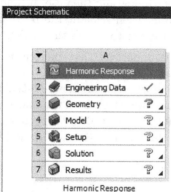

- Double-click on row 3 Geometry to start DESIGNMODELER.
- Select the desired length unit as Meter, then click the OK button.
- Move the mouse cursor over the Z axis on the triad in the lower right corner of the screen and click on the +Z axis so that the X-Y Plane is shown.

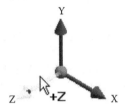

Alternatively, you can right-click on the XYPlane in the tree outline and select Look at.

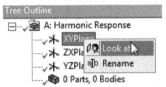

- Left-click the New Sketch icon to create a new sketch on the XYPlane.
- Click on the Sketching tab to open the Sketching Toolboxes window. Make sure that the point and edges selection filters are both active, which is indicated by the icons with a sunken appearance. If they are not active, you can click on the icons or type with the keyboard <Ctrl> p for points and <Ctrl> e for edges.

- Left-click the mouse button in the Graphics window and then use the middle mouse wheel to scroll the view until the dimension scale at the bottom of the screen shows 1.000m. This is to make it easier to sketch the model.
- Before commencing to sketch objects, it is important that the automatic constraints feature is turned on so that the cursor will "snap" to points and edges. In the Sketching Toolbox window, click on the Constraints tab. To scroll through the Constraint menu options, click on the downward pointing triangle next to the Setting tab until the Auto Constraints option is visible, then left-click on it. Click in the box next to Cursor: to activate the generation of automatic constraints.

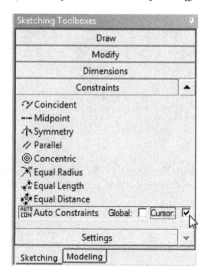

- Click on the Drawing tab in the Sketching Toolboxes windows and then click on Arc by Center. At the bottom of the window you should see Arc by Center -- Click, or Press and Hold, for center of circle.

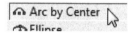

- Move the mouse cursor so that it hovers over the origin and the letter P is shown over the origin, which indicates that the cursor will snap to a coincident point at the origin, then left-click with the mouse button.

- Move the mouse cursor along the X axis so that the letter C is shown, which indicates that the cursor is coincident with the edge, and a circle is drawn that moves with the mouse cursor. Move the mouse cursor until the dimension in the bottom right corner is approximately 1. The exact dimension is not important as this will be defined in a later step. Left-click when you are happy with the placement of the start of the arc.

- Move the mouse cursor so that it is over the Y-axis and the letter C is shown and an arc is displayed, then left-click the mouse button to draw the arc.

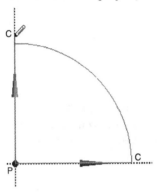

- Repeat these steps to draw another arc with a radius of 0.1 m.

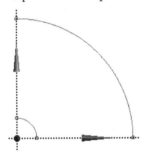

- Click on the Line tool under the Draw tab in the Sketching Toolboxes window. In the status bar at the bottom of window you should see Line -- Click, or Press and Hold, for start of line.

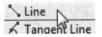

- Move the mouse cursor so that it hovers over the end point of the inner arcs, and a letter P is shown, which indicates that the cursor will snap to the coincident point, then left-click the mouse button to start drawing the line.

- Move the mouse cursor so that it hovers over the end point of the outer arc and a letter P is shown on the point and the letter H is shown on the X-axis to indicate a horizontal line.

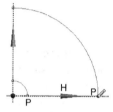

- Repeat this process to draw a vertical line that connects the ends of the inner and outer arcs, noting that a letter V will be displayed over the Y-axis to indicate a vertical line.

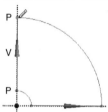

- Click on the Dimensions tab in the Sketching tab in the Sketching Toolboxes window.

Modify
Dimensions
Constraints
Settings
Sketching Modeling

- Click on the Horizontal dimension tool.

- Move the mouse cursor so that it hovers over the Y-axis and a highlighted line which will be drawn on the axis but is difficult to see. The mouse cursor will also change to a box. Left-click the mouse button to define the start of the dimension line.

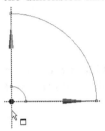

- The status bar at the bottom of the screen will change to Horizontal -- Select second point or 2D Edge for Horizontal dimension. Move the mouse cursor so that it hovers over the point on the X-axis for the inner arc, and the mouse cursor will change to a box with a dot in the upper left corner, then left-click the mouse button to define the end point for the dimension.

- Move the mouse cursor below the model and then left-click the mouse button to place the dimension. A symbol H1 will be displayed.

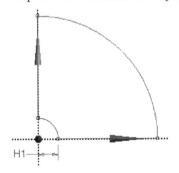

- Repeat this process to define the horizontal dimension from the Y-axis to the outer arc and a symbol H2 will be displayed.

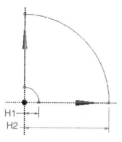

- In the Details View window under the Dimensions branch, there should be the horizontal dimensions H1 and H2 listed.

Details View	
Details of Sketch1	
Sketch	Sketch1
Sketch Visibility	Show Sketch
Show Constraints?	No
Dimensions: 2	
H1	0.17952 m
H2	1.0532 m

- Click in the box to the left of H1 and a dialog window will open. In the entry box for Parameter Name, type in piston_r and then click the OK button.

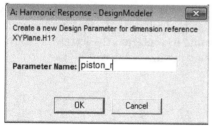

- Repeat this process for the H2 dimension and name the parameter baffle_r.
- Once this is completed, the dimensions H1 and H2 should have the letter D next to them in the Details View window. The letter D indicates that the dimensions are Design Parameters and will be listed in the Parameter Manager window when the Parameters icon is clicked, or selected from the main menu Tools | Parameters.

Details View	
⊟ **Details of Sketch1**	
Sketch	Sketch1
Sketch Visibility	Show Sketch
Show Constraints?	No
⊟ **Dimensions: 2**	
D H1	0.17952 m
D H2	1.0532 m

- Now is a good time to save your work. Click on File | Save Project and enter an appropriate filename such as piston_baffle_axisym.

- You may notice that in the Project Schematic window there is a new box labeled Parameter Set, which will be used later to define the dimensions of the model and the mesh density.

- Click in the window for DESIGNMODELER, and in the Tree Outline window, left-click on the XYPlane. Click on the New Sketch icon which will insert a Sketch2 branch under the XYPlane branch.

- Click on Sketch2, then click on the Sketching tab. Click on the Arc by Center tool and draw an arc starting at the origin that overlaps the inner arc, making sure that when the end points are selected the letter P is shown at the mouse cursor to indicate that the existing point will be used.

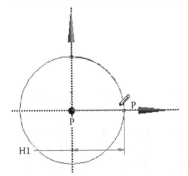

- Select the Line tool and draw radial lines that connect the origin to the points on the inner arc, so that a quarter sector is formed. When the lines are drawn you should notice that the letter H will appear for the horizontal line, and a letter V will appear for a vertical line.

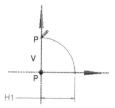

- Click on the Modeling tab, then click on the Sketch1 branch in the Tree Outline. From the menu bar, click on Concept | Surfaces from Sketches.

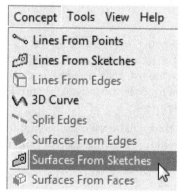

- In the Details View window, change the row Operation to Add Frozen, then click the Generate icon with the lightning bolt.

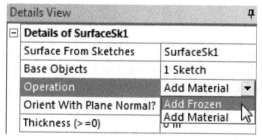

There are two states for bodies in DESIGNMODELER which are *Active* and *Frozen*. When a body is *Active* it can be altered by modeling operations. When the body is *Frozen* it is not affected by modeling operations (except for slicing). For further discussion see the ANSYS online help manual [33].

- Repeat these steps to generate a *frozen* Surface Body for Sketch2. In the Tree Outline view you should notice that there are now two Surface Body entries listed beneath the 2 Parts, 2 Bodies branch.

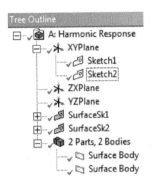

- Join the edges of the two surface bodies together by clicking on Tools | Joint and selecting the two areas. This can be done by activating the Faces selection filter by pressing <Ctrl> f on the keyboard, then holding down the <Ctrl> key on the keyboard and clicking each of the two areas, then click the Apply button, and then click the Generate icon.

- The next steps involve the creation of a part that will be used for the infinite acoustic radiation boundary, so that outward propagating waves will be absorbed. Select Concept | Lines From Edges.

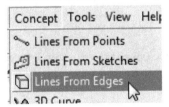

- Make sure that the Edge filter is selected and left-click on the outer arc so that it is selected, click the Apply button in the row labeled Edges, change the Operation to Add Frozen, then click the Generate icon. A new frozen Line Body part will be created.

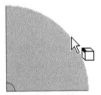

- A dummy cross section will be defined for the Line Body, but it will not be used in the analysis. The dummy cross section is needed so that ANSYS Mechanical will accept the model created in DESIGNMODELER. On the menu bar at the top, click on Concept | Cross Section | Rectangular and click the Generate icon. The dimensions of the cross-section are not important.

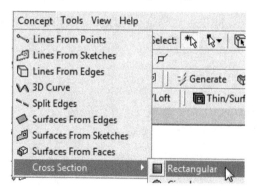

- Click on the Line Body entry under the branch 3 Parts, 3 Bodies and you will notice that the row Cross Section is highlighted in yellow. Click in the cell that has Not selected and a drop-down menu will appear. Click on the entry Rect1 cross-section definition.

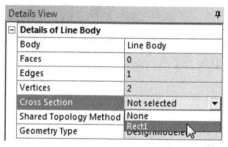

- To make it easier to refer to various parts of the model, we will rename the Surface Body and Line Body entries. In the Tree Outline under the branch labeled 3 Parts, 3 Bodies, click on the first Surface Body entry, which should highlight the larger of the two quarter sectors. Right-click on the Surface Body, which will open a context menu, and select Rename. Alternatively, press <F2> on the keyboard. Change the name to freefield_body.

- Repeat this process and rename the smaller quarter sector Surface Body to piston_body, and rename Line Body to radiation_boundary.

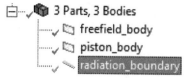

- The next step is to merge these three bodies so that the nodes on the boundary between the bodies will be shared. Left-click on the top freefield_body in the branch, hold down the <Ctrl> key on the keyboard, and left-click the piston_body and radiation_boundary bodies so that they are all highlighted in blue. While all three entries are highlighted, right-click and in the context menu that opens, left-click on Form New Part.

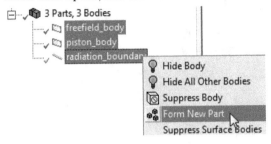

- You should notice that the Tree Outline window now lists 1 Part, 3 Bodies.

- As we want the finite element model to share the nodes at the boundaries between the bodies, click on the icon Share Topology, then click on the Generate icon.

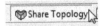

- In the menu bar, click on File | Save Project.

That completes the definition of the solid model for the baffled piston. The next stage is to set up the finite element model. You can keep the DESIGN-MODELER window open and you will be able to check that the dimensions of the model have been updated once they are defined in the Parameter Set.

- Click in the window for the ANSYS Workbench Project Schematic, and double-click the Parameter Set box which will display the tables for the parameters.

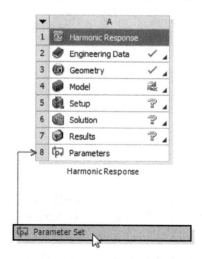

- The next steps are to define the dimensions of the model and some variables which will be used in the analysis. The learning outcome from these steps is to see how variables or parameters can be defined to create a parametric model which might be useful for conducting optimization or design of experiment studies for more complex analyses. The main tables that will be used in this example are in the `Outline` and `Properties` windows. The `Table of Design Points` can be used to define a set of parameters to be analyzed. For example, studies can be conducted to determine how the sound pressure level varies as the mesh density is changed.

- In the table labeled `Outline`, left-click in cell `C4`, which corresponds to `P1` `piston_r`, and type in the value `0.1`.

	A	B	C	D
	ID	Parameter Name	Value	Unit
1				
2	☐ Input Parameters			
3	☐ 🔊 Harmonic Response (A1)			
4	⎧p P1	piston_r	0.1	
5	⎧p P2	baffle_r	1.0532	
*	⎧p New input parameter	New name	New expression	
7	☐ Output Parameters			
*	p↲ New output parameter		New expression	
9	Charts			

Outline: No data ▾ ⊓ ✗

- Repeat these steps to define `baffle_r` as `1.0`.

- In the toolbar, click on the `Save` icon and then the `Refresh Project` icon. If you now return to the DESIGNMODELER and click on `Sketch1` under the

XYPlane tree in the Tree Outline window, you will see that the dimensions
for H1 = 0.1 m and H2 = 1 m are listed in the Details View window.

Details View	
⊟ **Details of Sketch1**	
Sketch	Sketch1
Sketch Visibility	Show Sketch
Show Constraints?	No
⊟ **Dimensions: 2**	
D H1	0.1 m
D H2	1 m

- You can close the DESIGNMODELER module, if you wish, as it will no longer
  be used in this example. DESIGNMODELER can be closed by clicking on the
  X in the top right corner of the window.

- Return to the view of the Parameter Set tables. Click in the cell labeled
  New Name.

Outline: No data

	A	B	C	D
1	ID	Parameter Name	Value	Unit
2	⊟ Input Parameters			
3	⊟ 🗛 Harmonic Response (A1)			
4	℗ P1	piston_r	0.1	
5	℗ P2	baffle_r	1	
*	℗ New input parameter	New name	New expression	

- Type in EPW as the acronym for elements per wavelength, then press the
  <Tab> key to move to the next cell for value in column C. Enter a value of
  12 and then press the <Enter> key on the keyboard.

Outline: No data

	A	B	C	D
1	ID	Parameter Name	Value	Unit
2	⊟ Input Parameters			
3	⊟ 🗛 Harmonic Response (A1)			
4	℗ P1	piston_r	0.1	
5	℗ P2	baffle_r	1	
6	℗ P3	EPW	12	
*	℗ New input parameter	New name	New expression	
8	⊟ Output Parameters			
*	℘ New output parameter		New expression	
10	Charts			

- Define another parameter 2ka and in the column for Value enter 16*PI, then
  press the <Enter> key on the keyboard. The expression will be evaluated as
  50.265.

	A	B	C	D
Outline: No data				
1	ID	Parameter Name	Value	Unit
2	⊟ Input Parameters			
3	⊟ 🅰 Harmonic Response (A1)			
4	ⓟ P1	piston_r	0.1	
5	ⓟ P2	baffle_r	1	
6	ⓟ P3	EPW	12	
7	ⓟ P4	2ka	50.265	
*	ⓟ New input parameter	New name	New expression	
9	⊟ Output Parameters			
*	ⓟ New output parameter		New expression	
11	Charts			

- We will return to the `Parameter Set` later to define expressions for the size of the mesh. For the moment, click on the `Return to Project` icon.

- In the `Project Schematic` window, double-click on row 4 `Model` to start ANSYS Mechanical. Once it has started, in the `Outline` window, click on the small plus sign next to `Geometry` to open the branch, and then click on the plus sign next to `Part`. You should notice that there are question marks next to the two `Surface Body` entries, which indicates that some information is missing.

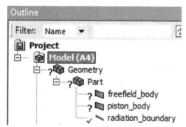

- Click on the `freefield_body` entry and you will notice that in the window `Details of "freefield_body"`, the row labeled `Thickness` is highlighted in yellow. Enter a dummy value of 1 for `Thickness`. The thickness value will not be used in this analysis as the element type used in this analysis will be an axi-symmetric 2D element.

- Repeat this step to define the thickness of `piston_body` as 1.

- You should notice that the icons next to the entries in the `Outline` window have changed from a question mark to a green tick.

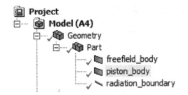

- The next steps involve specifying that 2D FLUID29 acoustic elements will be used in the model by inserting some APDL code. The element type has several options available, which are set using the KEYOPT settings, and define whether the element has displacement degrees of freedom active or inactive, and whether it is planar or axi-symmetric. In this case the elements are defined as without displacement DOFs (structure absent) and as axi-symmetric. In addition, the material properties will be defined for the speed of sound, and density of the gas. It is necessary to define the element type and material properties for each body. Right-click on the freefield_body entry and in the context menu that opens select Insert | Commands.

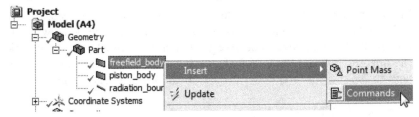

- In the window labeled Commands, some text will be displayed.

```
!   Commands inserted into this file will be executed just after material definitions
        in /PREP7.
!   The material number for this body is equal to the parameter "matid".

!   Active UNIT system in Workbench when this object was created:  Metric
        (m, kg, N, s, V, A)  with temperature units of C
!   NOTE:  Any data that requires units (such as mass) is assumed to be in the
!               consistent solver  unit system.  See Solving Units in the help system
                for more information.
```

At the end of this text you can insert the following APDL code that is contained in the file command_obj_solid_geom_01.txt and is included with this book.

```
 1   !-------------------------------------------------
 2   ! Units are m, kg
 3   !
 4   ! In the Details view, under "Input Arguments",
 5   ! please enter the following:
 6   !
 7   ! ARG1 = density
 8   ! ARG2 = speed of sound
 9   !-------------------------------------------------
10
11   !-------------------------------------------------
12   ! Change to acoustic element (purely acoustic)
13   !-------------------------------------------------
```

```
14   ! Standard acoustic element, no displacement DOF
15   ET,MATID,FLUID29
16   KEYOPT,MATID,2,1     ! 1=structure absent, 0=structure present
17   KEYOPT,MATID,3,1     ! 1=axi symmetric, 0=planar
18
19   !---------------------------------------------------,---
20   ! Define material properties
21   !------------------------------------------------------
22   MPDELE,ALL,MATID
23   MP,DENS,MATID,ARG1
24   MP,SONC,MATID,ARG2
25   MP,MU   ,MATID,1
```

- In the window labeled `Details of "Commands (APDL)"`, enter the value `1.21` in the cell next to ARG1, and `343` in the cell next to ARG2, which define the density and speed of sound of the fluid, respectively. Note that it would be possible to link these parameters for ARG1 and ARG2 back to the main `Parameter Set` in the `Project Schematic` window if desired, but will not be done for this simple example.

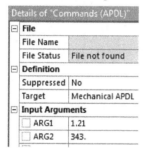

- Insert another command object for the `piston_body` and insert the following APDL code into the `Command` window. These commands are contained in the file `command_obj_solid_geom_02.txt` that is included with this book.

```
1    ! Attach these commands to the solid part under geometry
2    ! where the piston_body is the small piston region.
3
4    !------------------------------------------------------
5    ! Units are m, kg
6    !
7    ! In the Details view, under "Input Arguments",
8    ! please enter the following:
9    !
10   ! ARG1 = density
11   ! ARG2 = speed of sound
12   !------------------------------------------------------
13
14   !------------------------------------------------------
15   ! Change to acoustic element (with displacement DOFs)
16   !------------------------------------------------------
17   ! Standard acoustic element, with displacements DOF
18   ET,MATID,FLUID29
19   KEYOPT,MATID,2,0     ! 1=structure absent, 0=structure present
20   KEYOPT,MATID,3,1     ! 1=axi symmetric, 0=planar
21
22   !------------------------------------------------------
23   ! Define material properties
24   !------------------------------------------------------
```

```
25   MPDELE ,ALL ,MATID
26   MP ,DENS ,MATID ,ARG1
27   MP ,SONC ,MATID ,ARG2
28   MP ,MU   ,MATID ,1
```

- In the window labeled Details of "Commands (APDL)", for the piston_body, enter the value 1.21 in the cell next to ARG1, and 343 in the cell next to ARG2.

- The next step is to specify that the outer arc of the model has an infinite acoustic radiation boundary condition, so that outgoing acoustic waves are absorbed and not reflected. The following APDL code specifies that the line body will comprise FLUID129 infinite acoustic elements. In the Outline window, right-click on radiation_boundary and in the context menu that opens, left-click on Insert | Commands. In the Commands window that appears, insert the following APDL code. These commands are contained in the file command_obj_solid_geom_03.txt that is included with this book.

```
1    ! Attach these commands to the geometry
2    ! where the radiation_boundary for the outer infinite edge
3    ! on the circumference of the circle.
4
5    !----------------------------------------------------
6    ! Units are m, kg
7    !
8    ! In the Details view, under "Input Arguments",
9    ! please enter the following:
10   !
11   ! ARG1 = density
12   ! ARG2 = speed of sound
13   ! ARG3 = baffle_r: Radius of the baffle
14   !----------------------------------------------------
15
16   !----------------------------------------------------
17   ! Change to infinite acoustic elements
18   !----------------------------------------------------
19   ! Infinite element for the circumference
20   ET ,MATID ,FLUID129
21   KEYOPT ,MATID ,3 ,1    ! 1=axi symmetric , 0=planar
22
23   ! Infinite element for the circumference
24   R ,MATID ,ARG3 ,0 ,0
25
26   !----------------------------------------------------
27   ! define material properties
28   !----------------------------------------------------
29   MPDELE ,ALL ,MATID
30   MP ,DENS ,MATID ,ARG1
31   MP ,SONC ,MATID ,ARG2
32   MP ,MU   ,MATID ,1
```

- In the window labeled Details of "Commands (APDL)" enter the value 1.21 in the cell next to ARG1, 343 in the cell next to ARG2, and 1 in the cell next to ARG3.

Details of "Commands (APDL)"	
File	
File Name	
File Status	File not found
Definition	
Suppressed	No
Target	Mechanical APDL
Input Arguments	
☐ ARG1	1.21
☐ ARG2	343.
☑ ARG3	1.

The next steps involve setting up the size of the mesh for the model. Those that have prior experience using Mechanical APDL know that it is relatively easy to define mapped meshes for circular sector objects (using the AMAP command). However, when using ANSYS Workbench, the default mesh pattern for these circular-shaped objects is different.

- For interest, right-click on the Mesh branch in the Outline window and select Show | Mappable Faces. In the graphics window you will notice that only the 4-sided freefield_body is highlighted, and not the 3-sided piston_body.

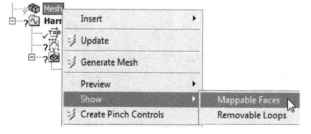

- Click on the Face selection filter in the toolbar, left-click on the area for the freefield_body, then right-click and select Insert | Sizing.

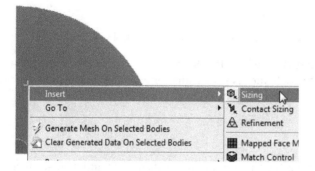

- In the window Details of "Face Sizing" - Sizing, click the box to the left of Element Size so that a letter P is shown to indicate it is a parameter.

Details of "Face Sizing" - Sizing	
⊟ **Scope**	
Scoping Method	Geometry Selecti‹
Geometry	1 Face
⊟ **Definition**	
Suppressed	No
Type	Element Size
P Element Size	Default
Behavior	Soft

- Use the Box Zoom tool to zoom onto the piston_body region to make it easier to select the edges.

- Click on the Edge selection filter, and then select the vertical line on the piston_body.

- Hold down the <Ctrl> key and select the horizontal edge of the piston_body. In the status bar at the bottom of the screen, you should see 2 Edges Selected: Length = 0.2m.

- Right-click with the mouse button and select Insert | Sizing.

- In the window for Details of "Edge Sizing" - Sizing, click in the box next to Element Size to define it as a parameter and a letter P will be shown in the box.

P Element Size	Default

- Define the Edge Sizing for the arc for the piston_body and also set it as a parameter.

- In the ANSYS Workbench Project Schematic window, double-click on the Parameter Set box. There will be three new entries for Face Sizing Element Size, Edge Sizing Element Size, and Edge Sizing 2 Element Size.

	A	B	C	D
Outline: No data				
	ID	Parameter Name	Value	Unit
1	ID	Parameter Name	Value	Unit
2	⊟ Input Parameters			
3	⊟ av Harmonic Response (A1)			
4	⏚ P1	piston_r	0.1	
5	⏚ P2	baffle_r	1	
6	⏚ P5	Face Sizing Element Size	0	m
7	⏚ P6	Edge Sizing Element Size	0	m
8	⏚ P7	Edge Sizing 2 Element Size	0	m
9	⏚ P3	EPW	12	
10	⏚ P4	2ka	50.265	
*	⏚ New input parameter	New name	New expression	

- Click in the cell C6 next to Face Sizing Element Size. In the window Properties of Outline C6: P5, click in the cell next to Expression with the entry 0 [m]. Type the following expression into the cell: 4*PI*P1/P4/P3*1 [m] remembering to include the square brackets around the m.

Expression	4*PI*P1/P4/P3*1 [m]

The element size has been defined in a convoluted manner for this example, so that it is compatible with the non-dimensional results presented in reference textbooks. Normally, one would just enter an appropriate numerical value for the element size.

The effective maximum analysis frequency is defined by the P4 parameter 2ka, which can be used to determine the wavelength of sound. As the number of elements per wavelength (EPW) has been defined, the size of each element can be calculated as follows. The number of elements per wavelength is the ratio of the wavelength λ divided by the element size ESIZE.

$$EPW = \frac{\lambda}{ESIZE}.$$ (8.30)

Hence the element size is

$$\text{ESIZE} = \frac{\lambda}{\text{EPW}}. \tag{8.31}$$

The minimum wavelength is calculated as

$$\lambda_{\min} = c_0/f_{\max}, \tag{8.32}$$

where $f_{\max}$ is the maximum frequency of the analysis, which is indirectly defined by the parameter $2ka$, where

$$2ka = \frac{2\omega a}{c_0} = \frac{4\pi f a}{c_0}. \tag{8.33}$$

By rearranging the terms for $2ka$, the wavelength is

$$\lambda = c_0/f = \frac{4\pi a}{(2ka)}, \tag{8.34}$$

which can be substituted into Equation (8.31) to give the element size

$$\text{ESIZE} = \left[\frac{4\pi a}{(2ka)}\right] \frac{1}{\text{EPW}}. \tag{8.35}$$

In terms of the `Input Parameter IDs`, the element size is

$$\text{ESIZE} = \left[\frac{4\pi \times \text{P1}}{\text{P4}}\right] \frac{1}{\text{P3}}, \tag{8.36}$$

and hence is entered into Workbench as the expression

$$4*\text{PI}*\text{P1}/\text{P4}/\text{P3}*1 \ [\text{m}]. \tag{8.37}$$

- Click in the cell to the right of `Edge Sizing Element Size` in the column labeled `Value` for parameter P6. In the window `Properties of Outline C7: P6`, click in the cell to the right of `Expression` and type P5. This will set the edge size to the same value as the face size defined for parameter P5.

Properties of Outline C7: P6		
	A	
1	Property	
2	⊟ General	
3	Expression	P5

- Repeat these steps to define the edge size for parameter P7 to be the same as P5.

- Click on the Save icon to save the project.

- Click on the Return to Project icon.

- Click on the Refresh Project icon.

- Double-click on row 5 Setup in the Project Schematic or open the Mechanical window.

- The next step is to change the options so that FLUID29 elements are used when the solid bodies are meshed. There is no method to directly specify which element type is used when using ANSYS Workbench. Instead, by changing various options the software will use the most appropriate element type. The FLUID29 elements do not have mid-side nodes, so by changing the options to "drop mid-side nodes," the software will use the FLUID29 elements when meshing. Left-click on the Mesh branch. In the window Details of "Mesh" click on the plus sign next to Advanced. Change the option for Element Midside Nodes to Dropped.

⊟ Advanced	
Shape Checking	Standard Mechanical
Element Midside Nodes	Program Controlled
Number of Retries	Program Controlled
Extra Retries For Assembly	Dropped
	Kept

- Right-click on the Mesh branch and select Generate Mesh. A reasonable mesh appears to have been generated. However, doing some hand calculations reveals the mesh size is inadequate.

- The learning outcome is to highlight the importance of confirming that the mesh size is adequate, rather than blindly proceeding with an analysis. Using the zoom tool, enlarge the area for the piston_body. It can be seen that there are 6 elements along each edge of the piston, which has a length of 0.1 m, hence the element size is $0.1/6 = 0.016$ m, whereas the required edge size to achieve EPW=12 is 0.002 m, and hence there should be about $0.1/0.002 = 50$ divisions along each edge.

- You could also visually inspect the element size by clicking on the Edge Sizing entry under the Mesh branch. You can see that there are dashed yellow lines, where the size of the dashes indicates the expected element size.

- To fix this problem, change the mesh sizing option from Soft to Hard for all three size definitions.

- Select the Face selection filter, left-click on Mesh, right-click and select Insert | Method. In the Geometry row, select the piston_body area in the Graphics window and then click the Apply button. Change the row labeled Method to MultiZone Quad/Tri. Change the row labeled Free Face Mesh Type to All Quad.

Details of "MultiZone Quad/Tri Method" - Method	
⊟ **Scope**	
Scoping Method	Geometry Selection
Geometry	1 Body
⊟ **Definition**	
Suppressed	No
Method	MultiZone Quad/Tri
Surface Mesh Method	Program Controlled
Element Midside Nodes	Use Global Setting
Free Face Mesh Type	All Quad

- Insert another Edge Sizing object for the 3 outside edges of the freefield_body, namely the vertical edge, outer arc, and the horizontal edge. Click in the cell to the left of Element Size to indicate that it is a parameter. Change the Behavior to Hard.

- Go to the Parameter Set and link the element size for the outer edges of the freefield_body to the main definition, by entering P5 in the cell for Expression for parameter P8, then click the Refresh Project icon.

Properties of Outline C9: P10		
	A	B
1	Property	Value
2	⊟ General	
3	Expression	P5
4	Description	
5	Error Message	
6	Expression Type	Derived
7	Usage	Input
8	Quantity Name	Length

- Return to the Mechanical module and in the menu bar click on `File | Save Project`.

- Right-click on the `Mesh` branch and select `Generate Mesh`. It will take a while to generate the mesh, so now would be a good time to have a break.

- After the mesh has been generated, zoom to the interfaces between `piston_body` and `freefield_body` and make sure that the mesh is continuous. Also check that the mesh between the `freefield_body` and the `radiation_boundary` is continuous. Another way to view the mesh is to click on the `Show Mesh` icon in the toolbar.

- Click on the `Mesh` and in the window `Details of "Mesh"`, click on plus sign next to the `Statistics` row. You will notice that the model has about 189,213 nodes. The exact number your model generates might vary slightly from this value.

The first set of results that will be reproduced is the pressure on the axis of the piston at a frequency corresponding to $2ka = 16\pi$, where the theoretical predictions are shown in Figure 8.14. We will define a parameter `f_max` for the maximum frequency of the analysis as

$$\text{f\_max} = [2ka] \times \frac{c_0}{4\pi a}. \tag{8.38}$$

- In the `Project Schematic` window, double-click on the box for the `Parameter Set`.

- Left-click in the cell labeled `New Name` and type `c_speed_sound` and then press the `<Tab>` key on the keyboard, which will cause the cursor to move to the next cell.

- In the cell for the value of `c_speed_sound`, type 343.

| 12 | ⚑ P9 | c_speed_sound | 343 | |

- Left-click in the cell labeled `New Name` and type `f_max` and then press the `<Tab>` key on the keyboard.

- In the window `Properties of Outline C12: P10`, click in the cell next to entry for `Expression` and enter the following formula

$$\text{P4*P9/(4*PI*P1)}. \tag{8.39}$$

The expression should evaluate to 13720 Hz. Unfortunately it is not possible to use the calculated maximum frequency as a varying parameter for the analysis frequency range in a harmonic analysis, and the value must be typed in manually.

- Click on the icon Refresh Project in the toolbar.

- Click on the icon Return to Project in the toolbar.

- Return to the Mechanical module, click on Analysis Settings. In the window for Details of "Analysis Settings", click in the cell for Range Maximum and type 13720.

- Change the Range Minimum to 13719.

- Change the Solution Intervals to 1.

- Change Solution Method to Full.

- Click on the plus sign next to Output Controls. Change Nodal Forces to Yes. In a later analysis we will attempt to use the results from nodal forces to calculate the equivalent nodal area. Change General Miscellaneous to Yes. This is needed to display the sound pressure level in decibels.

- Click on the plus sign next to Analysis Data Management and in the row for Save MAPDL db change to Yes. The Mechanical APDL database is needed when post-processing the results to plot the acoustic pressure field.

Details of "Analysis Settings"	
Options	
Range Minimum	13719 Hz
Range Maximum	13720 Hz
Solution Intervals	1
Solution Method	Full
Variational Technology	Program Controlled
Output Controls	
Stress	Yes
Strain	Yes
Nodal Forces	Yes
Calculate Reactions	Yes
General Miscellaneous	Yes
Damping Controls	
Analysis Data Management	
Solver Files Directory	C:\Users\cqhoward\Desk...
Future Analysis	None
Scratch Solver Files Directory	
Save MAPDL db	Yes
Delete Unneeded Files	Yes
Solver Units	Active System
Solver Unit System	mks

- Right-click on the Harmonic Response (A5) branch and select Insert | Commands. Right-click on this new object and left-click on Rename, or alternatively press the <F2> function key on the keyboard. Change the name of the object to Commands (APDL) PressureAxisA5.

- In the Commands window, enter the commands contained in the file command_obj_harmonic_A5_01.txt that is included with this book. This APDL code:
  - · turns on the Fluid-Structure-Interaction (FSI) flag for the elements that are on horizontal line representing the piston face,
  - · defines the harmonic displacement of the piston to be 1 micron in the Y-direction, and
  - · constrains the motion of the piston so that it cannot move in the X direction.

  When calculating the radiation impedance in a subsequent analysis, this APDL code will be replaced with a different block of code.

- In the window for Details of "Commands (APDL) PressureAxisA5", click in the cell next to ARG1 and type 0.1 for the piston radius.

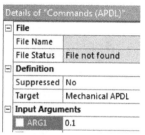

Details of "Commands (APDL)"	
⊟ **File**	
File Name	
File Status	File not found
⊟ **Definition**	
Suppressed	No
Target	Mechanical APDL
⊟ **Input Arguments**	
☐ ARG1	0.1

  Note that it is also possible to link ARG1 to the definition of the piston radius in the table of the Parameter Set, however it will not be done in this example.

- Right-click on Solution (A6) and select Insert | Commands. Right-click on this new object and left-click on Rename, or alternative press <F2> on the keyboard. Change the name of the object to Commands (APDL) PressureAxisA6.

- Click in the Command window and paste the code from the file command_obj_solution_A6_01.txt which is shown below.

```
1   !---------------------------------------
2   ! Export the pressure on axis of piston
3   !---------------------------------------
4
5   FINISH
6
7   ! Change the format of the output listings
8   ! to get rid of header information.
9   /HEADER,OFF,OFF,OFF,OFF,OFF,OFF
10  /PAGE,,,100000,1000
11
12  ! Enter the post-processor
13  /POST1
14
15  !---------------------------------------
16  ! Export the REAL pressure
```

```
17   !-----------------------------------------
18   ! Select the first set of results
19   ! and look at the REAL values
20   SET ,      1, 1      ,       ,REAL
21   NSEL ,S,LOC ,X ,0
22   /OUTPUT ,axis ,p_re.txt
23   PRNSOL ,PRES
24   /OUTPUT
25
26   !-----------------------------------------
27   ! Export the IMAGINARY pressure
28   !-----------------------------------------
29   ! Select the first set of results
30   ! and look at the IMAGINARY values
31   SET ,      1, 1      ,       ,IMAG
32   /OUTPUT ,axis ,p_im.txt
33   PRNSOL ,PRES
34   /OUTPUT
35
36   !-----------------------------------------
37   ! write the nodal coordinates
38   !-----------------------------------------
39   FINISH
40   /PREP7
41   NWRITE ,axis ,node.txt
42   FINISH
43   /POST1
44
45   ALLS                ! Select all nodes and elements
46
47   FINISH              ! Exit the post processor
48   !--------------------------
49   ! End of script
50   !--------------------------
```

This APDL code will do the following:

- Export a file called `axis.p_re.txt` that contains the real part of the complex acoustic pressure at the nodes along the vertical axis of the piston at $x = 0$.

- Export a file called `axis.p_im.txt` that contains the imaginary part of the complex acoustic pressure at the nodes along $x = 0$.

- Export a file called `axis.node.txt` that contains the nodal coordinates along the vertical axis at $x = 0$.

These files can be found in a sub-directory from where the project file is stored, such as `xxxx\piston_baffle_axisym_files\dp0\SYS\MECH`.

- Left-click on `Solution (A6)` and the ACT Acoustics extension toolbar will appear. In the ACT Acoustics toolbar, left-click on `Results | Acoustic Pressure`.

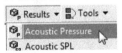

- In the `Details of "Acoustic Pressure"` window, left-click in the cell to the right of Geometry that has written All Bodies. Make sure the Face selection filter is active and then left-click on the area for the piston_body, and while the mouse button is still held down, move the mouse cursor onto the area for the freefield_body. The status bar at the bottom of the window should say 2 Faces Selected: Surface Area (approx.) = 0.78273m². Click the Apply button and the cell next to Geometry should say 2 Faces.

- The other results that are interesting to view are the sound pressure level in decibels, and the absolute value of pressure. For 3D analyses it is possible to use the ACT Acoustics extension by inserting Results | Acoustic SPL, however this does not work for a 2D analysis. Instead, click on the icon User Defined Result.

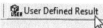

- In the Outline window, click on Solution A6 | User Defined Result. In the window Details of "User Defined Result", click in the cell next to Expression and type NMISC4. This cryptic code is a reference to a miscellaneous result output from the analysis. A list of the "element output definitions" for the FLUID29 element can be found in the ANSYS help manual [34, FLUID29, Table 29.2]. The expression NMISC4 is the sound pressure level of the root-mean-squared (RMS) pressure in decibels referenced to 20 μPa.

Details of "User Defined Result"	
⊟ **Scope**	
Scoping Method	Geometry Selection
Geometry	All Bodies
⊟ **Definition**	
Type	User Defined Result
Expression	= NMISC4

Note: the NMISCx and SMISCx results are not displayed in the Worksheet and the results can only be accessed by typing the appropriate expression into the cell next to Expression.

The learning outcome is to show how to extract specialized results from an analysis and to make use of the ANSYS Help Manual, which describes the details of acoustic elements.

- In the Outline window, right-click on the User Defined Result listed under Solution (A6) and in the context menu that opens, left-click on Rename Based on Definition.

- In the menu bar, click on File | Save Project.

That completes the pre-processing, setup of the analysis settings, and an initial request for results. The next stage is to solve the model, which involves one step.

- Click the `Solve` icon.

The model should solve after a while and there will be a green tick next to the `Solution (A6)` branch. The next stage is to examine the results and do further post-processing.

- Once the results have been calculated, click on the `Acoustic Pressure` branch under `Solution (A6)` to show the acoustic pressure field.

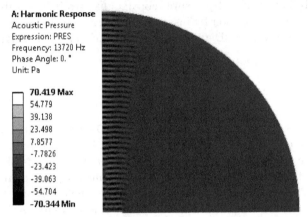

- Another method to visualize the results is to animate the contour plot. Click on the red triangle icon in the `Animation` toolbar in the `Graph` window to see a movie of the pressure wave radiating from the piston face.

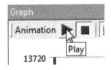

- As we want to plot the acoustic pressure versus axial distance, we need to know the locations of the nodes. In the `Outline` window, left-click on `Solution (A6)`. Click on the `Worksheet` icon in the toolbar which will reveal a table of `User Defined Result Expressions`.

- If the desired results are listed in this table, one would right-click on the row of interest and select `Create User Defined Result`. The result that we want to extract is the coordinate of the nodes along the Y-axis. Looking at the table we can see that in the column for `Expression`, this is `LOCY`.

LOC	Nodal	Scalar	X	LOCX	Displacement
LOC	Nodal	Scalar	Y	LOCY	Displacement
LOC	Nodal	Scalar	Z	LOCZ	Displacement
LOC_DEF	Nodal	Scalar	X	LOC_DEFX	Displacement
LOC_DEF	Nodal	Scalar	Y	LOC_DEFY	Displacement
LOC_DEF	Nodal	Scalar	Z	LOC_DEFZ	Displacement

- Right-click on this row for LOCY, and select Create User Defined Result.

- Click on the LOCY branch listed under the Solution (A6) tree. Click in the cell next to Geometry. Make sure that the Edge selection filter is active, then select the two vertical edges along the Y-axis. Click the Apply button.

- Insert another User Defined Result. In the window for Details of "User Defined Result 2", select the 2 vertical edges along the Y-axis for the Geometry. In the row for Expression type in PRES.

- Right-click on this entry for User Defined Result 2 and select Rename. Type in the name PRES_real.

  Note: The ANSYS help manual describes the mathematical operations that are supported in ANSYS, SAS IP, Inc. [35].

  Warning: The abs command effectively removes the sign from a numerical value. It does *not* calculate the magnitude of a complex number.

- Insert another User Defined Result. In the window for Details of "User Defined Result 3", select the 2 vertical edges along the Y-axis for the Geometry. In the row for Expression type in PRES. In the cell next to Phase Angle, type -90, which will calculate the imaginary part of the complex pressure.

- Right-click on this entry for User Defined Result 3 and select Rename. Type in the name PRES_imag.

- Right-click on one of the solution entries, such as LOCY, and select Evaluate All Results.

- Once the results have been calculated, click on the branch for NMISC4, which corresponds to the sound pressure level in decibels.

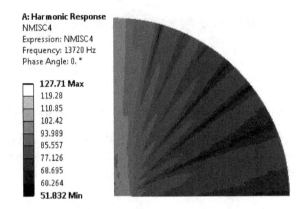

A: Harmonic Response
NMISC4
Expression: NMISC4
Frequency: 13720 Hz
Phase Angle: 0. °

127.71 Max
119.28
110.85
102.42
93.989
85.557
77.126
68.695
60.264
51.832 Min

- Right-click on PRES_real and select Export.

- Type in an appropriate filename such as axial_p_real.txt.

- Repeat these steps to export the imaginary pressure PRES_imag and name the file axial_p_imag.txt, and the nodal coordinates LOCY and name the file axial_node_loc.txt.

The results in the text files can then be processed using MATLAB or a spreadsheet to

- calculate the magnitude of the complex acoustic pressure and

- plot the magnitude of the acoustic pressure versus axial distance.

Pressure on the Axis of the Piston

The APDL code that was entered under Solution (A6) branch exports the results of the real and imaginary parts of the pressure along the axis of the piston, and also the nodal coordinates. The finite element model was created with 12 elements per wavelength (EPW). These results from the ANSYS analysis are compared with the theoretical predictions using the MATLAB model as shown in Figure 8.20. The results show that the pressure close to the piston face predicted using ANSYS is slightly higher than the theoretical values. These results can be improved by increasing the number of elements close to the piston face.

The model was then re-meshed at around 20 EPW (element size approximately 1.25 mm) and with a biased mesh around the piston. This was achieved

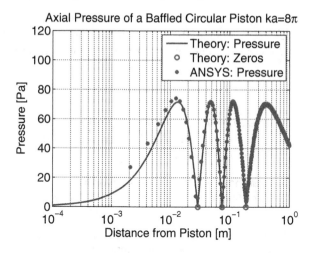

FIGURE 8.20

Pressure on axis of a baffled circular piston in an infinite plane baffle at an excitation frequency of $ka = 8\pi$ predicted theoretically using MATLAB and ANSYS, where the model has 12 EPW.

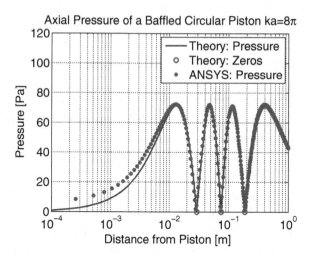

FIGURE 8.21

Pressure on axis of a baffled circular piston in an infinite plane baffle at an excitation frequency of $ka = 8\pi$ predicted theoretically using MATLAB and ANSYS, where the model had 20 EPW.

by using MultiZone Quad/Tri Method for the quarter circle around the piston, specifying the number of divisions as 100 on the three sides of a quarter circle around the piston and biasing the divisions toward the origin. This resulted in a mesh with 200,495 nodes and 199,613 elements. The sound pressure on the axis of the piston with increasing distance is shown in Figure 8.21.

As can be seen by comparing Figures 8.20 and 8.21, the pressure close to the face of the piston predicted using ANSYS approaches the theoretical values as the mesh density is increased. The far-field pressure (and hence the radiated power) are the same.

Acoustic Impedance

The next set of results that will be calculated is the acoustic impedance of the piston. This will involve replacing some of the APDL code and conducting the harmonic analysis over a range of frequencies.

- Start by right-clicking on Commands (APDL) PressureAxisA6 under the Solution (A6) branch and select Suppress. This will add a small blue cross next to the icon.

- Right-click on Solution (A6) and select Insert | Commands to add a new command object that will have the label Commands (APDL) 2.

- Right-click on the branch Commands (APDL) 2 and select Rename. Type the new name for the branch as Commands (APDL) ImpedanceA6.

- Copy the contents of the file command_obj_solution_A6_02.txt into this command object. The file is included with this book.

- In the window Details of "Commands (APDL) ImpedanceA6", click in the cell next to ARG1, ARG2, ARG3, ARG4 and type 1.21, 343, 0.1, and 1, respectively.

- Under the tree branch for Harmonic Response (A5), Suppress the Commands (APDL) PressureAxisA5.

- Right-click on Harmonic Response (A5) and select Insert | Commands.

- Right-click on the new command object and select Rename. Type the new name for the branch as Commands (APDL) ImpedanceA5 .

- Copy the contents of the file command_obj_harmonic_A5_02.txt into this command object. The file is included with this book.

- In the window Details of "Command (APDL) ImpedanceA5", click in the cell next to ARG1 and type 0.1 for the piston radius.

- Next we want to change the analysis frequency range to the same as in Figure 8.15, which has a maximum value of $2ka = 14$. Hence the maximum frequency for this analysis will be

$$f_{\max} = \frac{14c_0}{2a \times 2\pi} = 3821 \text{ Hz}. \tag{8.40}$$

Alternatively, you could have changed the value of P4 in the Parameter Set to 14 to calculate that the Parameter Name f_max = 3821 Hz.

Click on the Analysis Settings branch. Change the Range Minimum value to 0 and the Range Maximum value to 3821. Change Solution Intervals to 50.

- Right-click on the Solution (A6) and select Insert | Frequency Response | Deformation. In the Details of "Frequency Response" window, click in the cell next to Geometry. Make sure the Vertex selection filter is active and select the vertex at origin of the model, then click the Apply button. In the row for Orientation, change it to Y Axis. Change the row for Display to Real and Imaginary.

That completes the setup of the analysis. The next stage is to solve the model. Click on the Solve icon. It will take a while to solve the model, so now would be a good time for a break.

After the analysis has completed, there will be a green tick next to the Solution (A6) branch. The next steps describe the post-processing of the results to calculate the impedance of the piston.

- The analysis should have created an object under Commands (APDL) ImpedanceA6 with the label Post Output. Click on this icon and the Worksheet window will show a graph of the impedance of the piston, which should resemble Figure 8.15.

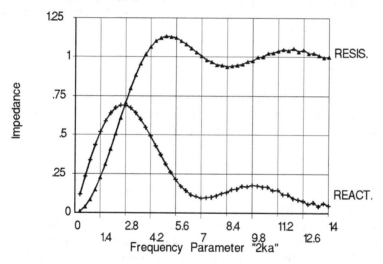

- Right-click on the entry for Frequency Response and select Export. Type in an appropriate filename such as piston_uy.txt.

The file piston_uy.txt contains 5 columns: the analysis frequency, magnitude, phase, real, and imaginary parts of the displacement of the node at the origin. This data can be processed using MATLAB or a spreadsheet to calculate the real and imaginary parts of the impedance.

Figure 8.22 shows the impedance calculated using ANSYS and the theoretical predictions using the MATLAB code are identical.

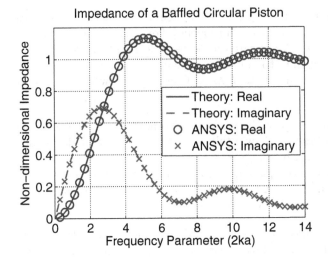

FIGURE 8.22

Real and imaginary normalized impedance of an oscillating piston installed in a rigid plane baffle, calculated theoretically using MATLAB and ANSYS, where the model had EPW=12.

Radiation Pattern

The next result that will be calculated is the radiation or beam pattern from the circular piston in an infinite plane baffle, at a frequency of $ka = 10$, which in this example corresponds to a frequency of

$$f = \frac{10c_0}{a \times 2\pi} = 5459 \text{ Hz}. \tag{8.41}$$

The intention of this analysis is to reproduce the theoretical results shown in Figure 8.16 using ANSYS.

The radiation pattern of the piston is determined by plotting the radiated acoustic pressure normalized to the maximum response that occurs on the axis of the piston. The directivity factor D_θ is calculated as the ratio of the intensity I_θ at an angle from the source, divided by the average intensity $<I>$ as [47, Eq. (5.123)]

$$D_\theta = I_\theta / <I> . \tag{8.42}$$

The directivity index DI is calculated as [47, Eq. (5.124)]

$$DI = 10 \log_{10} D_\theta . \tag{8.43}$$

The results are generated at a single frequency. Hence, it does not matter whether the piston is driven with a harmonic force or a harmonic displacement. For this example, we will drive the piston with a harmonic force with peak amplitude of 1×10^{-3}N.

- In the branch Harmonic Response (A5), right-click on Commands (APDL) PressureAxisA5 and select Duplicate.

- Rename the newly generated command object and give a name such as Commands (APDL) DirectivityA5.

- Right-click on Commands (APDL) DirectivityA5 and select Unsuppress. There should be a green tick next to this icon.

- Suppress the command object Commands (APDL) ImpedanceA5.

- Suppress the result Frequency Response.

- Suppress the result PRES_real, which calculated the real part of the acoustic pressure on the axis of the piston.

- Suppress the result PRES_imag, which calculated the imaginary part of the acoustic pressure on the axis of the piston.

- Suppress the result LOCY, which calculated the y-axis coordinate of the nodes on the axis of the piston.

- Four new User Defined Result objects are required for the nodes on the free-field boundary (the edge for the outer arc):
    - x coordinate of node
    - y coordinate of node
    - real part of the complex pressure
    - imaginary part of the complex pressure

  Although we want the sound pressure level in decibels at the nodes, this result is not available for the FLUID129 elements. Also note that the sound pressure level that is available for many of the acoustic elements (available with the NMISC4 expression), is based on the element result, not the nodal pressure result. Hence it is necessary to extract the real and imaginary nodal pressures and convert them to sound pressure level in decibels.

- Create a User Defined Result, in the Expression field type LOCX, for the x coordinate of the node. In the Geometry row, select the outer edge for the free-field boundary. Rename this User Defined Result to LOCX-boundary.

- Create a similar User Defined Result for LOCY, for the y coordinate of the node on the free-field boundary. Rename this User Defined Result to LOCY-boundary.

- Create a User Defined Result, in the Expression field type, PRES, for the *real* pressure at the node. In the Geometry row, select the edge for the free-field boundary. Rename this User Defined Result to P_real-boundary.

- Create a User Defined Result, in the Expression field type, PRES. In the row for Phase Angle type -90, to calculate the *imaginary* pressure at the node. In the Geometry row, select the edge for the free-field boundary. Rename this User Defined Result to P_imag-boundary.

- Click on Analysis Settings. In the window for Details of "Analysis Settings", click in the cell for Range Maximum and type 5459 as calculated in Equation (8.41). Change the Range Minimum to 5458. Change the Solution Intervals to 1.

- In the menu bar click on File | Save Project.

Now solve the model by clicking on the Solve icon. The model should not take too long to solve as the results at only 1 frequency have been requested.

Once the model has been solved, there will be a green tick next to the Solution (A6) branch. The post-processing of the results involve exporting the nodal coordinates, real and imaginary pressures.

- Right-click on LOCX-boundary and select Export. Type an appropriate filename such as LOCX-boundary.txt.

- Repeat these steps to export the data for LOCY-boundary, P_real-boundary, P_imag-boundary.

The data can be imported into a spreadsheet or MATLAB. The relative angle of the nodes on the boundary can be calculated using the MATLAB arc-tangent function atan2. The sound pressure level can be calculated by combining the real and imaginary parts of the acoustic pressure by using the expression

$$L_p = 20 \log_{10} \left[\frac{\sqrt{P_{\text{real}}^2 + P_{\text{imag}}^2}}{\sqrt{2} \times (20 \times 10^{-6})} \right] \text{ dB re } 20 \ \mu\text{Pa.} \qquad (8.44)$$

where the $\sqrt{2}$ is necessary to convert from a peak to RMS value, and 20×10^{-6} Pa is the reference pressure. The relative sound pressure level is the difference between the sound pressure level on the axis of the piston to the sound pressure level at a particular angle. The polar radiation pattern can then be plotted using the dirplot function in MATLAB that is included with this book.

Figure 8.23 shows the theoretical polar radiation pattern of the piston at a frequency of $ka = 10$ (solid line) calculated using Equation (8.23), pressure zeros at 22.5° and 44.6° (dashed lines) calculated by solving for the roots of Equation (8.24), and using ANSYS (× markers). The results show that the predictions using ANSYS are identical to theory.

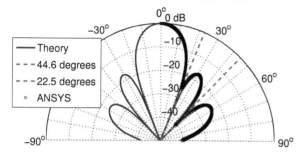

Beam Pattern of a Baffled Circular Piston

FIGURE 8.23
Radiation pattern of an oscillating piston installed in a rigid plane baffle at a frequency corresponding to $ka = 10$, calculated using MATLAB and ANSYS.

Radiated Sound Power

The next result that will be calculated is the sound power radiated from the piston. This will involve determining the effective area of each node on the boundary of the acoustic domain.

The sound power can be calculated by integrating the sound intensity over the surface of an imaginary hemisphere, which in this case is formed by the FLUID129 acoustic absorbing elements. The trap here is that the model comprises *line* elements and the area of the *hemisphere* is to be calculated. The effective area of each node can be calculated using three methods:

1. Exporting the coordinates of each node for the FLUID129 elements, calculating the effective nodal lengths, using the first theorem of Pappus [148] for the area of a body of revolution to calculate the effective nodal area.

2. Swapping the FLUID129 elements for SHELL61 elements, and using the APDL command ARNODE.

3. Swapping the FLUID129 elements for SHELL61 elements, applying a 1 Pascal load to each node, fixing the displacement of the nodes, conducting a static stress analysis, then exporting the reaction forces at the nodes to calculate the effective nodal area.

For this example, the first method will be used. The sound intensity at each node is calculated as

$$I_n = \frac{P_n^2}{2\rho_0 c_0}, \tag{8.45}$$

where P_n is the amplitude of the nodal pressure, ρ_0 is the density of the fluid, and c_0 is the speed of sound of the fluid. The sound power is calculated as

$$W = \sum_{n=1}^{N_a} I_n A_n \,, \tag{8.46}$$

where N_a is the number of nodes associated with the boundary surface, and A_n is the nodal area. The sound power level can be written in decibels as

$$L_w = 10 \log_{10} \left(\frac{W}{10^{-12}} \right) \text{dB re } 10^{-12} \text{ W.} \tag{8.47}$$

ANSYS Workbench Instructions

- Under the branch for Harmonic Response (A5), right-click on Commands (APDL) DirectivityA5 and select Suppress.

- Right-click on Commands (APDL) DirectivityA5 and select Duplicate. Rename the new entry to Commands (APDL) PowerA5. Right-click on this object and select Unsuppress and make sure there is an icon with a green tick next to it.

- Right-click on Solution (A6) and select Insert | Commands. Rename this new command object to Commands (APDL) PowerA6. Insert the APDL code from the file command_obj_solution_A6_03.txt into this command object. This APDL code will select the FLUID129 elements and nodes on the boundary of the acoustic domain, and export coordinates of the nodes as well as the real and imaginary nodal pressures.

- In the ANSYS Workbench Project Schematic window, double-click on the Parameter Set cell.

- Change the value for P4 2ka to 14, which should evaluate P10 f_max 3821.3, and hence the maximum analysis frequency will be 3821 Hz. Notice that the required element size for EPW = 12 is now 7 mm.

- Click on the Refresh Project icon and then click on the Return to Project icon.

- Return to the Mechanical window.

- Click on the Analysis Settings icon and change the analysis frequency range to 0 to 3821 Hz, using 20 Solution Intervals.

- In the menu bar, click on File | Save Project.

- Right-click on Mesh and select Generate Mesh.

- Once the meshing is completed, inspect the mesh by zooming on different parts of the model to make sure that the entire model was meshed, there is no significant distortion, that a predominantly quadrilateral mesh was used, and the size of the elements is as expected.

- In the menu bar, click on File | Save Project, again. Note that it is recommended to save the project before meshing as a separate program (thread) is used that might crash. Similarly, it is recommended to save the project before solving.

- Click on the Solve icon.

The results from the analysis are in the directory .\piston_baffle_axisym_files\dp0\SYS\MECH.

Results

The MATLAB code power_freefield_2Dhemisphere.m included with this book can be used to post-process the results exported from ANSYS for the nodal pressures, node and element data, and material properties, to calculate the sound power radiated into the free-field.

The results calculated using ANSYS can be compared with theoretical predictions by using the following MATLAB commands:

```
1  ansys=power_freefield_2Dhemisphere('fluid129');
2  baffled_piston;
3  hold on
4  p2=plot(2*(2*pi*ansys.f)/343*0.1,10*log10(ansys.power/1e-12),'ro');
```

Figure 8.24 shows the sound power radiated from the piston when it is oscillating with a peak displacement of 1×10^{-6} m calculated using the theory in Equation (8.22) and using ANSYS. The results show that they are identical, which is to be expected.

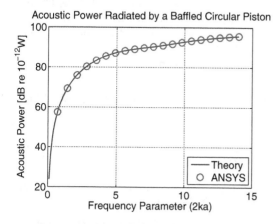

FIGURE 8.24

Sound power radiated from a piston with radius $a = 0.1$ oscillating with a peak displacement 1×10^{-6} m calculated theoretically using MATLAB and numerically using ANSYS.

8.4.5   ANSYS Mechanical APDL

This section describes how to create the same finite element model of the circular piston in an infinite plane baffle using ANSYS Mechanical APDL code. The model also makes use of 2D axi-symmetric FLUID29 acoustic elements.

The ANSYS Mechanical APDL code baffled_piston.inp included with this book can be used to calculate the mechanical impedance of a baffled circular piston. The code will plot the mechanical (radiation) impedance versus frequency, and will generate a figure similar to Figure 7.5.2 in Kinsler et al. [102, p. 187].

Figure 8.25 shows the mechanical impedance that was calculated and plotted using ANSYS Mechanical APDL. These results were exported to a text file and compared with the theoretical predictions, as shown in Figure 8.26. The figure shows that results predicted using ANSYS Mechanical APDL are similar to the theoretical predictions, which is to be expected.

The ANSYS Mechanical APDL code p_vs_d.inp included with this book can be used to calculate and plot the on-axis sound pressure level versus distance of a baffled circular piston, similar to Figure 7.4.2 in Kinsler et al. [102, p. 181], only the y-axis of the graph is the absolute value of pressure, rather than non-dimensionalized pressure as in Kinsler et al.

Note that the mathematical operations that are valid in ANSYS Mechanical APDL are listed in the ANSYS help manual under the *SET command [36]. Appendix D.2.1.3 contains further details about mathematical operations using APDL commands.

The number of elements per wavelength along lines can be calculated by typing LLIS at the command prompt, which will list all lines, their lengths, and number of divisions NDIV. Copy the output listing from this command into a spreadsheet or MATLAB, and calculate the element size along the mesh lines as

$$\mathtt{esize}_{\text{(meshed lines)}} = \frac{\mathtt{LENGTH}}{\mathtt{NDIV}}, \tag{8.48}$$

and then calculate the elements per wavelength (epw) along the meshed lines as

$$\mathtt{epw}_{\text{meshed lines}} = \frac{\lambda_{\min}}{\mathtt{esize}_{\text{(meshed lines)}}} \tag{8.49}$$

$$= \frac{c}{f_{\max}} \times \frac{1}{\mathtt{esize}_{\text{(meshed lines)}}}. \tag{8.50}$$

For this meshed model the element-per-wavelengths along the lines are listed in Table 8.3, and shows that they are close to EPW= 12, which was the desired mesh density.

This analysis was conducted at a frequency of 13,760 Hz. It is sometimes incorrectly suggested that finite element analysis cannot be used for high-frequency analyses. However, this analysis has shown that provided the model can be meshed with an adequate number of elements per wavelength, the

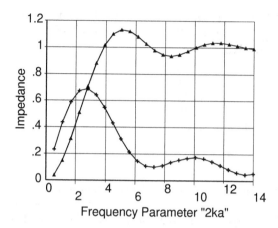

FIGURE 8.25
Real (triangle markers) and imaginary (cross markers) non-dimensional mechanical impedance of a circular piston installed in an infinite plane baffle calculated using ANSYS Mechanical APDL.

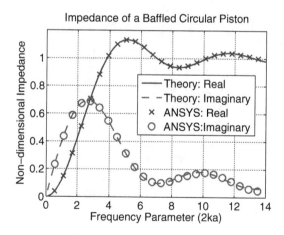

FIGURE 8.26
Real and imaginary normalized mechanical impedance of a circular piston installed in an infinite plane baffle calculated theoretically using MATLAB and numerically using ANSYS Mechanical APDL.

TABLE 8.3
Elements Per Wavelength (EPW) for the Meshed Lines for the Baffled Circular
Piston Example Using ANSYS Mechanical APDL are Close to EPW = 12

Line No.	EPW
1	11.96
4	12.06
7	11.97
8	11.97
9	11.97
10	11.97

results are accurate at "high frequencies" above 10 kHz. However, for 3D
models with large dimensions, the number of nodes and elements may become
large, and the computational resources required to solve these large models
may be demanding.

Figure 8.27 shows the nodal pressure along the axis of the circular piston,
oscillating with a peak displacement of 1 micron, calculated theoretically using Equation (8.16) in MATLAB and numerically using ANSYS Mechanical
APDL, at EPW = 12 and EPW = 20. The figure shows there is good correlation between the theoretical results and ANSYS Mechanical APDL.

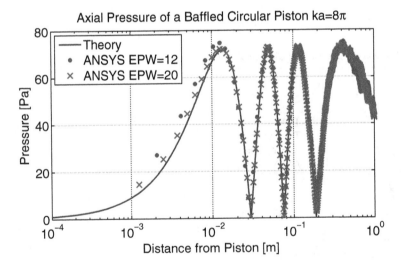

FIGURE 8.27
Pressure on the axis of a circular piston in an infinite plane baffle oscillating
with a peak displacement of 1 micron, calculated theoretically using MATLAB
and numerically using ANSYS Mechanical APDL, at EPW = 12 and EPW
= 20.

8.5 Scattering

The ANSYS software can be used to calculate the acoustic field scattered by an object due to an incident sound field. Some examples of where this type of analysis might be used are for sonar target strength estimates and diffraction of sound around an object such as an acoustic barrier.

Figure 8.28 shows a schematic of the typical arrangement of a finite element model used to investigate the acoustic scattering of an incident acoustic wave due to the presence of an object. The acoustic domain is bounded by a region of Perfectly Matched Layer (PML) elements that act to absorb outgoing acoustic waves. The object within the acoustic domain that causes the scattering of the incident acoustic wave field can be either rigid, where it is modeled by the absence of nodes and elements, or elastic, where it is modeled with structural elements. The acoustic source can be either external or internal to the finite element mesh.

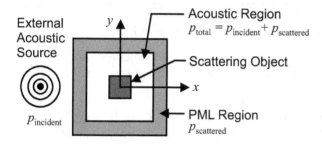

FIGURE 8.28
Configuration of a finite element model used to investigate acoustic scattering with an external acoustic source.

An analysis can return results of the total or scattered acoustic pressure fields, which is controlled by the APDL command HFSCAT, or in the ACT Acoustics extension object Excitation | Wave Sources in the row Pure Scattering Options. The value for this setting changes depending on the location where the incident external acoustic excitation is applied.

- When total sound pressure is calculated (APDL command HFSCAT,TOTAL), the incident acoustic excitation is applied on the exterior surface of the finite element model;

- When scattered pressure is calculated (APDL command HFSCAT,SCAT), the incident acoustic excitation is applied on the surface that interfaces the PML region and the regular (non-PML) acoustic elements.

- When scattering formulation is turned off (APDL command HFSCAT,OFF), which is the default setting, the incident acoustic excitation is applied on the surface that interfaces the PML region and the regular (non-PML) acoustic elements.

ANSYS has a useful feature called an `Equivalent Source Surface` (see Section 2.8.4.8), where acoustic results such as sound pressure level, sound power level, acoustic directivity, and others can be calculated beyond the finite element model.

In the following section, an example is described for the acoustic scattering of an incident plane wave striking a rigid, infinitely long cylinder.

8.6   Example: Scattering from a Cylinder

8.6.1   Learning Outcomes

The learning outcomes from this example when using ANSYS Workbench are:

- how to apply PML elements in a finite element model;

- how to define and evaluate the scattered pressure results are using the `Pure Scattering Options` in the ACT Acoustics extension object `Excitation | Wave Sources`;

- how to set up a cylindrical coordinate system that can be used to request acoustic particle velocity results along a radial direction, as opposed to the more commonly used Cartesian coordinate system;

- how to verify that the infinite acoustic domain has been created correctly by removing scattering objects from the model and confirming that the magnitude of the incident acoustic excitation source is correct and propagating as expected in a qualitative sense.

8.6.2   Theory

This section describes the theory to predict the sound field around an infinitely long rigid cylinder due to an impinging plane wave, as shown in Figure 8.29. It is assumed that the plane wave is incident normal to the axis of the cylinder and travels from left to right in Figure 8.29—the plane wave originates from the $-x$ axis ($\phi = 180°$) and propagates along the $+x$ axis. The cylinder has radius a and the pressure is measured at a location described in cylindrical coordinates at (ϕ, r).

The theory described here is from Morse and Ingard [117, Chapter 8, p. 400]. The theory will be implemented using MATLAB code that can be used to reproduce directivity plots shown in Faran [96, Figs. 9 and 13].

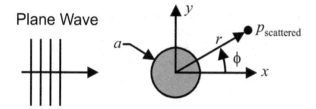

FIGURE 8.29
Schematic of an infinitely long rigid cylinder of radius a with an impinging plane wave normal to the axis of the cylinder.

The total sound field is a combination of the incident and scattered pressure fields. For an incident plane wave traveling normal to the axis of the cylinder the pressure is

$$p_p(k, x) = P_0 e^{ik(x - c_0 t)}, \tag{8.51}$$

where the subscript p represents plane wave, P_0 is the pressure amplitude of the plane wave, k is the wavenumber, x is the coordinate along the x-axis, c_0 is the speed of sound, and t is time. This can be expressed in terms of cylindrical waves as [117, Eq. (8.1.1)]

$$p_p(k, r) = P_0 \left[J_0(kr) + 2 \sum_{m=1}^{\infty} i^m \cos(m\phi) J_m(kr) \right] e^{-2\pi i \omega t}, \tag{8.52}$$

where J_m is the Bessel function of the first kind of order m, r is the radial distance from the center of the cylinder to the measurement location, and ω is the frequency in radians per second.

The radial velocity of this incident plane wave is

$$u_{pr}(k, r) = \frac{P_0}{\rho_0 c_0} \Big[i J_1(kr)$$

$$+ \sum_{m=1}^{\infty} i^{m+1} \left[J_{m+1}(kr) - J_{m-1}(kr) \right] \cos(m\phi) \Big] e^{-2\pi i \omega t}, \tag{8.53}$$

where the subscript r indicates the velocity in the radial direction, and ρ_0 is the density of the acoustic medium.

The scattered pressure from an incident plane wave striking the cylinder is

$$p_s(k, r) = \sum_{m=0}^{\infty} A_m \cos(m\phi) \left[J_m(kr) + i N_m(kr) \right] e^{-2\pi i \omega t}, \tag{8.54}$$

where the subscript s indicates the scattered pressure, and $N_m(kr)$ is the

Neumann function, which is the same as the Bessel function of the second kind of order m. The expression inside the square brackets $[J_m(kr)+iN_m(kr)]$ is the Hankel function of the first kind $H_m^1(kr)$ [1, p. 358], [116, p. 1373]. Note that there is an inconsistency in the equation in Morse and Ingard [117, Chapter 8, p. 401] which is discussed in detail in Appendix D.1.2. The terms in Equation (8.54) are [117, Eq. (8.1.2)]

$$A_m = -\epsilon_m P_0 i^{m+1} e^{-i\gamma m} \sin\gamma_m \tag{8.55}$$

$$\epsilon_m = \begin{cases} 1 & \text{for } m = 0, \\ 2 & \text{for } m > 1 \end{cases} \tag{8.56}$$

$$P_0 = \sqrt{\rho_0 c_0 I_p} \tag{8.57}$$

$$\tan\gamma_0 = -\left[\frac{J_1(ka)}{N_1(ka)}\right] \tag{8.58}$$

$$\tan\gamma_m = -\left[\frac{J_{m-1}(ka) - J_{m+1}(ka)}{N_{m-1}(ka) - N_{m+1}(ka)}\right], \tag{8.59}$$

where I_p is the sound intensity of the plane wave, and ϵ_m is called the Neumann factor [96]. The scattered pressure can be converted to a sound pressure level in decibels as

$$L_s = 20\log_{10}\left[\frac{|p_s|}{\sqrt{2} \times (20 \times 10^{-6})}\right] \text{ dB re } 20 \ \mu\text{Pa.} \tag{8.60}$$

The scattered acoustic particle velocity in the *radial* direction is

$$\begin{aligned}
u_{sr} = \frac{1}{\rho_0 c_0}\Big\{ &iA_0[J_1(kr) + iN_1(kr)] \\
&+ \frac{i}{2}\sum_{m=1}^{\infty} A_m\cos(m\phi)[J_{m+1}(kr) - J_{m-1}(kr) \\
&+ iN_{m+1}(kr) - iN_{m-1}(kr)]\Big\}e^{-2\pi i\omega t} \ .
\end{aligned} \tag{8.61}$$

The scattered (real) intensity in the radial direction can be calculated as [47, Eq. 1.72, p. 35]

$$I_{sr} = \frac{1}{2}\text{Re}(p_s u_{sr}^*), \tag{8.62}$$

where u_{sr}^* is the complex conjugate of the scattered radial acoustic particle velocity.

At *large* distances from the cylinder ($kr \gg 1$), an asymptotic expressions for the scattered pressure can be derived as [117, p. 402]

$$p_s = -\sqrt{\frac{2a\rho_0 c_0 I_p}{\pi r}}\ \psi_s(\phi)e^{ik(r-ct)}, \tag{8.63}$$

where

$$\psi_s(\phi) = \frac{1}{\sqrt{ka}} \sum_{m=0}^{\infty} \epsilon_m \sin(\gamma_m) e^{-i\gamma_m} \cos(m\phi) . \qquad (8.64)$$

The corresponding scattered radial acoustic particle velocity is

$$u_{sr} = \frac{p_s}{\rho_0 c_0} . \qquad (8.65)$$

The sound intensity of the scattered part of the acoustic field at a point (r, ϕ) at a large distance from the cylinder $(kr \gg 1)$ is [117, [Eq. (8.1.3), p. 402]

$$I_s \approx \frac{I_p a}{\pi r} |\psi_s(\phi)|^2 , \qquad (8.66)$$

where

$$|\psi_s(\phi)|^2 = \frac{1}{ka} \sum_{m,n=0}^{\infty} \epsilon_m \epsilon_n \sin\gamma_m \sin\gamma_n \cos(\gamma_m - \gamma_n) \cos(m\phi) \cos(n\phi) . \qquad (8.67)$$

The following section contains a description of the MATLAB code used in the analysis of this system.

8.6.3 MATLAB

The MATLAB code called `cylinder_plot_scattered_pressure.m` included with this book can be used to calculate the:

- incident acoustic pressure of a plane wave using Equation (8.52);
- scattered sound pressure level using Equations (8.54), (8.60), and (8.63);
- scattered acoustic particle velocity using Equations (8.61) and (8.65);
- scattered sound intensity using Equations (8.62) and (8.66).

The parameters used in the analysis are listed in Table 8.4.

TABLE 8.4
Parameters Used in the Analysis of the Scattering of an Acoustic Plane Wave by an Infinitely Long Rigid Cylinder.

Description	Parameter	Value	Units
Radius	a	1.0	m
Length	L	∞	m
Speed of sound	c_0	343.24	m/s
Density	ρ_0	1.2041	kg/m^3
Plane wave amplitude	P_0	1.0	Pa

The MATLAB code `cylinder_plot_scattered_pressure.m` was used to calculate the scattered sound pressure at a distance of $r = 50$ m, for a cylinder diameter of $a = 1$ m, and wavenumbers of $k = 3.4$ and $k = 5$. Hence, as $kr = 170$ and $kr = 250$ is much greater than 1, it is appropriate to use Equation (8.63) to calculate the scattered sound pressure at a *large* distance from the cylinder. These calculated results can be compared with published results such as Faran [96] to verify that the MATLAB code correctly calculates the scattered sound pressure. Faran calculated the normalized scattered sound pressure by a rigid cylinder for several frequencies where he plotted

$$\frac{1}{2} \left| \sum_{n=0}^{\infty} \epsilon_m \sin(\gamma_m) \exp(j\gamma_m) \cos(m\theta) \right| . \tag{8.68}$$

These results can be re-scaled to the *absolute* value of the scattered pressure, that is [96, Eq. (26)]

$$|p_s|_{r \to \infty} \to P_0 \left[\frac{2}{\pi k r} \right]^{1/2} \left| \sum_{m=0}^{\infty} \epsilon_m \sin(\gamma_m) \exp(j\gamma_m) \cos(m\theta) \right| . \tag{8.69}$$

Hence, it is necessary to multiply the results in Faran's figures (where he plotted Equation (8.68)) by

$$\left[P_0 \sqrt{\frac{2}{\pi k r}} \right] \times 2 . \tag{8.70}$$

Figures 8.30 and 8.31 show the absolute value of the scattered sound pressure calculated using the MATLAB code and compared with the predictions by Faran [96], at $ka = 3.4$ and $ka = 5$, respectively, at distance of $r = 50$ m. The results show that the MATLAB code calculates the same values as shown in Faran [96].

As shown in the theory section, Bessel, Neumann, and Hankel functions are used to calculate the acoustic wave field in cylindrical coordinates and MATLAB has functions to calculate them. However, there is a range of values where these functions will evaluate correctly, and beyond this range numerical errors will occur. These types of numerical errors can occur when one is attempting to evaluate the *scattered* pressure at a large distance from the center of the cylinder ($kr \gg 1$), and will also require a large number of m terms to accurately model the incident acoustic plane wave described by Equation (8.52).

For this problem of the scattering of a plane wave by a cylinder, the Hankel function of the first kind $H_m^1(kr)$ is evaluated, where m is the order, and kr is the wavenumber k, times the radial distance r from the center of the cylinder to the location where the scattered pressure is to be calculated. The corresponding MATLAB command is `besselh(m,1,kr)`. The values for m and kr that are used will depend on whether MATLAB is able to calculate the result. For example, the value of the Hankel function for $H_{1000}^1(250) = 2.9673\times$

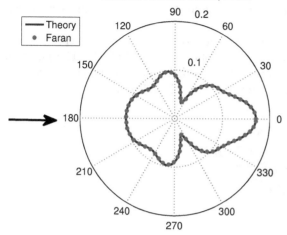

FIGURE 8.30
Scattered sound pressure, in units of Pascals, due to an incident plane wave at $ka = 3.4$, $r = 50$ m, striking an infinitely long rigid cylinder calculated using the theory described here and compared with the predictions by Faran [96].

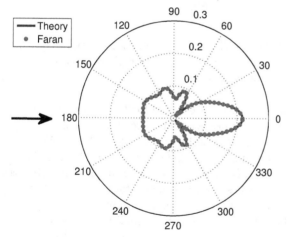

FIGURE 8.31
Scattered sound pressure, in units of Pascals, due to an incident plane wave at $ka = 5.0$, $r = 50$ m, striking an infinitely long rigid cylinder calculated using the theory described here and compared with the predictions by Faran [96].

$10^{-478} - j1.1078 \times 10^{474}$, and MATLAB is not able to calculate this value. In this case, it is necessary to reformulate the equations to prevent numerical errors.

Figure 8.32 shows the range of parameters for m and kr where the function `besselh(m,1,kr)` will evaluate a number using MATLAB. The non-shaded region indicates where `besselh` returns a finite number and the shaded region indicates conditions where it returns Not-a-Number (NaN).

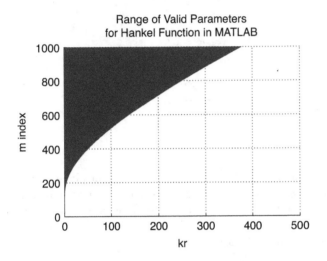

FIGURE 8.32

Range of valid parameters for the Hankel function that is evaluated using the `besselh` function in MATLAB, where the shaded region indicates where the function returns to Not-a-Number (NaN).

Similarly, Figure 8.33 shows the valid range for the Neumann function of the first kind $N_m^1(kr)$ (or sometimes written in reference books as $Y_m^1(kr)$) which is evaluated using the `bessely(m,kr)` command in MATLAB.

The Bessel function of the first kind $J_m^1(kr)$, is evaluated using the MATLAB command `besselj(m,kr)`. For large values of the input parameters m and kr, the Bessel function approaches zero, and hence for the range of values for m and kr considered here it will always evaluate a number.

Usually a large number of m terms is not required to model the scattered pressure field. However, at large distances, a large number of m terms is required to accurately model an incident plane wave. This calculation involves the use of Bessel functions that approaches zero for large values of m. One way to address the differing number of terms required in the summations to evaluate the incident and scattered pressure fields, is to use a different number of m terms for each pressure field.

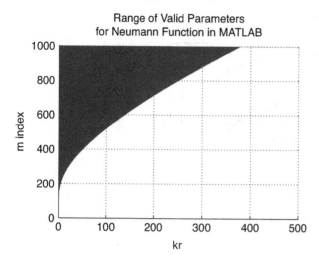

FIGURE 8.33

Range of valid parameters for the Neumann function that is evaluated using the bessely function in MATLAB, where the shaded region indicates where the function returns to Not-a-Number (NaN).

The following section describes an ANSYS Workbench model that can be used to predict the scattered sound pressure distribution around a cylinder due to an incident plane wave.

8.6.4 ANSYS Workbench

An ANSYS Workbench model was created to model the system shown in Figure 8.29, where an incident acoustic plane wave strikes an infinitely long rigid cylinder at an angle normal to the axis of the cylinder, and the scattered pressure is calculated. Figure 8.34 shows a schematic of the ANSYS Workbench model used to model the system. Only half the system has been modeled as it can be assumed to be symmetric about the ZX-plane. The acoustic domain is bounded by a region of Perfectly-Matched-Layer (PML) elements that act to absorb outgoing acoustic waves. The rigid cylindrical body of radius a is modeled by the *absence* of nodes and elements in the semi-circular region. The results along an arc will be exported and compared with theoretical predictions. The acoustic plane wave originates from outside the finite element mesh from the -X axis and propagates to the +X axis. The system will be modeled in 3D but will only have a thickness of 0.1 m and 1 element division along the Z-axis. The finite element model of the full system that will be generated for this example has 193,471 nodes and 26,675 elements (only half these values for the half model) and will exceed the node limit of 32,000 for teaching licenses of ANSYS.

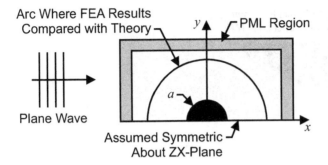

FIGURE 8.34
Schematic of the ANSYS Workbench model where an incident plane wave
strikes a rigid cylinder in an infinite domain. The system is assumed to be
symmetric about the ZX-plane.

Previous examples in this book included step-by-step instructions to create
ANSYS Workbench models. It is assumed that by this stage the reader will
have gained sufficient skills to enable creation of acoustic models using the
ACT Acoustics extension. For this example, an overview of the model will
be described and only the important features and steps will be described in
detail.

The completed ANSYS Workbench archive file of the system cylinder_
scattering.wbpz, which contains the .wbpj project file, is included with this
book.

Workbench Instructions

- Start ANSYS Workbench.

- Use DESIGNMODELER to create extruded blocks 0.1 m thick, as shown in
  Figure 8.34.

- The dimensions of the model are defined in the Parameter Set, in the
  Project Schematic window, using the values listed in Table 8.4. The analysis
  wavelength was selected as $\lambda = (2/5)\pi a$, which was used in [117, Fig. 8.1, p.
  402]. Hence, the analysis frequency is $f_{max} = c_0/\lambda = 343.024 \times (5/2)/(\pi a) =$
  273 Hz.

- Start ANSYS Mechanical.

- The model was created using FLUID220 elements that include mid-side nodes,
  which is the default element type when using the Acoustic Body object in
  the ACT Acoustics extension. A parameter for the elements per wavelength
  was set to (EPW=) 10, and is used to define the global mesh size based on
  the analysis wavelength. The global mesh size is $\lambda/\text{EPW} = 0.125$ m, and is
  used in the object Model (A4) | Mesh | Body Sizing.

- A cylindrical coordinate system has been defined to enable the calculation of the acoustic particle velocity in the radial direction. This can be created in the `Outline` window by right-clicking on `Model (A4) | Coordinate Systems` and left-click on `Insert | Coordinate System`.

- In the window `Details of "Coordinate System"`, change the row `Definition | Type` to `Cylindrical`. Change the row `Origin | Define By` to `Global Coordinates`. The `Geometry` window will show the orientation of the cylindrical coordinate system with the origin at the global origin, which is also at the center of the cylinder.

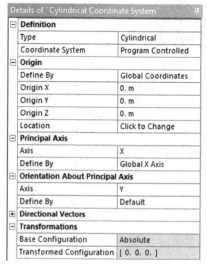

- Rename this coordinate system from the default name of `Coordinate System` to `Cylindrical Coordinate System`. In the `Outline` window, right-click on `Model (A4) | Coordinate Systems | Coordinate System` and left-click on `Rename`. Type in the new name as `Cylindrical Coordinate System`. The window `Details of "Cylindrical Coordinate System"` should look like the following image.

Details of "Cylindrical Coordinate System"	
Definition	
Type	Cylindrical
Coordinate System	Program Controlled
Origin	
Define By	Global Coordinates
Origin X	0. m
Origin Y	0. m
Origin Z	0. m
Location	Click to Change
Principal Axis	
Axis	X
Define By	Global X Axis
Orientation About Principal Axis	
Axis	Y
Define By	Default
Directional Vectors	
Transformations	
Base Configuration	Absolute
Transformed Configuration	[0. 0. 0.]

- In the Outline window, under the branch Model (A4) | Mesh, several objects were defined. A Body Sizing object was inserted and is used to define the global size of the elements, and was linked to a parameter called P5 Body Sizing Element Size in the Parameter Set window. The mesh method that was used for the acoustic domains was set to Multizone, shown in the following figure, which causes a predominantly hexagonal mesh (remember this is a 3D analysis) to be formed without the need for slicing the bodies into small regions.

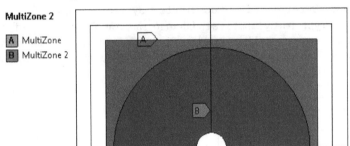

- A finite element mesh is shown in the following figure that has 94,741 nodes and 13,334 elements. This model cannot be solved using teaching licenses of ANSYS, which are limited to 32,000 nodes.

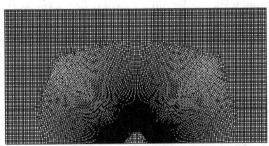

- The analysis frequency can be set in the branch Harmonic Response (A5) | Analysis Settings. The analysis will be conducted at $k = 5$, which corresponds to a frequency of $f = kc/(2\pi) = 5 \times 343.24/(2\pi) = 272.95$ Hz. The row Options | Range Minimum was set to 272, Options | Range Minimum was set to 273, Options | Solution Intervals was set to 1, and Options | Solution Method was set to Full. The options under the branch Output Control were all set to Yes.

- The ACT Acoustics extension was used to define four Acoustic Body objects as shown in the following figure, which have the following properties:

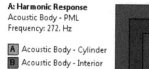

- Acoustic Body - Cylinder is a cylindrical body with a radius of 1 m. The acoustic body can be activated (Unsuppressed) to calculate the incident pressure field in the absence of the cylinder, or can be deactivated (Suppressed) so that the scattered pressure from the cylinder can be analyzed. Remember, if the cylinder acoustic body is to be suppressed, it is also necessary to suppress the corresponding solid geometry, otherwise the finite element model will contain SOLID186 structural elements.

- Acoustic Body - Interior, is the main region of interest. This region has been divided with a circular arc at a radius of 6.5 m where the acoustic pressure and acoustic particle velocity will be evaluated and compared with theoretical predictions.

- Acoustic Body - Buffer, which acts as a buffer region between the region of interest and the region of Perfectly Matched Layer elements. In this case, it is not necessary to include the buffer region. However, it is good practice to include a buffer region, especially if the Acoustic Far Field feature, where the sound pressure level and other results can be plotted as a function of angle to obtain directivity plots, were to be used. To use this feature, an Equivalent Source Surface must be defined, where the surface must fully enclose the acoustic radiator or scattering object, but does not need to be defined on areas of symmetry. In this example, the cylindrical body is modeled by the *absence* of nodes and elements. Hence, it is not possible to enclose the scattering object and therefore it is not possible to use the Acoustic Far Field feature to obtain directivity plots.

- Acoustic Body - PML, which is a layer of Perfectly Matched Layer elements used to absorb outgoing pressure waves. The PML feature is activated by clicking on the Acoustic Body - PML, and changing the row Definition | Perfectly Matched Layers (PML) to On, and change the following row PML Options to On - 3D PML.

• The bodies defined with the PML option also require that a boundary condition is defined such that the pressure is zero on the external faces. This

can be achieved by clicking on Boundary Conditions in the ACT Acoustics
extension menu bar and selecting Acoustic Pressure.

The rows beneath Definition have the default values of Pressure (Real)
0 [Pa], and Pressure (Im) 0 [Pa] and can be left unaltered. Change the
filter selection to Face and select the 4 (thin) faces on the exterior of the
PML bodies, but not the faces on the ZX-plane. In the window Details of
"Acoustic Pressure", in the row Scope | Geometry, click the Apply button.

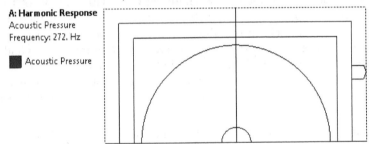

- The next step is to define the acoustic plane wave source. In the ACT Acoustics extension menu, select Excitation | Wave Sources (Harmonic).

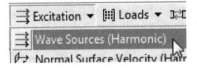

- In the window Details of "Acoustic Wave Sources", change the row Wave
  Type to Planar Wave, Excitation Type to Pressure, Source Location to
  Outside Model, Pressure Amplitude to 1, Angle Phi (From X Axis Toward
  Y Axis) to 180, Angle Theta (From Z Axis Toward X Axis) to 90, leave
  the density and speed of sound as the default values of 1.2041 and 343.24
  respectively, and change the row Pure Scattering Options to On (Output
  Scattered Pressure).

Details of "Acoustic Wave Sources"	
Definition	
Wave Number	1
Wave Type	Planar Wave
Excitation Type	Pressure
Source Location	Outside The Model
Pressure Amplitude	1 [Pa]
Phase Angle	0 [°]
Angle Phi (From X Axis Toward Y Axis)	180 [°]
Angle Theta (From Z Axis Toward X Axis)	90 [°]
Mass Density Of Environment Media	1.2041 [kg m^-1 m^-1 m^-1]
Sound Speed In Environment Media	343.24 [m sec^-1]
Pure Scattering Options	On (Output Scattered Pressure)

- Several acoustic results will be calculated as listed under the Solution (A6) branch.

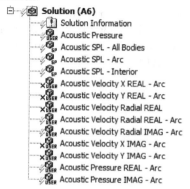

Results will be calculated for the sound pressure level (Acoustic SPL) at nodes on the arc of radius 6.5 m. The real and imaginary acoustic particle velocities are also calculated on the arc in the X and Y directions, and also in the radial direction.

- The steps used to request results along the X and Y directions has been described previously. The steps to request results in the radial direction will be described in detail as follows. Click on Results | Acoustic Velocity X from the ACT Acoustics extension menu, in the row Scope | Geometry, select the two edges for the arc on the radius of 6.5 m on the Z = 0 plane (the face nearest the origin, which can be determined by changing to an isometric view), and change the row Definition | Coordinate System to Cylindrical Coordinate System, which was defined earlier. Rename this object Acoustic Velocity X to Acoustic Velocity Radial REAL - Arc.

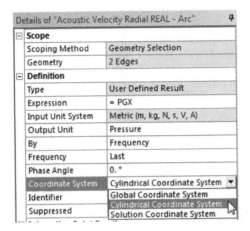

- It is also necessary to calculate the imaginary part of the complex acoustic particle velocity in the radial direction. This can be achieved by right-clicking on Acoustic Velocity Radial REAL - Arc and in the context menu that opens, left-click on Duplicate Without Results.

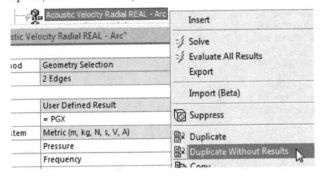

- Rename the new object that was created to Acoustic Velocity Radial IMAG - Arc. In the window Details of "Acoustic Velocity Radial IMAG - Arc", change the row Definition | Phase Angle to -90, which will calculate the imaginary component of the acoustic particle velocity.
- The first analysis that will be conducted will be used to verify that the scattered acoustic pressure is zero when there is no object inside the acoustic domain to cause scattering. Right-click in the Geometry window and left-click on Unsuppress All Bodies.
- Click the File | Save Project in the menu bar.
- Click the Solve icon.
- After the computations have completed, click on Solution (A6) | Acoustic Pressure to inspect the sound pressure. The following figure shows the *scat-*

tered sound pressure is zero throughout the acoustic domain, when there was no cylinder to cause scattering, which is to be expected.

- The next verification that will be conducted is to examine the *total* sound pressure level when there is no cylinder present. The expected result is the sound pressure level due to the incident acoustic plane wave. Change the scattering option in the window `Details of "Acoustic Wave Sources"` by changing the row `Pure Scattering Options` to `Off`.

- Click the `File | Save Project` in the menu bar.

- Click the `Solve` icon.

- After the computations have completed, click on `Solution (A6) | Acoustic SPL - Interior` to inspect the sound pressure level. The following figure shows the *total* sound pressure level throughout the acoustic domain, which is the sum of the incident and the scattered acoustic pressure, when there was no cylinder to cause scattering.

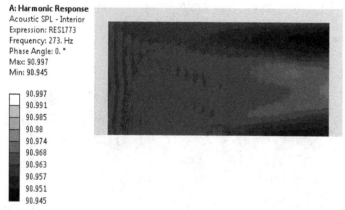

The sound pressure level in the acoustic domain is around 90.9 dB re 20 μPa. The theoretical sound pressure level for a plane wave of amplitude 1 Pa is

$$L_p = 20 \log_{10} \left[\frac{1}{\sqrt{2} \times (20 \times 10^{-6})} \right]$$
$$= 90.969 \text{ dB re } 20 \ \mu\text{Pa} \,,$$

hence the ANSYS Workbench result agrees with theoretical predictions.

Conducting these two analyses is useful to verify that the model has been constructed correctly. This is particularly important when using an Equivalent Source Surface where inadvertently defining the surfaces and bodies incorrectly can cause unexpected sound pressure level results.

The next step is to calculate the scattered acoustic pressure caused by the rigid cylinder.

- Change the filter section to Body and select the two bodies that form the 1 m radius cylinder. Right-click in the Geometry window and left-click on Suppress Body. Ensure there is a cross next to the object Acoustic Body - Cylinder under the branch Harmonic Response (A5), otherwise right-click on the object and left-click on Suppress.

- Click on the object Acoustic Wave Sources and in the window Details of "Acoustic Wave Sources", change the row Pure Scattering Options to On (Output Scattered Pressure).

- Click File | Save Project from the main menu, and then click on the Solve icon.

- The next step is to check that the results appear reasonable. Once the results have been calculated, in the branch Solution (A6), click on the object Acoustic SPL - Interior. The following figure shows the scattered sound pressure level at 273 Hz (note that the contour plot of the Acoustic SPL indicates the frequency is 0 Hz, which is a bug with Release 14.5 of the software). Reader may recognize this figure is used on the front cover of this book.

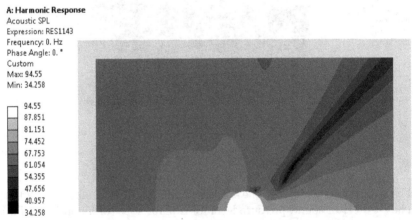

A: Harmonic Response
Acoustic SPL
Expression: RES1143
Frequency: 0. Hz
Phase Angle: 0. °
Custom
Max: 94.55
Min: 34.258

94.55
87.851
81.151
74.452
67.753
61.054
54.355
47.656
40.957
34.258

The contour plot indicates that: sound is being diffracted around the cylindrical body; there is a region of low sound pressure levels at an angle of about $+45°$ from the x-axis; and the sound pressure levels on the perimeter of the acoustic domain are lower than near the scattering object (the cylinder), which indicates that the sound is "spreading" with increasing distance from the scattering object. As these sound pressure level results appear reasonable, the next step is to export the results at the nodes on the arc of radius 6.5 m.

- Right-click on the following entries listed in the Solution (A6) branch, then left-click on Export, and type appropriate filename for each result:

  - · sound pressure level: Acoustic SPL - Arc

  - · real and imaginary pressure: Acoustic Pressure REAL - Arc , Acoustic Pressure IMAG - Arc

  - · real and imaginary acoustic particle velocity in the radial direction: Acoustic Velocity Radial REAL - Arc, Acoustic Velocity Radial IMAG - Arc

Results

The results exported from ANSYS can be post-processed using a spreadsheet or MATLAB and compared with the theoretical predictions using the theoretical model described in Section 8.6.3.

Figure 8.35 shows the real and imaginary parts of the scattered pressure, for a wavelength of $\lambda = (2/5)\pi a$ at a radius of 6.5 m from the center of the

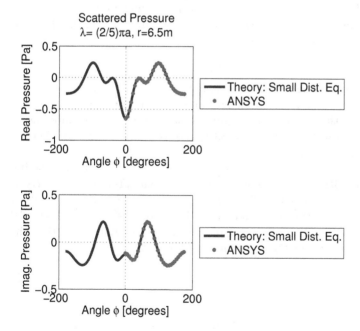

FIGURE 8.35

Scattered real and imaginary parts of the complex pressure at 6.5 m from a rigid cylinder due to an incident plane wave at a wavelength of $\lambda = (2/5)\pi a$ corresponding to 273 Hz, calculated using theory and ANSYS Workbench.

cylinder. The figure shows the results calculated using theoretical predictions from Equation (8.54) and from the ANSYS Workbench model compare favorably. Figure 8.36 shows the corresponding scattered sound pressure level and again the results from the ANSYS Workbench model overlay the theoretical predictions.

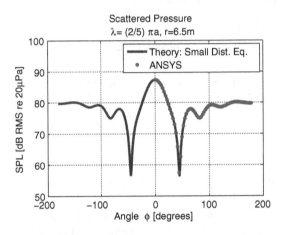

FIGURE 8.36

Scattered sound pressure level at 6.5 m from a rigid cylinder due to an incident plane wave at a wavelength of $\lambda = (2/5)\pi a$ corresponding to 273 Hz, calculated using theory and ANSYS Workbench.

Figure 8.37 shows the real and imaginary parts of the scattered *radial* acoustic particle velocity, for a wavelength of $\lambda = (2/5)\pi a$ at a radius of 6.5 m from the center of the cylinder. The figure shows the results calculated using theoretical predictions from Equation (8.61) and from the ANSYS Workbench model compare favorably.

Figure 8.38 shows the scattered sound pressure level at 273 Hz for the full acoustic domain with a full cylinder, where the assumption of symmetry is not implied. The mesh had 193,471 nodes and 26,675 elements. The contour plot for the full model can be compared with the results for the half-model to confirm that the results are the same, and hence the assumption that the system is symmetric about the ZX-plane is valid.

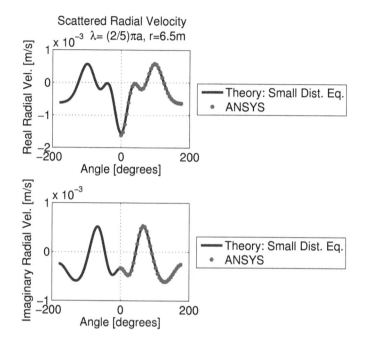

FIGURE 8.37

Scattered real and imaginary parts of the complex radial velocity at 6.5 m from a rigid cylinder due to an incident plane wave at a wavelength of $\lambda = (2/5)\pi a$ corresponding to 273 Hz, calculated using Equation (8.54) and ANSYS Workbench.

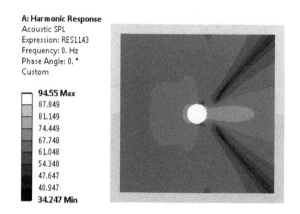

FIGURE 8.38

Scattered sound pressure level from a rigid cylinder due to an incident plane wave at a wavelength of $\lambda = (2/5)\pi a$ corresponding to 273 Hz, using the full model.

9

Fluid–Structure Interaction

9.1 Learning Outcomes

The learning outcomes of this chapter are:

- an ability to conduct a fluid–structure interaction analysis using ANSYS,

- understanding of the need to couple the displacement degrees of freedom of the nodes associated with structural elements with the displacement degrees of freedom of the nodes of acoustic elements, and

- introduction of modal-coupling methods for fluid–structure interaction analyses.

9.2 Fluid–Structure Interaction Using ANSYS

9.2.1 Introduction

The theory of fluid–structure interaction (FSI) using a finite element formulation was described in Section 2.4. In this chapter, two examples are described of how to conduct an FSI analysis using ANSYS and the ACT Acoustics extension.

The first example is used to provide an overview of the steps involved in setting up an FSI analysis in ANSYS Workbench. The example is the transmission loss of a simply supported plate in a duct.

The second example is the vibro-acoustic response of a flexible plate attached to an acoustic cavity. This example is used to introduce the modal-coupling method for solving vibro-acoustic problems. A theoretical model that uses model coupling is implemented in MATLAB and the example is solved in ANSYS using several methods.

9.2.2   Example: Transmission Loss of a Plate in a Duct

Figure 9.1 shows a schematic of an infinite duct that is divided by a simply supported plate. The parameters of the duct and plate are listed in Table 9.1.

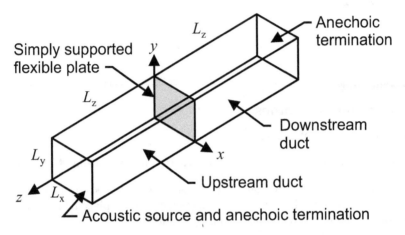

FIGURE 9.1
Schematic of an infinite duct divided by a simply supported plate.

TABLE 9.1
Parameters Used in the Analysis of an Infinite Acoustic Duct
Divided by a Simply Supported Thin Plate

Description	Parameter	Value	Units
Duct width	L_x	0.5	m
Duct height	L_y	0.5	m
Duct length	L_z	2.0	m
Plate width	L_x	0.5	m
Plate height	L_y	0.5	m
Plate thickness	h	0.002	m
Plate material		Structural Steel	
Speed of sound of air	c_0	343.24	m/s
Density of air	ρ_0	1.2041	kg/m^3

Each end of the duct is defined with acoustic anechoic terminations so that each duct can be considered as a semi-infinite duct. A simply supported thin plate is mounted in the duct. An acoustic excitation source is defined on the upstream end face of the duct so that sound will propagate along the duct until it strikes the simply supported panel. Some of the sound will be reflected upstream by the panel, and some will cause the panel to vibrate and generate sound in the upstream and downstream duct section. As the downstream duct section has an anechoic termination, the sound pressure level in the downstream section should remain constant along the axis of the duct, as there is no absorption mechanism or change in acoustic impedance that would cause a backward traveling wave.

Plane wave conditions will exist in the duct at frequencies below "cut-on," which can be calculated using Equation (3.17) as

$$f_{\text{cut-on: rectangular}} = \frac{c_0}{2H} = \frac{343.24}{2 \times 0.5} = 343 \text{ Hz}. \tag{9.1}$$

The following section describes how this example can be modeled using ANSYS Workbench.

ANSYS Workbench

The learning outcomes for this example are as follows:

- The steps involved in conducting a fluid–structure interaction analysis using ANSYS Workbench.

- How to model an anechoic termination in a duct.

- Demonstrate that an anechoic termination of a duct means there is no acoustic dissipation or acoustic modal response.

- Care must be taken when investigating the transmission loss of structures comprising shell elements as it is possible to inadvertently couple the pressure degree of freedom of the acoustic elements on each side of the structural elements, thereby by-passing the structure, and hence there would be no transmission loss.

Instructions

The completed ANSYS Workbench project file square_duct_plate.wbpz, which contains the .wbpj file, is available with this book.

- Start ANSYS Workbench.

- Make sure the ACT Acoustics extension is loaded by clicking on Extensions | Manage Extensions. There should be a tick in the column Load for the row ExtAcoustics.

- From the Analysis Systems window, double-click on Harmonic Response.

- Double-click on row 3 Geometry to start DESIGNMODELER.

- Define a square on the XYPlane that has equal length constraints. Define the edge length as 0.5 m.

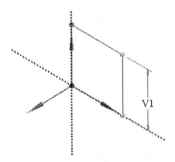

- Extrude the square 2 m to create an upstream duct.

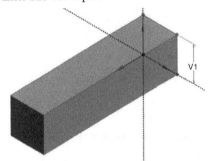

- Click on Sketch1 again so that it is highlighted.

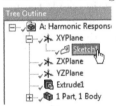

- Now we will generate the downstream duct. Click on the Extrude icon, making sure that the sketch of the square is highlighted in yellow. Click on the Apply button on the Geometry row, in the row labeled Direction change it from Normal to Reversed, make sure the FD1, Depth (>0) is still 2 m, change the row Operation to Add Frozen, then click on Generate. There should now be 2 volumes that are connected by a face, which will be the location of the plate.

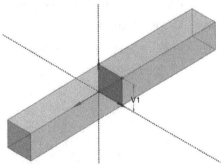

- Click on Sketch1 again, then select Concept | Surfaces from Sketches.

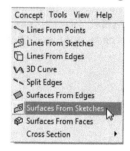

- In the row Base Objects, click Apply to use Sketch1, change the row Operation to Add Frozen, and change the Thickness to 0.002 m. Click on the Generate icon.

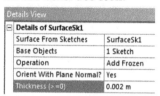

In the Outline window there should be 3 Parts, 3 Bodies, which indicates that they are all separate entities, which is correct. The reason we want 3 separate bodies is that the coupling of the degrees of freedom between the bodies will be manually defined.

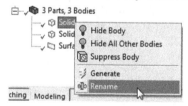

- The previous step renamed the solids, but it would also be useful to create named selections which can be used in the ANSYS Workbench Mechanical module. In the next few steps we will create named selections of the faces and bodies in the solid model, as shown in the following figure, to make it easier for defining boundary conditions, contact pairs, and fluid–structure interaction regions.

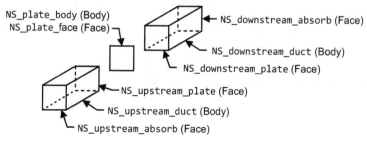

- In the tree branch for the parts and bodies, select the first Solid, or select the upstream (left side) block, then right-click to open a context menu and select Rename. Name the solid upstream_duct. Repeat this process to name the downstream_duct, and the plate. Click on the upstream_duct and then select Tools | Named Selection.

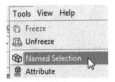

- Change the row Named Selection to NS_upstream_duct. In the Geometry make sure that the solid body of the upstream duct is selected, then click Apply. Click on the Generate icon to create the named selection.

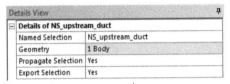

- Repeat these steps to define named selections for the bodies of the downstream duct and name it NS_downstream_duct, and the plate and name it NS_plate_body. Next define a named selection for the plate where the Geometry is the face belonging to the plate and name it NS_plate_face. The face of the plate can be selected by hiding the bodies for the upstream and downstream ducts. This last named selection will be used in the definitions of the contacts between the acoustic bodies and the plate.

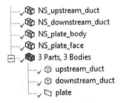

- Click on the downstream_duct solid body and then right-click and select Hide All Other Bodies.

- Change the selection filter method to Faces, then select the face on the downstream duct that is in contact with the plate.

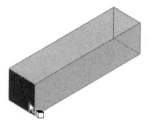

- Right-click and select Named Selection. Change the row Named Selection to NS_downstream_plate, click on Apply next to Geometry, then click Generate.

- Rotate the view of the downstream_duct so that you can clearly see the far end of the duct that is farthest from the plate. Define a named selection for the far end of the downstream duct that will have an acoustic wave-absorbing end condition and call it NS_downstream_absorb.

- Click on the upstream_duct solid body and then right-click and select Hide All Other Bodies.

- Define named selections for the upstream far end (farthest from the plate) that will have an acoustic wave-absorbing end condition, and define it NS_upstream_absorb.

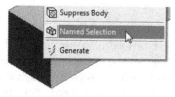

- Repeat this process to define a named selection for the face of the upstream_duct that touches the plate and call it NS_upstream_plate.

- In the Graphics window, right-click and select Show All Bodies. There should be 8 named selections that have been defined.

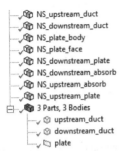

- In the main menu click on File | Save Project.
- Close DESIGNMODELER.

That completes the creation of the solid model. The next stage involves the setup of the finite element model using ANSYS Workbench Mechanical.

- Start ANSYS Workbench Mechanical by double-clicking on row 4 Model in the project view. The model should look like the figure below.

- Start by defining the acoustic bodies.

- You can define them by either using (a) geometry selection by clicking on the upstream duct and apply, or (b) change Scoping Method to Named Selection and then on the row for Named Selection, click on the drop-down list and select NS_upstream_duct. Change the row Acoustic-Structural Coupled Body Options to Coupled With Unsymmetric Algorithm. Selecting this option activates the displacement degrees of freedom for the acoustic elements to enable the fluid–structure interaction coupling. The "unsymmetric" algorithm was the first method that was developed in ANSYS for FSI analyses. You could also select Coupled With Symmetric Algorithm, and should arrive at the same answers as using the option Coupled With Unsymmetric Algorithm. Just make sure that whichever option you select, it is consistent for all definitions of the acoustic bodies in the model.

Details of "Acoustic Body"	
⊟ **Scope**	
Scoping Method	Named Selection
Named Selection	NS_upstream_duct
⊟ **Definition**	
Temperature Dependency	No
Mass Density	1.2041 [kg m^-1 m^-1 m^-1]
Sound Speed	343.24 [m sec^-1]
Dynamic Viscosity	0 [Pa sec]
Thermal Conductivity	0 [W m^-1 C^-1]
Heat Coefficient Cp	0 [J kg^-1 C^-1]
Heat Coefficient Cv	0 [J kg^-1 C^-1]
Specific Heat C	0 [J kg^-1 C^-1]
Equivalent Fluid of Perforated Material	No
Reference Pressure	2E-05 [Pa]
Reference Static Pressure	101325 [Pa]
Acoustic-Structural Coupled Body Options	Coupled With Unsymmetric Algorithm ▼
Perfectly Matched Layers (PML)	Off

- Repeat the process to define the downstream duct as an acoustic body, and make sure that you change the row `Acoustic-Structural Coupled Body Options` to `Coupled With Unsymmetric Algorithm`.

- Now check the contact conditions. Click on the `Connections` branch and then open the `Contacts` tree. Click on the first contact `Contact Region`.

- In the windows `Details of "Contact Region"`, you should notice that the row `Contact Bodies` is set to `upstream_duct` and the row `Target Bodies` is set to `downstream_duct`. This definition means that there is a direct connection between the nodes in the upstream duct and the downstream duct, and therefore by-passes the plate, which is not what we want.

Details of "Contact Region"	
⊟ **Scope**	
Scoping Method	Geometry Selection
Contact	1 Face
Target	1 Face
Contact Bodies	upstream_duct
Target Bodies	downstream_duct

What we want are two contact definitions between (1) the end face of the upstream duct to the plate, and (2) the plate and the face at the start of the downstream duct. To achieve this we will use the named selections that have been defined for the faces that are in contact with the plate.

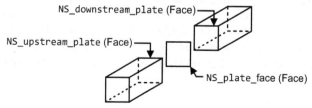

- First delete the unwanted contact pair by right-clicking on the first `Contact Region` and in the context menu, left-click on `Delete`.

- Left-click on the object labeled `Contact Region 2` and in the window `Details of "Contact Region 2"` you will see that the row `Contact Bodies` is `upstream_duct` and the row `Target Bodies` is `plate`, which is what we want.

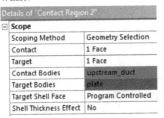

You can leave this as is or you can explicitly define the contact pair by changing the rows `Scoping Method` to `Named Selection`, `Contact` to `NS_upstream_plate`, and `Target` to `NS_plate_face` (note that you may need to scroll down the list of named selections to select this object). Change the row `Behavior` to `Symmetric`, which means that there is a double set of contact elements that are created (plate to duct and duct to plate) to couple the two systems. Right-click on `Contact Region 2` and in the context menu, left-click on `Rename Based on Definition`. The window `Details of "Bonded - NS_upstream_plate To NS_plate_face"` should look like the following figure.

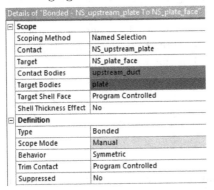

- Repeat this process to alter the definition for `Contact Region 3` to define a bonded contact between the face of the downstream duct and the face belonging to the plate by changing the row `Scoping Method` to `Named Selection`, `Contact` to `NS_downstream_plate`, `Target` to `NS_plate_face`, and `Definition | Behavior` to `Symmetric`. Right-click on `Contact Region 3` and in the context menu left-click on `Rename based on Definition`. The window `Details of "Bonded - NS_downstream_plate To NS_plate_face"` should look like the following figure.

Details of "Bonded - NS_downstream_plate To NS_plate_face"	
Scope	
Scoping Method	Named Selection
Contact	NS_downstream_plate
Target	NS_plate_face
Contact Bodies	downstream_duct
Target Bodies	plate
Target Shell Face	Program Controlled
Shell Thickness Effect	No
Definition	
Type	Bonded
Scope Mode	Manual
Behavior	Symmetric
Trim Contact	Program Controlled
Suppressed	No

- Right-click on the Mesh branch and select Insert | Sizing.

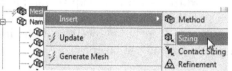

- Change the selection method to Edges. Change the selection filter to Box Select.

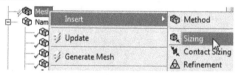

- Draw a box around the end square face on the upstream duct, then hold down the <Ctrl> key, draw another selection window around the plate region, and keeping the <Ctrl> key held down, draw a selection rectangle over the face at the far end of the downstream duct. There should be three squares that are highlighted. Click on the Apply button in the row with Geometry and there should be 20 Edges selected. Note that there are 3 overlapping squares at the plate region, a square at the upstream end, and a square at the downstream end, for a total of 5 squares with 4 sides each; hence there are 20 edges.

- Change the Type to Number of Divisions. Change the Number of Divisions to 10. Change the Behavior to Hard.

Details of "Edge Sizing" - Sizing	
Scope	
Scoping Method	Geometry Selection
Geometry	20 Edges
Definition	
Suppressed	No
Type	Number of Divisions
Number of Divisions	10
Behavior	Hard
Bias Type	No Bias

- The next step is to define the number of divisions of the edges along the axis of the duct. right-click on Mesh, select Insert | Sizing. With the Edges and the Box Selection active, start the selection window by left-clicking and holding the mouse button at the bottom right of the upstream duct and moving the mouse cursor upward to the left of the upstream duct. This will draw a selection window with marks through the middle of each square, which means that the selection filter will include objects that the lines of the selection box crosses. Release the left mouse button.

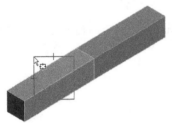

- The 4 axial edges on the upstream duct will be highlighted.

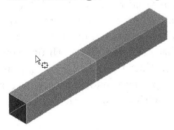

- Hold down the <Ctrl> key and select the 4 axial edges on the downstream duct. Then click on the Apply button in the row with Geometry and there should be 8 Edges that were selected.

- Change the Type to Number of Divisions. In the row with Number of Divisions, change it to 20. Change the behavior to Hard.

- Change the selection method to Single Select, and Body. Right-click on Mesh and select Insert | Method. Click in the row with Geometry and select the upstream and downstream ducts. Click on the Apply button.

- Right-click on Mesh and select Generate Mesh. This will generate a mapped mesh of brick elements throughout the blocks, and rectangular elements on the plate.

- Now add an acoustic mass source by clicking on Excitation | Mass Source (Harmonic).

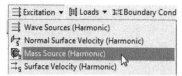

Other acoustic excitation sources that could also be used are Normal Source Velocity (Harmonic), Source Velocity (Harmonic). For this example, we would not use a Boundary Conditions | Acoustic Pressure, as this would fix the pressure at the inlet and we want to demonstrate that pressure interaction occurs in the upstream duct due to the combination of the incident and reflected waves.

- Change the Scoping Method to Named Selection. In the Named Selection row, select NS_upstream_absorb. Change the Amplitude Of Mass Source to 1.

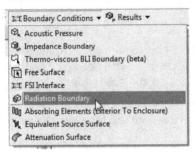

- Now add the absorbing boundaries to create the anechoic terminations at both ends of the duct. Select Boundary Conditions | Radiation Boundary.

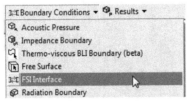

- Change Scoping Method to Named Selection and choose the NS_upstream_absorb.

- Repeat the process and define a radiation boundary condition for NS_downstream_absorb.

- The next step is to define the fluid–structure interaction (FSI) interfaces between the end of the upstream duct and the plate, and the end of the downstream duct and the plate. This step will activate the displacement degrees of freedom in the acoustic elements and enable them to be coupled to the displacement degrees of freedom in the structural elements. In the ACT Acoustics extension menu click on Boundary Conditions | FSI Interface.

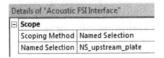

- Change Scoping Method to Named Selection, and change the Named Selection row to NS_upstream_plate.

Details of "Acoustic FSI Interface"	
Scope	
Scoping Method	Named Selection
Named Selection	NS_upstream_plate

- Repeat the process to define a FSI Interface on NS_downstream_plate.

- The next step is to set up the analysis options. Click on Analysis Settings under Harmonic Response (A5). Change the Range Minimum to 99, the Range

Maximum to 100, the Solution Intervals to 1, and Solution Method to Full. This will solve the model at the single frequency of 100 Hz, which is less than the cut-on frequency of this duct of 343 Hz.

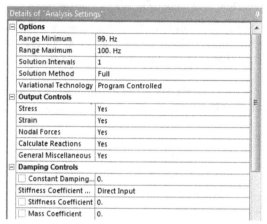

- Click on Results | Acoustic Pressure. Using the geometry selection method, click on the upstream duct and the downstream duct as the 2 acoustic bodies for which acoustic pressure results will be calculated.

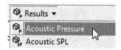

- Repeat this process and insert a Results | Acoustic SPL under the Solution (A6) branch, and again, only select the upstream and downstream acoustic bodies.

- Define the edges of the plate to have a simply supported boundary condition. Under Model (A4) | Geometry, right-click on plate and select Hide All Other Bodies.

- Right-click on Harmonic Response (A5) and select Insert | Simply Supported. Change the selection filter to Edges. In the graphics window right-click and left-click Select All. Then click the Apply button in the Geometry row. There should only be 4 edges that were selected.

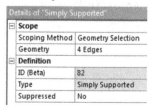

- Right-click on Solution (A6) and select Insert | Deformation | Total. Change the Scoping Method to Named Selection. Change the Named Selection to NS_plate_body. This will enable us to view the displacement of the plate.

- In the graphics window, right-click and select Show All Bodies.

- Click on the Harmonic Response (A5) and confirm the location of the applied loads and boundary conditions. There should be an Acoustic Mass Source on the upstream end of the duct, an Acoustic Radiation Boundary condition on the upstream end of the duct, two Acoustic FSI Interface conditions at the plate region, an Acoustic Radiation Boundary condition on the far end of the downstream duct, and a Simply Supported condition on the edges of the plate.

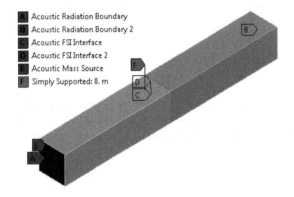

That completes the steps involved for the model set up. The next thing to do is save the project and then click the Solve icon. Wait for the model to solve.

The next stage involves investigating the results that were calculated.

- Once the computations have completed, click on Acoustic Pressure under Solution (A6) to see the acoustic pressure in the model. Note that the legend has written that Frequency: 0 Hz, whereas it should be at 100 Hz, as discussed in Section 2.8.5.2. This is a limitation with Release 14.5 of the ACT Acoustics extension in ANSYS. The frequency of the result that is requested is defined in the window Details of "Acoustic Pressure" under the row labeled Frequency.

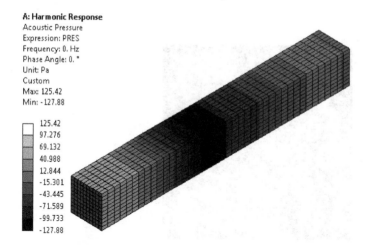

A: Harmonic Response
Acoustic Pressure
Expression: PRES
Frequency: 0. Hz
Phase Angle: 0. °
Unit: Pa
Custom
Max: 125.42
Min: -127.88

125.42
97.276
69.132
40.988
12.844
-15.301
-43.445
-71.589
-99.733
-127.88

- Click on the Acoustic SPL result, right-click on the numbers next to the color legend, and left-click on Adjust to Visible. Notice that the sound pressure level in the upstream duct varies with distance. Sound has reflected off the plate and traveled back upstream and interacted with the incident acoustic wave, causing a reduction in sound pressure level in one region of the duct.

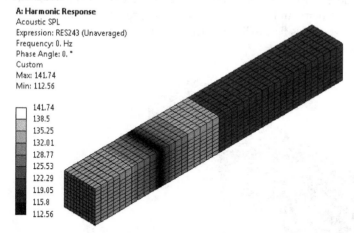

A: Harmonic Response
Acoustic SPL
Expression: RES243 (Unaveraged)
Frequency: 0. Hz
Phase Angle: 0. °
Custom
Max: 141.74
Min: 112.56

141.74
138.5
135.25
132.01
128.77
125.53
122.29
119.05
115.8
112.56

- Under the Geometry branch, right-click on plate and select Hide All Other Bodies. Click on Total Deformation under Solution (A6) to see the displacement of the simply supported plate.

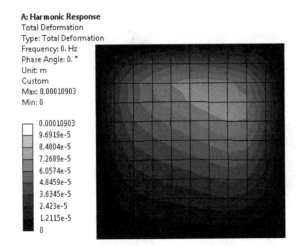

- One result worth highlighting is the acoustic pressure in the downstream duct. Under `Geometry`, right-click `downstream_duct` and select `Show All Bodies`, then right-click again and select `Hide All Other Bodies`. Under `Solution (A6)` select `Acoustic Pressure`. On the legend, right-click and select `Adjust to Visible`. You should notice that there is a pressure variation throughout the downstream duct as the vibrating plate has excited the air in contact with the plate and generated sound that propagates downstream in the duct. Note that this figure displays the *real* part of the *complex* pressure.

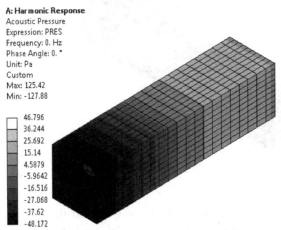

- Now click on `Acoustic SPL`. Right-click on the color legend, and in the context menu left-click on `Adjust to Visible`, which will change the range values in the legend to cover the range of sound pressure levels in the downstream duct body. You should notice that the SPL in the downstream duct is about 124 dB re 20 μPa and does not vary along the length of the duct. This is because a wave-absorbing end condition was applied to the end of the downstream duct which effectively creates a semi-infinite duct, so that there are no reflections from the end of the duct and there is no modal response. In addition, as there are no energy loss mechanisms, one would expect the SPL to be constant. Remember that the `Acoustic SPL` is a logarithmic value of the *magnitude* of the *complex* pressure value. Even though the real and imaginary parts of the complex pressure vary along the length of the duct, the magnitude of the complex pressure remains constant.

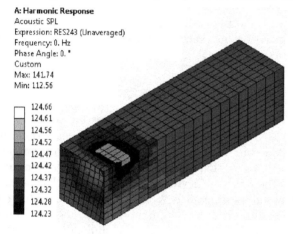

A: Harmonic Response
Acoustic SPL
Expression: RES243 (Unaveraged)
Frequency: 0. Hz
Phase Angle: 0. °
Custom
Max: 141.74
Min: 112.56

124.66
124.61
124.56
124.52
124.47
124.42
124.37
124.32
124.28
124.23

9.3 Fluid–Structure Interaction Using Modal Coupling

9.3.1 Introduction

The previous example of fluid–structure interaction analysis using ANSYS involved the computation of the mass, stiffness, and damping matrices, inverting a large matrix, and then multiplying this matrix by a vector of forces, to calculate the resulting pressures and displacements at each node. For harmonic analysis, the multiplication is conducted at every analysis frequency. For large models, these computations can take a long time. When conducting optimizations, where the configuration of the system changes slightly and the acoustic and structural performance is re-evaluated, this can take an excessively long time.

Another method of calculating the coupled vibro-acoustic response of a system is by using modal coupling theory. The mode shapes and natural frequencies of the structural and acoustic systems are calculated separately, and then mathematically combined to calculate the coupled vibro-acoustic response. Using this technique is computationally faster than conducting fluid–structure interaction full harmonic analyses. This technique was used in Howard and co-workers [82, 81, 79, 80, 78] for conducting optimizations where tuned mass dampers were attached to a structure, and Helmholtz resonators were inserted into an acoustic cavity, with the goal of minimizing the acoustic potential energy inside the acoustic cavity.

For one-off analyses, the time taken to learn how to conduct a modal coupling analysis is likely to be longer than using the computationally intensive approach of a fluid–structure interaction full harmonic analysis. However for large optimization problems, this modal coupling approach can reduce the computation times significantly.

The following sections describe the theory for fluid–structure interaction using modal coupling, followed by an example of a simply supported plate attached to one wall of a rectangular cavity.

9.3.2   Theory

Fahy and Gardonio [66, Sec. 7.6, p. 418] describe equations for determining the coupled structural–acoustic displacement response of a system, $w(\mathbf{r}_S)$, at some location, $\mathbf{r}_S$, on the structure, in terms of the combination and summation of structural and acoustic mode shapes. The structural mode shapes are evaluated by assuming that the structure is vibrating in a vacuum. The acoustic mode shapes of the enclosure surrounded by the structure are evaluated by assuming that the surrounding structure is infinitely rigid. The structural displacement at frequency, ω, is described in terms of a summation over the in-vacuo normal modes as:

$$w(\mathbf{r}_S, \omega) = \sum_{m=1}^{N_s} w_m(\omega)\phi_m(\mathbf{r}_S) , \qquad (9.2)$$

where the sinusoidal time dependency term, $e^{j\omega t}$, has been omitted from both sides of the equation. The quantity, $\phi_m(\mathbf{r}_S)$, is the mode shape of the mth structural mode at arbitrary location, $\mathbf{r}_S$ on the surface of the structure, and $w_m(\omega)$ is the modal participation factor of the mth mode at frequency, ω. Theoretically, the value of N_s should be infinity, but this is not possible to implement in practice, so N_s is chosen such that the highest-order mode considered has a natural frequency between twice and four times that of the highest frequency of interest in the analysis, depending on the model being solved and the accuracy required. The N_s structural mode shapes and natural frequencies can be evaluated using finite element analysis software, and the nodal displacements for a mode are described as a vector ϕ_m and then collated into a matrix $[\phi_1, \phi_2, ...\phi_{Ns}]$ for all the modes.

The acoustic pressure at frequency, ω, is described in terms of a summation of the acoustic modes of the fluid volume with rigid boundaries as:

$$p(\mathbf{r}, \omega) = \sum_{n=0}^{N_a} p_n(\omega)\psi_n(\mathbf{r}) , \qquad (9.3)$$

where the time dependency term has been omitted as it is not used in the analysis. The quantity, $\psi_n(\mathbf{r})$, is the acoustic mode shape of the nth mode at arbitrary location, $\mathbf{r}$ within the volume of fluid, and p_n is the modal participation factor of the nth mode. Theoretically, the value of N_a should be infinity, but this is not possible to implement in practice, so N_a is chosen such that the highest-order mode considered has a natural frequency between twice and four times that of the highest frequency of interest in the analysis, depending on the model being solved and the accuracy required. Note that the $n = 0$ mode is the acoustic bulk compression mode of the cavity that must be included in the summation. When conducting a modal analysis using finite element analysis software, the bulk compression mode of the cavity is the pressure response at 0 Hz. The N_a acoustic mode shapes and natural frequencies can be evaluated using finite element analysis software, where the nodal pressures for a mode, n, are described as a vector ψ_n and then collated into a matrix $[\psi_1, \psi_2, ...\psi_{Na}]$ for all the modes from 1 to N_a. The equation for the undamped coupled response of the structure for structural mode, m, is [66, Eq. (7.43)]

$$\ddot{w}_m + \omega_m^2 w_m = \frac{S}{\Lambda_m} \sum_{n=1}^{N_a} p_n C_{nm} + \frac{F_m}{\Lambda_m}, \tag{9.4}$$

where the frequency dependence of the pressures, forces, and displacements is implicit; that is, these quantities all have a specific and usually different value for each frequency, ω. The quantity, ω_m is the structural natural frequency for the mth mode, Λ_m is the modal mass calculated as

$$\Lambda_m = \int_S M(\mathbf{r}_0)\psi_m^2(\mathbf{r}_0)dS, \tag{9.5}$$

$M(\mathbf{r}_0)$ is the mass of structure at location $\mathbf{r}_0$, F_m is the modal force applied to the structure for the mth mode, S is the surface area of the structure, and C_{nm} is the dimensionless coupling coefficient between structural mode, m, and acoustic mode, n, given by the integral of the product of the structural, ϕ_m, and acoustic, ψ_n, mode shape functions over the surface of the structure, given by [66, Eq. (7.45)]

$$C_{nm} = \frac{1}{S} \int_S \psi_n(\mathbf{r}_s)\phi_m(\mathbf{r}_s)\mathrm{d}S. \tag{9.6}$$

The left-hand side of Equation (9.4) is a standard expression to describe the response of a structure in terms of its modes. The right-hand side of Equation (9.4) describes the forces that are applied to the structure in terms of modal forces. The first term describes the modal force exerted on the structure due to the acoustic pressure acting on the structure. The second term describes the forces that act directly on the structure. As an example, consider a point force F_a acting normal to the structure at nodal location (x_a, y_a) for which the mode shapes and natural frequencies have been evaluated using

FEA. As the force acts on the structure at a point, the modal force, F_m, at frequency, ω, for mode, m, is

$$F_m(\omega) = \psi_m(x_a, y_a)F_a(\omega) , \tag{9.7}$$

where $\psi_m(x_a, y_a)$ is the mth mode shape at the nodal location (x_a, y_a). Tangential forces and moment loadings on the structure can also be included in $F_m(\omega)$ and the reader is referred to Soedel [141] and Howard [72] for more information.

The dimensionless coupling coefficient C_{nm} may be calculated from finite element model results as

$$C_{nm} = \frac{1}{S} \sum_{i=1}^{J_s} \psi_n(\mathbf{r}_i)\phi_m(\mathbf{r}_i)S_i , \tag{9.8}$$

where S is the total surface area of the structure in contact with the acoustic fluid, S_i is the nodal area of the ith node on the surface (and hence the summation of all the individual nodal surface areas is equal to the total surface area $S = \sum_{i=1}^{i=J_s} S_i$), J_s is the total number of nodes on the surface, $\psi_n(\mathbf{r}_i)$ is the acoustic mode shape for the nth mode at node location $\mathbf{r}_i$, and $\phi_m(r_i)$ is the mode shape of the mth structural mode at node location $\mathbf{r}_i$. The area associated with each node of a structural finite element is sometimes available and if so, can be readily extracted from the software. The nodal areas can also be calculated by using the nodal coordinates that form the elements.

The equation for the undamped coupled response of the fluid (mode n) is given by Fahy and Gardonio [66, Eq. (7.44)]

$$\ddot{p}_n + \omega_n^2 p_n = -\left(\frac{\rho_0 c_0^2 S}{\Lambda_n}\right) \sum_{m=1}^{N_s} \ddot{w}_m C_{nm} + \left(\frac{\rho_0 c^2}{\Lambda_n}\right) \dot{Q}_n , \tag{9.9}$$

where the frequency dependence of p, w, and Q_n is implicit. The quantity, ω_n, represents the natural frequencies of the cavity, ρ_0 is the density of the fluid, c_0 is the speed of sound in the fluid, Λ_n is the modal volume defined as the volume integration of the square of the mode shape function

$$\Lambda_n = \int \psi_n^2(\mathbf{r})\mathrm{d}V , \tag{9.10}$$

and $\dot{Q}_n$ is a modal volume acceleration defined as

$$\dot{Q}_n(\omega) = \phi_n(x_b, y_b)\dot{Q}_b(\omega) , \tag{9.11}$$

where $\dot{Q}_b$ is the complex amplitude of the volume acceleration at nodal location (x_b, y_b), and $\phi_n(x_b, y_b)$ is the nth mode shape at the nodal location (x_b, y_b). A common definition for an acoustic source has units of volume velocity, which in this case is Q_b, and hence the time derivative of this expression is the source volume acceleration $\dot{Q}_b$.

An important point to note is that because the acoustic mode shapes used in the structural–acoustic modal coupling method are for a rigid-walled cavity, corresponding to a normal acoustic particle velocity at the wall surface equal to zero, the acoustic velocity at the surface resulting from the modal coupling method is incorrect [94]. However the acoustic pressure at the surface is correct, and this is all that is required for correctly coupling the acoustic and pressure modal equations of motion. The structural modal velocity is correct though.

For simple systems such as rectangular, rigid-walled cavities and simple plates it is possible to write analytical solutions for the mode shapes and natural frequencies. Anything more complicated than these simple structures nearly always involves the use of a discretized numerical model such as a finite element analysis, in which case, it is necessary to extract parameters from the finite element model to enable the calculation of the coupled response.

Cazzolato [52] described a method to calculate the acoustic and structural modal masses from a finite element model. When using finite element analysis software to evaluate the acoustic pressure mode shapes, the vectors returned by the software can be normalized to either unity or to the mass matrix. By normalizing the mode shapes to the mass matrix, the modal volume of the cavity can be obtained directly; that is

$$\mathbf{\Psi}_n^{\mathrm{T}}[\mathbf{M}_{\mathrm{fea}}]\mathbf{\Psi}_n = 1 \,, \tag{9.12}$$

where $\mathbf{\Psi}_n$ is the mass normalized mode shape function vector for the nth mode and $[\mathbf{M}_{\mathrm{fea}}]$ is the fluid element mass matrix defined as:

$$[\mathbf{M}_{\mathrm{fea}}] = \frac{1}{c_0^2} \int_{V_e} [\mathbf{N}][\mathbf{N}]^{\mathrm{T}} \mathrm{d}V_e \,, \tag{9.13}$$

where $[\mathbf{N}]$ is the shape function for the acoustic element with a single pressure degree of freedom and V_e is the volume of the element. If the mode shape vectors are normalized to unity; that is, the maximum value in the vector is 1, then

$$\hat{\mathbf{\Psi}}_n^{\mathrm{T}}[\mathbf{M}_{\mathrm{fea}}]\hat{\mathbf{\Psi}}_n = \frac{\Lambda_n}{c_0^2} \,, \tag{9.14}$$

where $\hat{\mathbf{\Psi}}_n$ is the mode shape vector normalized to unity for the nth mode and Λ_n is the modal volume of the nth mode. It can be shown that the relationship between the mass normalized mode shape vector $\mathbf{\Psi}_n$ and the unity normalized mode shape vector $\hat{\mathbf{\Psi}}_n$ is [62]

$$\mathbf{\Psi}_n = \frac{c_0}{\sqrt{\Lambda_n}}\hat{\mathbf{\Psi}}_n \,. \tag{9.15}$$

The advantage of this approach is that the mass matrix is not required. Given that the maximum value of the unity normalized mode shape vector $\hat{\mathbf{\Psi}}_n = 1$, the modal volume may be determined by rearranging Equation (9.15) to give

$$\Lambda_n = \frac{c_0^2}{\max(\mathbf{\Psi}_n^2)} \,. \tag{9.16}$$

Hence, to extract the acoustic modal volume of a system using finite element analysis software, an acoustic modal analysis is conducted and the results are normalized to the mass matrix. Then Equation (9.16) is used to calculate the acoustic modal volumes for each mode. The unity normalized mode shapes necessary for the modal coupling in Equations (9.2) to (9.11) can be calculated as

$$\hat{\mathbf{\Psi}}_n = \frac{\mathbf{\Psi}_n}{\max(\mathbf{\Psi}_n)} . \tag{9.17}$$

Equations (9.4) and (9.9) can form a matrix equation as:

$$\begin{bmatrix} \Lambda_m(\omega_m^2 - \omega^2) & -S[\mathbf{C}_{nm}]^{\mathrm{T}} \\ (-\omega^2)S[\mathbf{C}_{nm}] & \dfrac{\Lambda_n}{\rho_0 c_0^2}(\omega_n^2 - \omega^2) \end{bmatrix} \begin{bmatrix} \mathbf{w}_m \\ \mathbf{p}_n \end{bmatrix} = \begin{bmatrix} \mathbf{F}_m \\ \dot{\mathbf{Q}}_n \end{bmatrix}, \tag{9.18}$$

where all the m structural and n acoustic modes are included in the matrices, so that the square matrix on the left-hand side of Equation (9.18) has dimensions $(m + n) \times (m + n)$, $\mathbf{w}_m$ is a vector of all the structural modal participation factors from Equation (9.2), and $\mathbf{p}_n$ is a vector of all the acoustic modal participation factors from Equation (9.3). The left-hand matrix in Equation (9.18) can be made symmetric by dividing all terms in the lower equation by ω^2. The structural, w_m, and acoustic, p_n, modal participation factors, which are frequency dependent, can be calculated by pre-multiplying each side of the equation by the inverse of the square matrix on the left-hand side. Once these factors are calculated, the vibration displacement of the structure can be calculated from Equation (9.2) and the acoustic pressure inside the enclosure can be calculated using Equation (9.3).

The method described above can be used to make predictions of the vibro-acoustic response of an enclosed system, but it does have limitations. One mistake that is commonly made is to make numerical calculations with an insufficient number of structural and acoustic modes. This problem affects all numerical methods involving the summation of modes to predict the overall response and has been known since the early 1970s. Cazzolato et al. [56] demonstrated the errors that can occur with modal truncation and how it can lead to erroneous conclusions. As a start, the analyst should consider including structural and acoustic modes that have natural frequencies up to two octaves higher than the frequency range of interest. Methods have been proposed to reduce the number of modes required to be included in the analysis by including the affects of the higher-order modes in a residue or pseudo-static correction term [144, 68, 151].

The modal coupling method described above is applicable to vibro-acoustic systems where there is "light" coupling, such as between air and a structure. If the vibro-acoustic response of a system is to be calculated where there is "heavy" coupling due to the fluid loading, such as between water and a structure, then this method will generate erroneous results because it does not account for the cross-fluid coupling terms; that is, the coupling between fluid modes.

The equations described above do not have damping terms. In practice, it is common to include damping in the structure by using a complex elastic modulus, and damping in the fluid by a complex bulk modulus. Another way of incorporating damping is to include a modal loss factor.

One of the main advantages of using the modal coupling method is that the time taken to solve the system of equations is significantly less than conducting a full fluid–structure interaction analysis using finite element analysis. This is very important if optimization studies are to be conducted that involve many FEA evaluations while converging to an optimum solution.

The following section describes an example of the use of this modal-coupling theory to calculate the vibro-acoustic response of a system.

9.4 Example: Flexible Plate Attached to an Acoustic Cavity

Figure 9.2 shows a schematic of a rectangular cavity with 5 rigid-walls, and a flexible simply supported plate on the end wall. A harmonic point force acts on the plate that causes it to vibrate. The vibrating plate generates acoustic pressure inside the cavity. The dimensions and properties of the system are listed in Table 9.2.

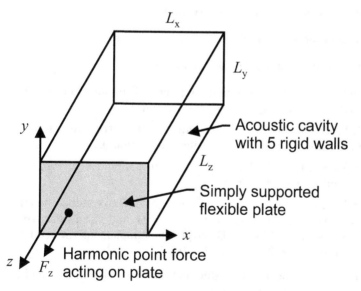

FIGURE 9.2

Schematic of the simply supported plate attached to the end of a rigid-walled rectangular cavity.

TABLE 9.2

Parameters Used in the Analysis of a Rectangular Acoustic Cavity with a Flexible Plate

Description	Parameter	Value	Units
Box width	L_x	0.5	m
Box height	L_y	0.3	m
Box depth	L_z	1.1	m
Plate width	L_x	0.5	m
Plate height	L_y	0.3	m
Plate thickness	h	0.003	m
Plate density	ρ_{plate}	2700	kg/m^3
Plate Young's modulus	E	70×10^9	Pa
Plate Poisson's ratio	ν	0.3	No units
Speed of sound of air	c_0	344	m/s
Density of air	ρ_0	1.21	kg/m^3
Driving force	F_z	1.0	N
Driving force location	(x_F, y_F, z_F)	$(0.10, 0.075, 0.00)$	(m,m,m)
Microphone location	(x_q, y_q, z_q)	$(0.125, 0.150, -0.875)$	(m,m,m)

Analyses will be conducted to calculate the

- sound pressure at a point inside the cavity,

- displacement of the plate at the driving point,

- acoustic potential energy inside the cavity, and

- structural kinetic energy of the plate.

These results will be calculated using four methods:

1. A theoretical model using modal coupling will be implemented in MATLAB, as described in Sections 9.4.1 and 9.4.2.

2. ANSYS Workbench is used to to conduct a fluid–structure interaction full harmonic analysis as described in Section 9.4.3.

3. ANSYS Mechanical APDL is used to conduct a fluid–structure interaction full harmonic analysis.

4. ANSYS Mechanical APDL is used in Section 9.4.4 to conduct modal analyses of the structure and acoustic cavity and the results are post-processed using MATLAB scripts that implement the modal coupling theory described in Section 9.3.2.

The results calculated using the four different methods should be identical and are demonstrated in the following sections.

9.4.1 Theory

The theoretical analysis of the vibro-acoustic response of a flexible plate attached to cavity has been investigated by numerous researchers [61, 70, 124, 140, 49, 132, 101]. The derivation of the theory follows these steps:

1. Calculate the natural frequencies of the simply supported plate (ω_m) covering the desired frequency range.

2. Sort the natural frequencies (ω_m) and modal indices (m_1, m_2) of the plate into order of increasing frequency.

3. Determine how many structural modes will be retained in the analysis (N_s).

4. Derive equations of motion of the plate in terms of its modal response.

5. Calculate the equivalent modal forcing function of the plate.

6. Follow similar steps in 1–5 for the acoustic cavity, where the natural frequencies of the rigid-walled cavity are calculated (ω_n).

7. Sort the natural frequencies (ω_n) and modal indices (n_1, n_2, n_3) of the cavity into order of increasing frequency.

8. Determine how many acoustic modes will be retained in the analysis (N_a).

9. Derive the equations for the acoustic response of the cavity in terms of its acoustic modes.

10. Calculate the vibro-acoustic coupling coefficient between the vibration of the plate and the acoustic response in the cavity.

11. Form a matrix equation of the coupled response of the system.

12. Calculate the modal participation factors for the structure and the acoustic cavity.

13. Use the modal participation factors to calculate the acoustic pressure in the cavity and the vibration response of the plate.

Modal Response of the Plate

The natural frequencies of a simply supported thin rectangular plate of dimensions L_x and L_y and thickness h can be calculated as [108, Eq. (4.20), p. 44, note that Leissa defines density ρ_{plate} as mass per unit area of the plate] [141, Eq. (6.18.8), p. 176] [132, Eq. (A1)]

$$\omega_m(m_1, m_2) = \sqrt{\frac{D}{h\rho_{\text{plate}}}} \left(\left[\frac{m_1\pi}{L_x}\right]^2 + \left[\frac{m_2\pi}{L_y}\right]^2 \right), \qquad (9.19)$$

and have units of radians / second, where m_1, m_2 are the modal indices along the axes x, y, respectively, the bending stiffness D is given by

$$D = \frac{Eh^3}{12(1 - \nu^2)} , \tag{9.20}$$

and E is the Young's modulus of the plate, h is the thickness, and ν is Poisson's ratio.

The natural frequencies of the plate are calculated for all permutations using an arbitrary large number for each mode index (i.e., $m_1 = 1 \cdots 50, m_2 = 1 \cdots 50$), and then sorted into increasing frequencies. The set of ordered natural frequencies is searched and counted to determine how many resonant modes are included in the analysis frequency range, which determines the number of structural modes that will be used in the analysis N_s.

The displacement of a point on the flexible plate can be described in terms of the summation of the structural modal participation factors w_m and its mode shapes ϕ_m as

$$w(x, y) = \sum_{m=1}^{N_s} \phi_m(x, y) \, w_m , \tag{9.21}$$

where the mode shapes of the simply supported plate are given by

$$\phi_{(m_1, m_2)}(x, y) = \sin\left(\frac{m_1 \pi x}{L_x}\right) \sin\left(\frac{m_2 \pi y}{L_y}\right) . \tag{9.22}$$

The modal mass of the simply supported plate is calculated using Equation (9.5) as

$$\Lambda_m = \int_{y=0}^{L_y} \int_{x=0}^{L_x} \rho_{\text{plate}} h \, \phi_m^2(x, y) \, dx \, dy \tag{9.23}$$

$$= \rho_{\text{plate}} h L_x L_y / 4 , \tag{9.24}$$

which is one quarter of the mass of the plate.

The modal response of a simply supported rectangular plate is given by Equation (9.4) as

$$\ddot{w}_m + \omega_m^2 w_m = \frac{S}{\Lambda_m} \sum_{n=1}^{N_a} p_n C_{n,m} + \frac{F_m}{\Lambda_m} , \tag{9.25}$$

where F_m is the modal force acting on the plate and is calculated as

$$F_m = F_z \phi_m(x_F, y_F) , \tag{9.26}$$

where F_z is the magnitude of the force applied at (x_F, y_F). Equation (9.25) can be rearranged as

$$\Lambda_m(\omega_m^2 - \omega^2) w_m - S \sum_{n=0}^{N_a} p_n C_{n,m} = F_m . \tag{9.27}$$

Modal Response of the Cavity

The natural frequencies of a rigid-walled rectangular acoustic cavity are calculated as [46, Eq. (6.13), p. 153] [132, Eq. (A5)]

$$\omega_a = \pi c_0 \sqrt{\left(\frac{n_1}{L_x}\right)^2 + \left(\frac{n_2}{L_y}\right)^2 + \left(\frac{n_3}{L_z}\right)^2},\tag{9.28}$$

and have units of radians/second, where n_1, n_2, n_3 are the modal indices along the x, y, z axes of the rectangular cavity. The natural frequencies of the cavity are calculated for all permutations using an arbitrary large number for each mode index (i.e., $n_1 = 0 \cdots 50, n_2 = 0 \cdots 50, n_3 = 0 \cdots 50$), and then sorted into increasing frequencies. The set of ordered natural frequencies is searched and counted to determine how many are included in the analysis frequency range, which determines the number of acoustic modes N_a that will be used in the analysis.

The acoustic pressure in the cavity can be described in terms of the summation of modal participation factors p_n and its mode shapes ψ_n as

$$p(x, y, z) = \sum_{n=0}^{N_a} \psi_n(x, y, z) \, p_n,\tag{9.29}$$

where the mode shapes of the acoustic cavity are given by [46, Eq. (6.13), p. 153]

$$\psi_{(n_1, n_2, n_3)}(x, y, z) = \cos\left(\frac{n_1 \pi x}{L_x}\right) \times \cos\left(\frac{n_2 \pi y}{L_y}\right) \times \cos\left(\frac{n_3 \pi z}{L_z}\right).\tag{9.30}$$

The modal volume of the rectangular cavity is calculated using Equation (9.10) as [46, p. 162]

$$\Lambda_n = \int_{z=0}^{L_z} \int_{y=0}^{L_y} \int_{x=0}^{L_x} \psi_n^2(x, y, z) \, dx \, dy \, dz\tag{9.31}$$

$$= V \times \varepsilon_{n1} \varepsilon_{n2} \varepsilon_{n3},\tag{9.32}$$

where $V = L_x \times L_y \times L_z$ is the volume of the cavity, $\varepsilon_n = 1$ if $n = 0$, and $\varepsilon_n = 1/2$ if $n > 0$.

The modal response of a rectangular cavity is given by Equation (9.9) as

$$\ddot{p}_n + \omega_n^2 p_n = -\left(\frac{\rho_0 c_0^2 S}{\Lambda_n}\right) \sum_{m=1}^{N_s} \ddot{w}_m C_{nm} + \left(\frac{\rho_0 c_0^2}{\Lambda_n}\right) \dot{Q}_n,\tag{9.33}$$

where $\dot{Q}_n$ is the modal acoustic source excitation of the cavity described in terms of modal volume acceleration and is calculated as

$$\dot{Q}_n = j\omega Q \psi_n(x_q, y_q, z_q),\tag{9.34}$$

where Q is the magnitude of the volume velocity applied at (x_q, y_q, z_q). For this example there is no acoustic source within the cavity, so the vector of the acoustic modal volume accelerations $\dot{\mathbf{Q}}_n$ is a vector of zeros of dimensions $[N_a \times 1]$.

Equation (9.33) can be rearranged as

$$\left(\frac{\Lambda_n}{\rho_0 c_0^2}\right)(\omega_n^2 - \omega^2)p_n + \frac{-\omega^2}{S}\sum_{m=1}^{N_s} C_{nm}w_m = \dot{Q}_n = 0. \tag{9.35}$$

Coupling Coefficient

The equation for the coupling coefficient is calculated by substituting the mode shape function for the plate (Equation (9.22)) and cavity (Equation (9.30)) into Equation (9.6).

For this example, the flexible plate is attached to the cavity in the xy plane at $z = 0$. Hence $z = 0$ can be substituted into Equation (9.30) for the mode shape function of the acoustic cavity and will reduce to

$$\psi_{n_1, n_2, n_3}(x, y, 0) = \cos\left(\frac{n_1 \pi x}{L_x}\right) \times \cos\left(\frac{n_2 \pi y}{L_y}\right). \tag{9.36}$$

Substituting Equation (9.36) and Equation (9.22) into Equation (9.6) and performing the double integration over the area of the plate ($x = 0 \cdots L_x$ and $y = 0 \cdots L_y$) gives

$$C_{n,m} = \frac{1}{\pi^2} \times \left[\frac{m_1 \times [(-1)^{n_1 + m_1} - 1]}{n_1^2 - m_1^2}\right] \times \left[\frac{m_2 \times [(-1)^{n_2 + m_2} - 1]}{n_2^2 - m_2^2}\right]. \tag{9.37}$$

References [140, 49, 132, 101] describe the analysis of a similar system where the flexible plate is mounted in the xy plane at $z = L_z$, and the coupling coefficient is given by

$$C_{n,m} = \frac{(-1)^{n_3}}{\pi^2} \times \left[\frac{m_1 \times [(-1)^{n_1 + m_1} - 1]}{n_1^2 - m_1^2}\right] \times \left[\frac{m_2 \times [(-1)^{n_2 + m_2} - 1]}{n_2^2 - m_2^2}\right]. \tag{9.38}$$

Matrix Equation of the Coupled System

The equations for motion of the plate in Equation (9.27) and the acoustic response of the cavity in Equation (9.35) can be formed into a matrix equation as shown in Equation (9.18) as

$$\underbrace{\begin{bmatrix} \mathbf{A}_{11} & \mathbf{A}_{12} \\ \mathbf{A}_{21} & \mathbf{A}_{22} \end{bmatrix}}_{\mathbf{A}} \begin{bmatrix} \mathbf{w}_m \\ \mathbf{p}_n \end{bmatrix} = \begin{bmatrix} \mathbf{F}_m \\ \dot{\mathbf{Q}}_n \end{bmatrix}, \tag{9.39}$$

where $\mathbf{A}_{11}$ is matrix of dimensions $[N_s \times N_s]$ with entries only on the diagonal defined as

$$\mathbf{A}_{11}(m, m) = \Lambda_m(\omega_m^2 - \omega^2), \tag{9.40}$$

$\mathbf{A}_{12}$ is matrix of dimensions $[N_s \times N_a]$ defined as

$$\mathbf{A}_{12} = -S[\mathbf{C}_{nm}]^{\mathrm{T}} , \qquad (9.41)$$

$\mathbf{A}_{21}$ is matrix of dimensions $[N_a \times N_s]$ defined as

$$\mathbf{A}_{21} = (-\omega^2)S[\mathbf{C}_{nm}] , \qquad (9.42)$$

$\mathbf{A}_{22}$ is matrix of dimensions $[N_a \times N_a]$ with entries only on the diagonal given by

$$\mathbf{A}_{22}(n, n) = \frac{\Lambda_n}{\rho_0 c_0^2} (\omega_n^2 - \omega^2) , \qquad (9.43)$$

where $\mathbf{w}_m$ is a vector of dimensions $[N_s \times 1]$ of the modal participation factors of the structural response, $\mathbf{p}_n$ is a vector of dimensions $[N_a \times 1]$ of the modal participation factors of the acoustic response, $\mathbf{F}_m$ is a vector of dimensions $[N_s \times 1]$ of the modal forces acting on the structure that were defined in Equation (9.7), and as defined previously $\dot{\mathbf{Q}}_n$ is a vector of dimensions $[N_a \times 1]$ comprising of zeros of the acoustic modal volume accelerations as there is no acoustic source in this example.

Calculation of Modal Participation Factors

The structural $\mathbf{w}_m$ and acoustic $\mathbf{p}_n$ modal participation factors are calculated by solving Equation (9.39) as

$$\left[\begin{array}{c} \mathbf{w}_m \\ \mathbf{p}_n \end{array} \right] = [\mathbf{A}]^{-1} \left[\begin{array}{c} \mathbf{F}_m \\ \mathbf{0} \end{array} \right] , \qquad (9.44)$$

where $\mathbf{A}$ is defined in Equation (9.39).

Calculation of Acoustic and Structural Responses

Once the structural and acoustic modal participation factors are calculated, the displacement of the structure can be calculated using Equation (9.21) and the acoustic pressure at points within the cavity can be calculated using Equation (9.29).

9.4.2 MATLAB

The theoretical model described in Section 9.4.1 was implemented in MAT-LAB and the file theory_coupled_plate_cavity.m is included with this book.

Results

Figure 9.3 shows the magnitude of the acoustic pressure in the cavity at the location $(0.125, 0.150, -0.875)$ over a frequency range from 0 Hz to 400 Hz. Figure 9.4 shows the magnitude of the displacement of the plate at the location of the driving force at $(0.10, 0.075, 0.00)$. These results will be compared with predictions using ANSYS in the following sections.

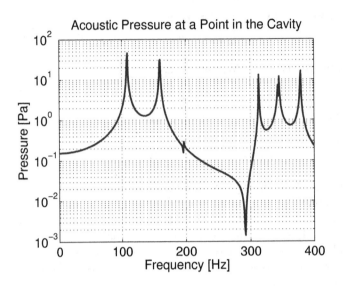

FIGURE 9.3
Magnitude of acoustic pressure at $(0.125, 0.150, -0.875)$ arising from a unit force at $(0.10, 0.075, 0.00)$ calculated using the MATLAB code theory_couple_plate_cavity.m.

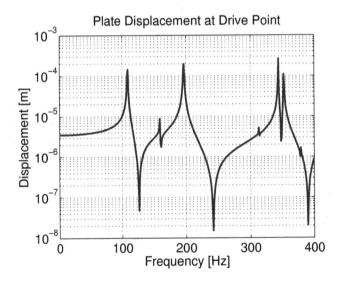

FIGURE 9.4
Magnitude of displacement at the node corresponding to the driving force at $(0.10, 0.075, 0.00)$ calculated using the MATLAB code theory_couple_plate_cavity.m.

9.4.3 ANSYS Workbench

The learning outcomes from this section are:

- how to use ANSYS to undertake a fluid–structure interaction analysis of a system;

- how to use the Worksheet to select a particular node in the finite element mesh as a Named Selection;

- how to apply a force to a node using ANSYS Workbench;

- how to request results at a particular node using ANSYS Workbench;

- how to select an appropriate element size for a fluid–structure interaction analysis needs to be based on the wavelength (or mode indices) in the structure, and the acoustic wavelength; and

- how to define custom material properties and assign them to a structure.

An ANSYS Workbench model will be created of the system shown in Figure 9.2. The default element types will be used for the acoustic domain, which are FLUID220 elements, and for the plate, which are SHELL181 elements. A harmonic analysis will be conducted using fluid–structure interaction coupling to predict the displacement of the plate at the location of the excitation force, and the sound pressure at a location within the cavity. The predictions using ANSYS Workbench analysis will be conducted using unsymmetric and symmetric formulations of the acoustic–structural coupling. The results from ANSYS Workbench will be compared with predictions using ANSYS Mechanical APDL and theoretical predictions using modal coupling methods.

Instructions

The completed ANSYS Workbench archive file fsi_plate_box.wbpz, which contains the .wbpj project file, is available with this book.

- Start ANSYS Workbench.

- Make sure the ACT Acoustics extension is loaded by clicking on Extensions | Manage Extensions. There should be a tick in the column Load for the row ExtAcoustics.

- From the Analysis Systems window, double-click on Harmonic Response.

- Double-click on row 3 Geometry to start DESIGNMODELER.

- Create a sketch in the XY Plane of a rectangle of dimensions $L_x = 0.5$ m $L_y = 0.3$ m.

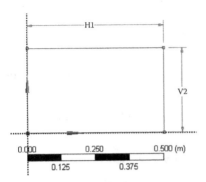

- Extrude the rectangle to a depth of 1.1 m in the -Z direction, by changing the Direction to Reversed. The reason we are extruding the area along the $-z$ axis is so that it is easier to see the plate when viewing the model from an isometric angle.

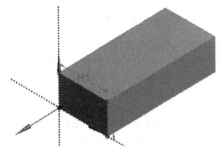

- In the next couple of steps we will create a surface body for the plate. Change the selection filter to Faces and left-click on the end wall of the box at $z = 0$.

- In the menu bar, select Concept | Surfaces from Faces, making sure that the end wall of the box is still selected, and in the window Details View, click the Apply button in the row Faces. Change the row Operation to Add Material.

- Click the Generate icon.

- In the Tree Outline window there should be 2 Parts, 2 Bodies. Click on the plus sign next to this entry to reveal the two bodies.

- Click on the object Surface Body in the tree list. In the window Details View, type the value 0.003 in the row Thickness.

Details of Surface Body	
Body	Surface Body
Thickness (> =0)	0.003 m
Thickness Mode	Manual
Surface Area	0.15 m²
Faces	1
Edges	4
Vertices	4
Fluid/Solid	Solid
Shared Topology Method	Automatic
Geometry Type	DesignModeler

- The next step is to define 3 Named Selections for (1) the Solid box, which will become the acoustic cavity, (2) the Surface Body, which will become the flexible plate, and (3) the edges of the plate that will have simple support boundary conditions. The important step here is that the named selection should be for Surface Body and not the face of the Solid body, so a couple of important steps are needed to ensure that the correct object is defined for the Named Selection. Right-click on the object Surface Body to open a context menu and select Hide All Other Bodies. This step ensures that it is not possible to accidentally select a face or the body of the box.

- Change the selection filter to Body. In the Model View window, left-click on the Surface Body. The status window at the bottom of the screen should say 1 Body: Area = 0.15 m².

- Right-click in the Model View window to open a context menu and left-click on Named Selection. Ensuring that the Surface Body is still selected, in the window Details View, click the Apply button in the row Geometry. In the row Named Selection, type the name NS_plate_body. Click the Generate button.

Details of NS_plate_body	
Named Selection	NS_plate_body
Geometry	1 Body
Propagate Selection	Yes
Export Selection	Yes

- Change the filter selection to Edges. Select the 4 Edges of the plate. The status bar at the bottom of the screen should indicate 4 Edges: Length = 1.6m. Right-click to open a context menu and left-click on Named Selection. Click the Generate icon. Change the name of the object to NS_plate_edges.

- In the Model View window, right-click to open a context menu and left-click on Show All Bodies.

- Make sure that the Body selection filter is still active. In the Model View window, left-click on the Solid box to select it. Right-click in the window Model view and left-click on Named Selection. In the window Details View, click the Apply button in the row Geometry. Change the name to NS_box_body. Click the Generate icon.

Details View	
⊟ **Details of NS_box_body**	
Named Selection	NS_box_body
Geometry	1 Body
Propagate Selection	Yes
Export Selection	Yes

- In the menu bar click File | Save Project and type in an appropriate filename such as fsi_plate_box.

- In the Project Schematic window, double-click on row 4 Model to start Mechanical.

- The first step is to define the mesh size, which will depend on the analysis frequency range. In this example we will be conducting a harmonic response analysis over 0 Hz to 400 Hz. Make sure the Body selection filter is active. Right-click on Model (A4) | Mesh and select Insert | Sizing. In the Geometry window, right-click and left-click on Select All. In the window Details of "Sizing" - Sizing, click the Apply button in the row Scope | Geometry. In the row Definition | Element Size, type the value 0.025. This will result in a mesh size of approximately $344/400/0.025 = 34$ elements per acoustic wavelength, which is much higher than the recommendation of 6 elements per wavelength for the FLUID220 acoustic elements that will be used. At 400 Hz, the maximum mode indices in the plate are $m = 3$ along the x-axis and $n = 2$ along the y-axis. If 6 elements per wavelength is used for the plate, there should be at least $3 \times 6 = 18$ divisions along the x-axis of the plate, which would result in an element size of $L_x/18 = 0.5/18 = 0.0278$ m. Hence, an element size of 0.025 is reasonable. It is advisable that the mesh density of the acoustic elements in the region of fluid–structure interaction should have a similar mesh density, and ideally coincident nodes, as the structural elements, so that the displacement of the nodes belonging to the structural elements will cause an identical displacement of the nodes belonging to the acoustic elements. The mesh density of the acoustic elements farther away from the fluid–structure interaction region need only have sufficient mesh density for the acoustic wavelength being investigated.

- In the menu bar click on File | Save Project.

- Right-click on Model (A4) | Mesh and left-click on Generate Mesh. As the bodies are regular shapes, a swept mesh will be created.

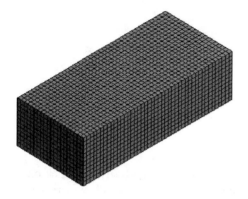

- The next step is to change the material properties of the Surface Body, which is for the flexible plate, from Structural Steel to Aluminium. In the Outline window, click on Model (A4) | Geometry | Surface Body.

- In the window Details of "Surface Body" in the row Material | Assignment, click in the cell for Structural Steel, which will reveal an arrow pointing to the right. Click on this arrow and select New Material....

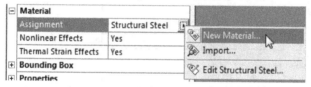

- Click the OK button on the warning dialog box.

- In the window Outline of Schematic A2: Engineering Data, left-click in the cell that has written Click here to add a new material and type myAluminium.

	A	B	C
1	Contents of Engineering Data		Source
2	⊟ Material		
3	🏷 Structural Steel		🖙 General_Materials.xml
4	?🏷 myAluminium		
*	Click here to add a new material		

Outline of Schematic A2: Engineering Data

- The material properties that need to be defined are Young's modulus, density, and Poisson's ratio. In the window Toolbox, under the branch Linear Elastic, double-click on Isotropic Elasticity, which add this material property to the window Properties of Outline Row 4: myAluminium.

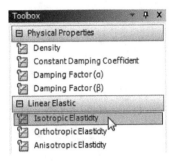

- In the window `Properties of Outline Row 4: myAluminium`, click on the plus sign in row 2 next to `Isotropic Elasticity`, which will reveal the properties that can be defined. In cell B4 for the value of `Young's Modulus`, type `70e9`. Enter a value of `0.3` for `Poisson's Ratio`.

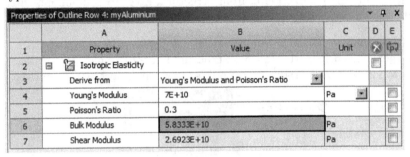

- In the window `Toolbox`, double-click on `Physical Properties | Density`. In the window `Properties of Outline Row 4: myAluminium`, enter the value of `2700` for `Density`.

- That completes the definition of the material properties for the plate. Click on the `Refresh Project` icon.

- Return to the `Mechanical` window and change the material property of the `Surface Body` to `myAluminium`.

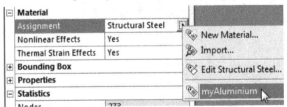

- The next step is to define `Named Selections` for the node locations where the excitation force is applied to the plate, and where the pressure will be measured within the acoustic cavity. This can be done using the `Worksheet`. Right-click on `Model (A4) | Named Selection` and left-click on `Insert | Named Selection`. In the window `Details of "Selection"`, change the row

Scope | Scoping Method to Worksheet. Move the mouse cursor into the window Worksheet, right-click in the table to open the context menu, and left-click on Add Row.

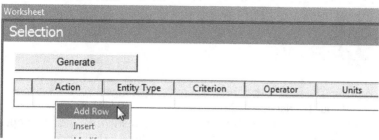

- The Worksheet can be used to select a particular object by using several selection filters. The first step is to select the body associated with the material myAluminium. Select all the nodes associated with this object, then filter the set of nodes to only those where $x = 0.1$ and then filter the remaining set to the node where $y = 0.075$. Change the entries in the Worksheet as per the following image. When finished, click on the Generate button. In the window Details of "Selection", the row Scope | Geometry should indicate 1 Node. Rename this Named Selection object from Selection to NS_plate_force_node.

Worksheet

Selection

Generate							
	Action	Entity Type	Criterion	Operator	Units	Value	Lower Bound
☑	Add	Body	Material	Equal	N/A	myAluminium	N/A
☑	Convert To	Mesh Node	N/A	N/A	N/A	N/A	N/A
☑	Filter	Mesh Node	Location X	Equal	m	0.1	N/A
☑	Filter	Mesh Node	Location Y	Equal	m	0.075	N/A

- Repeat this process to define a Named Selection object for the node at location $(0.125, 0.150, -0.875)$. Rename the Named Selection object to NS_mic_node.

Worksheet

NS_mic_node

Generate						
	Action	Entity Type	Criterion	Operator	Units	Value
☑	Add	Body	Named Selection	Equal	N/A	NS_box_body
☑	Convert To	Mesh Node	N/A	N/A	N/A	N/A
☑	Filter	Mesh Node	Location X	Equal	m	0.125
☑	Filter	Mesh Node	Location Y	Equal	m	0.15
☑	Filter	Mesh Node	Location Z	Equal	m	-0.875

- With the node selected, the harmonic force will be applied to the structure. In the window Outline, left-click on Harmonic Response (A5). In the Environment toolbar, click on Direct FE | Nodal Force.

- In the window Details of "Nodal Force", change the rows Scope | Scoping Method to Named Selection, Scope | Named Selection to NS_plate_force_node, and Definition | Z Component to 1. This will apply a force of 1 N that acts normal to the plate at the selected node.

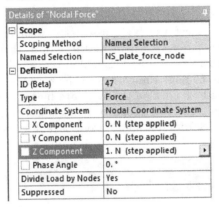

- In the Acoustics toolbar, click on Acoustic Body | Acoustic Body. In the window Details of "Acoustic Body", change the rows Scope | Scoping Method to Named Selection, Scope | Named Selection to NS_box_body, Definition | Mass Density to 1.21, and Definition | Sound Speed to 344.

- In order to couple the vibration of the plate to an acoustic response it is necessary to change the option Definition | Acoustic-Structural Coupled Body Options to Coupled With Unsymmetric Algorithm or Coupled With Symmetric Algorithm. As an exercise, the reader could leave the setting as Uncoupled and the results will show that no sound is generated in the cavity.

Details of "Acoustic Body"	
Scope	
Scoping Method	Named Selection
Named Selection	NS_box_body
Definition	
Temperature Dependency	No
Mass Density	1.21 [kg m^-1 m^-1 m^-1]
Sound Speed	344 [m sec^-1]
Dynamic Viscosity	0 [Pa sec]
Thermal Conductivity	0 [W m^-1 C^-1]
Specific Heat Cv	0 [J kg^-1 C^-1]
Specific Heat Cp	0 [J kg^-1 C^-1]
Equivalent Fluid of Perforated Material	No
Reference Pressure	2E-05 [Pa]
Reference Static Pressure	101325 [Pa]
Acoustic-Structural Coupled Body Options	Coupled With Unsymmetric Algorithm ▼
Perfectly Matched Layers (PML)	Off

- The next step is to define the FSI interface between the plate and the acoustic cavity. This can be done automatically provided that ANSYS Workbench created a contact region listed under the Model (A4) | Connections | Contacts | Contact Region. Click on this entry. In the window Details of "Contact Region", the rows Scope | Contact Bodies should indicate Solid and Scope | Target Bodies should indicate Surface Body. In the Acoustic toolbar, click on Tools | Automatically create FSI condition based on contacts. An object should be created under Harmonic Response (A5) | Acoustic FSI Interface.

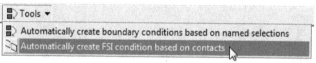

- In the Environment toolbar, click on Supports | Simply Supported. Change the rows Scope | Scoping Method to Named Selection, and Scope | Named Selection to NS_plate_edges.

- In the Acoustics toolbar, click on Results | Acoustic SPL. This will create a display of the sound pressure levels within the acoustic body.

- In the window Details of "Acoustic SPL", change the rows Scope | Scoping Method to Named Selection, and Scope | Named Selection to NS_box_body.

- In the Acoustics toolbar, click on Results | Acoustic Time_Frequency Plot.

- In the window Details of "Acoustic Time_Frequency Plot", change the rows Scope | Scoping Method to Named Selection, and Scope | Named Selection to NS_mic_node. This object will plot a spectrum of the sound pressure at the select node in the cavity versus frequency.

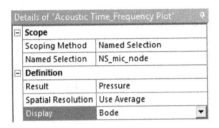

- We would also like to view a frequency response plot of the displacement of the plate where the point normal force is applied. In the Outline window, left-click on Solution (A6). In the Solution toolbar, click on Frequency Response | Deformation. In the window Details of "Frequency Response" change the rows Scope | Scoping Method to Named Selection, and Scope | Named Selection to NS_plate_force_node. Change the row Definition | Orientation to Z Axis. Rename this object to Frequency Response Plate Disp.

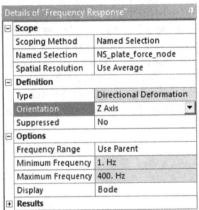

- In the window Outline, left-click on Harmonic Response (A5) | Analysis Settings. Change the rows Range Minimum to 1, Range Maximum to 400, and Solution Intervals to 399. Note that no response is calculated at the lower end of the frequency range. As a first step to demonstrate the technique of undertaking a fluid–structure interaction analysis, the reader may wish to change this value to 4, so that results are obtained only at 4 frequencies, rather than 399 frequencies, which can take several hours to solve and is done for the comparison of the various analysis methods. Change row Solution Method to Full. Click on Analysis Data Management | Save MAPDL db and change this setting to Yes, which enables the calculation of results during the post-processing stage.

- Click on File | Save Project.

This completes the setup of the project. Click the Solve icon. Once the computations have completed, which could take several hours, click on each of the three results to inspect them.

Results

- Right-click on Solution (A6) | Acoustic Time_Frequency Plot and left-click on Export. Type in an appropriate filename such as ansys_wb_mic_pressure.txt. The results predicted using ANSYS Workbench are shown in Figure 9.5.

- Right-click on Solution (A6) | Frequency Response Plate Disp and left-click on Export. Type in an appropriate filename such as ansys_wb_plate_disp.txt. The results predicted using ANSYS Workbench are shown in Figure 9.6.

There is an option to use a symmetric formulation of the matrices for fluid–structure interaction. The following steps describe how to set this option and recalculate the results. The aim is to show that identical results are calculated in a shorter time compared to using the unsymmetric formulation.

- In ANSYS Workbench Mechanical, in the Outline window, click on Harmonic Response (A5) | Acoustic Body.

- In the window Details of "Acoustic Body", click in the cell next to the row Acoustic - Structural Coupled Body Options and change the setting to Coupled With Symmetric Algorithm.

Details of "Acoustic Body"	
Scope	
Scoping Method	Named Selection
Named Selection	NS_box_body
Definition	
Temperature Dependency	No
Mass Density	1.21 [kg m^-1 m^-1 m^-1]
Sound Speed	344 [m sec^-1]
Dynamic Viscosity	0 [Pa sec]
Thermal Conductivity	0 [W m^-1 C^-1]
Specific Heat Cv	0 [J kg^-1 C^-1]
Specific Heat Cp	0 [J kg^-1 C^-1]
Equivalent Fluid of Perforated Material	No
Reference Pressure	2E-05 [Pa]
Reference Static Pressure	101325 [Pa]
Acoustic-Structural Coupled Body Options	Coupled With Symmetric Algorithm
Perfectly Matched Layers (PML)	Uncoupled
	Coupled With Unsymmetric Algorithm
	Coupled With Symmetric Algorithm

- To ensure that the results from the previous analysis using the unsymmetric formulation have been erased, right-click on Solution (A6) and left-click on

Clear Generated Data, then click the Yes button to confirm that the data
from the previous analysis should be erased.

- In the menu bar click on File | Save Project.

- Click the Solve icon.

Once the computations have been completed, which could take several
hours, the results can be exported as described earlier and compared with the
results where the unsymmetric matrix formulation was used.

Figure 9.5 shows the magnitude of the acoustic pressure at $(0.125, 0.150,$
$-0.875)$ calculated using ANSYS Workbench with unsymmetric and symmetric formulations for the acoustic-structural coupling and the results are identical.

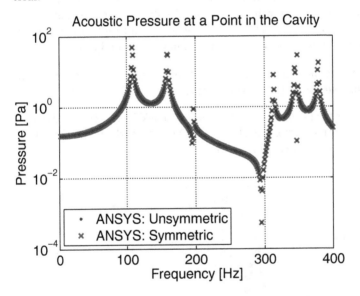

FIGURE 9.5
Acoustic   pressure   at $(0.125, 0.150, -0.875)$ calculated   using
ANSYS Workbench for an FSI full harmonic analysis with unsymmet-
ric and symmetric formulations for the acoustic–structural coupling.

Figure 9.6 shows the magnitude of displacement at the node at
$(0.10, 0.075, 0.00)$ (where the driving force was applied) calculated using
ANSYS Workbench with unsymmetric and symmetric formulations for the
acoustic-structural coupling, and the results are identical.

The main difference between these two solution methods is that the computation time taken to complete the analysis with the symmetric formulation

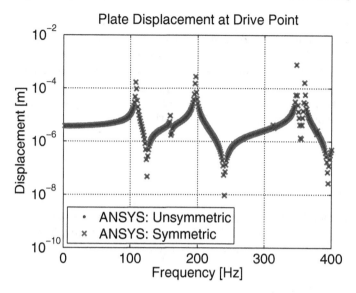

FIGURE 9.6
Magnitude of displacement at the node corresponding to the driving force of 1 N at $(0.10, 0.075, 0.00)$ calculated using ANSYS Workbench for an FSI full harmonic analysis with unsymmetric and symmetric formulations for the acoustic–structural coupling.

is significantly less than using an unsymmetric formulation. The unsymmetric formulation took about 8.8 hours and the symmetric formulation took 4.5 hours on a laptop computer (Microsoft Windows 7 64-bit, Intel Core i5 M540 2.53 GHz, 4 GB RAM) in 2014. The actual computation times will vary depending on numerous factors such as processor type, processor speed, operating system, memory, hard disk type, and so on. Hence it is not possible to estimate how long these analyses will take on other system configurations.

The next section describes how to conduct a similar analysis using ANSYS Mechanical APDL.

9.4.4 ANSYS Mechanical APDL

A finite element model will be created using ANSYS Mechanical APDL of the system shown in Figure 9.2. The acoustic cavity is modeled using FLUID30 acoustic elements and the structural plate is modeled using SHELL181 elements. The acoustic elements only have the pressure degree of freedom active, except for the elements in contact with the structural plate, which have pressure and 3 displacement degrees of freedom. The edges of the plate have simply supported boundary conditions. A point force is applied to a node on the plate in a direction normal to the plate.

For this example, where there is only a point force acting on the plate, it is acceptable that the nodes for the structural plate are shared with the nodes for the adjacent acoustic elements. However, if one were attempting to apply an external harmonic excitation pressure loading to the plate or trying to examine the transmission loss of the plate, as the nodes are shared, the pressure would also be directly applied to the acoustic elements and hence there would be no transmission loss across the plate. In order to investigate an external pressure loading, it would be necessary to have separate nodes for the plate and acoustic elements and couple the displacement degrees of freedom.

Instructions

- Start ANSYS Mechanical APDL.

- Type the command /INPUT,box_plate,inp   This will create the model, apply the loads, and perform a fluid–structure interaction full harmonic analysis. The finite element model that is created is shown in the following image.

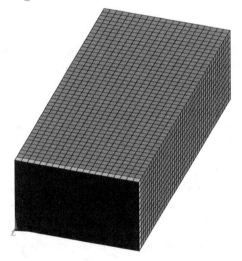

The element size is 0.025 m, which results in line divisions of $(20, 12, 44)$ along the (L_x, L_y, L_z) edges. As described in Section 9.4.3 for the analysis using ANSYS Workbench, the selection for the element size was based on the structural wavelength in the plate at 400 Hz.

- Once the calculations have been completed, the graph in Figure 9.7 will be displayed.

- The next step is to export the results from this harmonic analysis so that they can be compared with the predictions using other analysis methods. Either type the following lines of APDL code into the command line, or open the file box_plate.inp and copy the code into the command line. Note that when using ANSYS Mechanical APDL in interactive (rather than batch) mode, when trying to redirect the output to a file, this command is ignored

and the output is displayed in a new window. The user then has to select
File | Save As and type in an appropriate filename.

```
1   ! Export the drive point displacement
2   /OUTPUT,d_node,txt
3   PRVAR,2
4   /OUTPUT
5
6   ! Export the pressure inside the cavity
7   /OUTPUT,p_node,txt
8   PRVAR,3
9   /OUTPUT
10
11  ! Export the displacement at all nodes of the plate
12  ! and at all frequencies.
13  ! This could take a while....
14  ESEL,S,ENAME,,SHELL181
15  NSLE,S,1
16  /INPUT,nodal_disp,inp
17  ALLS
18
19  ! Export the pressure at all nodes in the cavity
20  ! at all frequencies.
21  ! This could take a while....
22  ESEL,S,ENAME,,FLUID30
23  NSLE,S,1
24  /INPUT,nodal_p,inp
25  ALLS
```

Results

Figure 9.7 shows the graph from ANSYS Mechanical APDL of the displacement of the plate versus frequency at the node where the normal force of
$F_z = 1$ N was applied, and hence this result is the drive point compliance
(inverse of stiffness). The first resonance peak in the graph occurs at 108 Hz.

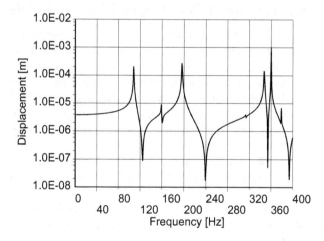

FIGURE 9.7
Displacement of the plate for a normal force $F_z = 1$ N, calculated using
ANSYS Mechanical APDL.

Figure 9.8 shows a contour plot of the sound pressure level (in dB RMS re 20 μPa) inside the cavity at 108 Hz, which corresponds to the first resonance peak in Figure 9.7.

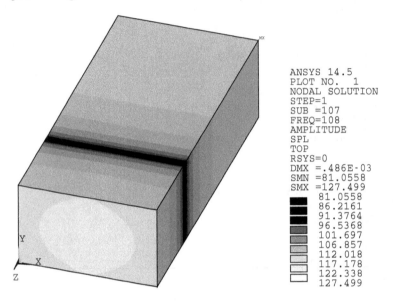

```
ANSYS 14.5
PLOT NO.   1
NODAL SOLUTION
STEP=1
SUB =107
FREQ=108
AMPLITUDE
SPL
TOP
RSYS=0
DMX =.486E-03
SMN =81.0558
SMX =127.499
      81.0558
      86.2161
      91.3764
      96.5368
      101.697
      106.857
      112.018
      117.178
      122.338
      127.499
```

FIGURE 9.8
Sound pressure level the inside cavity at 108 Hz.

9.4.5   MATLAB Code for Modal Coupling of ANSYS Models

This section describes the use of MATLAB functions to calculate the vibro-acoustic response from structural and acoustic modal analyses conducted using ANSYS and using the modal coupling theory described in Section 9.3.2.

The code essentially performs the same mathematical operations as were done in Section 9.4.2, but the theoretically calculated mode shapes, natural frequencies, etc., are replaced with values calculated using the results from ANSYS. The use of these scripts enables the analysis of complex shaped systems.

Table 9.3 lists the filename of each function along with a brief description of what it does. Further documentation about the use of these functions can be found in Appendix C.

Comparisons can be made between the results calculated using ANSYS Mechanical APDL where a fluid–structure interaction full harmonic analysis was conducted, and the theoretical modal coupling method. At Release 14.5 of ANSYS, there is no in-built feature to conduct a harmonic response using modal summation of a coupled vibro-acoustic system. However, this capability does exist in Release 15.0.

TABLE 9.3

MATLAB Functions Used to Calculate the Vibro–Acoustic Response
Using Modal Coupling Theory

Filename	Description
loadmodel.m	Loads the ANSYS results created with the ANSYS macro extract_modes.mac.
loadstr.m	Called by loadmodel.m and loads the results from the structural analysis.
loadcav.m	Called by loadmodel.m and loads the results from the acoustic analysis.
bli.m	Called by loadmodel.m and calculates the coupling coefficients.
plotmodel.m	Plots the structural or acoustic model.
plotmode.m	Plots the structural or acoustic mode shapes.
createloadcase.m	Defines the structural and acoustic loads.
coupled_response_fahy.m	Performs the vibro-acoustic coupling and calculates the modal participation factors that enables calculation of other results.
cav_pressure.m	Calculates the acoustic pressure at a node.
str_displacement.m	Calculate the structural displacement at a node.
acousticpotentialenergy.m	Calculate the acoustic potential energy of the cavity.
structuralkineticenergy.m	Calculate the kinetic energy of the structure.

Instructions

• Assuming that the ANSYS Mechanical APDL structural and acoustic model
  has already been created using the instructions in Section 9.4.4, type the
  following commands

```
1  FINI
2  ALLS
3  extract_modes,'boxplate',1600
```

These commands will execute the macro extract_modes.mac that is included
with this book. The natural frequencies and mode shapes of the structure
and acoustic cavity are calculated up to 1600 Hz and the results are exported
to a number of files that will be post-processed using MATLAB scripts.

The next step is to process the exported ANSYS results using MATLAB scripts. Start MATLAB and change the working directory to the path where the ANSYS results and MATLAB scripts are stored. The MATLAB script compare_modal_coupling_vs_full_FSI_press.m included with this book can be used to calculate the coupled modal response and display the acoustic pressure at a point within the cavity, and compare it with the results from the full FSI analysis. The script contains the following MATLAB commands.

```
1   % Define a frequency vector in Hertz
2   freq=[1:400];
3   % Load the modal results from Ansys
4   m=loadmodel('boxplate');
5   % Plot the shape of the acoustic cavity
6   plotmodel(m.c);
7   % Plot the shape of the structure
8   plotmodel(m.s);
9   % Define the load case - a point force on the plate at node 373 of force 1N.
10  m=createloadcase(m,'point_force',freq,373,1,[],[],[],[]);
11  % Calculate the modal participation factors for the coupled response.
12  m=coupled_response_fahy(m,'point_force');
13  % Use the participation factors to calculate the pressure at a node
14  [cavp,f]=cav_pressure(m,'point_force',5414);
15  % Import the Ansys results from the FSI full harmonic analysis
16  ansys_p_5414=importdata('p_node.txt',' ',7);
17  % Plot the theoretical modal coupling and Ansys results
18  p1h=semilogy(f,abs(cavp),ansys_p_5414.data(:,1),ansys_p_5414.data(:,2),'.')
19  xlabel('Frequency [Hz]')
20  ylabel('Pressure [Pa]')
21  legend('ANSYS MAPDL Modal Coupling','Ansys MAPDL Full FSI',  ...
22          'location','southwest')
```

Figure 9.9 shows the shape of the (a) acoustic cavity and the (b) structure, respectively, using the MATLAB script plotmodel.m, which imports the data generated by the ANSYS script extract_modes.mac. The arrows in Figure 9.9(b) indicate the direction normal to the node, which is important as only motion normal to the surface of the structure causes pressure excitation in the acoustic fluid. These two plots can be used to check that the MATLAB model of the system has been interpreted correctly.

Figure 9.10 shows the magnitude of the (complex) acoustic pressure versus frequency at a node in the cavity at the coordinate $(0.125, 0.150, -0.875)$, calculated using the modal coupling theory in Section 9.4.1, using ANSYS Mechanical APDL and the modal coupling theory described in Section 9.4.5, and using ANSYS Mechanical APDL for a full harmonic analysis, which was described in Section 9.4.4. Although the three sets of results generally have good correlation, it can be seen that at the anti-resonances in the response (i.e., the troughs), the results from the theoretical modal coupling do not precisely align with the results from the ANSYS FSI full harmonic analysis. This is caused by modal truncation, where only modes up to 1600 Hz were used to calculate the vibro-acoustic response using the modal coupling theory. If a greater number of modes were included in the analysis, the modal-coupling results would approach the results from the FSI full harmonic analysis.

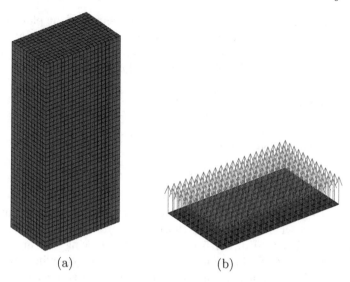

(a)   (b)

FIGURE 9.9

Plot of the finite element model using the MATLAB script `plotmodel.m` showing the (a) acoustic cavity, and (b) structure, where the arrows indicate the direction of the vector normal to the node.

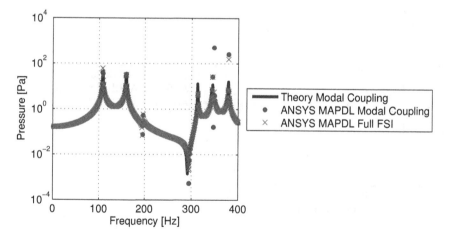

FIGURE 9.10

Acoustic pressure at $(0.125, 0.150, -0.875)$ calculated using modal coupling theory, ANSYS Mechanical APDL using modal coupling, and ANSYS Mechanical APDL for a FSI full harmonic analysis.

Similar comparisons can be made for the displacement at the driving point on the structure. The MATLAB script `compare_modal_coupling_vs_full_FSI_disp.m` included with this book can be used, after running the script `compare_` modal_coupling_vs_full_FSI_press.m, to plot the displacement of the plate at the driving point, calculated using the modal coupling theory, and compared with the results from the full harmonic fluid–structure interaction analysis. The script contains the following MATLAB commands.

```
1   % Use the participation factors to calculate the displacement at a node
2   [str_disp,f]=str_displacement(m,'point_force',373);
3   % Import the Ansys results from the full FSI harmonic analysis
4   ansys_d_373=importdata('d_node.txt',' ',7);
5   % Plot the results of theoretical modal coupling and Ansys
6   semilogy(f,abs(str_disp),ansys_d_373.data(:,1),ansys_d_373.data(:,2),'.')
7   xlabel('Frequency [Hz]')
8   ylabel('Displacement [m]')
9   legend( 'ANSYS MAPDL Modal Coupling','Ansys MAPDL Full FSI ',      ...
10          'location','southwest')
```

Figure 9.11 shows the magnitude of the (complex) displacement versus frequency at the node where the driving harmonic force was applied on the plate at the coordinate $(0.10, 0.075, 0.00)$, calculated using the modal coupling theory, using ANSYS Mechanical APDL and modal coupling, and ANSYS Mechanical APDL for a full harmonic analysis. As described previously, the reason for the small discrepancies in the anti-resonances is due to an insufficient number of modes included in the analyses.

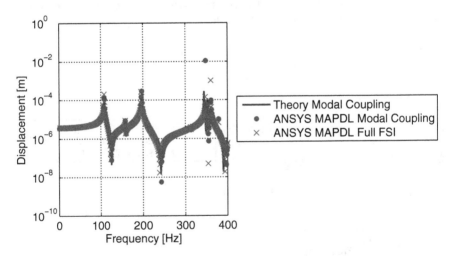

FIGURE 9.11
Magnitude of displacement at the node corresponding to the driving force at $(0.10, 0.075, 0.00)$ calculated using modal coupling theory, ANSYS Mechanical APDL using modal coupling, and ANSYS Mechanical APDL for a FSI full harmonic analysis.

The following instructions describe the steps to calculate the acoustic potential energy using the modal coupling theory from Section 2.10.3 and Equation (2.54). These modal coupling results are compared with the results from the FSI full harmonic analysis where the real and imaginary pressure at every node and every frequency was exported. The results are processed using the following MATLAB commands that call the MATLAB script ape_from_ansys.m, which is included with this book.

```
1  % Calculate the acoustic potential energy from modal coupling results
2  [ape,f]=acousticpotentialenergy(m,'point_force');
3  % Need to define a vector for the analysis frequencies that were used in
4  % the FSI full analysis
5  ansys_freq_full=[2:400];
6  % Post-process the Ansys results from the FSI full harmonic analysis
7  % to calculate the acoustic potential energy
8  ape_from_ansys;
9  % Plot the results of theoretical modal coupling and Ansys
10 p2h=semilogy(f,ape,ansys_freq_full,ansys_ape_long_way,'.');
11 xlabel('Frequency [Hz]')
12 ylabel('Acoustic Potential Energy [J/m^2]')
13 legend('Theory','ANSYS','location','northwest')
```

Figure 9.12 shows the acoustic potential energy calculated using the modal coupling theory and Equation (2.55) and from the results from the ANSYS FSI full harmonic analysis. The results are identical, which is to be expected.

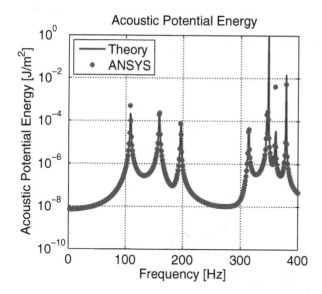

FIGURE 9.12
Acoustic potential energy in the cavity calculated using modal coupling theory and Equation (2.55) and ANSYS Mechanical APDL for a FSI full harmonic analysis.

The following instructions describe the steps to calculate the structural kinetic energy using the modal coupling theory from Section 2.10.5 and Equation (2.64). These modal coupling results are compared with the results from the FSI full harmonic analysis where the real and imaginary displacement at every node and every frequency was exported. The results are pro-

cessed using the following MATLAB commands that call the MATLAB script
ske_from_ansys.m , which is included with this book.

```
1   % Calculate the structural kinetic energy from modal coupling results
2   [ske,f]=structuralkineticenergy(m,'point_force');
3   % Need to define several parameters
4   % a vector for the analysis frequencies that were used in
5   % the FSI full analysis
6   ansys_freq_full=[2:400];
7   % density of the plate
8   rho_structure=2700;
9   % thickness of the plate
10  thick_structure=0.003;
11  % Calculate the structural kinetic energy from the FSI full harmonic analysis
12  ske_from_ansys;
13  % Plot the results of theoretical modal coupling and Ansys
14  p3h=semilogy(f,ske,ansys_freq_full,ansys_ske_long_way,'.');
15  xlabel('Frequency [Hz]')
16  ylabel('Structural Kinetic Energy [J]')
17  title('Structural Kinetic Energy')
18  legend('Theory','ANSYS','location','southeast')
```

Figure 9.13 shows the structural kinetic energy calculated using the modal
coupling theory and Equation (2.64) and from the results from the ANSYS
FSI full harmonic analysis. The results are identical, which is to be expected.

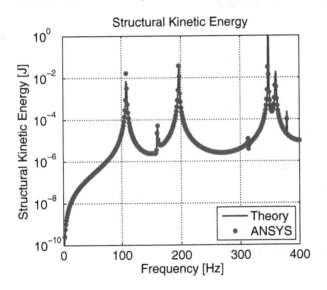

FIGURE 9.13
Structural kinetic energy of the plate calculated using modal coupling theory
and Equation (2.64) and ANSYS Mechanical APDL for an FSI full harmonic
analysis.

9.5 Example: Transmission Loss of a Simply Supported Panel

9.5.1 Learning Objectives

The learning objectives of this section are:

- learn how to model a system that incorporates fluid–structure interaction and some of the subtleties, and

- learn how to calculate the sound power that radiates out of an imaginary hemisphere in ANSYS.

9.5.2 Theory

Roussos [130] describes theory to calculate the transmission loss of a simply supported panel using the modal summation method. The method involves

1. calculating the modal force that is applied to a panel due to an incident plane-wave striking the panel at an arbitrary angle,

2. applying this modal excitation force to the panel and calculating the vibration response,

3. calculating the pressure, intensity, and radiated power from the panel, and

4. calculating the transmission loss of the panel by using the ratio of the incident sound power striking the panel and the radiated power from the panel.

The theoretical model makes an assumption of "weak" coupling where the fluid-loading of the air that is contact with the panel is ignored.

Incident Acoustic Excitation

Consider a plane wave incident on a simply supported panel as shown in Figure 9.14. Figure 9.15 shows a cross-sectional view of the acoustic plane wave striking the panel.

The incoming pressure wave has an amplitude P_i and strikes the panel at angles θ_i normal to the panel and ϕ_i in the plane of the panel. The pressure that is incident on the panel $p_i(x, y)$ is given by

$$p_i(x, y) = P_i \exp\left[j(\omega t - kx \sin\theta_i \cos\phi_i - ky \sin\theta_i \sin\phi_i) \right], \qquad (9.45)$$

where $k = \omega/c_0 = 2\pi f/c_0$ is the wavenumber, and ω is the circular frequency (radians/s). When the incident acoustic wave is reflected by the surface of the panel, the pressure at the surface of the panel is doubled [47, Sec. 5.9.2,

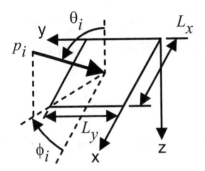

FIGURE 9.14
Coordinate system for a plane wave striking a simply supported rectangular panel.

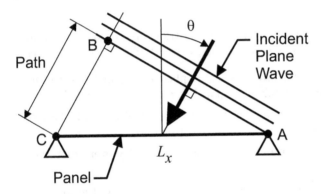

FIGURE 9.15
Cross-sectional view of plane wave striking a simply supported panel.

p. 221] and is called the blocking pressure $p_b = 2p_i$, which is the excitation pressure acting on the panel.

The next step is to examine the dynamics of the simply supported panel and determine the natural frequencies and modes shapes. Once the equation for the mode shapes is defined, it is possible to write Equation (9.45) as a modal force that is applied to the plate.

Plate Dynamics

For the simply supported panel under consideration here, the displacement w of the panel can be written in terms of an infinite sum of its vibration modes multiplied by the modal participation factor for each mode $w_{m,n}$ as

$$w(x,y) = \sum_{m=1}^{\infty}\sum_{n=1}^{\infty} w_{m,n}\sin\left(m\pi x/L_x\right)\sin\left(n\pi y/L_y\right) = \mathbf{w}\boldsymbol{\Psi}, \qquad (9.46)$$

where L_x, L_y are the lengths of the panel along the x and y axes, $w_{m,n}$ are the modal participation factors, $\mathbf{w}$ is the corresponding vector of all the modal participation factors, $\boldsymbol{\Psi}$ is the corresponding matrix of mode shapes, and a particular combination of m, n indices is called the sth index where the natural frequency of the m, n mode has been sorted into increasing natural frequencies. The natural frequencies of a simply supported panel $\omega_{m,n}$ are given by

$$\omega_{m,n}^2 = \omega_s^2 = \frac{D\pi^4}{h\rho_{\text{panel}}}\left[\left(\frac{m}{L_x}\right)^2 + \left(\frac{n}{L_y}\right)^2\right]^2 \tag{9.47}$$

$$D = \frac{Eh^3}{12(1-\nu^2)}, \tag{9.48}$$

where D is the bending stiffness of the panel, E is the Young's modulus, h is the thickness, ρ_{panel} is the density of the panel, and ν is the Poisson's ratio.

The equation of motion of the panel can be written in a modal summation format as [66, p. 95, Eq. (2.55a)]

$$\Lambda_{m,n}(\omega_{m,n}^2 - \omega^2 + j2\zeta\omega_{m,n}\omega)w_{m,n} = p_{m,n}, \tag{9.49}$$

where $\Lambda_{m,n}$ is the modal mass given in Equation (9.24), $\zeta = C_D/(2M_{m,n}\omega_{m,n})$ is the modal damping ratio, C_D is the modal viscous damping coefficient, and $p_{m,n}$ is the modal forcing function given by

$$p_{m,n} = 8P_i\bar{I}_m\bar{I}_n, \tag{9.50}$$

where

$$\alpha = kL_x \sin\theta_i \cos\phi_i \tag{9.51}$$
$$\beta = kL_y \sin\theta_i \sin\phi_i \tag{9.52}$$

$$\bar{I}_m = \begin{cases} -\dfrac{j}{2}\text{sign}(\sin\theta_i\,\cos\phi_i) & \text{if } (m\pi)^2 = \alpha^2 \\ (m\pi)\dfrac{1-(-1)^m e^{-j\alpha}}{(m\pi)^2 - \alpha^2} & \text{if } (m\pi)^2 \neq \alpha^2 \end{cases} \tag{9.53}$$

$$\bar{I}_n = \begin{cases} -\dfrac{j}{2}\text{sign}(\sin\theta_i\,\sin\phi_i) & \text{if } (n\pi)^2 = \beta^2 \\ (n\pi)\dfrac{1-(-1)^n e^{-j\beta}}{(n\pi)^2 - \beta^2} & \text{if } (n\pi)^2 \neq \beta^2. \end{cases} \tag{9.54}$$

Equation (9.49) can be re-arranged to calculate the modal participation factors for the panel as

$$w_{m,n} = \frac{p_{m,n}}{\Lambda_{m,n}(\omega_{m,n}^2 - \omega^2 + j2\zeta\omega_{m,n}\omega)}. \tag{9.55}$$

Now that the participation factors have been calculated, which means that the displacement and velocity of the panel is known, these can be used to calculate the pressure that is generated by the plate on the "receiver" side.

Radiated Acoustic Power and Transmission Loss

The transmitted pressure at a point remote from the panel due to the vibration of the panel is calculated using the Rayleigh integral and can be written as [146, 66]

$$p_{m,n}^t = j\omega(j\omega w_{m,n})\rho_0 \frac{e^{-jkr}}{2\pi r} L_x L_y I_m I_n \,, \tag{9.56}$$

where

$$I_m = \begin{cases} -\dfrac{j}{2}\operatorname{sign}(\sin\theta_i \,\cos\phi_i) & \text{if } (m\pi)^2 = \alpha^2 \\[2mm] (m\pi)\dfrac{1-(-1)^m e^{-j\alpha}}{(m\pi)^2 - \alpha^2} & \text{if } (m\pi)^2 \neq \alpha^2 \end{cases} \tag{9.57}$$

$$I_n = \begin{cases} -\dfrac{j}{2}\operatorname{sign}(\sin\theta_i \,\sin\phi_i) & \text{if } (n\pi)^2 = \beta^2 \\[2mm] (n\pi)\dfrac{1-(-1)^n e^{-j\beta}}{(n\pi)^2 - \beta^2} & \text{if } (n\pi)^2 \neq \beta^2 \,. \end{cases} \tag{9.58}$$

The transmitted intensity is calculated as

$$I^t = \left| \sum_m \sum_n p_{m,n}^t \right|^2 / (2\rho_0 c_0) \,. \tag{9.59}$$

The total power Π^t that is radiated by the panel is calculated as the integral of the sound intensity over an imaginary far-field hemisphere as

$$\Pi^t = \int_{\phi_t=0}^{2\pi} \int_{\theta_t=0}^{\pi/2} I^t r^2 \sin\theta_t \, d\theta d\phi \,. \tag{9.60}$$

The power that is incident on the panel is given by

$$\Pi^i = (|P_i|^2 L_x L_y \cos\theta_i)/(2\rho_0 c_0) \,. \tag{9.61}$$

Finally, the transmission loss TL for a plane-wave striking the panel is given by

$$\text{TL} = 10\log_{10}(\tau(\theta_i, \phi_i)) = 10\log_{10}(\Pi^i/\Pi^t) \,. \tag{9.62}$$

This is the extent of the analysis that will be investigated in this book, where only a single incident plane wave that strikes the panel will be examined. A transmission loss test of a panel can be conducted using source and receiver acoustic reverberation chambers. It is feasible to simulate this test configuration using ANSYS, however the computational resources required to solve the model will be large. In a real source reverberation room, the acoustic field is characterized by a diffuse field that can be described by an infinite number of uncorrelated plane-waves. The transmission loss for a diffuse field is calculated as [69]

$$\text{TL}_{\text{diffuse}} = \frac{\int_0^{2\pi} \int_0^{\pi/2} \tau(\theta_i, \phi_i) \sin \theta_i \cos \theta_i \, d\theta_i d\phi_i}{\int_0^{2\pi} \int_0^{\pi/2} \sin \theta_i \cos \theta_i \, d\theta_i d\phi_i}$$

$$= \frac{\int_0^{2\pi} \int_0^{\pi/2} \tau(\theta_i, \phi_i) \sin 2\theta_i \, d\theta_i d\phi_i}{2\pi}. \qquad (9.63)$$

This diffuse field formulation and an enhanced model by Roussos [130] was used to study the transmission loss of a panel with an array of lumped masses and tuned vibration absorbers attached to a panel in Howard [72, 73].

Roussos' theoretical model for the transmission loss of a finite panel can also be compared with predictions based on infinite panel theory. The transmission loss of an infinite isotropic panel subject to an acoustic wave at incident angle θ_i to the normal of the surface of the panel is [46, p. 286, Eq. (9.85)]

$$\text{TL}_{\theta i} = 10 \log_{10} \left[1 + \left| \frac{h \rho_{\text{panel}} \, \omega \cos \theta_i}{2 \rho_0 c_0} \times \left\{ 1 - \left(\frac{\omega}{\omega_c} \right)^2 \sin^4 \theta_i \right\} \right|^2 \right], \qquad (9.64)$$

where ω is the excitation frequency (radian/s), ρ_{panel} is the panel density (kg/m^3), h is the thickness of the panel (m), ρ_0 is the density of air (kg/m^3), c_0 is the speed of sound of air (m/s), and ω_c is the coincidence or critical frequency of the panel (radians/s) given by

$$\omega_c = c_0^2 \sqrt{\frac{h \rho_{\text{panel}}}{D}}, \qquad (9.65)$$

where D is the bending stiffness defined in Equation (9.48), E is the Young's modulus of the panel (Pa), and ν is the Poisson's ratio of the panel.

Table 9.4 lists the parameters used to model the transmission loss of the panel, which are obtained from Roussos [130].

9.5.3 MATLAB

The MATLAB script TL_panel_roussos.m included with this book can be used to calculate the transmission loss of a finite panel using the theory of Roussos [130] described in Section 9.5.2. Figure 9.16 shows the theoretical transmission loss of the panel calculated using the MATLAB script, which attempts to reproduce the results shown in [130, Fig. 7], and the transmission loss of an infinite panel using Equation (9.64).

9.5.4 ANSYS Mechanical APDL

The ANSYS Mechanical APDL script incident_pressure.inp included with this book can be used to model the transmission loss of a rectangular panel in

TABLE 9.4
Parameters Used in the Analysis of the Transmission Loss of a
Panel

Description	Parameter	Value	Units
Panel length	L_x	0.38	m
Panel width	L_y	0.15	m
Panel thickness	h	8.1×10^{-4}	m
Panel Young's Modulus	E	70×10^9	Pa
Panel density	ρ_{panel}	2700	kg/m^3
Panel Poisson's ratio	ν	0.33	
Panel damping ratio	ζ	0.1	
Air speed of sound	c_0	343	m/s
Air density	ρ_0	1.21	kg/m^3
Angle of incidence	θ_i, ϕ_i	$60°, 0°$	
Incident pressure	P_i	1.0	Pa

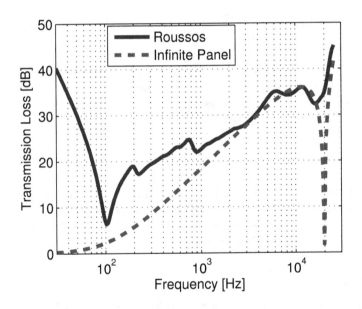

FIGURE 9.16
Transmission loss of a 0.38m $\times$ 0.15m $\times$ 8.1×10^{-4}m aluminium panel for an
incident plane-wave at $\theta_i = 60°$ and $\phi_i = 0°$, calculated using the theory from
Roussos [130, compare with Fig. 7] and an infinite panel using Equation (9.64).

an infinite baffle due to an incident plane wave striking the surface of the panel
at an arbitrary angle. This ANSYS model will require a significant amount
of computational resources to solve. You will need at least 30 GB of disk

space free for the results files, plenty of free swap space to solve the model, which has 76,324 nodes, and 216,360 elements, which is more than permitted by the teaching licenses of ANSYS. The finite element model that is created comprises a

- *rectangular panel* that is excited by nodal forces that are equivalent to the pressure from an incident acoustic plane wave, and

- *hemispherical acoustic region* to model an infinite planar baffle in which the panel is mounted.

Figure 9.17 shows the solid model of the system created by the APDL script and Figure 9.18 shows the finite element model. A rectangular panel is created that is centered about the origin of the global Cartesian system using SHELL181 elements. The damping of the panel was specified as a material property, using the APDL command MP,DMPR,1,value, which defines the critical damping ratio. A hemispherical acoustic volume is created in the $-z$ direction comprising FLUID30 linear acoustic elements. The acoustic free-field is achieved by inserting FLUID130 infinite acoustic elements on the exterior surface of the hemisphere, that act to absorb outgoing acoustic waves. The acoustic elements in the model had a mesh density of approximately 12 elements per wavelength. However, near the surface of the panel, the mesh density was higher to correspond with the mesh density of the panel. The dimensions of the panel were 0.38 m × 0.15 m and it was divided into 120 × 80 divisions, which is more than required. As described in Section 2.11, a sufficient mesh density is required to model the bending waves in the panel, up to the frequency range of interest. However, if a modal summation method is employed to calculate the harmonic response of the structure, it is recommended that natural frequencies and mode shapes up to twice the analysis frequency range are included in the analysis to avoid modal truncation issues.

It can be seen in Figure 9.17 that there are 4 rectangular volumes near the origin, that are located adjacent to the panel. These volumes are meshed with FLUID30 acoustic elements that have their displacement (UX, UY, UZ) degrees of freedom activated at the nodes to enable the coupling between the displacement of the panel and the displacement of the acoustic nodes. The 4 areas of the panel are different from the areas that form the volumes adjacent to the panel. Hence, when these bodies are meshed, there are no common nodes between the panel and the acoustic domain. The mesh pattern in the panel and the volume adjacent to the panel are identical, so the nodes are coincident. The nodes from the FLUID30 elements are coupled to the displacement degrees of freedom in the nodes belonging to the SHELL181 elements used to model the panel using the APDL command CPINTF. The reason why the nodes in the panel and acoustic domains are different is because if a pressure load were applied to the panel and the nodes in the panel and acoustic domain were the same, the pressure load would also be directly applied to the acoustic domain which effectively by-passes the transmission loss provided by the panel.

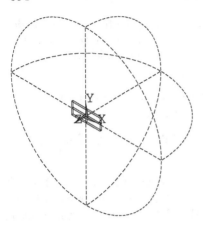

FIGURE 9.17
Lines of the solid model of a panel installed in a planar infinite baffle, with a hemispherical acoustic free-field.

FIGURE 9.18
Finite element model of a panel installed in a planar infinite baffle, with a hemispherical acoustic free-field with FLUID130 infinite acoustic elements on the exterior surface used to absorb outgoing waves.

The acoustic free-field is simulated using FLUID130 infinite acoustic elements which must be placed on the surface of a spherical body. ANSYS recommends that the surface of the spherical region should be separated from the nearest object by at least 0.2 times the acoustic wavelength. These two requirements dictate the size of the finite element model. For this example, the minimum frequency of analysis is 100 Hz, which corresponds with the maximum wavelength of $\lambda_{max} = c_0/f_{max} = 343/100 = 3.43$ m. Hence the separation between the edge of the panel and the FLUID130 elements should be at least $0.2\lambda_{max} = 0.2 \times 3.43 = 0.686$ m. The maximum radius of the panel is $r_{panel} = \sqrt{(0.38/2)^2 + (0.15/2)^2} = 0.204$ m. Hence the radius of the hemispherical region should be at least $r > (0.204 + 0.686) = 0.890$ m. For this model the radius of the hemisphere was selected as $r = 1$ m.

The acoustic excitation of the panel caused by the incident plane wave is applied as forces to the nodes in the panel in the $-z$ direction. The pressure load at a node on the panel can be calculated using Equation (9.45) and the nodal area A_{node} is determined using the APDL command ARNODE, and then the nodal excitation force is calculated as $F_{node} = p_i(x, y)A_{node}$.

The nodes on the edge of the panel were constrained to provide a simply supported boundary condition. A subtle point is that the nodes in the acoustic domain that are adjacent to the nodes on the boundary of the panel should also be constrained from motion; if they are not constrained, then they are free to move and spurious results can occur.

That completes the description of the finite element model. In the following text the results from the finite element analysis are presented and compared with the results from theoretical predictions.

Results

Figure 9.19 shows a contour plot of the sound pressure level (dB re 20 μPa) at 400 Hz calculated using ANSYS Mechanical APDL, where the region $y > 0$ of the finite element model has been hidden to show the sound radiation pattern on a plane through the center of the panel. The incident plane wave at $\theta_i = 60°$ and $\phi_i = 0°$ travels toward the panel in the $-z$ direction, strikes the panel causing it to vibrate, and then re-radiates sound into the hemispherical free-field region. Figure 9.19(a) shows the model using an isometric projection and Figure 9.19(b) shows the contour plot of the sound pressure level on the XZ plane.

Figure 9.20 shows the sound pressure (Pa) calculated theoretically and using ANSYS Mechanical APDL. The pressure results calculated using ANSYS are slightly below the theoretical values. The reason for this is due to the lower vibration levels of the panel predicted using ANSYS, which is discussed below.

Figure 9.21 shows a contour plot of the amplitude of the nodal displacement of the panel at 400 Hz in the z direction, calculated using ANSYS Mechanical APDL. The figure shows that the displacement of the panel is not symmetric about the y axis, which is to be expected because the plane wave strikes the panel at an oblique angle of incidence. The nodal displacements can be used to calculate the structural kinetic energy of the panel.

Figure 9.22(a) shows the structural kinetic energy of the panel calculated theoretically and using ANSYS Mechanical APDL where there was no fluid

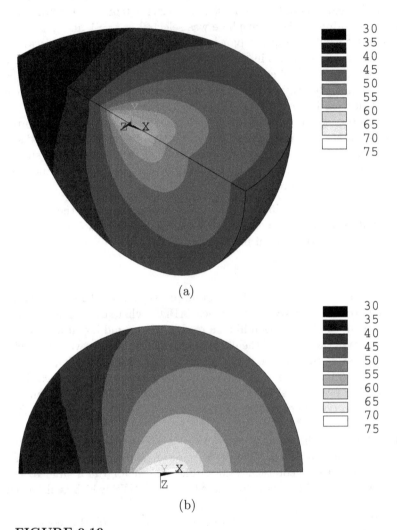

(a)

(b)

FIGURE 9.19
Sound pressure level (dB re 20 μPa) at 400 Hz calculated using
ANSYS Mechanical APDL, where the region $y > 0$ of the finite element model
has been hidden to show the sound radiation pattern, displayed (a) using an
isometric projection, and (b) looking at the XZ plane.

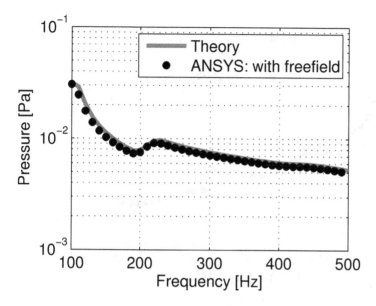

FIGURE 9.20
Sound pressure at 1 m from the center of the panel $(0, 0, -1)$ calculated theoretically and using ANSYS Mechanical APDL.

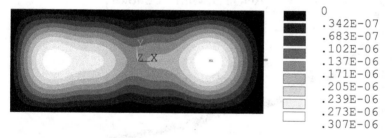

FIGURE 9.21
Contour plot of the nodal displacement (m) of the panel in the Z direction at 400 Hz for the coupled system.

and no fluid–structure interaction. The kinetic energy was evaluated using Equation (2.63) and it can be seen that the results are identical.

Figure 9.22(b) shows the structural kinetic energy for the system under investigation where the panel was installed in an infinite planar baffle and radiates sound into a hemispherical acoustic free-field. It can be seen that the structural kinetic energy is now slightly less than the theoretical results, as the ANSYS model now includes bi-directional fluid–structure interaction, where the air in contact with the panel provides additional mass and inertia that reduces the motion of the panel. Hence these two results show

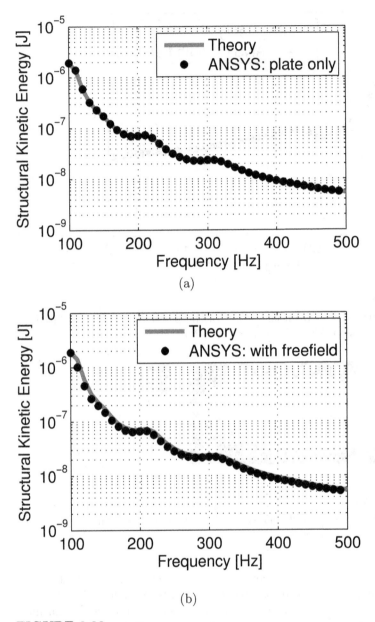

(a)

(b)

FIGURE 9.22
Structural kinetic energy of the panel calculated theoretically and using
ANSYS Mechanical APDL, when there was (a) no fluid and no fluid–structure
interaction coupling, and (b) when the panel was installed in a planar baffle
and radiated sound into a hemispherical acoustic free-field.

- that when the fluid is not included in the model, the kinetic energy of the panel calculated theoretically and using ANSYS are nearly identical, which suggests that the excitation and the responses of the panel are correct; and

- that when the fluid is included in the model, the displacement of the panel is reduced slightly, due to the small fluid-loading caused by the air acting on the panel.

The slight difference between the theoretical and ANSYS results for the response of the panel will have consequences for the radiated acoustic power and the transmission loss results that are presented below.

Figure 9.23 shows the acoustic power radiated by the panel, calculated using Equation (9.60). The calculation of the radiated acoustic power using the ANSYS Mechanical APDL model involves the following:

- Exporting the real and imaginary parts of the pressure for the nodes of the FLUID130 elements that are on the hemispherical surface of the acoustic domain.

- Determining the effective areas for the nodes of the FLUID130 elements that are on the hemispherical surface of the acoustic domain. This is accomplished by exporting the coordinates of the nodes using the APDL command NWRITE, and the nodes belonging to each elements using the APDL command EWRITE, then mathematically calculating the nodal areas. Alternatively, one can temporarily swap the FLUID130 acoustic elements for SHELL181 structural elements and use the APDL command ARNODE to determine the nodal areas.

- Using the MATLAB script power_freefield_hemisphere.m that imports the complex nodal pressures and areas, and calculates the radiated sound power.

The radiated acoustic power by the panel calculated using ANSYS Mechanical APDL is slightly below the theoretical predictions, which is a consequence of the slightly lower panel vibration, as shown in Figure 9.22.

Figure 9.24 shows the transmission loss of the panel due to an incident plane wave at $\theta_i = 60°$ and $\phi_i = 0°$, calculated theoretically and using ANSYS Mechanical APDL. The results presented in Figure 9.24 are intended to reproduce the results in Roussos [130, Fig. 7]. It can be seen that the transmission loss results calculated using ANSYS Mechanical APDL are slightly *greater* than the theoretical values. It was shown in Figure 9.23 that the (radiated) transmitted power by the panel calculated using ANSYS is slightly *less* than the theoretical predictions. As the transmission loss is the ratio of the incident power divided by transmitted power, as shown in Equation (9.62), the lower transmitted power causes a higher value of transmission loss. As mentioned previously, the difference in the response of the panel is because the theoretical model assumes "weak" coupling and neglects the fluid-loading by the air on the panel, whereas the ANSYS Mechanical APDL model utilizes

bi-directional fluid–structure interaction between the fluid and the panel and
includes the fluid-loading on the panel.

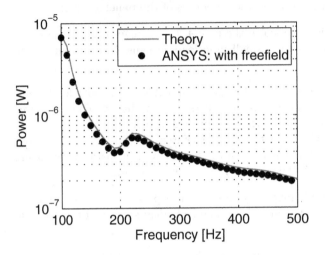

FIGURE 9.23
Acoustic power from the vibrating panel calculated by integrating the in-
tensity over the surface of a hemisphere, from the theoretical model and
ANSYS Mechanical APDL.

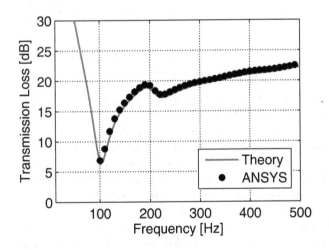

FIGURE 9.24
Transmission loss of the panel due to an incident plane wave at $\theta_i = 60°$
and $\phi_i = 0°$, calculated using the theoretical model and ANSYS Mechanical
APDL.

A

Files Included with This Book

A.1   Table of Files Included with This Book

The following table lists the MATLAB scripts, ANSYS Workbench archive
files, and ANSYS Mechanical APDL scripts that are included with this book.

Section, Page No.	Filename / Description
Chapter 2 Background	
Sec 2.12, p. 98	impedance_driven_closed_pipe.m This MATLAB script is used to calculate the mechanical impedance of a piston attached to the end of a closed duct. The script is used to highlight the scaling that is needs to be applied to a 1/4 acoustic finite model to calculate the results for a full model.
Chapter 3 Ducts	
Sec 3.3.1, p. 107	res_freqs_duct.wbpz ANSYS Workbench model of a circular duct used to calculate the resonance frequencies.
Sec 3.3.2, p. 130	res_freqs_duct_FLUID220.wbpz ANSYS Workbench model of a circular duct used to calculate the resonance frequencies, meshed with FLUID220 elements.
Sec 3.3.4, p. 138	driven_duct_pres_dist.wbpz ANSYS Workbench model of a circular duct with a normal surface velocity excitation at one end, which simulates a piston, and a rigid termination at the other, that is meshed with FLUID30 elements, and is used to calculate the sound pressure levels and acoustic particle velocities along the duct.

Section, Page No.	Filename / Description
Sec 3.3.5, p. 144	spl_along_duct_4pole.m This MATLAB script can be used to calculate the sound pressure and acoustic particle velocity along a circular duct using the four-pole transmission matrix method.
Sec 3.3.7.2, p. 149	radiation_open_duct.wbpz ANSYS Workbench model of a duct with a piston on one end and a hemispherical volume on the other end that is used to simulate a plane baffle (hemispherical free-field). The solid model contains the whole geometry, however many of the bodies are Suppressed, so that only 1/4 of the model is shown and analyzed. The piston face is driven with a harmonic force of 1×10^{-3}N and the resulting displacement of the piston is calculated by the harmonic response analysis. The results can be exported to MATLAB and used to calculate the piston velocity, mechanical impedance, and mechanical power, and compared with theoretical predictions with the MATLAB script radiation_end_of_pipe.m.
Sec 3.3.7.3, p. 161	radiation_end_of_pipe.m Script to calculate the mechanical impedance of a piston attached to a duct that radiates into a plane baffle (hemispherical free-field). The script calculates the real and imaginary parts of the mechanical impedance and the mechanical power into the piston.
Sec 3.3.7.4, p. 162	freq_depend_impedance.wbpz This ANSYS Workbench archive file contains a model of a duct with a piston on one end and an impedance on the other end that varies with frequency and is used to simulate radiation into a plane baffle (hemispherical free-field). The impedance is implemented using the APDL command SF,,IMPD inside a command object within the Workbench model.

Chapter 4 Sound Inside a Rigid-Walled Cavity

Chapter 5 Introduction to Damped Acoustic Systems

Section, Page No.	Filename / Description
Sec 5.5.5, p. 273	code_ansys_impedance_tube.txt This ANSYS Mechanical APDL script is used to generate an FE model of a lined duct using the 2D FLUID29 acoustic elements with a real admittance at one end, as well as conduct a harmonic analysis using the full method. The impedance is activated using SF,,IMPD,1 and the admittance is defined using MP, MU, , Admittance. The file also includes additional code to write the ANSYS results to a text file, as well as the code for generating plots of key results.
Sec 5.6.2, p. 280	impedance_surf153.m This MATLAB script is used to analyze the response of a one-dimensional waveguide with an arbitrary boundary impedance opposite a plane-wave source. The script calculates the pressure response in the duct and uses the two-microphone method to estimate the surface impedance, complex reflection coefficient and sound absorption coefficient. The script also analyzes the results from ANSYS Mechanical APDL.
Sec 5.6.2, p. 280	code_ansys_surf153.txt This ANSYS Mechanical APDL script is used generate a FE model of a lined duct using the 2D FLUID29 acoustic elements with an arbitrary complex impedance at one end, as well as conduct a harmonic analysis using the full method. The impedance is achieved using the SURF153 surface effect element. The file also includes additional code to write the ANSYS results to a text file, as well as the code for generating plots of key results.
Sec 5.9.3, p. 306	plane_wave_viscous_losses.m This MATLAB script is used to calculate the expected classical attenuation in a duct. Using the expressions derived in Section 5.9.1 the attenuation per unit length is calculated using Equations (5.49) and (5.47). The script also analyzes the results from ANSYS Mechanical APDL.

Section, Page No.	Filename / Description
Sec 5.9.4, p. 306	`Visco-thermal.wbpj` This ANSYS Workbench project file contains a model of a hard-walled duct with the visco-thermal losses activated. The file conducts a harmonic analysis using the full method. The acoustic attenuation per meter due to classical absorption is calculated.
Sec 5.9.5, p. 313	`code_ansys_visco_thermal.txt` This ANSYS Mechanical APDL script is used to generate an FE model of a rigid-walled duct using the quadratic `FLUID220` acoustic elements; the appropriate material properties are defined, a mass source boundary condition is applied to one end, and an anechoic termination is applied to the other. A harmonic analysis is performed and the results are exported to a text file.
Sec 5.10.3, p. 316	`rigid_wall_cavity_damping.m` This MATLAB script is used to calculate the effect of spectral (global) damping on a rigid-walled rectangular cavity. The script is is a modified version of the script `rigid_wall_cavity.m` presented previously. The script also analyzes the results from ANSYS Workbench and ANSYS Mechanical APDL.
Sec 5.10.4, p. 316	`rigid_cavity_modal_super_damped.inp` This ANSYS Mechanical APDL script is used to generate an FE model of a rigid-walled rectangular cavity to which various forms of spectral damping are applied. The pressure response is calculated using harmonic analysis with the full option, and the results are exported to a text file for analysis by the MATLAB script `rigid_wall_cavity_damping.m`. This analysis is based on ANSYS Mechanical APDL file `rigid_cavity_modal_super.inp`.

Chapter 6 Sound Absorption in a Lined Duct

Sec 6.5.1, p. 338	`lined_duct.m` This MATLAB script is used define the key parameters defined in Table 6.2, then using the expressions derived in Section 6.4.2 the attenuation per unit length is calculated using Equations (6.9), (6.10), and (6.11). The script also analyzes the results from ANSYS Workbench and ANSYS Mechanical APDL.
Sec 6.6.1, p. 364	`scott.m`

Section, Page No.	Filename / Description
	This MATLAB function is called by lined_duct.m when solving the transcendental equation used to calculate the absorption by a bulk reacting liner made from porous media.
Sec 6.5.2.1, p. 338	Lined_Duct.wbpj
	This ANSYS Workbench project file contains a model of a duct, with a silencer section and anechoic termination. The section is modeled as both a locally reacting impedance using the SF,,IMPD command as well as a bulk reacting liner using the Johnson–Champoux–Allard equivalent fluid model. The file conducts a harmonic analysis using the full method. Inputs are both acoustic FLOW and mass sources, and output quantities of interest are sound pressures, sound pressure levels, and particle velocities.
Sec 6.5.3, p. 357	code_ansys_lined_duct.txt
	This ANSYS Mechanical APDL script is used generate a finite element model of a lined duct, as well as conduct a harmonic analysis using the full method. The file also includes additional code to write the ANSYS results to a text file, as well as the code for generating plots of key results.

Chapter 7 Room Acoustics

Sec 7.4.1.1, p. 379	Sabine.m
	This MATLAB script can be used to calculate the modal density, modal overlap, frequency bounds, and reverberation time using the equations in Section 7.3 as well as post-process the ANSYS Workbench and ANSYS Mechanical APDL results.
Sec 7.4.1.2, p. 380	Sabine.wbpj
	This ANSYS Workbench project file contains a model of a reverberation room, comprised of a rigid-walled rectangular acoustic cavity that is lined on one surface with a sound-absorbing material. The file conducts undamped and damped model analysis, a harmonic analysis using the full method, as well as a transient analysis using the full method. Inputs are both acoustic FLOW and Mass Source, and output quantities of interest are sound pressures and sound pressure levels.

Section, Page No.	Filename / Description
Sec 7.4.1.3, p. 393	code_ansys_sabine.txt This ANSYS Mechanical APDL script is used to model a reverberation room, comprised of a rigid-walled rectangular acoustic cavity that is lined on one surface with a sound-absorbing material. The file conducts undamped and damped model analysis, a harmonic analysis using the full method, as well as a transient analysis using the full method. Inputs are both acoustic FLOW and Mass Source, and output quantities of interest are sound pressures and sound pressure levels.

Chapter 8 Radiation and Scattering

Sec 8.3, p. 438	acoustic_sources_PML.wbpz This ANSYS Workbench archive file is used to plot the directivity of Acoustic Wave Sources, Monopole, Dipole, back-enclosed loudspeaker, and Bare Loudspeaker.
Sec 8.3.1, p. 443	monopole_spl_vs_angle.m This MATLAB code can be used to calculate the sound pressure level versus angle for a monopole source radiating into a free-field.
Sec 8.3.4, p. 457	dipole_spl_vs_angle.m This MATLAB code can be used to calculate the sound pressure level versus angle for a dipole source radiating into a free-field.
Sec 8.4.3, p. 462	pressure_on_axis.m This MATLAB code can be used to calculate the pressure on the axis of symmetry radiated from a baffled circular piston.
Sec 8.4.3, p. 462	baffled_piston.m This MATLAB code can be used to calculate the normalized impedance of a vibrating baffled circular piston.
Sec 8.4.3, p. 462	struve.m This MATLAB function can be used to calculate the Struve function.
Sec 8.4.3, p. 464	radiation_pattern_baffled_piston.m This MATLAB code can be used to calculate the beam pattern or directivity of a vibrating baffled circular piston.

Section, Page No.	Filename / Description
Sec 8.4.4, p. 467	piston_baffle_axisym.wbpz This ANSYS Workbench archive file contains a model of an axi-symmetric piston radiating into an infinite plane baffle.
Sec 8.4.4, p. 482	command_obj_solid_geom_01.txt This file contains APDL code that can be copied into a command object in the ANSYS Workbench model of an axi-symmetric piston radiating into an infinite plane baffle.
Sec 8.4.4, p. 483	command_obj_solid_geom_02.txt This file contains APDL code that can be copied into a command object in the ANSYS Workbench model of an axi-symmetric piston radiating into an infinite plane baffle.
Sec 8.4.4, p. 484	command_obj_solid_geom_03.txt This file contains APDL code that can be copied into a command object in the ANSYS Workbench model of an axi-symmetric piston radiating into an infinite plane baffle.
Sec 8.4.4, p. 493	command_obj_harmonic_A5_01.txt This file contains APDL code that can be copied into a command object in the ANSYS Workbench model of an axi-symmetric piston radiating into an infinite plane baffle.
Sec 8.4.4, p. 500	command_obj_harmonic_A5_02.txt This file contains APDL code that can be copied into a command object in the ANSYS Workbench model of an axi-symmetric piston radiating into an infinite plane baffle.
Sec 8.4.4, p. 493	command_obj_solution_A6_01.txt This file contains APDL code that can be copied into a command object in the ANSYS Workbench model of an axi-symmetric piston radiating into an infinite plane baffle.
Sec 8.4.4, p. 500	command_obj_solution_A6_02.txt This file contains APDL code that can be copied into a command object in the ANSYS Workbench model of an axi-symmetric piston radiating into an infinite plane baffle.

Section, Page No.	Filename / Description
Sec 8.4.4, p. 506	command_obj_solution_A6_03.txt This file contains APDL code that can be copied into a command object in the ANSYS Workbench model of an axi-symmetric piston radiating into an infinite plane baffle.
Sec 8.4.4, p. 507	power_freefield_2Dhemisphere.m This MATLAB code can be used to post-process the results exported from ANSYS of the sound pressure of FLUID129 elements on a circular arc to calculate the sound power radiated into a free-field. The MATLAB script will read the exported nodal pressures, node and element data, and material properties.
Sec 8.4.5, p. 508	baffled_piston.inp This ANSYS Mechanical APDL input file is used to generate an axi-symmetric (2D) model of a rigid circular piston vibrating in an infinite plane baffle. The code will calculate and plot the mechanical radiation impedance versus frequency of the piston. The code will reproduce Figure 7.5.2, p. 187 in Kinsler et al. [102].
Sec 8.4.5, p. 508	p_vs_d.inp This ANSYS Mechanical APDL input file is used to generate an axi-symmetric (2D) model of a rigid circular piston vibrating in an infinite plane baffle. The code will calculate and plot the on-axis sound pressure level versus distance from the piston. The graph is similar to Figure 7.4.2, p. 181 in Kinsler et al. [102], only the y-axis of the graph is the absolute value of pressure, rather than non-dimensionalized pressure.
Sec 8.6.3, p. 515	cylinder_plot_scattered_pressure.m This MATLAB code can be used to plot the scattered sound pressure level from an incident plane wave striking a cylinder.
Sec 8.6.4, p. 520	cylinder_scattering.wbpz This ANSYS Workbench archive file can be used to calculate the scattered sound pressure level from an infinitely long cylinder when struck by an acoustic plane wave at an angle normal to the axis of the cylinder.

Section, Page No.	Filename / Description

Section, Page No.	Filename / Description
Sec 9.4.5, p. 582	compare_modal_coupling_vs_full_FSI_press.m This MATLAB script is used to process the results from the ANSYS Mechanical APDL analyses using the modal coupling theory described in Section 9.3.2 to calculate the sound pressure at a point within the cavity. This is compared with the results from conducting a full fluid–structure interaction harmonic analysis.
Sec 9.4.5, p. 583	compare_modal_coupling_vs_full_FSI_disp.m This MATLAB script is used to calculate the displacement of the plate at the driving point from ANSYS Mechanical APDL analyses using the modal coupling theory described in Section 9.3.2. This result is compared with the results from conducting a full fluid–structure interaction harmonic analysis. This script should only be run after using the MATLAB script compare_modal_coupling_vs_full_FSI.m.
Sec 9.4.5, p. 585	ape_from_ansys.m This MATLAB function is used to calculate acoustic potential energy from results exported by the ANSYS Mechanical APDL function box_plate.inp. The acoustic pressure at every node associated with FLUID30 elements and at every analysis frequency is imported and used to calculate the acoustic potential energy of the acoustic cavity.
Sec 9.4.5, p. 586	ske_from_ansys.m This MATLAB function is used to calculate structural kinetic energy from results exported by the ANSYS Mechanical APDL function box_plate.inp. The displacement at every node associated with SHELL181 elements and at every analysis frequency is imported and used to calculate the kinetic energy of the structure.
Sec 9.3, p. 581	loadmodel.m This MATLAB function is part of a group of functions to calculate the coupled vibro-acoustic response using modal coupling theory. This function loads the data that was exported from the ANSYS Mechanical APDL macro export_modes.mac.

Section, Page No.	Filename / Description
Sec 9.3, p. 581	`loadstr.m` This MATLAB function is part of a group of functions to calculate the coupled vibro-acoustic response using modal coupling theory. This function is called by `loadmodel.m` and loads the results from the structural analysis that was exported from the ANSYS Mechanical APDL macro `export_modes.mac`.
Sec 9.3, p. 581	`loadcav.m` This MATLAB function is part of a group of functions to calculate the coupled vibro-acoustic response using modal coupling theory. This function is called by `loadmodel.m` and loads the results from the acoustic analysis that was exported from the ANSYS Mechanical APDL macro `export_modes.mac`.
Sec 9.3, p. 581	`bli.m` This MATLAB function is part of a group of functions to calculate the coupled vibro-acoustic response using modal coupling theory. This function is used to calculate the coupling coefficient between the acoustic modes and the structural modes. It is called by the function `loadmodel.m`.
Sec 9.3, p. 581	`plotmodel.m` This MATLAB function is part of a group of functions to calculate the coupled vibro-acoustic response using modal coupling theory. This function can be used to plot the shape of the acoustic and structural models.
Sec 9.3, p. 581	`plotmode.m` This MATLAB function is part of a group of functions to calculate the coupled vibro-acoustic response using modal coupling theory. This function can be used to plot the mode shape of the acoustic and structural models.
Sec 9.3, p. 581	`createloadcase.m` This MATLAB function is part of a group of functions to calculate the coupled vibro-acoustic response using modal coupling theory. This function is used to define the structural and acoustic loads that are applied to the system.

Section, Page No.	Filename / Description
Sec 9.5.3, p. 591	TL_panel_roussos.m This MATLAB function is used to calculate the transmission loss of a rectangular panel due to an incident plane wave striking the surface of the panel at an arbitrary angle using the theory presented in Roussos [130].

Appendix D Errors

Sec D.1.2, p. 632	cylinder_plot_scattered_pressure_junger_feit.m This MATLAB code can be used to calculate the scattered sound pressure level from a plane wave striking an infinitely long rigid cylinder, using the theory by Junger and Feit [97, p. 322].

Appendix E Export of Nodal Area from ANSYS

Sec E.1, p. 650	power_freefield_hemisphere.m This MATLAB function is used to calculate the sound power radiating through a hemi- or spherical surface comprising FLUID130 infinite elements. The coordinates of the nodes and the pressure at the FLUID130 elements are exported from ANSYS and this script imports the results and calculates the radiated sound power.

B

Advice for Using ANSYS

B.1 Recommended Practice

One of the challenges when starting to do acoustic analyses (or any sort for that matter) using ANSYS is to have a pre-conceived idea about what the results should look like. This is difficult for new users as they often lack experience or knowledge. For this reason it is vital to "crawl, walk, run." The general advice is:

crawl: Start with the simplest analysis cases you can find, preferably with a known solution, or use one of the cases from the ANSYS verification manual. This "crawling" step cannot be emphasized enough, and many people, even experienced analysts, think this is below them—it is not, and it will save you numerous wasted hours of frustration.

walk: Once you have "crawled," extend the simple problem to something slightly more complicated, but related. Slightly change the geometry, boundary conditions, loads, analysis frequency range, etc., to see what happens.

run: Once you have "crawled" and "walked," only then should you consider conducting the full analysis of your problem.

Some further advice:

- Use a 64-bit operating system, especially for Microsoft Windows operating systems. If you intend to conduct analyses with large models, then you will have problems if you use a 32-bit operating system that has a limit on the maximum permissible size of disk files and also with memory allocation.

- Get as much RAM as practical.

- Have lots of free disk space for the swap files, solution matrices, and results. Sometimes peculiar errors occur that seem like an unexplained crash, but are simply caused by running out of disk space, even though the log files do not indicate "Disk Full" errors.

- Never run analyses using network file storage disks, where data is transferred via a wired or wireless network. Always run it from a local disk.

- Save Before Solve: It is suggested that you always save your model before clicking the Solve icon in ANSYS Workbench Mechanical. There is an option that will automatically do this for you. In a Workbench Mechanical analysis, click on Tools | Options. In the window that opens, click on the plus sign next to Mechanical, click on Miscellaneous, and change the row Save Project Before Solution to Yes.

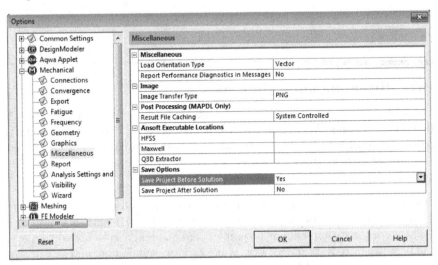

- Export Setup: If you are frequently exporting nodal results and want to export the nodal locations at the same time, from the Mechanical window, select Tools | Options | Mechanical | Export and change Include Node Location to Yes.

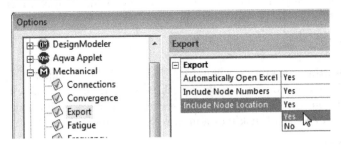

C

MATLAB Functions for Modal Coupling

C.1 MATLAB Functions for Modal Coupling

Table 9.3 lists several MATLAB scripts that are used to conduct a vibro-acoustic analysis using the modal-coupling theory described in Section 9.3.2. This section contains further details about each of these scripts.

General Steps

The general steps involved in the use of the modal coupling functions are:

1. Construct a structural-acoustic finite element model in ANSYS. The acoustic cavity is meshed with FLUID30 elements and the structure is meshed with SHELL181 elements. The ANSYS model should have coincident nodes for the structural and acoustic elements at the Fluid-Structure-Interaction (FSI) interface.

2. Calculate the resonance frequencies and mode shapes of the cavity and structure. This is done by the ANSYS macro extract_mode.mac.

3. Export the coordinates of the nodes and elements. This is done by the ANSYS macro extract_mode.mac.

4. Load the model and ANSYS results into MATLAB. This is done using the MATLAB function loadmodel.m, which calls the function loadcav.m that reads the acoustic results, and loadstr.m that reads the structural results. The functions calculate the normals, areas associated with each node, the modal volumes, and modal masses. The function also calls the MATLAB function bli.m that calculates the coupling coefficients between the structural and acoustic modes.

5. Apply loads to the model using the function createloadcase.m. The loads are stored in a cell array, and are created by the file createloadcase.m. Several load cases can be associated with the model. Each load case contains the frequency range whose results are to be evaluated, the structural forces and acoustic volume velocities applied to the model, and the cavity and structural damping.

6. Conduct the modal-coupling analysis using the function coupled_

response_fahy.m. This function implements the modal-coupling theory in the frequency variable of the specified load case. The results are the structural and acoustic participation factors that are stored for the specified load case.

7. Use the modal participation factors to calculate acoustic and structural results. There are four additional functions that can be used to calculate: (1) the acoustic pressure at a node cav_pressure.m, (2) the structural displacement at a node str_displacement.m, (3) the acoustic potential energy of the cavity acousticpotentialenergy.m, and (4) the kinetic energy of the structure structuralkineticenergy.m.

Example Use of the Scripts

Assuming that the ANSYS Mechanical APDL model of the structure and acoustic system has been created properly using SHELL181 and FLUID30 elements, the following command is typed into the command line:
extract_modes,'boxplate',1600
This command will execute the ANSYS macro extract_modes.mac and will calculate the modes below 1600 Hz from the cavity and the structure. Change the frequency range for the modes as required. The results are exported to several text files with the prefix boxplate.

Once the ANSYS results have been exported, the results can be imported into MATLAB using the command
m=loadmodel('boxplate');

A MATLAB *structure array* called m contains the following

m =

      name: 'boxplate'
         s: [1x1 struct]
         c: [1x1 struct]
     bli: [100x19 double]

Unfortunately, this terminology might cause confusion, as MATLAB has *structure arrays* and should not be confused with the physical structure that vibrates.

The fields of the *structure array* m are as follows:

- name is a text string that is a label used to describe the contents of the MATLAB *structure array.*

- s is a MATLAB *structure array* that contains data about the finite element model of the structure.

- c is a MATLAB *structure array* that contains data about the finite element model of the acoustic cavity.

- bli is the coupling matrix.

The MATLAB *structure array* m.s contains the information about the finite element model of the structure that was modeled and has several fields as listed below:

```
m.s =

     n: [1x1 struct]
     e: [1x1 struct]
     f: [1x19 double]
     p: [273x19 double]
    mm: [1x19 double]
    id: 'str'
```

The fields of m.s are as follows:

- n and e are MATLAB *structure arrays* for the nodes and elements data.
- f is a vector of the natural frequencies of the structure in Hertz.
- p is a matrix of the normal mode shape matrix psi.
- mm is the modal mass of the structure.
- id is a string used to label the MATLAB *structure array*.

The MATLAB *structure array* m.s.n contains information about the nodes associated with the structure of the finite element model and contains the following fields:

```
m.s.n =

     num: [273x1 double]
       x: [273x1 double]
       y: [273x1 double]
       z: [273x1 double]
     idx: [546x1 double]
    area: [273x1 double]
    norm: [273x3 double]
```

The fields of m.s.n are as follows:

- num are the node numbers from the ANSYS model.
- x, y, z are vectors of the nodal coordinates.
- idx is a vector that references the node numbers from their index. For example, m.s.n.area(m.s.n.idx(373)) will return the area associated with node 373.

- area is a vector containing the area associated with each structural node. This is calculated using geometric methods, rather than the method involving fixing the displacement degrees of freedom, applying a 1 Pascal pressure to the node, and using the reaction force from a static analysis as the equivalent nodal area.

- norm is a matrix containing a unit vector in the direction normal to the surface at the node.

The MATLAB *structure array* m.s.e contains information about the elements associated with the structure of the finite element model and contains the following fields:

```
m.s.e =

     i: [240x1 double]
     j: [240x1 double]
     k: [240x1 double]
     l: [240x1 double]
   mat: [240x1 double]
  type: [240x1 double]
  real: [240x1 double]
   num: [240x1 double]
   idx: [10800x1 double]
```

The fields of m.s.e are as follows:

- i, j, k, l are vectors of the node numbers associated with an element. Only the first 4 nodes associated with an element indices are imported; even if the element contains more than 4 nodes, they are ignored. This means that elements with mid-side nodes, such as the SHELL99 composite element, are automatically accommodated.

- num is a vector of the element numbers from the ANSYS model.

- mat, type and real are vectors containing the indices from the MAT, TYPE, and REAL definitions associated with the element as defined in the ANSYS model. These vectors can be used to select elements.

- idx is a vector that references the element numbers from their index.

The MATLAB *structure array* m.c contains the following fields:

```
m.c =

    n: [1x1 struct]
    e: [1x1 struct]
    f: [1x100 double]
    p: [12285x100 double]
   mv: [1x100 double]
```

```
      c: 344
   dens: 1.2100
     id: 'cav'
```

The fields of m.c are as follows:

- n and e are MATLAB *structure arrays* containing data about the node and elements associated with the acoustic cavity.
- f is a vector of the resonance frequencies of the cavity.
- p is a matrix of the mode shape matrix phi.
- mv is a vector of the modal volume of the cavity.
- c and dens are constants for the speed of sound and density of the fluid, respectively.
- id is a string used to label the MATLAB *structure array*.

The MATLAB *structure array* m.c.n contains information about the nodes associated with the acoustic cavity of the finite element model and contains the following fields:

m.c.n =

```
   num: [12285x1 double]
     x: [12285x1 double]
     y: [12285x1 double]
     z: [12285x1 double]
   idx: [12285x1 double]
   vol: [12285x1 double]
```

The fields of m.c.n are as follows:

- x, y, z are the nodal coordinates.
- num is the node numbers from the ANSYS model.
- idx is a vector that references the node numbers from their index.
- vol is the volume associated with each node.

The MATLAB *structure array* m.c.e contains information about the elements associated with the acoustic cavity of the finite element model and contains the following fields:

m.c.e =

```
     i: [10560x1 double]
     j: [10560x1 double]
     k: [10560x1 double]
     l: [10560x1 double]
```

```
    m: [10560x1 double]
    n: [10560x1 double]
    o: [10560x1 double]
    p: [10560x1 double]
  mat: [10560x1 double]
 type: [10560x1 double]
 real: [10560x1 double]
  num: [10560x1 double]
```

The fields of m.c.e are as follows:

- i, j, k, l, m, n, o, p are vectors of the element indices. These are the node numbers of the element corners for the volume element.

- num are the element numbers from the ANSYS model.

- mat, type and real are vectors containing the indices from the MAT, TYPE, and REAL definitions associated with the element as defined in the ANSYS model. These vectors can be used to select elements.

Apply Load to the Model

Once the model has been loaded into MATLAB, structural and acoustic loads can be defined using the MATLAB function createloadcase.m. The function is used with the following input parameters

```
createloadcase(model, unique_id_string, frequency, str_nodes, ...
    force, str_damping, cav_nodes, volume_velocity, cav_damping)
```

where the input parameters are

- model is the MATLAB *structure array* (not the model structure) to which the load will be applied.

- unique_id_string is a text string used to identify the load case.

- frequency is a vector of the analysis frequencies to be used in the computations.

- str_nodes is a vector of the structural node numbers where the normal point forces listed in the vector force will be applied.

- force is a vector of normal point forces that will be applied to the str_nodes.

- str_damping is the structural damping loss factor.

- cav_nodes is a vector of cavity node numbers where acoustic volume velocities listed in the vector volume_velocity will be applied.

- volume_velocity is a vector of acoustic volume velocities that will be applied at the nodes listed in cav_nodes.

- cav_damping is the cavity damping loss factor.

If an input parameter is not used in the load case, then it should be replaced with a null matrix [].

An example of single point force load on a node is

```
freq=[1:400];
m=createloadcase(m,'point_force',freq,373,1,[],[],[],[]);
```

which results in the following *structure array* for m:

```
m =

    name: 'boxplate'
       s: [1x1 struct]
       c: [1x1 struct]
     bli: [100x19 double]
      lc: {[1x1 struct]}
```

and the MATLAB *structure array* m.lc contains the following fields

```
m.lc{1}=

    name: 'point_force'
    freq: [1x400 double]
      sn: 373
      sf: 1
      sl: []
      cn: [0x1 double]
      cq: []
      cl: []
```

where the fields are

- name is a text string to identify the load case.
- freq is the frequency vector of the analysis.
- sn is a vector of structural node numbers where the point force will be applied.
- sf is a vector of normal forces that will be applied at the nodes.
- sl is the structural damping loss factor.
- cn is a vector of cavity node number where acoustic volume velocities will be applied.
- cq is a vector of volume velocities that will be applied at cavity nodes.
- cl is the acoustic damping loss factor.

An example of a point acoustic volume velocity source is

```
freq=[1:400];
m=createloadcase(m,'point_q',freq,[],[],0.1,4886,1,[]);
```

An example of a more complicated load involves varying the load at each frequency. The structural loads force or the acoustic volume velocities volume_velocity can be defined as a matrix with the number of rows corresponding to the length of the vector str_nodes or cav_nodes, as appropriate, and number of columns corresponding to the length of the frequency vector frequency.

```
m=createloadcase(m,'rand_force',freq,m.s.n.num, ...
rand(length(m.s.n.num),length(freq)),0.0001,[],[],0.0001)
```

```
m =

    name: 'boxplate'
       s: [1x1 struct]
       c: [1x1 struct]
     bli: [100x19 double]
      lc: {[1x1 struct]  [1x1 struct]}
```

```
m.lc{2} =

    name: 'rand_force'
    freq: [1x400 double]
      sn: [273x1 double]
      sf: [273x400 double]
      sl: 1.0000e-04
      cn: [0x1 double]
      cq: []
      cl: 1.0000e-04
```

It is not possible to overwrite a load case. Instead, the load case must be deleted and then the replacement load case is defined. To delete all load cases, use the following commands:

```
m=rmfield(m,'lc');
```

Calculate the Coupled Response

The coupled response is calculated using the MATLAB function coupled_response_fahy.m. This function uses the modal-coupling theory to evaluate the modal participation factors for the structure and the cavity for the selected load case.

The MATLAB function coupled_response_fahy.m has the following input parameters coupledresponse(model,load_case_id) where

- model is the MATLAB *structure array* containing the model and
- load_case_id is either the string used when defining the load case unique_id_string, or an index number (1,2,...).

An example of the use of the MATLAB function `coupled_response_fahy.m` is

```
m=coupled_response_fahy(m,'point_force')

m =

      name: 'boxplate'
         s: [1x1 struct]
         c: [1x1 struct]
       bli: [100x19 double]
        lc: {[1x1 struct]  [1x1 struct]}
```

The function will calculate the participation factors for the cavity (e.g., `m.lc{1}.cp`) and structure (e.g., `m.lc{1}.sp`) at each frequency and store the results in a MATLAB *structure array* for the specified load case.

```
m.lc{1} =

      name: 'point_force'
      freq: [1x400 double]
        sn: 373
        sf: 1
        sl: []
        cn: [0x1 double]
        cq: []
        cl: []
        sp: [19x400 double]
        cp: [100x400 double]
```

These participation factors (`sp` and `cp`) can be used to calculate structural and acoustic results.

Calculating Acoustic and Structural Results

There are four MATLAB functions available for calculating acoustic and structural results

- `cav_pressure.m` is used to calculate the acoustic pressures at nodes.

- `str_displacement.m` is used to calculate structural displacements at nodes.

- `acousticpotentialenergy.m` is used to calculate the acoustic potential energy inside the cavity.

- `structuralkineticenergy.m` is used to calculate the kinetic energy of the structure.

Cavity Pressure

The MATLAB function `cav_pressure.m` can be used to calculate the acoustic pressure at nodes within the cavity and has the following input parameters:

`[pressure,freq]=cav_pressure(m,lcindex,node)`

where

- `m` is the MATLAB *structure array* that contains the model and results,
- `lcindex` is the number or text string for the load case (e.g., `'point_force'`), and
- `node` is a vector containing node numbers where the pressures are to be calculated (e.g `[node1, node2, node3]`).

An example of the use of this MATLAB function is shown below.

```
>> [cavp,f]=cav_pressure(m,'point_force',5414);
>> size(cavp)

ans =

   400     1

>> [cavp,f]=cav_pressure(m,'point_force',[5414,5413]);
>> size(cavp)

ans =

   400     2
```

Structural Displacement

The MATLAB function `str_displacement.m` can be used to calculate the displacements at nodes on the structure and has the following input parameters:

`[str_disp,freq]=str_displacement(m,lcindex,node)`

where

- `m` is the MATLAB *structure array* that contains the model and results,
- `lcindex` is the number or text string for the load case (e.g., `'point_force'`), and
- `node` is a vector containing node numbers where the displacements are to be calculated (e.g `[node1, node2, node3]`).

An example of the use of this MATLAB function is shown below.

`[str_disp,f]=str_displacement(m,'point_force',373)`

Acoustic Potential Energy

The MATLAB function acousticpotentialenergy.m can be used to calculate the acoustic potential energy in the cavity and has the following input parameters:

[ape,freq]=acousticpotentialenergy(m,lcindex)

where

- m is the MATLAB *structure array* that contains the model and results.
- lcindex is the number or text string for the load case (e.g., 'point_force').

Note that for this function there is no input parameter for nodes as it is assumed that all nodes associated with the acoustic cavity participate in the calculation of the total acoustic potential energy.

An example of the use of this MATLAB function is shown below.

[ape,f]=acousticpotentialenergy(m,'point_force')

Structural Kinetic Energy

The MATLAB function structuralkineticenergy.m can be used to calculate the kinetic energy of the structure and has the following input parameters:

[ske,freq]=structuralkineticenergy(m,lcindex)

where

- m is the MATLAB *structure array* that contains the model and results.
- lcindex is the number or text string for the load case (e.g., 'point_force').

Note that for this function there is no input parameter for nodes as it is assumed that all nodes associated with the structure participate in the calculation of the total structural kinetic energy.

An example of the use of this MATLAB function is shown below.

[ske,f]=structuralkineticenergy(m,'point_force')

Displaying the Model

The MATLAB function plotmodel.m can be used to display the acoustic and structural models. The usage of the function is

[p]=plotmodel(s)

where the input parameter s is either

- m.s, which will display the structure, or
- m.c, which will display the cavity.

An example use of the function to plot the acoustic cavity is

```
>> plt=plotmodel(m.c)

plt =

    f: 1
    a: 173.0011
    e: 175.0011
```

where 3 fields are returned:

- f figure number
- a axes handle
- e patch handle

An example use of the function to plot the structure is

```
>> plt2=plotmodel(m.s)

plt2 =

    f: 2
    a: 349.0011
    n: 350.0011
    e: 355.0011
```

The image of the structure will include arrows on the nodes indicating the normal direction, which is from the data in m.s.n.norm. The arrows can be removed (and added again) from the image using the MATLAB function togglenorm.m. An example of the use of this function is

```
>> togglenorm(plt2)
```

The mode shapes can be plotted with the MATLAB function plotmode.m. The following example shows how to plot the fifth acoustic mode shape with the following MATLAB commands:

```
m.c.pl=plotmodel(m.c);
plotmode(m.c,5);
```

The following example shows how to plot the fifth structural mode shape with the following MATLAB commands:

```
m.s.pl=plotmodel(m.s);
plotmode(m.s,5);
togglenorm(m.s.pl); % turn off the arrows for the node normals
```

D

Errors

D.1 Errors Relating to References

D.1.1 Definition of Power

There is a small typographical error in the equation for power in Kinsler et al. [102, p. 276] which incorrectly has written that the power is calculated as

$$\text{Power} = \frac{|\tilde{F}|^2 R_{m0}}{2Z_{m0}^2}, \tag{D.1}$$

when the correct equation has the magnitude of the impedance in the denominator as

$$\text{Power} = \frac{|\tilde{F}|^2 R_{m0}}{2|Z_{m0}|^2}, \tag{D.2}$$

which is consistent with the derivation by Fahy and Gardonio [66, Eq. (2.5), p. 77]. Impedance is a complex value and $Z_{m0}^2 = (R + jX)^2 = R^2 + j2RX - X^2$ is not the same as $|Z_{m0}|^2 = R^2 + X^2$.

D.1.2 Equation for Scattered Pressure by a Cylinder

The equation for the scattered wave field from an incident plane wave striking a rigid cylinder is written in Morse and Ingard [117, Chapter 8, p. 401] as

$$p_s = \sum_{m=1}^{\infty} A_m \cos(m\phi)[J_m(kr) + iN_m(kr)]e^{-2\pi i\nu t}, \tag{D.3}$$

where the summation is over $m = 1 \cdots \infty$. However, in Morse [115, p. 348] and Morse and Feshbach [116, p. 1377, Eq. (11.2.28)], the summation is over $m = 0 \cdots \infty$. This is also consistent with expressions by Junger and Feit [97, p. 322], and Skudrzyk [139, p. 446, Eq. (144)]. Hence Equation (8.54) in this book uses the summation over $m = 0 \cdots \infty$.

This correction, where the summation was altered to $m = 0 \cdots \infty$, was verified by two methods—the scattered sound pressure level calculated using the corrected formula shown in Equation (8.54) was compared with

1. predictions using ANSYS Workbench, as shown in Figure 8.36, and

2. with theoretical predictions using the Junger and Feit theory [97, p. 322] that was programmed in MATLAB, as shown in Figure D.1.

Both of these comparisons showed identical results and hence confirm that the summation should be $m = 0 \cdots \infty$.

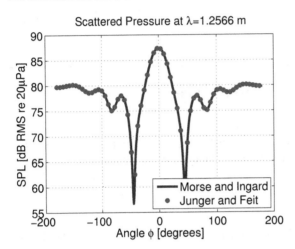

FIGURE D.1
Scattered sound pressure level from a plane wave striking an infinitely long rigid cylinder. The results were calculated using the theories from Morse and Ingard [115, p. 348] and Junger and Feit [97, p. 322].

The MATLAB code `cylinder_plot_scattered_pressure_junger_feit.m` included with this book can be used to calculate the scattered sound pressure level from a plane wave striking an infinitely long rigid cylinder, using the theory by Junger and Feit [97, p. 322].

Another discrepancy occurs with the definitions for the incident plane wave. Morse [115, p. 347] defined the amplitude of the incident plane wave as

$$P_0 = \sqrt{2\rho_0 c_0 I_p} \, , \tag{D.4}$$

where I_p is the plane wave intensity. However, in Morse and Ingard [117, p. 401] the equation is defined as

$$P_0 = \sqrt{\rho_0 c_0 I_p} \, , \tag{D.5}$$

where the factor of $\sqrt{2}$ has been removed. Hence, take care when selecting the appropriate units for the plane wave intensity.

Another inconsistency occurs with the equation for the scattered pressure at *large* distances ($kr \gg 1$) from the cylinder. The analyses using ANSYS described in this book involve the comparison of the predicted sound level in regions relatively close to the cylinder. Morse and Feshbach [116, p. 1378] describe the asymptotic form of this equation as the radius from the origin to the measurement location approaches infinity $r \to \infty$.

Morse has defined several variations for Equation (8.63) as listed in Table 9.1.

TABLE 9.1
Variation of Equations Listed in Books by Morse for the Scattered Pressure from an Incident Plane Wave Striking an Infinitely Long Rigid Cylinder at Large Distances

Reference	Equation
Morse [115, p. 349]	$p_s = -\sqrt{\dfrac{4a\rho_0 c_0 I_p}{\pi r}}\,\psi_s(\phi)e^{ik(r-ct)}$
Morse and Ingard [117, p. 402]	$p_s = -\sqrt{\dfrac{2a\rho_0 c_0 I_p}{\pi r}}\,\psi_s(\phi)e^{ik(r-ct)}$
Morse and Feshbach [116, p. 1378, Eq. (11.2.29)]	$p_s = -\sqrt{\dfrac{i\,2a\rho_0 c_0 I_p}{\pi r}}\,\psi_s(\phi)e^{ik(r-ct)}$

D.1.3 Temperature Gradient in a Duct

Section 3.6.1 describes a theoretical model for the calculation of the pressure and acoustic particle velocity in a duct that uses a four-pole transmission matrix, based on the theory by Sujith [142]. The equations presented in the paper were an extension of the work by Sujith et al. [143]. It was found that there were several errors in the equations by Sujith [142]. The equations were re-derived and the corrected equations are shown in Equations (3.72) to (3.75) and published in Howard [74].

D.2 Issues Relating to ANSYS

ANSYS provides excellent capabilities for conducting acoustic simulations and the software is in a state of continual improvement to implement new features and address issues. This section contains a description of issues that have been identified during the course of writing this book, which may also affect readers of this book trying to conduct simulations. This section is divided into subsections that deal with ANSYS Mechanical APDL and ANSYS Workbench, the ACT Acoustics extension, and issues with the ANSYS help documentation.

The subsections dealing with ANSYS products are divided further into three subsections:

Issues are incorrect implementation of features in ANSYS Workbench, ANSYS Mechanical APDL, and the ACT Acoustics extension.

Traps are things that a user might unintentionally misunderstand, leading to what is perceived to be an incorrect result, when really the user has not appreciated the way ANSYS has implemented a feature.

Limitations are features that would be good to have (or should be) in ANSYS, but have to be implemented another way.

Lastly, there is a subsection that describes the error messages that are generated by ANSYS and explanations of what they mean and how to address them.

D.2.1 ANSYS Mechanical APDL and ANSYS Workbench

D.2.1.1 Issues

Transient analysis using a Mass source: When using a Mass Source in ANSYS Mechanical APDL, issued using the command BF,,JS, unless a KBC,1 command is also issued only the last time step will have a non-zero result. In other words, the default ramp input KBC,0 does not work. If using a F,,FLOW command instead, then either the default ramp input KBC,0 or KBC,1 works. This has been rectified in ANSYS Release 15.0.

SPL and transient analysis solutions: When reviewing transient results in ANSYS Mechanical APDL there is an error when viewing SPLs using any of the following three commands: NSOL/PRNSOL/PLNSOL,SPL. ANSYS does not return SPL = 10*log10((PRES)^2/(PRES REF)^2) but instead returns a value = 10*log10((PRES)^2/STEPSIZE). According to ANSYS, the SPL results are not meant to be available in transient analyses and an attempt to use the APDL commands NSOL/PRNSOL/PLNSOL,SPL should display an error message. The ANSYS Help [37, // Theory Reference // 8. Acoustics // 8.6. Acoustic Output Quantities] does not contain a discussion of obtaining SPL results from a transient analysis, which is an indication that the feature is not available.

Impedance sheet: The impedance sheet (implemented with the ANSYS APDL command BF,,IMPD,RESIS,REACT) is not fully functional in ANSYS Release 14.5. The reactive part of the impedance does not work. This has been rectified in ANSYS Release 15.0.

Naming of objects: The names of objects in the Outline window are not necessarily retained upon closing and reopening the ANSYS Workbench file.

D.2.1.2 Traps

The following text describes situations that a user might encounter that can generate unexpected outcomes.

PRCPLX,1 , POST26, and SPL: When using the time-frequency post-processor /POST26 in ANSYS Mechanical APDL, if the format of the results have been set as magnitude and phase with the command PRCPLX,1 then an attempt to store the sound pressure level using the command NSOL,,,SPL will give incorrect results. Instead, one must use the real and imaginary format by issuing the command PRCPLX,0, which is the default setting. This has been fixed in ANSYS Release 15.0.

Johnson–Champoux–Allard model: In ANSYS Releases 14.5 and 15.0 the Johnson–Champoux–Allard Equivalent Fluid Model implements a porous media model of the fluidic phase in the pores of the rigid-walled media. It is important to note that it is not the homogenous fluid with equivalent bulk properties, which requires dividing the complex density and bulk modulus of the fluidic phase by the porosity. Consequently, surface impedances at the boundaries of such elements will be out by a factor equal to the porosity of the material. The implementation of the Johnson–Champoux–Allard Equivalent Fluid Model will be changed in ANSYS Release 16.0 so it uses the equivalent fluid rather than the fluidic phase.

Using VGET to get SPLs: Consider the situation in /POST26 when a sound pressure level had been stored in the 3rd variable by issuing the command, NSOL,3,,SPL. Any attempt to move the variable into an array parameter variable using a command of the form VGET,SPL,3 will fail. The reason is that SPL values are never really variables but are calculated when issuing the PRVAR or PLVAR command. If you want to manipulate SPL variables you need to use the real and imaginary pressure and recompute SPL as a real variable. See the ANSYS Mechanical APDL code below, which can be used to calculate the sound pressure level.

```
1   PRCPLX,0
2   CURR = 0
3   PREF = 20e-6
4   *DO,AR30,1,NODENUM
5   CURR = NDNEXT(CURR)
6   NSOL,3,CURR,PRES
7   REALVAR, 4, 3, , ,
8   REAL IMAGIN, 5, 3, , ,
9   IMAG PROD, 6, 4, 4
10  PROD, 7, 5, 5
11  ADD,8,6,7
12  FACT = 1/(2*PREF**2)
13  CLOG, 9, 8, , , SPL, , , FACT, 10
14  *IF,AR30,EQ,1,THEN
15  ADD, 10, 9
16  *ELSE
17  ADD, 10, 2, 9
18  *ENDIF
```

```
19   ADD,2,10
20   *ENDDO
21   PROD, 11, 10, , ,SPL, , , 1/NODENUM
```

Impedance sheet: In the ANSYS online Help manual [37, // Command Reference // III. B Commands // BF] there is little information on how the impedance is implemented for the BF,,IMPD command. It should be noted that the command does not work the same way as the SF command, which is discussed in // Command Reference // XX. S Commands // SF.

With BF,,IMPD, the first term is the resistance and the second term is the reactance, regardless of the sign of these terms. Whereas with SF,,IMPD, for acoustic harmonic response analyses, the first value is the resistance in Ns/m^3 if > 0 and is conductance in mho if < 0. The second value is the reactance in Ns/m^3 if > 0 and is the product of susceptance and angular frequency if < 0. In acoustic transient analyses, the second value is not used. It is expected that in ANSYS Release 16.0, there will be a check for negative values of resistance. Furthermore, the impedance sheet behaves as an acoustic side-branch and not as a structural impedance sheet. For the latter, ANSYS has created a Trim element that will be available from ANSYS Release 15.0.

Numerical damping: The amount of numerical damping used in transient analyses is set by the parameter GAMMA (> 0), which is the Amplitude Decay Factor for second-order transient integration. In ANSYS Mechanical APDL the default value is 0.005. However, in ANSYS Workbench the default value for GAMMA is 0.1, which can significantly alter the response of lightly damped vibro-acoustic systems.

Symmetric FSI option: In the model radiation_open_duct.wbpj, if Acoustic Body and Acoustic Body 2, which are for the duct and the 1/8 spherical regions, have the Acoustic – Structural Coupled Body Options changed to Coupled With Symmetric Algorithm, as shown below, the results are incorrect.

Acoustic-Structural Coupled Body Options	Coupled With Symmetric Algorithm ▼
Perfectly Matched Layers (PML)	Uncoupled
	Coupled With Unsymmetric Algorithm
	Coupled With Symmetric Algorithm

Figure D.2 shows that the real part of the impedance is incorrect. However, Figure D.3 shows that the imaginary part of the impedance is correct.

If the symmetric FSI formulation is used, then all the elements in the model must use the symmetric formulation.

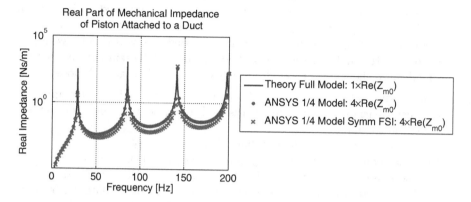

FIGURE D.2

Real part of mechanical impedance of a piston attached to a 3 m circular duct that radiates into a baffled plane, calculated theoretically and using ANSYS Workbench with a 1/4 model, showing that using the Symmetric FSI formulation causes incorrect results.

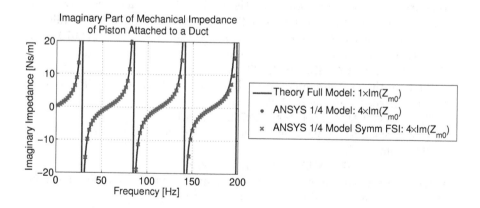

FIGURE D.3

Imaginary part of mechanical impedance of a piston attached to a 3 m circular duct that radiates into a baffled plane, calculated theoretically and using ANSYS Workbench with a 1/4 model.

Thermal conductivity of fluid: A feature exists for the FLUID30, FLUID220, and FLUID221 elements where the temperature at the nodes can be defined using the ANSYS Mechanical APDL command BF,node,TEMP,value, which applies the temperature as a nodal body force load. Note that these acoustic elements do not have temperature degrees of freedom, so it is not possible to conduct a thermal analysis using these elements. The temperature is used to calculate the speed of sound and density of the fluid at each node.

When using ANSYS Release 14.5, an issue can occur if one attempts to conduct a harmonic analysis if the thermal conductivity of the gas is defined using the command
MP,KXX,1,0.0257   ! [W/m.K] thermal conductivity
then the acoustic particle velocity is not calculated. It is necessary to delete the thermal conductivity definition for the gas using the command
MPDELE,KXX,matid before solving the model in order to calculate the acoustic particle velocity. This issue has been fixed in ANSYS Release 15.0.

Sound pressure level results for FLUID29 and FLUID30: In versions of ANSYS before Release 14.0, the results from a harmonic analysis could return the sound pressure level (in decibels) within an element using the command PLESOL,NMISC,4, as is still the case for the 2D FLUID29 elements. The sound pressure level is now obtained by requesting that complex valued results are displayed as amplitude using the APDL command SET,,,,AMPL, and then plotting the sound pressure level with the command PLNSOL,SPL , or listing the results with the command PRNSOL,SPL.

Pressure gradient and velocity: Users should note that the way to retrieve results for the FLUID29 and FLUID30 elements are slightly different and should consult the help manual for the specifics. For the FLUID29 2D acoustic element, the APDL command:
ESOL, 3, ELEM_NUM, NODE_NUM, PG, X, pg_x
will return the *pressure gradient* at the node. To obtain the *particle velocity* the user needs to use the APDL command:
ESOL, 4, ELEM_NUM, NODE_NUM, SMISC, 3, vel_x.
However, when using 3D acoustic elements such as FLUID30, then the same APDL command:
ESOL, 3, ELEM_NUM, NODE_NUM, PG, X, pg_x
will return the *particle velocity* for modal and harmonic analyses, and pressure gradient in the X direction for transient analyses.

D.2.1.3   Limitations

Modal superposition using transient analysis: ANSYS Release 14.5 does not support modal superposition method (MSUP) for transient analysis of damped acoustic systems. Therefore, when there is damping or absorption in the model, it is necessary to use the Full analysis. It is possible to use

MSUP with an undamped acoustic model, but as of ANSYS Release 14.5 this feature is not supported.

Contact elements in acoustic analysis: Contact regions do not support acoustics elements (for pressure DOF) so it is necessary to use connected meshes (that share common nodes). Alternatively, it is possible to couple the displacement degrees of freedom of a fluid (if activated).

Johnson–Champoux–Allard model: The flow resistivity, σ, the viscous characteristic length, Λ, and the thermal characteristic length, Λ', are all temperature dependent. Since ANSYS does not provide a tabular form of Johnson–Champoux–Allard model data, the only way currently (in ANSYS Release 14.5) to model the effects of varying temperature across an absorbent is to have individual element layers in the absorbent and to use different JCA parameter values in each.

Mass source and tabular data: In ANSYS Release 14.5 the acoustic Mass Source does not support tabular data. This is to be rectified in Release 15.0. As an alternative, one can use an acoustic FLOW source, which does support tabular data.

Spectral damping and full harmonic analysis: The pressure-formulated acoustic elements (FLUID30, etc) do not support spectral damping (Rayleigh damping or structural damping ratios) when conducting a full harmonic analysis. These are supported when using modal superposition.

Imaginary results from a harmonic analysis: When using the ANSYS Workbench model freq_depend_impedance.wbpj, if the starting analysis frequency range is set to 0 in the ARG4 parameter as part in the Commands (APDL) command object, the imaginary results for the UZ displacement are zero for all analysis frequencies, which is incorrect. If the starting analysis frequency is non-zero, say 2 Hz, then the results are calculated correctly.

APDL mathematical operations: The ANSYS online help manual describes the mathematical operations that can be done using APDL commands under sections:

Parametric expressions // ANSYS Parametric Design Language Guide // 3. Using Parameters // 3.7. Parametric Expressions

Parametric functions // ANSYS Parametric Design Language Guide // 3. Using Parameters // 3.8. Parametric Functions

APDL Functions // Command Reference // XX. S Commands // *SET

A frustrating lack of capability in ANSYS APDL is the inability to perform mathematical operations on complex numbers. For example, an attempt to evaluate aa=SQRT(-1) in ANSYS Mechanical APDL will generate the error
*** ERROR *** Value= -1 is outside function range SQRT.

Hence, when attempting to evaluate expressions with complex numbers, it is necessary to separate the real and imaginary components and perform the operations for each part.

The /POST26 post-processing module in ANSYS Mechanical APDL has some capabilities for performing mathematical operations with complex valued results. The APDL command CFACT defines complex scaling factors that can be used with some operations in /POST26 such as ADD, PROD, QUOT, etc. The full list of /POST26 operations are listed in ANSYS, SAS IP [38, Table 2.109]

D.2.2   ACT Acoustics Extension

D.2.2.1   Issues

Acoustic Time_Frequency Plot: From time to time, all the Acoustic Time_Frequency Plots will disappear and the worksheet will disappear. For some reason the default setting with the worksheet activated for the plots gets unselected. You can get the results back by clicking on the Worksheet button. If this fails, then it is necessary to clear the result and generate it again.

Imaginary results: When plotting complex valued results such pressures and particle velocities using the ACT Acoustics extension Version 8, there is the option for retrieving results at a specified phase angle. A phase of $0°$ corresponds to the real component. A phase of $90°$ should correspond to the imaginary component, however, the ACT Acoustics extensions returns the conjugate of the imaginary result. Hence, to obtain the imaginary value one needs to request results at a phase angle of $-90°$. This behavior can be confirmed by viewing the (real and imaginary) results in a Acoustic Time_Frequency Plot.

Using node selection with ACT Acoustics mass source: When using the ACT Acoustics extension Version 8, if a Mass Source is defined where it is applied to a node using a Named Selection,

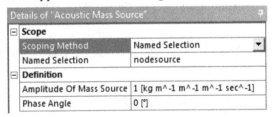

then warning messages will be generated when solving such as

```
*** WARNING ***              CP =       1.825   TIME= 06:38:22
Node 1 on element 1 is unselected.
```

Many of these warning messages will be generated that eventually causes an error once the number of warning messages exceeds the defined limit. The workaround is to change the Scope | Scoping Method back to Geometry selection so that the object is selected, instead of using the Named Selection.

Details of "Acoustic Mass Source"	
Scope	
Scoping Method	Geometry Selection
Geometry	1 Node
Definition	
Amplitude Of Mass Source	1 [kg m^-1 m^-1 m^-1 se...
Phase Angle	0 [°]

The row Scope | Geometry will still indicate that 1 Node is selected. It is then possible to solve the model.

D.2.2.2 Limitations

This section describes some of the feature limitations of ANSYS at Release 14.5.

FLOW source in ACT Acoustics extension: In ANSYS Release 14.5 the acoustic FLOW source is not implemented in the ACT Acoustics extension and a Command object must be used instead.

Acoustic pressure results: When selecting an Acoustic Pressure result in ANSYS Workbench using the ACT Acoustics extension toolbar, one sees a little P next to the icon indicating pressure. However, when these are inserted into the results tree in the OUTLINE window, it reverts back to the default (which is User). The reason is the ACT Acoustics extension uses a User Defined Result for Acoustic Pressure. The same does not happen for SPL which employs a custom code. The reason is that it is only possible to modify the icons of customized results. However customized results are a little slower to evaluate, and a User Defined Result is used instead.

User-defined result identifier: This is not currently available for the SPL results. This is only for the native results such as pressure and velocity.

Averaging SPLs: When plotting SPLs using the Acoustic Time_Frequency Plot it is possible to display the average of all selected nodes. It should be noted that this is a linear average of the nodal SPLs and not a logarithmic average. In other words, for N measurements this process would return

$$\bar{L}_p = \frac{1}{N} \sum_{i=1}^{N} L_{p,i} \,, \tag{D.6}$$

where $L_{p,i}$ is the sound pressure level at the ith location, which is not correct. To correctly average the SPLs for the microphones, the following expression is required

$$\bar{L}_p = 10 \log_{10} \left(\frac{1}{N} \sum_{i=1}^{N} 10^{L_{p,i}/10} \right) . \tag{D.7}$$

D.2.3   Other

D.2.3.1   ANSYS Documentation

- The following documentation is incorrect // Technology Demonstration Guide // 34. Analysis of a Piezoelectric Flextensional Transducer in Water // 34.7. Results and Discussion
  The line Expression: SPL = 20*log10(Pmag/20e-06) should actually be SPL = 20*log10(Prms/20e-06).

- In the help manual // Fluids Analysis Guide // II. Acoustics // 1. Acoustics // 1.1. Types of Acoustic Analysis it has written

  > The program assumes that the fluid is compressible, but allows only relatively small pressure changes with respect to the mean pressure. Also, the fluid is assumed to be non-flowing and inviscid (that is, viscosity causes no dissipative effects).

  This is no longer the case. See the Theory Manual under section // Theory Reference // 8. Acoustics // 8.1. Acoustic Fundamentals. for the correct information.

- For the 2D FLUID29 element, ANSYS automatically adjusts the value of the absorption coefficient MU to accommodate the effects of temperature in the element. This feature is documented in the FLUID29 help. However, for the 3D acoustic elements it is not clear from the ANSYS help if the surface absorption coefficient,SF,,CONV, is also automatically adjusted for temperature. In other words, if one were to apply CONV,1 the resistance will remain $\rho_0 c_0$ at the nominal temperature or the modified average nodal temperature when issuing a BFE,ALL,TEMP,,TMP_HOT command. The attenuation surface is adapted according to the applied temperature and static pressure, and therefore one will obtain the same results with applied temperature and manually modified sound speed and density.

- The acoustic wave equation given by Eq. (8-1) in the ANSYS Help manual shows the presence of viscous losses, however it does not show thermal losses. Both viscous and thermal losses are available in ANSYS Release 14.5, however, the thermal losses are undocumented. This is rectified in Release 15.0 where visco-thermal losses are fully documented.

D.2.4   ANSYS Errors Messages

There are many error messages that can be generated by ANSYS and it is often unclear what they mean or how to address them. Below are listed some error messages, descriptions of what they mean, and how they can be addressed.

```
*** ERROR ***                    CP =        2.090    TIME= 03:46:42
PML element 10071 is not in the 3-D PML region.
```

According to the help manual, "You construct a block about the origin of the global Cartesian coordinate system or a local Cartesian coordinate system. You align the edges of the 3-D PML region with the axes of the Cartesian coordinate system." If you try to create a PML with faces of the volume that are not aligned with the Cartesian axes, you can get these errors.

Another point to note from the ANSYS help is that acoustic excitation sources are not permitted in the PML region, as described, otherwise the following error message will be generated.

```
*** ERROR ***                    CP =       20.826    TIME= 14:49:42
The mass source defined by the BF command is not allowed in the
PML element 13171 (KEYOPT(4)=1).
```

This error can occur if the mass source that was defined happens to touch one of the PML elements, which can occur on the boundary between a non-PML region and a PML region, and a mass source has been defined that extends onto the common vertex, line, or face between the two regions.

The solution is to create another non-PML region between the geometry where the mass source was applied and the PML region.

Negative Pivot: this often means that your model does not have sufficient boundary conditions and it can move freely along one or more axes. Consider applying more boundary conditions to limit the movement of the model.

```
The program is unable to open file file.LN09. If the suggestions
in the associated error messages immediately preceding and/or
following this error message do not help resolve the file issue,
please send the data leading to this operation to your technical
support provider, along with the system error code of -1, for
help in determining the possible reasons as to why this file
could not be opened.
```

When using Microsoft Windows 7, it is possible to set up a process monitor to continuously log the amount of free disk space. Although the log file showed that there was 17 GB free at all times, this was insufficient. After freeing a large amount of disk space the model could be solved, and the log file showed that the solution required 55 GB of disk space. The moral of the story is make sure you have *much* more free disk available than you think you will ever need.

```
*** ERROR ***                    CP =      6.864   TIME= 16:17:08
Node no. 465 of the FLUID130 Acoustic Absorbing element 131337
should lie on a SPHERE of radius 4; instead lies at a radius of
3.99951169. Ensure that the radius input through the element Real
Constant matches with that of the absorbing boundary.
```

Absorbing Elements must be placed on the outside of a spherical or hemi-spherical surface. A peculiar error can occur where the numerical precision of the placement of the nodes is inaccurate. This has to be manually corrected using the following Command Object, which can be placed after the definition of the Absorbing Elements in the (Harmonic Response) analysis branch.

The APDL code below is intended to move the location of all nodes associated with the FLUID130 elements to a radius of 4 m. The user should use this code judiciously to ensure that nodes in other parts of the model are not unintentionally moved.

Change the value used for the radius in the code below as required.

```
1   /PREP7                      ! Change to the preprocessor
2   CSYS,2                      ! Change to a spherical coordinate system
3   ESEL,S,ENAME,,130
4   NSLE,S
5   !NSEL,S,LOC,X,3.99,10       ! OR select all nodes on a radius of 3.99 to 10
6   NMOD,ALL,4.000              ! Change all these nodes to have a radius of 4.000
7   NSEL,ALL                    ! SELECT ALL THE NODES AGAIN
8   ESEL,ALL
9   CSYS,0                      ! CHANGE BACK TO A CARTESIAN COORDINATE SYSTEM
10  /SOLU                       ! Go back to the solution module
```

The display settings are windows Aero and image capture might not work. please change them to another theme.

This is a problem that occurs with Microsoft Windows Vista and Windows 7 operating systems. The way to fix this problem is to disable "Enable Desktop Composition" or select the Windows Classic theme instead of Aero.

Method 1:

1. Open Control Panel > System Properties > Advanced System Settings > Performance Settings > Visual Effects.

2. Choose Custom.

3. Uncheck Enable Desktop Composition.

4. Click Apply.

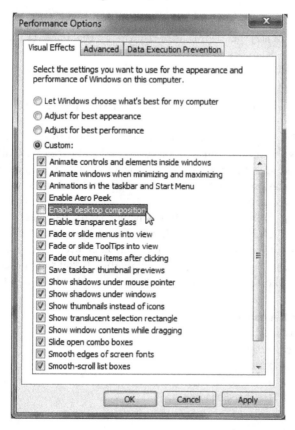

Method 2:

1. Open Control Panel > Appearance and Personalization > Change the theme.

2. Select Windows Classic.

3. Click Apply.

It is expected that this issue will be addressed in ANSYS Release 16.0.

Warning: The Analysis Type (2D/3D) cannot be changed after the first attach.

If you wanted to create a 2D model but inadvertently started Mechanical in the 3D mode and were working, but then decided to change the mode from 3D to 2D, you will see this warning message. The only way to correct this is to start a new analysis from the Workbench Project Schematic.

A warning window can appear when opening a project with the text:

The files shown in Details are missing from the project.

See "Project File Management" in the Workbench help for more information. Files that may be repairable in the Files pane (View > Files):

...\driven_duct_pres_dist_files\dpall\global\AdvancedAddin Package\log.html

The Extension menu in the Project window is written in gray and not available for a particular project.

Click on View | Files to show the files related to the project. There might be a missing log.html file, which is of type "ACT log file", which is normally found in the sub-directory from the location of the project file .wbpj. For example, for the project driven_duct_pres_dist, the log file would be located in the sub-directory

.\driven_duct_pres_dist_files\dpall\global\AdvancedAddinPackage

If you were to open ANSYS Mechanical, you might find that existing entries for ACT Acoustics extension options have an icon with a red circle with a white bar, indicating that the option is not available. If you were to click on one of these entries, the details window, such as Details of "Acoustic Body", would be blank.

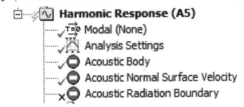

To fix this issue do the following steps:

- Make sure that you installed the ACT Acoustic Extension package previously, by opening an existing file and checking that you can access the ACT Acoustic Extensions. If this does not work, then you might need to install the ACT Acoustic Extensions add-in package.

- From the menu bar, select File | Save As, and save the project with a new filename, such as replacement.wbpj.

- Right-click on the row with log.html and select Remove log.html from List. You will get a Warning dialog box that says

  Any reference to the selected file(s) will be permanently removed from Workbench? Do you want to continue? Click Yes.

- Click the Save icon.

- Now you need to re-open the project. Unfortunately there is no option in Workbench to close a project, so instead click on the New icon (or File | New).

- Click on File | Open and select the project that you just created (e.g., replacement.wbpj).

- You should notice that the Extension menu is now written in black.

```
*** ERROR ***                    CP =      2.886    TIME= 05:35:55
The dipole axis from the positive to the negative is not defined
correctly for the wave 1.   The AWAVE command is ignored.
```

This error can occur if the user did not specify the orientation of the dipole by defining the components of the unit vector.

X Component Of Unit Dipole Vector	0 [m]
Y Component Of Unit Dipole Vector	0 [m]
Z Component Of Unit Dipole Vector	0 [m]

The default values for the unit vector are $(0,0,0)$ and an error will be generated unless values are defined. Note that when defining the components of the vector, it is *not* necessary to define the components of the vector so that the magnitude ($\sqrt{x^2 + y^2 + z^2}$) equals 1—the equivalent vector of unit length will be calculated by ANSYS.

```
*** ERROR ***                    CP =      1.638    TIME= 10:36:44
Data file FILE.DB does not exist for RESUME.
```

This error can be generated when using ANSYS Workbench and the ACT Acoustics extension if you forgot to change the option to retain the MAPDL db file. The solution is to click on Harmonic Response (A5) | Analysis Settings, and in the window Details of "Analysis Settings", change the row Analysis Data Management | Save MAPDL db to Yes.

E

Export of Nodal Area from ANSYS

E.1   Calculation of Nodal Area

There are three suggested methods for calculating the area associated with nodes.

1. Nodal Area Using APDL Command ARNODE

The APDL command ARNODE(node) can be used to determine the area associated with a node. Only the nodes associated with the element should be selected before using this command.

An example use of this command is shown below for determining the area associated with a node belonging to a planar SHELL181 element.

```
1  ! Select the elements associates with SHELL181
2  ESEL,S,ENAME,,SHELL181
3  !Select only the nodes associated with the SHELL181
4  NSLE,S,1
5  ! Find a node that belongs to the SHELL181
6  node=NDNEXT(0)
7  narea=ARNODE(node)
```

Note the ARNODE command will work for structural elements such as SHELL181, but does not work for FLUID129, FLUID130, or MESH200 elements. Hence, to determine the area associated with nodes attached to FLUID130 elements, it is necessary to either temporarily re-assigned the elements as SHELL181, export the nodal areas using the ARNODE command, and then return the elements to FLUID130; or use one of the other methods listed below.

2. Nodal Area from Reaction Force of 1Pa Load

A method that can be used for structural elements is as follows:

- Fix all the degrees of freedom of all the nodes.

- Apply a 1 Pa nodal load to the face of the structure.

- Conduct a static structural analysis.

- Examine the reaction forces at all the nodes. Pressure is calculated as force divided by area, and because a unit pressure of 1 Pa was applied, the reaction force is effectively the nodal area.

3. Mathematically Calculating Nodal Area from Nodal Coordinates

Another method of determining the nodal areas involves exporting the coordinates of all the nodes and elements and calculating the area using geometry. This method is used in the MATLAB functions `loadstr.m` and `power_freefield_hemisphere.m` that are included with this book. Examples of the use of this function are shown in Sections 9.4.5 and 9.5.4.

References

[1] Milton Abramowitz and Irene A. Stegun, editors. *Handbook of Mathematical Functions with Formulas, Graphs, and Mathematical Tables*. Dover Publications, New York, 1972.

[2] Jean Allard and Noureddine Atalla. *Propagation of Sound in Porous Media: Modelling Sound Absorbing Materials*. John Wiley and Sons, second edition, November 2009.

[3] ANSI/ASA S1.11-2004 (R2009) Octave-Band and Fractional-Octave-Band Analog and Digital Filters, 2004.

[4] ANSYS, SAS IP, Inc. *ANSYS 14.5 Help*. ANSYS Help System // Mechanical APDL // Mechanical Equations // Theory Reference // 17. Analysis Procedures // 17.3. Mode-Frequency Analysis.

[5] ANSYS, SAS IP, Inc. *ANSYS 14.5 Help*. ANSYS Help System // Mechanical APDL // Mechanical Equations // Theory Reference // 15. Analysis Tools // 15.15. Eigenvalue and Eigenvector Extraction.

[6] ANSYS, SAS IP, Inc. *ANSYS 14.5 Help*. ANSYS Help System // Mechanical APDL // Theory Reference // 17. Analysis Procedures // 17.4. Harmonic Analysis // 17.4.2. Description of Harmonic Analysis.

[7] ANSYS, SAS IP, Inc. *ANSYS 14.5 Help*. ANSYS Help System // Mechanical APDL // Theory Reference // 15. Analysis Tools // 15.9. Mode Superposition Method.

[8] ANSYS, SAS IP, Inc. *ANSYS 14.5 Help*. ANSYS Help System // Mechanical APDL // Theory Reference // 17. Analysis Procedures // 17.4. Harmonic Analysis // 17.4.5. Harmonic Analysis Solution // 17.4.5.2. Mode Superposition Method.

[9] ANSYS, SAS IP, Inc. *ANSYS 14.5 Help*. ANSYS Help System // Mechanical APDL // Structural Analysis Guide // 5. Transient Dynamic Analysis.

[10] ANSYS, SAS IP, Inc. *ANSYS 14.5 Help*. ANSYS Help System // Mechanical APDL // Theory Reference // 8. Acoustics // 8.4. Acoustic Fluid-Structural Interaction.

[11] ANSYS, SAS IP, Inc. *ANSYS 14.5 Help.* ANSYS Help System // Mechanical APDL // Theory Reference // 8. Acoustics // 8.4. Acoustic Fluid-Structural Interaction // 8.4.2. Coupled Acoustic Fluid-Structural System with Symmetric Matrix Equation for Full Harmonic Analysis.

[12] ANSYS, SAS IP, Inc. *ANSYS 14.5 Help.* ANSYS Help System // Mechanical APDL // Verification Manual // I. Verification Test Case Descriptions // VM242.

[13] ANSYS, SAS IP, Inc. *ANSYS 14.5 Help.* ANSYS Help System // Mechanical APDL // Theory Reference // 8. Acoustics // 8.3. Propagation, Radiation, and Scattering of Acoustic Pressure Waves // 8.3.4. Acoustic Excitation Sources // 8.3.4.3. Analytic Wave Sources.

[14] ANSYS, SAS IP, Inc. *ANSYS 14.5 Help.* ANSYS Help System // Mechanical APDL // Theory Reference // 8. Acoustics // 8.3. Propagation, Radiation, and Scattering of Acoustic Pressure Waves // 8.3.4. Acoustic Excitation Sources // 8.3.4.2. Mass Source in the Wave Equation.

[15] ANSYS, SAS IP, Inc. *ANSYS 14.5 Help.* ANSYS Help System // Mechanical APDL // // Theory Reference // 8. Acoustics // 8.3. Propagation, Radiation, and Scattering of Acoustic Pressure Waves // 8.3.5. Sophisticated Acoustic Media.

[16] ANSYS, SAS IP, Inc. *ANSYS 14.5 Help.* ANSYS Help System // Mechanical APDL // // Theory Reference // 8. Acoustics // 8.3. Propagation, Radiation, and Scattering of Acoustic Pressure Waves // 8.3.5. Sophisticated Acoustic Media // 8.3.5.3. Impedance Sheet Approximation.

[17] ANSYS, SAS IP, Inc. *ANSYS 14.5 Help.* ANSYS Help System // Mechanical APDL // Theory Reference // 8. Acoustics // 8.3. Propagation, Radiation, and Scattering of Acoustic Pressure Waves // 8.3.1. Acoustic Boundary Conditions.

[18] ANSYS, SAS IP, Inc. *ANSYS 14.5 Help.* ANSYS Help System // Mechanical APDL // High-Frequency Electromagnetic Analysis Guide.

[19] ANSYS, SAS IP, Inc. *ANSYS 14.5 Help.* ANSYS Help System // Mechanical APDL // Element Reference // I. Element Library // FLUID30, FLUID30 Input Data.

[20] ANSYS, SAS IP, Inc. *ANSYS 14.5 Help.* ANSYS Help System // Mechanical APDL // Theory Reference // 8. Acoustics // 8.1. Acoustic Fundamentals, Eq. (8.1).

[21] ANSYS, SAS IP, Inc. *ANSYS 14.5 Help.* ANSYS Help System // Mechanical APDL // Technology Demonstration Guide // 31. Acoustic Analysis of a Small Speaker System // 31.5. Boundary Conditions and Loading // 31.5.2. Acoustic Loads and Boundary Conditions.

[22] ANSYS, SAS IP, Inc. *ANSYS 14.5 Help*. ANSYS Help System // Mechanical APDL // Theory Reference // 19. Postprocessing // 19.15. POST1 and POST26 Complex Results Postprocessing.

[23] ANSYS, SAS IP, Inc. *ANSYS 14.5 Help*. ANSYS Help System // Mechanical APDL // Theory Reference // 8. Acoustics // 8.6. Acoustic Output Quantities.

[24] ANSYS, SAS IP, Inc. *ANSYS 14.5 Help*. ANSYS Help System // Mechanical APDL // Structural Analysis Guide // 1. Overview of Structural Analyses // 1.4. Damping.

[25] ANSYS, SAS IP, Inc. *ANSYS 14.5 Help*. ANSYS Help System // Mechanical APDL // Theory Reference // 15. Analysis Tools // 15.3. Damping Matrices.

[26] ANSYS, SAS IP, Inc. *ANSYS 14.5 Help*. ANSYS Help System // Mechanical APDL // Mechanical Equations // Theory Reference // 8. Acoustics // 8.1. Acoustic Fundamentals // 8.1.1. Governing Equations.

[27] ANSYS, SAS IP, Inc. *ANSYS 14.5 Help*. ANSYS Help System // Mechanical APDL // Theory Reference // 8. Acoustics // 8.3. Propagation, Radiation, and Scattering of Acoustic Pressure Waves // 8.3.5.2 Equivalent Fluid of Perforated Materials.

[28] ANSYS, SAS IP, Inc. *ANSYS 14.5 Help*. ANSYS Help System // Mechanical APDL // Fluids Analysis Guide // II. Acoustics // 1. Acoustics // 1.5. Applying Loads and Obtaining the Solution // 1.5.4.3 Load Types.

[29] ANSYS, SAS IP, Inc. *ANSYS 14.5 Help*. ANSYS Help System // Mechanical Applications // Mechanical User Guide // Specifying Geometry // Named Selections // Defining Named Selections // Specifying Named Selections using Worksheet Criteria // Adjusting Tolerance Settings for Named Selections by Worksheet Criteria.

[30] ANSYS, SAS IP, Inc. *ANSYS 14.5 Help*. ANSYS Help System // Mechanical Applications // Mechanical User Guide // Configuring Analysis Settings // Steps and Step Controls for Static and Transient Analyses // Guidelines for Integration Step Size.

[31] ANSYS, SAS IP, Inc. *ANSYS 14.5 Help*. ANSYS Help System // Mechanical APDL // Mechanical Equations // Theory Reference // 17. Analysis Procedures // 17.2. Transient Analysis.

[32] ANSYS, SAS IP, Inc. *ANSYS 14.5 Help*. ANSYS Help System // Mechanical APDL // Theory Reference // 8. Acoustics // 8.3. Propagation, Radiation, and Scattering of Acoustic Pressure Waves // 8.3.2. Absorbing Boundary Condition (ABC).

[33] ANSYS, SAS IP, Inc. *ANSYS 14.5 Help.* ANSYS Help System // DesignModeler User Guide // 3D Modeling // Advanced Features and Tools // Freeze.

[34] ANSYS, SAS IP, Inc. *ANSYS 14.5 Help.* ANSYS Help System // Mechanical APDL // Element Reference // I. Element Library // FLUID29, Table 29.2: FLUID29 Item and Sequence Numbers.

[35] ANSYS, SAS IP, Inc. *ANSYS 14.5 Help.* ANSYS Help System // Mechanical Applications // Mechanical User Guide // Using Results // User Defined Results // User Defined Result Expressions.

[36] ANSYS, SAS IP, Inc. *ANSYS 14.5 Help.* ANSYS Help System // Mechanical APDL // Command Reference // XX. S Commands // *SET.

[37] ANSYS, SAS IP, Inc. *ANSYS 14.5 Help.*

[38] ANSYS, SAS IP, Inc. *ANSYS 14.5 Help.* ANSYS Help System // Mechanical APDL // Command Reference // 2. Command Groupings // 2.8. POST26 Commands, Table 2.109 Operations.

[39] ASTM C384 - 04(2011) Standard test method for impedance and absorption of acoustical materials by impedance tube method. Technical report, ASTM, 2011.

[40] ASTM C423-08a. Standard test method for sound absorption and sound absorption coefficients by the reverberation room method.

[41] ASTM C522 - 03(2009)e1 Standard test method for airflow resistance of acoustical materials. Technical report, ASTM, 2009.

[42] N. Atalla and R.J. Bernhard. Review of numerical solutions for low-frequency structural-acoustic problems. *Applied Acoustics*, 43(3):271–294, 1994. doi:10.1016/0003-682X(94)90050-7.

[43] Juha Backman. A model of open-baffle loudspeakers. In *Proceedings of the 107th Convention of the Audio Engineering Society*, New York, USA, 24–27 September 1999. Audio Engineering Society, Audio Engineering Society. Preprint 5025. URL: http://www.aes.org/e-lib/browse.cfm?elib=8155.

[44] J.S. Bendat and A.G. Piersol. *Random Data Analysis and Measurement Procedures*. Wiley-Interscience, second edition, 1986.

[45] Leo L. Beranek, editor. *Noise and Vibration Control*. Institute of Noise Control Engineering, second edition, 1988.

[46] Leo L. Beranek and István L. Vér, editors. *Noise and Vibration Control Engineering: Principles and Application*. Wiley Interscience, New York, USA, 1992.

[47] David A. Bies and Colin H. Hansen. *Engineering Noise Control: Theory and Practice.* Spon Press, London, UK, fourth edition, August 12 2009.

[48] David A. Bies, Colin H. Hansen, and Gareth E. Bridges. Sound attenuation in rectangular and circular cross-section ducts with flow and bulk-reacting liner. *Journal of Sound and Vibration*, 146(1):47–80, 1991.

[49] V.B. Bokil and U.S. Shirahatti. Technique for the modal analysis of sound-structure interaction problems. *Journal of Sound and Vibration*, 173(1):23–41, 26 May 1994. doi:10.1006/jsvi.1994.1215.

[50] R. Bossart, N. Joly, and M. Bruneau. Hybrid numerical and analytical solutions for acoustic boundary problems in thermo-viscous fluids. *Journal of Sound and Vibration*, 263(1):69–84, 2003. doi:10.1016/S0022-460X(02)01098-2.

[51] Ben S. Cazzolato and Justin Ghan. Frequency domain expressions for the estimation of time-averaged acoustic energy density. *The Journal of the Acoustical Society of America*, 117(6):3750–3756, 2005. doi:10.1121/1.1567273.

[52] Benjamin S. Cazzolato. *Sensing systems for active control of sound transmission into cavities.* PhD thesis, School of Mechanical Engineering, The University of Adelaide, South Australia, April 1999. URL: http://hdl.handle.net/2440/37893.

[53] Benjamin S. Cazzolato and Colin H. Hansen. Errors arising from three-dimensional energy density sensing in one-dimensional sound fields. *Journal of Sound and Vibration*, 236(3):375–400, 2000. doi:10.1006/jsvi.1999.2992.

[54] Benjamin S. Cazzolato and Colin H. Hansen. Active control of enclosed sound fields using three-axis energy density sensors: Rigid-walled enclosures. *International Journal of Acoustics and Vibration*, 8(1):39–51, 2003.

[55] Benjamin S. Cazzolato, Carl Q. Howard, and Colin H. Hansen. Finite element analysis of an industrial reactive silencer. In *ICSV5: Proceedings of the 5th International Congress of Sound and Vibration*, volume 3, pages 1659–1668, Adelaide, South Australia, Australia, 15–18 December 1997. IIAV.

[56] Benjamin S. Cazzolato, Cornelis D. Petersen, Carl Q. Howard, and Anthony C. Zander. Active control of energy density in a 1D waveguide: A cautionary note. *The Journal of the Acoustical Society of America*, 117(6):3377–3380, 2005. doi:10.1121/1.1920213.

[57] J.Y. Chung and D.A. Blaser. Transfer function method of measuring in-duct acoustic properties. i. theory. *Journal of Acoustical Society of America*, 68(3):907–913, 1980.

[58] A. Craggs. The transient response of a coupled plate-acoustic system using plate and acoustic finite elements. *Journal of Sound and Vibration*, 15(4):509–528, 1971. doi:10.1016/0022-460X(71)90408-1.

[59] M.J. Crocker, editor. *Encyclopedia of Acoustics*. John Wiley & Sons, 1997.

[60] Olivier Doutres, Yacoubou Salissou, Noureddine Atalla, and Raymond Panneton. Evaluation of the acoustic and non-acoustic properties of sound absorbing materials using a three-microphone impedance tube. *Applied Acoustics*, 71(6):506–509, 2010. doi:10.1016/j.apacoust.2010.01.007.

[61] E.H. Dowell and H.M. Voss. The effect of a cavity on panel vibration. *AIAA Journal*, 1(2):476–477, 1963. doi:10.2514/3.1568.

[62] David J. Ewins. *Modal Testing: Theory, Practice and Application*. Wiley, second edition, August 2001. (Mechanical Engineering Research Studies: Engineering Dynamics Series).

[63] Frank J. Fahy. *Sound and Structural Vibration: Radiation, Transmission and Response*. Academic Press, San Diego, California, USA, 1994.

[64] Frank J. Fahy. *Sound Intensity*. E & FN Spon, and imprint of Chapman & Hall, London, UK, second edition, 1995.

[65] Frank J. Fahy. *Foundations of Engineering Acoustics*. Academic Press, London, UK, 2001. URL: http://www.sciencedirect.com/science/book/9780122476655.

[66] Frank J. Fahy and Paolo Gardonio. *Sound and Structural Vibration: Radiation, Transmission and Response*. Academic Press, San Diego, California, USA, second edition, 2007.

[67] Justin Ghan, Benjamin S. Cazzolato, and Scott D. Snyder. Expression for the estimation of time averaged acoustic energy density using the two-microphone method. *The Journal of the Acoustical Society of America*, 113(5):2404–2407, 2003. doi:10.1121/1.1567273.

[68] Jianmin Gu, Zheng-Dong Ma, and Gregory M. Hulbert. Quasi-static data recovery for dynamic analyses of structural systems. *Finite Elements in Analysis and Design*, 37(11):825841, October 2001. doi:10.1016/S0168-874X(01)00070-1.

[69] C. Guigou-Carter and M. Villot. Modelling of sound transmission through lightweight elements with stiffeners. *Building Acoustics*, 10(3):193–209, 2003. doi:10.1260/135101003322662005.

[70] R.W. Guy and M.C. Bhattacharya. The transmission of sound through a cavity-backed finite plate. *Journal of Sound and Vibration*, 27(2):207223, 22 March 1973. doi:10.1016/0022-460X(73)90062-X.

[71] J.P. Holman. *Heat Transfer, SI Metric Edition*. McGraw Hill, Singapore, 1989.

[72] Carl Q. Howard. Transmission loss of a panel with an array of tuned vibration absorbers. *Acoustics Australia*, 36(3):98–103, December 2008.

[73] Carl Q. Howard. Transmission loss of a panel with an array of tuned vibration absorbers. In *Proceedings of Acoustics 2008*, Geelong, Victoria, Australia, 24–26 November 2008. Australian Acoustical Society. Paper 12.

[74] Carl Q. Howard. The corrected expressions for the four-pole transmission matrix for a duct with a linear temperature gradient and an exponential temperature profile. *Open Journal of Acoustics*, 3(3):62–66, September 2013. doi:10.4236/oja.2013.33010.

[75] Carl Q. Howard. Transmission matrix model of a quarter-wave-tube with gas temperature gradients. In *Proceedings of Acoustics 2013*, Victor Harbor, South Australia, Australia, 9–11 November 2013. Australian Acoustical Society. Paper 71.

[76] Carl Q. Howard, Benjamin S. Cazzolato, and Colin H. Hansen. Exhaust stack silencer design using finite element analysis. *Noise Control Engineering Journal*, 48(4):113–120, 2000.

[77] Carl Q. Howard and Richard A. Craig. Noise reduction using a quarter wave tube with different orifice geometries. *Applied Acoustics*, 76:180–186, February 2014. doi:10.1016/j.apacoust.2013.08.006.

[78] Carl Q. Howard, Colin H. Hansen, and Anthony C. Zander. Multivariable optimisation of a vibro-acoustic system using a distributed computing network. In *ICSV12: Proceedings of the 12th International Congress of Sound and Vibration*, Lisbon, Portugal, July 2005. International Institute of Acoustics and Vibrations. Paper No. 665.

[79] Carl Q. Howard, Colin H. Hansen, and Anthony C. Zander. Noise reduction of a rocket payload fairing using tuned vibration absorbers with translational and rotational dofs. In *Proceedings of Acoustics 2005*, pages 165–171, Busselton, Western Australia, Australia, 9–11 November 2005. Australian Acoustical Society.

[80] Carl Q. Howard, Colin H. Hansen, and Anthony C. Zander. Optimisation of design and location of acoustic and vibration absorbers using a distributed computing network. In *Proceedings of Acoustics 2005*, pages 173–178, Busselton, Western Australia, Australia, 9–11 November 2005. Australian Acoustical Society.

[81] Carl Q. Howard, Colin H. Hansen, and Anthony C. Zander. Vibro-acoustic noise control treatments for payload bays of launch vehicles: Discrete to fuzzy solutions. *Applied Acoustics*, 66(11):1235–1261, November 2005. doi:10.1016/j.apacoust.2005.04.009.

[82] Carl Q. Howard, Rick C. Morgans, Colin H. Hansen, and Anthony C. Zander. A tool for the optimisation of vibro-acoustic systems using a parallel genetic algorithm and a distributed computing network. *Noise Control Engineering Journal*, 53(6):256–267, November 2005.

[83] Uno K. Ingard. *Noise Reduction Analysis*. Jones and Bartlett Publishers, Sudbury, Massachusetts, USA, 2010.

[84] Daniel J. Inman. *Engineering Vibration*. Prentice Hall, third edition, May 19 2007. ISBN: 0132281732.

[85] ISO 35-2003 Acoustics: Measurement of sound absorption in a reverberation room.

[86] ISO 3741:2010 Acoustics – Determination of sound power levels and sound energy levels of noise sources using sound pressure – Precision methods for reverberation test rooms.

[87] ISO 10534-1:1996 Acoustics – Determination of sound absorption coefficient and impedance in impedance tubes – Part 1: Method using standing wave ratio. Technical report, ISO, 1996.

[88] ISO 10534-2:1998 Acoustics – Determination of sound absorption coefficient and impedance in impedance tubes – Part 2: Transfer-function method. Technical report, ISO, 1998.

[89] ISO 266:1997 Acoustics – Preferred frequencies. Technical report, ISO, 1997.

[90] Finn Jacobsen. The sound field in a reverberation room. Note 31261, Technical University of Denmark, 2011.

[91] Finn Jacobsen and Peter Møller Juhl. *Fundamentals of General Linear Acoustics*. John Wiley and Sons., West Sussex, UK, 2013.

[92] L. Jaouen. Acoustical porous material recipes. Website. URL: http://apmr.matelys.com/.

[93] L. Jaouen and F.-X. Becot. Acoustical characterization of perforated facings. *Journal of the Acoustical Society of America*, 123(3):1400–1406, 2011.

[94] V. Jayachandran, S. Hirsch, and J. Sun. On the numerical modelling of interior sound fields by the modal expansion approach. *Journal of Sound and Vibration*, 210(2):243–254, 1998. doi:10.1006/jsvi.1997.1328.

[95] Z.L. Ji. Acoustic length correction of closed cylindrical side-branched tube. *Journal of Sound and Vibration*, 283(35):1180–1186, 2005. doi: 10.1016/j.jsv.2004.06.044.

[96] James J. Faran Jr. Sound scattering by solid cylinders and spheres. *The Journal of the Acoustical Society of America*, 23(4):405–418, July 1951. doi:10.1121/1.1906780.

[97] Miguel C. Junger and David Feit. *Sound Structures, and Their Interaction*. Acoustical Society of America, through the American Institute of Physics, Woodbury, NY, USA, second edition, 1993.

[98] Colin D. Kestell, Benjamin S. Cazzolato, and Colin H. Hansen. Active noise control in a free field with virtual sensors. *The Journal of the Acoustical Society of America*, 109(1):232–243, 2001. doi:10.1121/1.1326950.

[99] M.S. Khan, C. Cai, and K.C. Hung. Acoustics field and active structural acoustic control modeling in ANSYS. In *International ANSYS Conference*, 2002.

[100] M.R.F. Kidner and C.H. Hansen. A comparison and review of theories of the acoustics of porous materials. *International Journal of Acoustics and Vibration*, 13(3):112–119, 2008.

[101] Sang-Myeong Kim and Michael J. Brennan. Active control of harmonic sound transmission into an acoustic enclosure using both structural and acoustic actuators. *Journal of the Acoustical Society of America*, 107(5):2523–2534, 2000. doi:10.1121/1.428640.

[102] Lawrence E. Kinsler, Austin R. Frey, Alan B. Coppens, and James V. Sanders. *Fundamentals of Acoustic*. John Wiley and Sons, New York, USA, fourth edition, 1999.

[103] S.-H. Ko. Sound attenuation in lined rectangular ducts with flow and its application to the reduction of aircraft engine noise. *Journal of Acoustical Society of America*, 50(6):1418–1432, 1971.

[104] U. Kurze and E. Riedel. Silencers. In G. Müller and M. Möser, editors, *Handbook of Engineering Acoustics*, chapter 11, pages 269–299. Springer-Verlag, Berlin Heidelberg, 2013. doi:10.1007/978-3-540-69460-1_11.

[105] U.J. Kurze and C.H. Allen. Influence of flow and high sound level on the attenuation in a lined duct. *Journal of the Acoustical Society of America*, 49(5):1643–1654, 1970.

[106] U.J. Kurze and I.L. Ver. Sound attenuation in ducts lined with non-isotropic material. *Journal of Sound and Vibration*, 24(2):177–187, 1972.

[107] Y.W. Lam. AEOF3/AEOF4 acoustics of enclosed spaces, 1995.

[108] Arthur W. Leissa. *Vibration of Plates*. Acoustical Society of America, American Institute of Physics, Woodbury, New York, 1993.

[109] Los Alamos National Laboratories. DeltaEC: Design environment for low-amplitude thermoacoustic energy conversion. Software, http://www.lanl.gov/thermoacoustics/DeltaEC.html, 18 February 2012. version 6.3b11 (Windows, 18-Feb-12).

[110] D.-Y. Maa. Potential of microperforated panel absorber. *Journal of the Acoustical Society of America*, 104(5):28682866, 1998.

[111] Steffen Marburg. Six boundary elements per wavelength: is that enough? *Journal of Computational Acoustics*, 10(1):25–51, 2002. doi:10.1142/S0218396X02001401.

[112] F.P. Mechel. Schalldämpfer. In M. Heckl and H.A. Müller, editors, *Taschenbuch der Technischen Akustik*, chapter 20. Springer, Berlin, Germany, 1994.

[113] Danielle J. Moreau, Justin Ghan, and Benjamin S. Cazzolato ANS Anthony C. Zander. Active noise control in a pure tone diffuse sound field using virtual sensing. *The Journal of the Acoustical Society of America*, 125(6):3742–3755, 2009. doi:10.1121/1.3123404.

[114] Philip M. Morse. The transmission of sound inside pipes. *Journal of the Acoustical Society of America*, 11:205–210, 1939. doi:10.1121/1.1916024.

[115] Philip M. Morse. *Vibration and Sound*. Acoustical Society of America, Woodbury, NY, USA, second edition, 1981.

[116] Philip M. Morse and Herman Feshbach. *Methods of Theoretical Physics*, volume 2. McGraw-Hill, New York, USA, 1953.

[117] Philip M. Morse and K. Uno Ingard. *Theoretical Acoustics*. Princeton University Press, Princton, New Jersey, USA, 1986.

[118] G. Müller and M. Möser, editors. *Handbook of Engineering Acoustics*. Springer, 2013.

[119] M.L. Munjal. *Acoustics of Ducts and Mufflers with Application to Exhaust and Ventilation System Design*. Wiley-Interscience, 1987.

[120] Harry Nyquist. Certain factors affecting telegraph speed. *Transactions of the American Institute of Electrical Engineering*, 47(2):617644, 1928. doi:10.1109/T-AIEE.1928.5055024.

[121] Harry Nyquist. Certain topics in telegraph transmission theory. *Proceedings of the IEEE*, 90(2):280–305, February 2002. doi:10.1109/5.989875.

[122] Roger Ohayon. Reduced symmetric models for modal analysis of internal structural-acoustic and hydroelastic-sloshing systems. *Computer Methods in Applied Mechanics and Engineering*, 190(24-25):3009–3019, 2 March 2001. doi:10.1016/S0045-7825(00)00379-0.

[123] X. Olny and R. Panneton. Acoustical determination of the parameters governing thermal dissipation in porous media. *Journal of the Acoustical Society of America*, 123(2):814–824, 2008.

[124] Jie Pan and David A. Bies. The effect of fluidstructural coupling on sound waves in an enclosure-Theoretical part. *Journal of the Acoustical Society of America*, 87(2):691–707, 1990. doi:10.1121/1.398939.

[125] R. Panneton. Comments on the limp frame equivalent fluid model for porous media. *Journal of the Acoustical Society of America*, 122(6):EL217–EL222, 2007. doi:10.1121/1.2800895.

[126] R. Panneton and X. Olny. Acoustical determination of the parameters governing viscous dissipation in porous media. *Journal of the Acoustical Society of America*, 119(4):2027–2040, 2006.

[127] K.C. Park. Course notes for ASEN 5022: Dynamics of aerospace structures, chapter 16 modeling for structural vibrations: FEM models, damping, similarity laws. 1998.

[128] Eugene T. Patronis Jr. Electroacoustics. In J.C. Whitaker, editor, *The Electronics Handbook*, chapter 1.3, pages 20–30. CRC Press, Taylor and Francis Group, second edition, 2005. doi:10.1201/9781420036664.

[129] Alan D. Pierce. *Acoustics: An Introduction to Its Physical Principles and Applications*. Acoustical Society of America, Woodbury, New York, USA, 1994.

[130] Louis A. Roussos. Noise transmission loss of a rectangular plate in an infinite baffle. Technical report, NASA, Langley Research Center, Hampton, Virginia, USA, March 1985. NASA technical paper TP2398.

[131] Daniel A. Russell, Joseph P. Titlow, and Ya-Juan Bemmen. Acoustic monopoles, dipoles, and quadrupoles: An experiment revisited. *American Journal of Physics*, 67(8):660–664, 1999. doi:10.1119/1.19349.

[132] F. Scarpa. Parametric sensitivity analysis of coupled acoustic-structural systems. *Journal of Vibration and Acoustics*, 122(2):109–115, April 2000. doi:10.1115/1.568447.

[133] Steven H Schot. Eighty years of Sommerfeld's radiation condition. *Historia Mathematica*, 19(4):385–401, 1992. doi:10.1016/0315-0860(92)90004-U.

[134] R.A. Scott. The propagation of sound between walls of porous materials. *Proceedings of the Physical Society*, 58:358–368, 1946.

[135] A.F. Seybert and D.F. Ross. Experimental determination of acoustic properties using a two-microphone random-excitation technique. *Journal of Acoustical Society of America*, 61(5):1362–1370, 1977.

[136] C.E. Shannon. Communication in the presence of noise. *Proceedings of the IRE*, 37(1):1021, January 1949. doi:10.1109/JRPROC.1949.232969.

[137] C.E. Shannon. Classic paper: Communication in the presence of noise. *Proceedings of the IEEE*, 86(2):447–457, February 1998. doi:10.1109/JPROC.1998.659497.

[138] Sarabjeet Singh, Colin H. Hansen, and Carl Q. Howard. A detailed tutorial for evaluating in-duct net acoustic power transmission in a circular duct with an attached cylindrical Helmholtz resonator using transfer matrix method. In *Acoustics 2008*, pages 1–9, Geelong, Victoria, Australia, 24–26 November 2008. Australian Acoustical Society, Australian Acoustical Society. Paper 9.

[139] Eugen Skudrzyk. *The Foundations of Acoustics, Basic Mathematics and Basic Acoustics*. Springer-Verlag, reprinted by the Acoustical Society of America, New York, USA, 1971. doi:10.1007/978-3-7091-8255-0.

[140] Scott Snyder and Nobuo Tanaka. On feedforward active control of sound and vibration using vibration error signals. *The Journal of the Acoustical Society of America*, 94(4):2181–2193, October 1993. doi:10.1121/1.407489.

[141] Werner Soedel. *Vibration of Shells and Plates*. Marcel Dekker, third edition, 2004.

[142] R.I. Sujith. Transfer matrix of a uniform duct with an axial mean temperature gradient. *The Journal of the Acoustical Society of America*, 100(4):2540–2542, October 1996. doi:10.1121/1.417362.

[143] R.I. Sujith, G.A. Waldherr, and B.T. Zinn. An exact solution for one-dimensional acoustic fields in ducts with an axial temperature gradient. *Journal of Sound and Vibration*, 184(3):389–402, 1995. doi:10.1006/jsvi.1995.0323.

[144] Michel Tournour and Noureddine Atalla. Pseudostatic corrections for the forced vibroacoustic response of structure-cavity system. *The Journal of the Acoustical Society of America*, 107(5):2379–2386, 2000. doi:10.1121/1.428624.

[145] D. Turo, J. Vignola, and A. Glean. Acoustic metrology—Chapter 5 modelling sound propagation in porous media. Course Notes, The Catholic University of America.

[146] C.E. Wallace. Radiation resistance of a rectangular panel. *The Journal of the Acoustical Society of America*, 51(3):946–952, 1972. part 2.

[147] C. Wassilieff. Experimental verification of duct attenuation models with bulk reacting linings. *Journal of Sound and Vibration*, 114(2):239–251, 1987.

[148] Eric W. Weisstein. Pappus's centroid theorem. From MathWorld– A Wolfram Web Resource. http://mathworld.wolfram.com/ PappussCentroidTheorem.html.

[149] David B. Woyak. Acoustics and fluid–structure interaction - A revision 5.0 tutorial // 3.2 analysis solution procedures (FLUID29 and FLUID30) // FLUID29 and FLUID30 boundary conditions and fluid loads. Technical Report 1, Swanson Analysis Systems Inc., Houston, PA, 15342, 11 June 1992. DN-T044:50, Jordan, Apostal, Ritter and Associates, Inc., North Kingstown, RI 02852.

[150] David B. Woyak. Acoustics and fluid–structure interaction: A revision 5.0 tutorial. Technical Report 1, Swanson Analysis Systems Inc., Houston, PA, 15342, 11 June 1992. DN-T044:50, Jordan, Apostal, Ritter and Associates, Inc., North Kingstown, RI 02852.

[151] You-qun Zhao, Su-huan Chen, San Chai, and Qing-wen Qu. An improved modal truncation method for responses to harmonic excitation. *Computers and Structures*, 80(1):99–103, January 2002. doi: 10.1016/S0045-7949(01)00148-1.

[152] O.C. Zienkiewicz, R.L. Taylor, and J.Z. Zhu. *The Finite Element Method: Its Basis & Fundamentals*. Elsevier, sixth edition, 2005.

Index